Carreau / De Kee / Chhabra
Rheology of Polymeric Systems

Pierre J. Carreau / Daniel C.R. De Kee / Raj P. Chhabra

Rheology of Polymeric Systems

Principles and Applications

Hanser Publishers, Munich Vienna New York

Hanser/Gardner Publications, Inc., Cincinnati

The Authors:
Prof. Pierre J. Carreau, Dept. of Chemical Engineering, École Polytechnique of Montreal, Montreal, Quebec H3C 3A7, Canada
Prof. Daniel C.R. De Kee, Dept. of Chemical Engineering, University of Sherbrooke, Quebec, Canada
Prof. Raj P. Chhabra, Dept. of Chemical Engineering, Indian Institute of Technology, Kanpur, India

Distributed in the U.S.A. and in Canada by
Hanser/Gardner Publications, Inc.
6600 Clough Pike, Cincinnati, Ohio 45244-4090, U.S.A.
Fax: (513) 527-8950
Phone: (513) 527-8977 or 1-800-950-8977
Internet: http://www.hansergardner.com

Distributed in all other countries by
Carl Hanser Verlag
Postfach 86 04 20, 81631 München, Germany
Fax: +49 (89) 98 12 64
Internet: http://www.hanser.de

The use of general descriptive names, trademarks, etc., in this publication, even if the former are not especially identified, is not to be taken as a sign that such names, as understood by the Trade Marks and Merchandise Marks Act, may accordingly be used freely by anyone.

While the advice and information in this book are believed to be true and accurate at the date of going to press, neither the authors nor the editors nor the publisher can accept any legal responsibility for any errors or omissions that may be made. The publisher makes no warranty, express or implied, with respect to the material contained herein.

Library of Congress Cataloging-in-Publication Data
Carreau, Pierre J.
Rheology of polymeric systems : principles and applications /
Pierre J. Carreau, Daniel C.R. De Kee, Raj P. Chhabra.
   p. cm.
Includes bibliographical references and index.
ISBN 1-56990-218-6
1. Polymers—Rheology. I. De Kee, D. (Daniel) II. Chhabra, R. P. III. Title.
TP1092.C37   1997
620.1′920423— DC20   96-36329

Die Deutsche Bibliothek - CIP-Einheitsaufnahme
Carreau, Pierre J.:
Rheology of polymeric systems : principles and applications /
Pierre J. Carreau/Daniel C. R. De Kee/Raj P. Chhabra.-
Munich ; Vienna ; New York : Hanser ; Cincinnati :
Hanser/Gardner, 1997
  ISBN 3-446-17182-7
NE: De Kee, Daniel C.R.:; Chhabra, Raj P.:

All rights reserved. No part of this book may be reproduced or transmitted in any form or by any means, electronic or mechanical, including photocopying or by any information storage and retrieval system, without permission in writing from the publisher.

© Carl Hanser Verlag, Munich Vienna New York, 1997
Typeset in England by Techset Composition Ltd., Salisbury
Printed and bound in Germany by Druckerei Hubert & Co., Göttingen

This book is dedicated to Professor R.B. Bird and Professor A.S. Lodge, both of the University of Wisconsin (Madison), who have had a strong influence on the education and research training of numerous students. They are a constant source of inspiration and motivation to us. Their unique contribution will continue to inspire generations to come.

Preface

Rheology is now recognized as an important field of research and finds applications in a variety of industrial sectors such as polymers, foods, cosmetics, paints, and pharmaceuticals. Some of the currently available books cover the new trends in research very well, while only a few books address the applications. This book intends to bridge the gap between fundamental concepts and applications. The bulk of the material presented here has been used successfully for many years in our respective courses.

This book is intended to be used as a textbook for a graduate or advanced undergraduate course in polymer rheology. The level is between that of introductory texts and of highly advanced monographs. We consider the introduction of a treatment of rheology at this level to be timely. This work aims to develop a systematic approach and a clear understanding of the envisaged applications. The reader is expected to be familiar with introductory transport phenomena, or equivalent fluid mechanics and heat transfer.

The organization of this book is as follows. The text introduces the subject of rheology via the description of unusual phenomena such as rod climbing, extrudate swell, stable bubble shapes, ... In Chapter 2, material functions are defined for a variety of flow situations. Generalized Newtonian fluid models are introduced, and their predictions are compared to typical data. Chapter 3 deals with the subject of rheometry. Measurements of viscosity, normal stress differences, yield stress, ... using capillary, concentric cylinder, cone-and-plate geometries, ... are reviewed in detail. Isothermal flows, as well as heat and mass transfer in simple geometries involving generalized non-Newtonian fluids, are dealt with in Chapter 4. The subject of linear viscoelasticity is discussed in Chapter 5, whereas Chapter 6 reviews the area of nonlinear deformation and the formulation of appropriate constitutive equations. The modern molecular approach is dealt with in an introductory fashion in Chapter 7, and topics dealing with the flow properties of non-Newtonian media involving bubbles, drops, and particles are discussed in Chapter 8, and mixing of complex fluids in Chapter 9. Finally, we present a substantial appendix dealing with tensor analysis, which is largely based on a text in preparation on advanced mathematics. The material presented in the first five chapters can be used as an introduction to the subject. A more advanced course would also encompass Chapters 6 and 7, and Appendix A, while Chapters 8 and 9 focus on major areas of applications.

In writing this book, we have been strongly inspired by *Transport Phenomena* (Bird, Stewart, and Lightfoot, 1960) and by *Dynamics of Polymeric Liquids*, especially Volume I (Bird, Armstrong, and Hassager, 1977, 1987). While we do not try to match the in-depth coverage of *Dynamics of Polymeric Liquids*, we present results of our extensive teaching and research experience in this field in a coherent manner. In this regard, this book has a distinct engineering flavor, covering topics such as mixing and flow of particulate systems, which are seldom discussed in other books on rheology. Statements such as "it can easily be shown" have carefully been avoided, in favor of a fair amount of detailed explanation. Several homework problems appear at the end of most chapters and at the end of Appendix A. These problems are labelled by a subscript a or b indicating the level of difficulty. The "b-problems" are the more demanding ones.

In preparing this book, we have made extensive use of the research literature and research performed in our respective laboratories by our graduate students and research associates. Special thanks go to Drs. C.F. Chan Man Fong and M. Grmela, who contributed to many facets of this book. We acknowledge also the devotion of Ms. D. Héroux who patiently typed and re-typed what must have seemed endless revisions of the text. Finally we are thankful to Mr. F. St-Louis and Ms. N. Chapleau for preparing the artwork for this book.

P.J. Carreau
D. De Kee
R.P. Chhabra

Contents

1 Introduction 1
   1.1 Definitions and Classification 1
      1.1-1 Purely Viscous or Inelastic Material 3
      1.1-2 Perfectly Elastic Material 3
      1.1-3 Viscoelastic Material 3
   1.2 Non-Newtonian Phenomena 3
      1.2-1 The Weissenberg Effect 4
      1.2-2 Entry Flow, Extrudate Swell, Melt Fracture, and Vibrating Jet 4
      1.2-3 Recoil 7
      1.2-4 Drag Reduction 7
      1.2-5 Hole Pressure Error 13
      1.2-6 Mixing 14
      1.2-7 Bubbles, Spheres, and Coalescence 15

2 Material Functions and Generalized Newtonian Fluid 18
   2.1 Material Functions 19
      2.1-1 Simple Shear Flow 19
      2.1-2 Sinusoidal Shear Flow 25
      2.1-3 Transient Shear Flows 26
      2.1-4 Elongational Flow 32
   2.2 Generalized Newtonian Models 35
      2.2-1 Generalized Newtonian Fluid 36
      2.2-2 The Power-Law Model (Ostwald, 1925) 37
      2.2-3 The Ellis Model (Bird, Armstrong, and Hassager, 1977) . 37
      2.2-4 The Carreau Model (1972) 38
      2.2-5 The Cross-Williamson Model (1965) 39
      2.2-6 The Four-Parameter Carreau Model (Carreau et al., 1979b) . 39
      2.2-7 The De Kee Model (1977) 40
      2.2-8 The Carreau-Yasuda Model (Yasuda, 1976) 41
      2.2-9 The Bingham Model (1922) 41
      2.2-10 The Casson Model (1959) 42
      2.2-11 The Herschel-Bulkley Model (1926) 42
      2.2-12 The De Kee-Turcotte Model (1980) 42
      2.2-13 Viscosity Models for Complex Flow Situations 43
   2.3 Thixotropy, Rheopexy, and Hysteresis 44
   2.4 Relations Between Material Functions 48

	2.5	Temperature, Pressure, and Molecular Weight Effects	50
		2.5-1 Effect of Temperature on Viscosity	50
		2.5-2 Effect of Pressure on Viscosity	52
		2.5-3 Effect of Molecular Weight on Viscosity	52
	2.6	Problems	57

3 Rheometry ... 61

	3.1	Capillary Rheometry	62
		3.1-1 Rabinowitsch Analysis	64
		3.1-2 End Effects or Bagley Correction	68
		3.1-3 Mooney Correction	72
		3.1-4 Intrinsic Viscosity Measurements	73
	3.2	Coaxial-Cylinder Rheometers	76
		3.2-1 Calculation of Viscosity	77
		3.2-2 End Effect Corrections	81
		3.2-3 Normal Stress Determination	82
	3.3	Cone-and-Plate Geometry	84
		3.3-1 Viscosity Determination	86
		3.3-2 Normal Stress Determination	88
		3.3-3 Inertial Effects	90
		3.3-4 Criteria for Transient Experiments	94
	3.4	Concentric Disk Geometry	98
		3.4-1 Viscosity Determination	99
		3.4-2 Normal Stress Difference Determination	101
	3.5	Yield Stress Measurements	103
	3.6	Problems	106

4 Transport Phenomena in Simple Flows ... 112

	4.1	Criteria for Using Purely Viscous Models	113
	4.2	Isothermal Flow in Simple Geometries	114
		4.2-1 Flow of a Shear-Thinning Fluid in a Circular Tube	114
		4.2-2 Film Thickness for the Flow on an Inclined Plane	116
		4.2-3 Flow in a Thin Slit	118
		4.2-4 Helical Flow in an Annular Section	119
		4.2-5 Flow in a Disk-Shaped Mold	122
	4.3	Heat Transfer to Non-Newtonian Fluids	126
		4.3-1 Convective Heat Transfer in Poiseuille Flow	126
		4.3-2 Heat Generation in Poiseuille Flow	134
	4.4	Mass Transfer to Non-Newtonian Fluids	138
		4.4-1 Mass Transfer to a Power-Law Fluid Flowing on an Inclined Plate	138
		4.4-2 Mass Transfer to a Power-Law Fluid in Poiseuille Flow	141
	4.5	Boundary Layer Flows	144
		4.5-1 Laminar Boundary Layer Flow of Power-Law Fluids over a Plate	144
		4.5-2 Laminar Thermal Boundary Layer Flow over a Plate	149
	4.6	Problems	152

5 Linear Viscoelasticity ... 162
- 5.1 Importance and Definitions ... 162
- 5.2 Linear Viscoelastic Models ... 163
  - 5.2-1 The Maxwell Model ... 164
  - 5.2-2 Generalized Maxwell Model ... 170
  - 5.2-3 The Jeffreys Model ... 178
  - 5.2-4 The Voigt-Kelvin Model ... 180
  - 5.2-5 Other Linear Models ... 182
- 5.3 Relaxation Spectrum ... 184
- 5.4 Time-Temperature Superposition ... 186
- 5.5 Problems ... 189

6 Nonlinear Viscoelasticity ... 194
- 6.1 Nonlinear Deformations ... 195
  - 6.1-1 Expressions for the Deformation and Deformation Rate ... 197
  - 6.1-2 Pure Deformation or Uniaxial Elongation ... 200
  - 6.1-3 Planar Elongation ... 204
  - 6.1-4 Expansion or Compression ... 205
  - 6.1-5 Simple Shear ... 205
- 6.2 Formulation of Constitutive Equations ... 208
  - 6.2-1 Material Objectivity and Formulation of Constitutive Equations ... 209
  - 6.2-2 Maxwell Convected Models ... 210
  - 6.2-3 Generalized Newtonian Models ... 215
- 6.3 Differential Constitutive Equations ... 220
  - 6.3-1 The De Witt Model ... 221
  - 6.3-2 The Oldroyd Models ... 222
  - 6.3-3 The White-Metzner Model ... 223
  - 6.3-4 The Marrucci Model ... 230
  - 6.3-5 The Phan-Thien-Tanner Model ... 232
- 6.4 Integral Constitutive Equations ... 234
  - 6.4-1 The Lodge Model ... 235
  - 6.4-2 The Carreau Constitutive Equation ... 239
  - 6.4-3 The K–BKZ Constitutive Equation ... 247
  - 6.4-4 The LeRoy-Pierrard Equation ... 254
- 6.5 Concluding Remarks ... 257
- 6.6 Problems ... 258

7 Constitutive Equations from Molecular Theories ... 263
- 7.1 Bead– and Spring–Type Models ... 264
  - 7.1-1 Hookean Elastic Dumbbell ... 265
  - 7.1-2 Finitely Extensible Nonlinear Elastic (FENE) Dumbbell ... 272
  - 7.1-3 Rouse and Zimm Models ... 276
- 7.2 Network Theories ... 284
  - 7.2-1 General Network Concept ... 284
  - 7.2-2 Rubber-Like Solids ... 286

		7.2-3	Elastic Liquids	288
		7.2-4	Recent Developments	290
	7.3	Reptation Theories		294
		7.3-1	The Tube Model	294
		7.3-2	The Doi-Edwards Model	296
		7.3-3	The Curtiss-Bird Kinetic Theory	300
	7.4	Conformation Tensor Rheological Models		304
		7.4-1	Basic Description of the Conformation Model	304
		7.4-2	FENE-Charged Macromolecules	307
		7.4-3	Rod-Like and Worm-Like Macromolecules	312
		7.4-4	Generalization of the Conformation Tensor Model	320
	7.5	Problems		327
8	**Multiphase Systems**			**329**
	8.1	Systems of Industrial Interest		330
	8.2	Rheology of Suspensions		331
		8.2-1	Viscosity of Dilute Suspensions of Rigid Spheres	332
		8.2-2	Rheology of Emulsions	334
		8.2-3	Rheology of Concentrated Suspensions of Non-Interactive Particles	343
		8.2-4	Rheology of Concentrated Suspensions of Interactive Particles	347
		8.2-5	Concluding Remarks	351
	8.3	Flow About a Rigid Particle		352
		8.3-1	Flow of a Power-Law Fluid Past a Sphere	352
		8.3-2	Other Fluid Models	356
		8.3-3	Viscoplastic Fluids	356
		8.3-4	Viscoelastic Fluids	357
		8.3-5	Wall Effects	357
		8.3-6	Non-Spherical Particles	359
		8.3-7	Drag-Reducing Fluids	360
		8.3-8	Behavior in Confined Flows	361
	8.4	Flow Around Fluid Spheres		362
		8.4-1	Creeping Flow of a Power-Law Fluid Past a Gas Bubble	362
		8.4-2	Experimental Results on Single Bubbles	363
	8.5	Creeping Flow of a Power-Law Fluid Around a Newtonian Droplet		366
		8.5-1	Experimental Results on Falling Drops	367
	8.6	Flow in Packed Beds		368
		8.6-1	Creeping Power-Law Flow in Beds of Spherical Particles: The Capillary Model	368
		8.6-2	Other Fluid Models	373
		8.6-3	Viscoelastic Effects	373
		8.6-4	Wall Effects	374
		8.6-5	Effects of Particle Shape	375
		8.6-6	Submerged Objects' Approach to Fluid Flow in Packed Beds: Creeping Flow	376

8.7	Fluidized Beds	377
	8.7-1	Minimum Fluidization Velocity	378
	8.7-2	Bed Expansion Behavior	380
	8.7-3	Heat and Mass Transfer in Packed and Fluidized Beds	382
8.8	Problems	383

9 Liquid Mixing ... 386

9.1	Introduction	387
9.2	Mechanisms of Mixing	388
	9.2-1 Laminar Mixing	389
	9.2-2 Turbulent Mixing	391
9.3	Scale-Up and Similarity Criteria	391
9.4	Power Consumption in Agitated Tanks	396
	9.4-1 Low Viscosity Systems	396
	9.4-2 High Viscosity Inelastic Fluids	397
	9.4-3 Viscoelastic Systems	412
9.5	Flow Patterns	414
	9.5-1 Class I Agitators	415
	9.5-2 Class II Agitators	416
	9.5-3 Class III Agitators	418
9.6	Mixing and Circulation Times	420
9.7	Gas Dispersion	423
	9.7-1 Gas Dispersion Mechanisms	423
	9.7-2 Power Consumption in Gas-Dispersed Systems	425
	9.7-3 Bubble Size and Holdup	428
	9.7-4 Mass Transfer Coefficient	429
9.8	Heat Transfer	430
	9.8-1 Class I Agitators	431
	9.8-2 Class II Agitators	432
	9.8-3 Class III Agitators	434
9.9	Mixing Equipment and its Selection	436
	9.9-1 Mechanical Agitation	436
	9.9-2 Extruders	436
9.10	Problems	439

Appendix A General Curvilinear Coordinate Systems and Higher Order Tensors ... 441

A.1	Cartesian Vectors and Summation Convention	442
A.2	General Curvilinear Coordinate Systems	445
	A.2-1 Generalized Base Vectors	445
	A.2-2 Transformation Rules for Vectors	449
	A.2-3 Tensors of Arbitrary Order	452
	A.2-4 Metric and Permutation Tensors	454
	A.2-5 Physical Components	458

Contents

A.3 Covariant Differentiation 462
    A.3-1 Definitions 462
    A.3-2 Properties of Christoffel Symbols 464
    A.3-3 Rules of Covariant Differentiation 465
    A.3-4 Grad, Div, and Curl 468
A.4 Integral Transforms 474
A.5 Isotropic Tensors, Objective Tensors and Tensor-Valued Functions 476
    A.5-1 Isotropic Tensors 476
    A.5-2 Objective Tensors 478
    A.5-3 Tensor-Valued Functions 480
A.6 Problems 483

Appendix B Equations of Change 487
  B.1 The Equation of Continuity in Three Coordinate Systems 487
  B.2 The Equation of Motion in Rectangular Coordinates (x, y, z) 487
    B.2-1 In Terms of σ 487
    B.2-2 In Terms of Velocity Gradients for a Newtonian Fluid with Constant ρ and μ 488
  B.3 The Equation of Motion in Cylindrical Coordinates (r, θ, z) 488
    B.3-1 In Terms of σ 488
    B.3-2 In Terms of Velocity Gradients for a Newtonian Fluid with Constant ρ and μ 489
  B.4 The Equation of Motion in Spherical Coordinates (r, θ, ϕ) 490
    B.4-1 In Terms of σ 490
    B.4-2 In Terms of Velocity Gradients for a Newtonian Fluid with Constant ρ and μ 490

References 492

Notation 503

Subject Index 513

1 Introduction

1.1 Definitions and Classification 1
   1.1-1 Purely Viscous or Inelastic Material. 3
   1.1-2 Perfectly Elastic Material 3
   1.1-3 Viscoelastic Material 3

1.2 Non-Newtonian Phenomena 3
   1.2-1 The Weissenberg Effect 4
   1.2-2 Entry Flow, Extrudate Swell, Melt Fracture, and Vibrating Jet. 4
   1.2-3 Recoil 7
   1.2-4 Drag Reduction 7
   1.2-5 Hole Pressure Error. 13
   1.2-6 Mixing 14
   1.2-7 Bubbles, Spheres, and Coalescence 15

Because science evolved and developed first through experimentation, it is appropriate to introduce the complex field of rheology by discussing some of the intriguing and paradoxical phenomena encountered with polymeric liquids and some particulate suspensions. A similar presentation can be found in most textbooks on rheology. For this reason, we have restricted the number of such examples in this chapter. Some definitions and a classification are presented first.

1.1 Definitions and Classification

Rheology is a science that deals with the deformation of materials as a result of an applied stress. It can therefore be considered part of continuum mechanics, although it is also possible to relate the stress to the deformation or to the rate of deformation via molecular kinetic theory.

Two physical laws dating back to the seventeenth century are very important in the present context. They are

(i) *Hooke's law*, describing the behavior of an elastic solid, given in shear by

$$\sigma_{yx} = -G \frac{du_x}{dy}, \tag{1.1-1}$$

where the shear stress σ_{yx} (see Chapter 2) is related to the deformation gradient du_x/dy via the constant elastic modulus G.

(ii) *Newton's law*, describing the behavior of a linear viscous fluid, given by

$$\sigma_{yx} = -\mu \frac{dV_x}{dy}, \tag{1.1-2}$$

where the shear stress σ_{yx} is related to the rate of deformation dV_x/dy, via the constant Newtonian viscosity μ.

We can consider two limiting cases of material response: that of a non-deformable body on the one hand and that of an inviscid fluid on the other hand. For a non-deformable body, the elastic modulus is infinite. For an inviscid fluid, the viscosity is zero.

The behavior of real materials falls between these limiting situations. Table 1.1-1 summarizes the observed rheological behavior. For instance, we can say that although most real materials have a finite viscosity, under certain conditions (e.g., flow of air over an aerofoil), the effect of viscosity is confined to a thin layer (boundary layer, see Section 4.5). Beyond this, the fluid behavior can be well-represented by an ideal inviscid fluid. However, in most confined flow situations, the effects of viscosity cannot be ignored. At the other extreme is an ideal elastic material which attains an equilibrium deformation when subjected to an external stress. For some materials, these limiting behaviors are easily observed. For instance, the viscosity of ice or the elasticity of water may go unnoticed! In between these two extremes, the fluid behavior gradually passes from inviscid ideal to viscous, to viscoelastic fluid-like, to solid-like, and then to elastic solids, as summarized schematically in Table 1.1-1.

Many of the terms used in the field of rheology have been carefully defined by Lodge (1964). We summarize here some of the important definitions.

Table 1.1-1 Summary of Rheological Behavior

			Examples
Continuum mechanics	Fluids	Inviscid fluid (ideal case with $\mu = 0$)	None
		Linear viscous fluid (Newtonian behavior)	Water
		Nonlinear viscous material (Generalized Newtonian behavior defined in Chapter 2)	Suspensions in Newtonian media
		Linear viscoelastic material	Polymer under small deformation
		Nonlinear viscoelastic material	Concentrated polymer solutions or plastics under large deformation
	Solids	Nonlinear elastic material	Rubber
		Linear elastic solid	Linear Hookean spring
		Nondeformable solid (ideal case with $G = \infty$)	None

1.1-1   Purely Viscous or Inelastic Material

A material is purely viscous or inelastic if following any flow or deformation history, the stresses in the material become instantaneously zero (or isotropic) as soon as the flow is stopped (deformation rate set to zero); or the deformation rate (in the absence of inertial effects) becomes instantaneously zero when the stresses are set equal to zero (or are isotropic).

1.1-2   Perfectly Elastic Material

A material is perfectly elastic if the equilibrium shape is attained instantaneously when non-isotropic stresses are applied, or if the stresses become non-isotropic as soon as the material is deformed. Hooke's law (equation 1.1-1) describes a perfectly linear elastic body, if the modulus G is considered constant. The behavior of a rubber band approximates closely that of a perfectly elastic body, but a highly non-linear one, since in this case the modulus changes with deformation.

1.1-3   Viscoelastic Material

Any material which obeys neither the purely viscous nor the perfectly elastic criteria is viscoelastic. The parts of the word, *viscous* and *elastic*, describe a rheological behavior between that of a purely viscous liquid and that of a perfectly elastic solid. In simple terms, a viscoelastic material will not deform instan-taneously when non-isotropic stresses are applied, or the stresses will not respond instantaneously to any imposed deformation or deformation rate. Typical examples are polymer solutions and plastics that are known to exhibit memory effects such as relaxation, described in Chapter 2. The phenomena described in Section 1.2 are mostly due to viscoelasticity.

1.2   Non-Newtonian Phenomena

Most polymer systems, as well as many other complex fluids, do not obey Newton's law of viscosity. These fluids generally exhibit a viscosity that decreases with increasing rate of deformation. This is referred to as pseudoplastic or shear-thinning behavior. Very large decreases in viscosity are observed in polymeric fluids, as illustrated in Chapter 2. Moreover, polymeric fluids have a viscoelastic character that is responsible for a number of spectacular phenomena not observed with Newtonian fluids. The design of many industrial processing operations requires taking into account several of these phenomena.

   We review here some of the more striking viscoelastic effects. An excellent presentation of rheological or non-Newtonian phenomena can be found in the books of Bird, Armstrong, and Hassager (Volume 1, 1977 or 1987) and in another book by Boger and Walters (1993). In writing the following sections we have been largely inspired by these authors.

4   Introduction

1.2-1 The Weissenberg Effect

This phenomenon is illustrated schematically in Fig. 1.2-1. If a rod is rotated in a beaker containing a Newtonian fluid such as water, the free surface is deformed by a centrifugal force, creating a vortex in the center. In contrast, if a rod is rotated in a polymer solution or melt, the fluid tends to climb the rod, and an inverted vortex is created. Weissenberg (1947) was able to explain this phenomenon in terms of unequal normal stresses present in such materials under steady shearing conditions (see Section 3.2-3).

The polymer molecules in a solution or melt form an entangled network, which, when deformed in one direction through the action of a rotating or moving surface, generates internal tensions in the flow direction as well as normal to the flow direction. These tensions are the normal stresses mentioned in the previous paragraph. In fact, if we could measure the pressure at point A on the rod and at point B on the beaker, we would observe, contrary to the Newtonian case, that with the polymeric fluid, $P_A > P_B$. This excess pressure is compensated by an extra hydrostatic head.

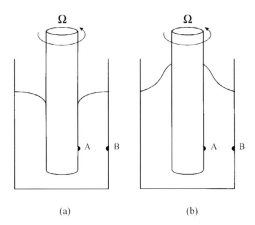

Figure 1.2-1 Shape of the liquid's free surface for a rotating rod in a reservoir
(a) Newtonian liquid
(b) viscoelastic liquid

1.2-2 Entry Flow, Extrudate Swell, Melt Fracture, and Vibrating Jet

Flow visualization of the entry flow in the case of a sudden contraction is illustrated in Fig. 1.2-2. Depending on the liquid rheology and flow conditions, two main patterns are observed. The pattern of Fig. 1.2-2a is observed for branched polymer melts such as low density polyethylene (LDPE), polystyrene (PS), and polymethyl methacrylate (PMMA). The pattern of Fig. 1.2-2b is typical of linear polymer melts such as high density polyethylene (HDPE) and linear low density polyethylene (LLDPE), as well as Newtonian fluids at low Reynolds numbers.

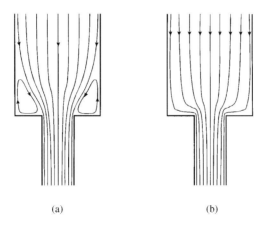

Figure 1.2-2 Main flow patterns in sudden contraction flow

The size of the vortices observed for LDPE, PS,... (Fig. 1.2-2a) increases first with flow rate and eventually becomes unstable. The unstable flow in the reservoir appears at the same moment as the helical distortion illustrated in Fig. 1.2-4. This has been reported by several authors (Den Otter, 1970, 1971; Ballenger et al., 1971; Boger and Ramamurthy, 1972). For linear polyethylenes, the corner vortices are usually not observed, and the flow pattern is that shown in Fig. 1.2-2b.

Another spectacular observation that is very important in the transformation of plastics is the swell of the extrudate as it emerges from a capillary. This is shown in Fig. 1.2-3. A Newtonian fluid (C) normally shows a small decrease ($<20\%$) in diameter as it emerges from the capillary. This is due to inertial effects. In contrast, a highly elastic fluid (A), such as a polymer melt, could show a 200% to 400% increase in diameter. This extrudate swell effect is very frequently referred to in the literature as the *die swell* effect. For obvious reasons this connotation should be avoided!

Qualitatively, we can explain this phenomenon through the presence of normal stresses (extra pressure) created at the wall of the capillary. As the polymeric fluid emerges from the capillary, this internal pressure is released, resulting in a lateral expansion. Another important contribution is due to memory effects in polymeric fluids that behave like rubbery materials. When entering a small capillary die from a large reservoir, the fluid is subjected to a rapid change of shape (large deformation), and as it emerges from the die, it tends through its rubbery nature to recover its initial shape (elastic recovery). For this reason polymeric fluids are often referred to as fluids with memory. Other effects, such as velocity changes at the exit and thermal gradients in the extrudate, also contribute to this phenomenon (Tanner, 1985).

Extrudate swell is thus associated with the elastic nature of the fluid, and its measurement is frequently used to characterize polymer melt elasticity in relation to its molecular structure, molecular weight, and molecular weight distribution. Extrudate swell is a phenomenon that has to be taken into account in fiber production operations. Critical velocity gradients are also complicating and may lead to melt fracture.

6 Introduction

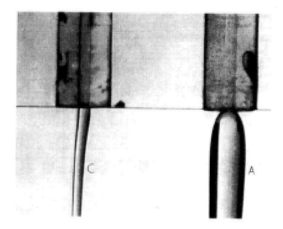

Figure 1.2-3 Fluid extrudates from capillary tubes
A stream of a Newtonian silicone fluid shows no diameter increase (C); a solution of 2.44 g of polymethyl methacrylate ($M_n = 10^6$ kg/kmol) in 100 mL of dimethylphthalate shows a 200% increase in diameter (A). Both fluids have a similar viscosity (From Lodge, 1964, with permission)

Melt fracture is observed as a polymer is extruded freely from a die at a rate exceeding a critical value. The diameter of the extrudate is no longer uniform and may exhibit various distortions, all referred to as melt fracture. Figure 1.2-4 illustrates various shapes of melt fracture encountered under different flow conditions.

- Defects known as *sharkskin* are shown in Fig. 1.2-4a and b. This is an often periodic instability, which depends on the flow rate, temperature, and properties of the polymer. In (a), the extrudate is a linear low density polyethylene, whereas in (b), it is a high density polyethylene.
- In some cases, we observe smooth surfaces followed by so-called sharkskin zones. This is referred to as a bamboo effect (attributed to the stick-slip phenomenon (c)).
- As the extrusion rate is increased, the sharkskin may disappear and the surface of the extrudate may again become smooth, as shown in (d).
- Helical or screw shapes are frequently encountered in the flow of polystyrene (e) or in the flow of polypropylene (f). The amplitude of the distortions increases with increasing flow rate. As the flow rate is further increased, polyethylene, polystyrene, and polypropylene exhibit chaotic distortions (g).

Other polymers may exhibit one or more of the distortions shown in Fig. 1.2-4.

Distortions similar to melt fracture have been observed in polymer processing, such as the calendering of polyvinylchloride (PVC). The film of PVC, usually transparent, becomes partly opaque, and the surface that is not in contact with the roller shows surface defects as the roller's velocity increases, or if the nip between the rollers becomes too narrow.

As a final example in this extrusion section, we show in Fig. 1.2-5 the behavior of a jet emerging from a nozzle subjected to a transverse vibration. The Newtonian fluid (a) breaks into droplets. A concentrated polymer solution (c) emerges as a structurally stable non-uniform wave. Dilute (and very dilute) polymer solutions (b) exhibit a behavior in between

1.2 Non-Newtonian Phenomena 7

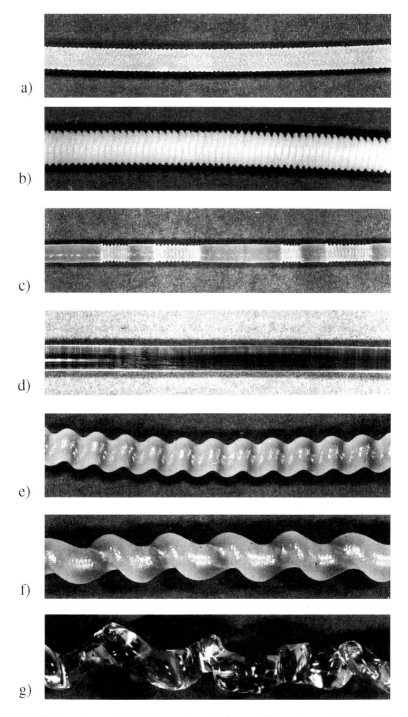

Figure 1.2-4 Various shapes of extrudates under melt fracture (From Agassant et al., 1991)

8   Introduction

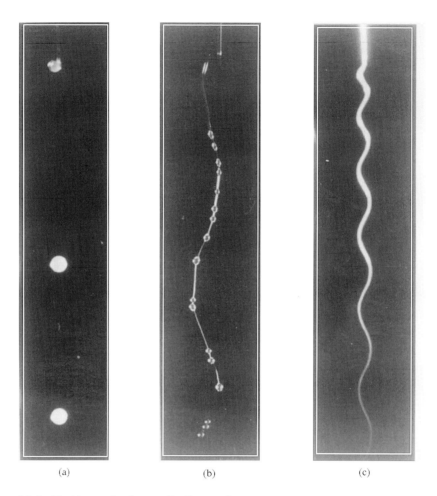

Figure 1.2-5 Liquid emerging from a vibrating nozzle
(a) Newtonian fluid
(b) dilute polymer solution
(c) concentrated polymer solution

that of (a) and (c); that is, drops are connected by a thread. Chan Man Fong et al. (1993) have presented an analysis of this problem, involving elongational as well as oscillatory flow.

1.2-3   Recoil

One experiment that may easily be performed to show the elastic nature of polymeric fluids is the following. If an elastic fluid is forced down a tube through a pressure gradient, the fluid will deform continuously, as indicated by the trace on Fig. 1.2-6. At a given time, the pressure gradient is set to zero, and the fluid starts to flow in the opposite direction. Recoil can be quite spectacular, as shown by Professor Lodge's experiment in Fig. 1.2-7.

1.2 Non-Newtonian Phenomena 9

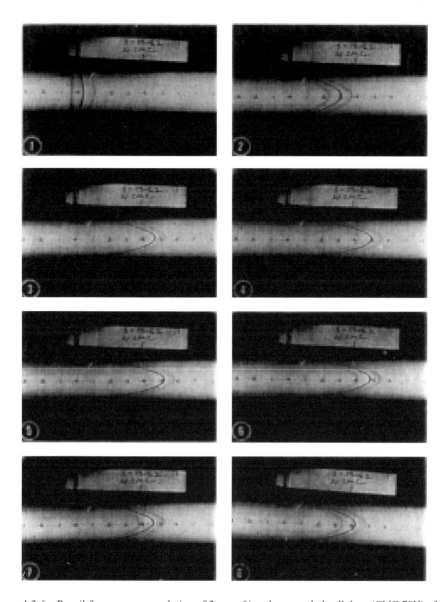

Figure 1.2-6 Recoil for an aqueous solution of 2 mass % carboxy methyl cellulose (CMC 70H) after the pressure gradient has been turned off (just before (f))
Recoil is shown by the charcoal tracer lines in the subsequent photos (From Fredrickson, 1964, with permission)

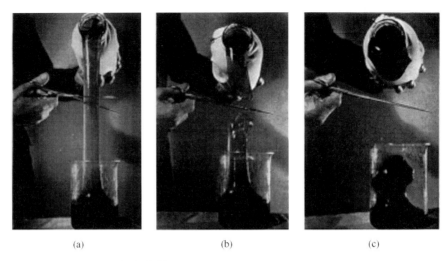

(a)   (b)   (c)

Figure 1.2-7   Recoil in an elastic fluid
An aluminum soap solution (aluminum dilaurate in decalin and m-cresol), is being poured from a beaker (a) and suddenly cut in midstream (b). In photo (c), we note that the liquid above the cut snaps back into the upper beaker (From Lodge, 1964, with permission)

This phenomenon is closely related to the behavior of an elastic band when released of its tension. For viscoelastic fluids, recoil is only partial and takes a finite time. Viscoelastic fluids are said to have a "fading memory", in the sense that they are more affected by recent deformation as opposed to those experienced in the more distant past. Moreover, the effect is strongly dependent on the rate of deformation.

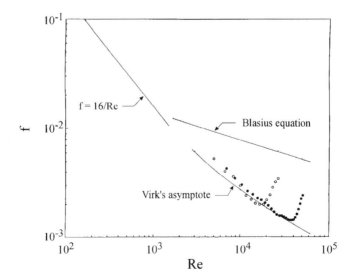

Figure 1.2-8   Typical drag reduction data for the turbulent flow of a 100 ppm PIB solution in cyclohexane (●) and a 100 ppm ODR solution in kerosene (○) (Adapted from Tiu et al., 1995)

1.2-4 Drag Reduction

One phenomenon that was of considerable interest in research in the seventies is the drag reduction obtained by adding a small quantity of high molecular weight, linear, soluble polymer to a fluid in turbulent motion. Figure 1.2-8 shows a conventional friction factor-Reynolds number plot obtained for the two polymers in turbulent tube flow. This friction factor and the Reynolds number are defined by

$$f = \frac{1}{4}\left(\frac{D}{L}\right)\frac{\Delta P}{\frac{1}{2}\rho\langle V\rangle^2} \qquad (1.2\text{-}1)$$

and

$$Re = \frac{D\langle V\rangle\rho}{\mu}, \qquad (1.2\text{-}2)$$

where D and L are the tube diameter and length respectively, ΔP is the pressure drop, $\langle V\rangle$ is the average fluid velocity, and ρ and μ are the fluid density and viscosity respectively. One fluid was a solution containing 100 ppm (mass parts per million) of polyisobutylene (PIB) of a very high molecular weight ($\approx 2 \times 10^6$ kg/kmol) in cyclohexane. The other fluid was a 100 ppm solution of a commercial organic drag reducer (ODR) in kerosene. The molecular weight of the polymer was about 4×10^6 kg/kmol. The figure also shows the theoretical laminar result ($f = 16/Re$), the empirical Blasius equation for the turbulent flow in a smooth pipe ($f = 0.0791/Re^{0.25}$), and the Virk (1975) asymptote for drag-reducing fluids. Both polymer solutions exhibit a substantial reduction of the friction in the turbulent flow regime up to critical values of the Reynolds number. A reduction by a factor of about 2 with respect to the Blasius result obtained for Newtonian solvents is observed. At critical Reynolds numbers, the data show an upward turn, suggesting polymer degradation.

Figure 1.2-9 compares the drag reduction with the heat transfer reduction in turbulent flow in a tube, obtained for a 100 ppm polyacrylamide solution in water. The shear viscosity of this solution was found to be constant and slightly larger than that of water

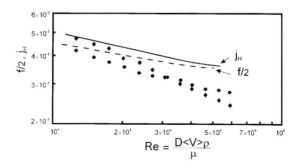

Figure 1.2-9 Heat transfer (———) and pressure drop data (– – –) for water in a circular pipe Friction factor (●) and heat transfer data (◆) for a 100 ppm aqueous polyacrylamide solution (From Del Villar et al., 1984)

($\mu = 1.2$ mPa·s). The pressure drop and heat transfer are reported in terms of the friction factor f and the heat transfer factor j_H defined by

$$j_H = \frac{Nu}{RePr^{\frac{1}{3}}} \qquad (1.2\text{-}3)$$

The Nusselt number, Nu, and the Prandtl number, Pr, are defined by

$$Nu = \frac{hD}{k} \qquad (1.2\text{-}4)$$

and

$$Pr = \frac{\hat{C}_p \mu}{k}, \qquad (1.2\text{-}5)$$

where h is the heat transfer coefficient, and k and $\hat{C}_p$ are the fluid thermal conductivity and the heat capacity per unit mass respectively.

Although the viscosity of the polyacrylamide solution is slightly larger than that of water, the friction factor for the polymer solution is considerably less. The heat transfer reduction when using the polymer solution is possibly more important at higher values of the Reynolds number. For highly turbulent flow, we expect the Chilton-Colburn (1934) analogy to be valid, that is

$$j_H = \frac{f}{2}. \qquad (1.2\text{-}6)$$

This is observed here for water only.

This unusual drag reduction phenomenon has initiated a series of industrial and military investigations. For example, some fire departments tried to make practical use of this drag reduction phenomenon, as illustrated in Fig. 1.2-10. The "rapid" water, containing a small amount of polyethylene oxide, turns out to be slippery, and creates safety problems. Well-known drag reducers are polyethylene oxide and polyacrylamide with molecular weight

Figure 1.2-10 Effect of drag reduction on fire hose range (Taken from Schowalter, 1978, with permission)

above 10^6 kg/kmol. Reductions in friction by a factor of 2 to 5 are possible, but applications of drag reduction to pipeline transportation and marine applications are severely jeopardized by a rapid mechanical degradation of the polymers.

There is no clear understanding of the mechanism of drag reduction. Some researchers have associated this effect with the elastic properties of the polymeric fluids. However, at low concentrations in the range of 10 to 100 ppm, the fluids exhibit hardly any measurable elastic properties. A more acceptable explanation is that those supermacromolecules have a large hydrodynamic volume in the fluid, suppressing a considerable number of sites for the formation of eddies, thereby reducing the turbulence intensity. Also, such large molecules may get trapped at the wall, as a result of the wall roughness conditions. The resulting new surface (wall plus polymer) may be smoother, reducing the pumping energy requirements.

1.2-5 Hole Pressure Error

This experiment illustrates non-Newtonian and viscoelastic effects associated with pressure measurements using pressure transducers. Typically, a wall pressure measurement, P_M, is made by taking a reading at the bottom of a well. As shown in Fig. 1.2-11, the measured pressure gives the correct result for a Newtonian fluid (N) but is too low for the polymer solution (P). As discussed in Chapter 3, the pressure measured at a wall surface for the flow of a viscoelastic fluid is the sum of the thermodynamic pressure, P, and a normal stress component, σ_{yy}. The pressure measured at the bottom of the well, as shown in the figure, will be lower than that measured by a transducer, flush-mounted at the wall of the flow section.

The hole pressure error for different geometries is related to shear and normal stresses. For example, for a circular hole, the hole pressure error is associated with the shear stress, the primary normal stress difference, and the secondary normal stress difference.

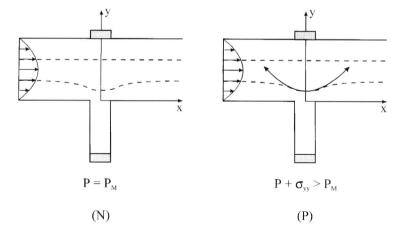

Figure 1.2-11 Hole pressure error
The arrows in the polymer solution indicate how an extra tension along a streamline tends to lift the fluid out of the cavity resulting in a low pressure reading (Adapted from Bird, Armstrong, and Hassager, 1987)

While the current consensus on the secondary normal stress difference seems to be that this quantity is about ten times smaller in magnitude than the primary normal stress difference, as well as being opposite in sign, its history has been quite turbulent. In 1950, Weissenberg postulated the secondary normal stress difference to be zero. Since then, experimenters have found the secondary normal stress difference to be positive, again zero, and now negative. The fact that the magnitude of this quantity is rather small is probably a major cause of the difficulties associated with its measurement. In addition, the secondary normal stress difference is believed to have little bearing on most viscoelastic phenomena.

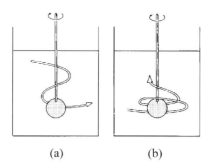

Figure 1.2-12 Flow patterns near a sphere rotating in a viscoelastic fluid
(a) inertial forces dominate
(b) elastic forces dominate (From Ulbrecht and Carreau, 1985)

Figure 1.2-13 Conical stagnant zone observed in a 2 mass % aqueous solution of sodium carboxy methyl cellulose
A decoloration process is used to determine the mixing time

However, one notable exception where the secondary normal stress difference plays an important role deals with the wire coating process. If, because of a disturbance, the wire finds itself off center, a force will act to bring the wire back into a central position, provided the secondary normal stress difference is negative.

1.2-6   Mixing

Non-Newtonian and elastic effects are also responsible for rather striking flow patterns associated with mixing operations. Figure 1.2-12 illustrates the difference in flow patterns in the vicinity of a sphere rotating in a viscoelastic solution.

Another striking as well as detrimental phenomenon is the one shown in Fig. 1.2-13. This figure illustrates the existence of stagnant zones when a polymer solution is mixed by a helical ribbon agitator. Because there is no macroscale mixing going on in a stagnant zone, a situation such as the one depicted in Fig. 1.2-13 could be associated with extremely long mixing times, before a homogeneous product results.

Keirstead et al. (1980) reported large differences in mixing effectiveness depending on the direction of rotation. They reported a difference in mixing time of one order of magnitude between rotations in the helix and counterhelix directions, when mixing ammonium nitrate gels at 200 rpm.

1.2-7   Bubbles, Spheres, and Coalescence

A variety of industrial phenomena rely on mass transfer resulting from liquid-gas contact. The shape of the bubbles is very much affected by the type of fluid. Figure 1.2-14 illus-

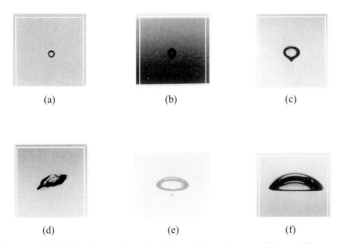

Figure 1.2-14   Shapes of bubbles in a polyacrylamide solution (1 mass % in a 50 mass % mixture of glycerine and water): (a) 0.01 mL; (b) 0.1 mL; (c) 1.0 mL; (e) 2.0 mL; (f) 10 mL; and in a Newtonian 40 mass % aqueous glycerine solution (d) 1.0 mL (From Dajan, 1985)

trates the shapes of bubbles of different volume in a viscoelastic polyacrylamide (PAA) solution and in a Newtonian glycerine solution. Different degrees of magnification were used in order to better portray the shapes, especially for the small volume bubbles. Note the striking difference in bubble shape between (c) and (d). They both portray a 1.0 mL air bubble, but rising in a viscoelastic solution in (c), and in a Newtonian fluid in (d). Bubble shapes, except for very small volume bubbles, are not stable in Newtonian fluids. Several snapshots of the same volume bubble would result in very different pictures (De Kee and Chhabra, 1988). In viscoelastic fluids, the shapes are stable and vary with increasing volumes, from a spherical shape, to a prolate teardrop, to an oblate cusped, and finally to a spherical cap shape (Chhabra and De Kee, 1992).

Figure 1.2-15 illustrates the coalescence phenomena of bubbles in a viscoelastic fluid. Photo (a) shows the simultaneous injection of two bubbles. Photos (b) and (c) illustrate the capture of the trailing bubble in the wake of the leading bubble, and then the film draining after the bubbles make contact. Photo (d) illustrates the tremendous deformations associated with bubble capture, shown here for the simultaneous injection of three bubbles. If the time required for the film to drain and thin after bubble contact is made exceeds the

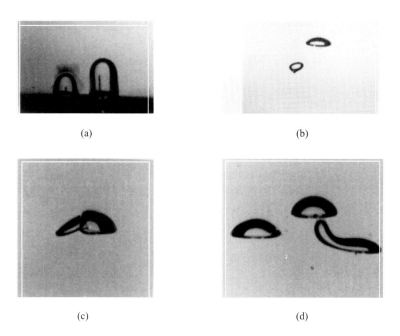

Figure 1.2-15 Bubble coalescence (From Dajan, 1985)
(a) Simultaneous injection of two air bubbles $V_1 = 3.5$ mL and $V_2 = 9.3$ mL in the 1.0 mass % PAA fluid of Fig. 1.2-13. The initial separation between the bubbles is 24 mm
(b) A 1.0 mL bubble moves into the wake of a 4.7 mL bubble. The initial separation between the bubbles was 9 mm. The fluid is again the 1.0 mass % PAA fluid
(c) Bubble contact for the system in frame b
(d) Bubble deformation and capture following a three bubble injection in a 1 mass % aqueous carboxy methyl cellulose solution. Each bubble had a volume of 7.5 mL and their initial separation was 30 mm

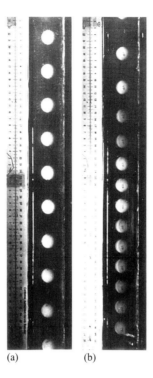

Figure 1.2-16 Motion of a sphere of radius 11.3 mm and density 8×10^3 kg/m^3 in a Newtonian fluid (a) of viscosity 3 Pa·s and in an elastic Boger fluid (b) of viscosity 3 Pa·s and relaxation time of 0.05 s (From Jones et al., 1994, with permission)

period of contact, coalescence will not occur. This is usually the case for equal volume bubbles.

Figure 1.2-16 illustrates the motion of a sphere falling in a Newtonian (a) and in an elastic (b) fluid.

We can observe the successive positions of the spheres. In the case of the Newtonian fluid, a constant velocity is obtained, whereas in the elastic fluid we observe a deceleration over the distance of the tank.

The above are only a few examples of the different behavior exhibited by polymeric materials as compared to Newtonian fluids. We could easily discuss several more of these effects. However, the idea is on the one hand to draw attention to the striking differences between the behaviors of Newtonian and non-Newtonian materials, and on the other hand to suggest that there is probably a variety of flow phenomena involving viscoelastic liquids that is still to be discovered and explained.

2 Material Functions and Generalized Newtonian Fluid

2.1 Material Functions 19
   2.1-1 Simple Shear Flow 19
        2.1-1.1 Steady-State Simple Shear Flow 21
            2.1-1.1.1 Non-Newtonian Viscosity 21
            2.1-1.1.2 Primary (or First) Normal Stress Coefficient 21
            2.1-1.1.3 Secondary Normal Stress Coefficient. 22
   2.1-2 Sinusoidal Shear Flow 25
   2.1-3 Transient Shear Flows 26
        2.1-3.1 Stress Growth Experiment 26
        2.1-3.2 Stress Relaxation Following Steady Shear Flow 29
        2.1-3.3 Stress Relaxation Following a Sudden Deformation. 31
   2.1-4 Elongational Flow. 32
        2.1-4.1 Uniaxial Elongation 32
        2.1-4.2 Biaxial Elongation. 34

2.2 Generalized Newtonian Models 35
   2.2-1 Generalized Newtonian Fluid 36
   2.2-2 The Power-Law Model (Ostwald, 1925) 37
   2.2-3 The Ellis Model (Bird, Armstrong, and Hassager, 1977). 37
   2.2-4 The Carreau Model (1972). 38
   2.2-5 The Cross-Williamson Model (1965) 39
   2.2-6 The Four-Parameter Carreau Model (Carreau et al., 1979b). 39
   2.2-7 The De Kee Model (1977). 40
   2.2-8 The Carreau-Yasuda Model (Yasuda, 1979) 41
   2.2-9 The Bingham Model (1922). 41
   2.2-10 The Casson Model (1959) 42
   2.2-11 The Herschel-Bulkley Model (1926) 42
   2.2-12 The De Kee-Turcotte Model (1980) 42
   2.2-13 Viscosity Models for Complex Flow Situations 43

2.3 Thixotropy, Rheopexy, and Hysteresis 44

2.4 Relations Between Material Functions 48

2.5 Temperature, Pressure and, Molecular Weight Effects. 50
   2.5-1 Effect of Temperature on Viscosity 50
   2.5-2 Effect of Pressure on Viscosity. 52
   2.5-3 Effect of Molecular Weight on Viscosity. 52
        2.5-3.1 Low Molecular Weight Polymers. 54
        2.5-3.2 High Molecular Weight Polymers 55

2.6 Problems 57
   $2.6\text{-}1_a$ Viscosity Data of a PIB Solution 57
   $2.6\text{-}2_a$ Viscosity Data of a CMC Solution 57
   $2.6\text{-}3_a$ The Ellis Model. 58
   $2.6\text{-}4_b$ Viscosity Data for a PS Solution 58

2.6-5_b Rheological Behavior of Drilling Muds 59
2.6-6_b The Cross-Williamson Model 59
2.6-7_b Viscosity-Molecular Weight Relationship 60

It has been shown that even for the most complicated constitutive equations for fluids, there are special flows for which the response functional manifests itself through three viscometric functions only (Coleman, Markovitz, and Noll, 1966). A constitutive equation relates the stress to the deformation or to the rate of deformation. One of these viscometric functions is a nonlinear (non-Newtonian) shear viscosity, the other two are differences of normal stresses. The answer to the question whether the viscometric functions are independent is of theoretical as well as practical value. In this chapter we define a variety of important material functions which we will encounter throughout the book. Frequently used viscosity models are presented, and some useful relations between material functions are given.

2.1 Material Functions

2.1-1 Simple Shear Flow

Simple shear (viscometric) flow is defined as follows: a fluid is contained between two infinite flat parallel plates as illustrated in Fig. 2.1-1

We can imagine the liquid to be composed of several thin sheets of fluid, parallel to the plates. Under static conditions (both plates are stationary), the velocity profile, assuming we can talk about a velocity profile under static conditions, is represented by a vertical line. If suddenly we decide to set the lower plate in motion in the positive x-direction, the velocity profile may be given by the same vertical line except for the layer in contact with the moving plate. Molecules in this layer now have the plate velocity, V, associated with their masses, and as such a different momentum.

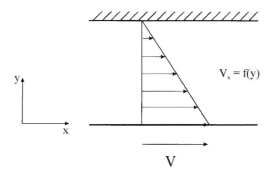

Figure 2.1-1 Sketch defining unidirectional shear flow

20   Material Functions and Generalized Newtonian Fluid

It is now feasible for molecules to jump from layer one into layer two and vice versa. Those molecules arriving in layer one will, because of the moving plate, instantly adopt the plate velocity. The molecules arriving in layer two (from layer one) will increase the momentum of layer two. Jumps occurring simultaneously in layers farther away from the moving plate are not yet affecting the net change in velocity profile at this stage. The jumping process from layer to layer will result in momentum being transported in the positive y-direction.

Eventually, provided the gap between the plates is small enough and the flow is laminar, a linear velocity profile will be established for which we can write

$$V_x = \dot{\gamma}_{yx} y; \quad V_y = V_z = 0, \tag{2.1-1}$$

where the shear rate is

$$\dot{\gamma}_{yx} = \frac{dV_x}{dy}. \tag{2.1-2}$$

The force per unit area required to keep the lower plate moving at a constant velocity V defines a shear stress σ_{yx}, which is directly proportional to the plate velocity and inversely proportional to the distance between the plates. That is,

$$\sigma_{yx} \propto \frac{dV_x}{dy} = \dot{\gamma}_{yx} \tag{2.1-3}$$

The interpretation of the subscripts yx has been given by Bird, Stewart, and Lightfoot (1960) as follows: σ_{yx} represents a shear stress exerted in the x-direction on a fluid surface of constant y by the fluid in the region of lesser y. It can also be interpreted as a flux of x-momentum transferred in the y-direction. The quantity on the right side of equation 2.1-3, $\dot{\gamma}_{yx}$, is a shear component of the rate-of-deformation tensor, $\underline{\underline{\dot{\gamma}}}$, defined as

$$\underline{\underline{\dot{\gamma}}} = \nabla \underline{V} + \nabla \underline{V}^+, \tag{2.1-4}$$

where $\nabla \underline{V}$ and $\nabla \underline{V}^+$ are the velocity gradient tensor and its transpose, respectively; $\underline{\underline{\dot{\gamma}}}$ thus represents nine components.

Labeling the axes x and y is of course arbitrary. In a more general way we can refer to the quantity on the left side of equation 2.1-3 as σ_{ij}, where both i and j can take on the values 1, 2, or 3. In the particular case of Cartesian coordinates, 1 refers to the x-direction, 2 to the y-direction, and 3 to the z-direction. σ_{ij} thus represents a quantity characterized by nine components. This quantity is a second-order tensor. We recall that a first-order tensor (or a vector) such as, for example, the velocity, requires three components to be defined (V_x, V_y, and V_z in Cartesian coordinates), and that a zero-order tensor (a scalar), such as temperature, requires only one numerical value to be completely defined.

The nine components of the stress tensor can be represented by a 3×3 matrix as follows:

$$\underline{\underline{\sigma}} = \begin{pmatrix} \sigma_{11} & \sigma_{12} & \sigma_{13} \\ \sigma_{21} & \sigma_{22} & \sigma_{23} \\ \sigma_{31} & \sigma_{32} & \sigma_{33} \end{pmatrix}. \tag{2.1-5}$$

Any component of the stress tensor can be interpreted as the component of a force per unit area acting on a specific surface of a material elementary volume as depicted in Fig. 2.1-2 for Cartesian coordinates. Let us consider surface (2), which is normal to the x_2-axis: $\underline{\sigma}_2$

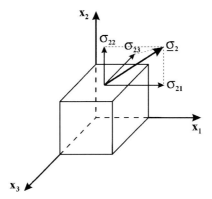

Figure 2.1-2 Decomposition of the force acting on a surface of a cubic element

represents the net force acting on the surface per unit area; its magnitude and orientation depend on the flow field. This force per unit area is a vector that can be decomposed into three components, σ_{21}, σ_{22}, and σ_{23}.

The same procedure can be followed for the other surfaces. Since this material element is in equilibrium with its surroundings, only three resulting forces are independent, that is $\underline{\sigma}_1$, $\underline{\sigma}_2$, and $\underline{\sigma}_3$ (Lodge, 1964). Nine independent components are thus generated: the first index in σ_{ij} refers to the surface considered and the second gives the direction of the force. Finally, in equilibrium, no resultant torque can be acting on the material element: hence, $\sigma_{12} = \sigma_{21}$, $\sigma_{13} = \sigma_{31}$, and $\sigma_{23} = \sigma_{32}$ (Lodge, 1964). The stress tensor is symmetric, and this reduces the number of independent stress components from nine to six.

Of particular interest in our context (shear flow) is the component σ_{21} (the shear stress), which by symmetry equals σ_{12}, and the components σ_{ii} on the diagonal. We will be mainly interested in differences among those normal stresses, as they explain a variety of rheological phenomena. As outlined next, the shear stress σ_{21} is related to the shear rate $\dot{\gamma}_{21}$. In this context, the second subscript (1) indicates the direction of flow, and the first subscript (2) indicates the direction in which the velocity changes.

2.1-1.1 Steady-State Simple Shear Flow

For steady shear flow, where the shear rate $\dot{\gamma}_{21}$ is constant, we define the following material functions (using $\dot{\gamma}$ for the shear rate and subscripts y and x in place of 2 and 1).

2.1-1.1.1 Non-Newtonian Viscosity

$$\eta(\dot{\gamma}) = -\frac{\sigma_{yx}}{\dot{\gamma}} \qquad (2.1\text{-}6)$$

2.1-1.1.2 Primary (or First) Normal Stress Coefficient

$$\psi_1(\dot{\gamma}) = -\left(\frac{\sigma_{xx} - \sigma_{yy}}{\dot{\gamma}^2}\right) \qquad (2.1\text{-}7)$$

2.1-1.1.3 Secondary Normal Stress Coefficient

$$\psi_2(\dot{\gamma}) = -\left(\frac{\sigma_{yy} - \sigma_{zz}}{\dot{\gamma}^2}\right) \quad (2.1\text{-}8)$$

The quantities $(\sigma_{xx} - \sigma_{yy})$ and $(\sigma_{yy} - \sigma_{zz})$ represent the primary normal stress difference N_1 and the secondary normal stress difference N_2 respectively. The relation between ψ_1 and ψ_2 is normally taken as $\psi_2 \approx -0.1\psi_1$. In the majority of flow situations, the secondary normal stress coefficient ψ_2 is not all that important. Figures 2.1-3 to 2.1-5 show typical viscosity-shear rate and primary normal stress-shear rate behavior for a variety of viscoelastic solutions, such as a 2.0 mass % solution of polyacrylamide (Separan AP 30) in a 50 mass % mixture of water and glycerine, and a 6.0 mass % solution of polyisobutylene (PIB) in Primol 355. Primol 355 is a pharmaceutical-grade white oil with a viscosity of 0.15 Pa·s at 298 K. Note the tremendous drop in viscosity over the shear rate range shown. This behavior is typical for viscoelastic solutions. The primary normal stress coefficient data of Fig. 2.1-4 show a similar trend. However, note that the limiting behavior at low shear rates is not accessible and that the drop with increasing shear rate is more severe.

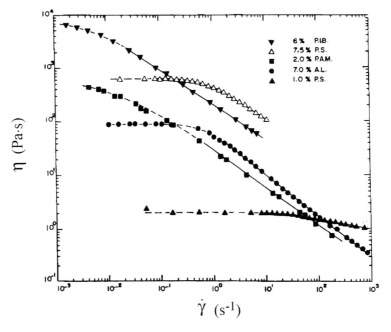

Figure 2.1-3 Viscosity-shear rate plots for typical viscoelastic solutions
The 1.0 and 7.5 PS are respectively 1.0 and 7.5 mass % solutions of narrow molecular weight polystyrene ($M_w = 860\,000$ kg/kmol) in Aroclor 1248 (Data from Ashare, 1968). Aroclor 1248 is a chlorinated diphenyl with a viscosity of 0.3 Pa·s at 298 K. The 7.0% AL is a 7.0 mass % solution of aluminum laurate in decalin and m-cresol (Data from Huppler, 1965). The 2.0% PAM is a 2.0 mass % solution of polyacrylamide (AP30 of Dow Chemical) in a 50 mass % mixture of water and glycerine. The 6% PIB is a 6.0 mass % solution of polyisobutylene ($M_w \sim 1.5 \times 10^7$ kg/kmol) in Primol 355 (Data from De Kee, 1977)

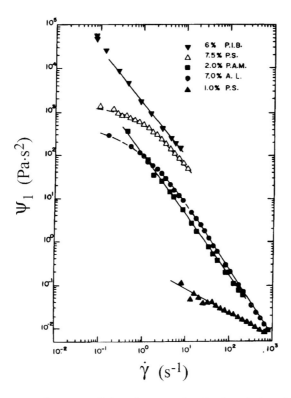

Figure 2.1-4 Primary normal stress coefficient-shear rate plots for the solutions described in Fig. 2.1-3

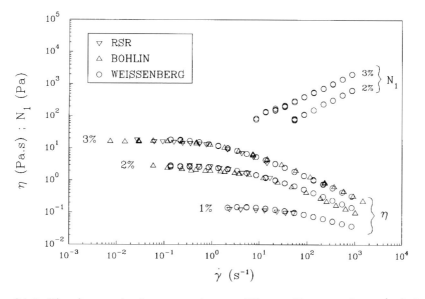

Figure 2.1-5 Viscosity η and primary normal stress difference N_1 versus shear rate $\dot{\gamma}$ for PEO ($M_v = 1.8 \times 10^6$ kg/kmol) solutions of different concentration: (a) in water

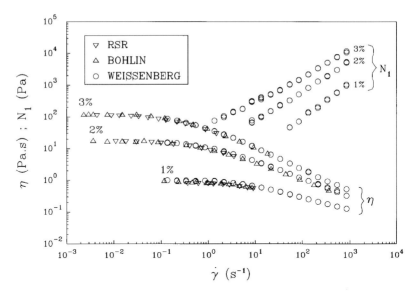

Figure 2.1-5 (Continued) (b) in 50 mass % mixture of water and glycerine (Data from Ortiz, 1992, see Ortiz et al., 1994)

In Fig. 2.1-5 we show typical data obtained using a Union Carbide (N-60 K) polyethylene oxide (PEO) with a molecular weight of 1.8×10^6 kg/kmol for solutions of 1 to 3 mass % in water (a) and in water and glycerine (b). Shear thinning becomes more important with increasing polymer concentration. At low shear rate we can observe a zero shear rate viscosity plateau which is more pronounced at higher concentration. The water-glycerine solvent produces viscosities and normal stresses of a higher magnitude than the aqueous solutions. The data obtained with different rheometers were within 15% (in the worst case) and superposed well. The three instruments used were a Weissenberg rheogoniometer, a Rheometrics controlled stress rheometer (RSR), and a Bohlin VOR rheometer.

Figure 2.1-6 reports the steady shear viscosity and primary normal stress difference for a so-called Boger fluid (Boger, 1977), which is a very dilute solution of a high molecular weight polymer in a very viscous solvent. The Boger fluid here is 0.244 mass % of a polyisobutylene in a mixed solvent consisting of 7 mass % of kerosene in polybutene, known as M1 (Sridhar, 1990). As shown in the figure, the viscosity is almost constant (very little shear thinning) and the primary normal stress difference is quadratic with respect to the shear rate ($\psi_1 =$ constant) for the lower values of the shear rate. Boger fluids that are non-shear-thinning but elastic are useful model fluids for investigating rheological effects in various flow situations (Boger and Walters, 1993).

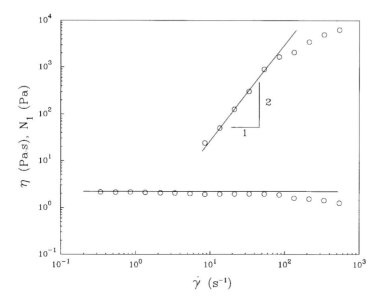

Figure 2.1-6 Steady shear viscosity and primary normal stress difference of the M1 fluid at 25 °C Data obtained on a Weissenberg rheogoniometer, model R-18

2.1-2 Sinusoidal Shear Flow

For small amplitude oscillatory shear flow, the lower plate in Fig. 2.1-1 would be required to oscillate sinusoidally in the x-direction, with small amplitude, under a variety of frequencies ω. This situation is illustrated in Fig. 2.1-7. We define a complex viscosity as follows:

$$\eta^* = \eta' - i\eta'' = -\frac{\sigma_{21}^0}{\dot{\gamma}^0}, \qquad (2.1\text{-}9)$$

with

$$\dot{\gamma}(t) = \text{Re}[\dot{\gamma}^0 e^{i\omega t}] \qquad (2.1\text{-}10)$$

and

$$\sigma_{21}(t) = \text{Re}[\sigma_{21}^0 e^{i\omega t}]. \qquad (2.1\text{-}11)$$

For small deformation (in the linear viscoelastic domain as discussed in Chapter 5), inertial effects can be ignored, and the stress response is a sine wave of the same frequency as the input function, but out of phase. Here Re [-] stands for the real part of [-]; $\dot{\gamma}^0$ and σ_{21}^0 represent the complex amplitudes of $\dot{\gamma}$ and σ_{21} respectively; η' is referred to as the dynamic viscosity and is associated with energy dissipation; while the coefficient of i, η'', represents an elastic contribution associated with energy storage, and which could be labeled dynamic rigidity.

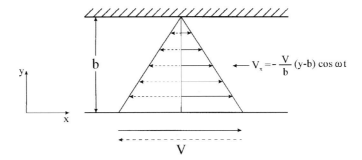

Figure 2.1-7 Sketch defining sinusoidal shear flow

It is also possible to work in terms of a quantity G^*, defined as

$$G^* = G' + iG'', \qquad (2.1\text{-}12)$$

where the storage modulus $G' = \omega \eta''$ and the loss modulus $G'' = \omega \eta'$. These material functions are used for material characterization, and they relate to molecular structure (Ferry, 1980).

Figure 2.1-8 illustrates the dependence of the dynamic viscosity η' and the storage modulus G' on the frequency ω, for 3.0 mass % PEO solutions of different molecular weights. We observe that at a given frequency, ω, both η' and G' increase with molecular weight, and at high frequency they both approach a simple power-law behavior, almost independent of molecular weight.

2.1-3   Transient Shear Flows

Transient or time-dependent shear flows are associated, for example, with the start-up of processes involving the displacement of viscoelastic materials. Under such initial flow conditions, stresses can reach magnitudes which are substantially more important than their steady-state values, achieved for the applied shear rate.

2.1-3.1   Stress Growth Experiment

For stress growth, after onset of steady simple shear (the lower plate in Fig. 2.1-1 starts moving in the positive *x*-direction), we have

$$\dot{\gamma}(t) = \dot{\gamma}_\infty h(t), \qquad (2.1\text{-}13)$$

where $\dot{\gamma}_\infty$ is the constant velocity gradient for $t > 0$, and $h(t)$ is the unit step function

$$h(t) = 0 \quad \text{for} \quad t < 0 \qquad (2.1\text{-}14\text{a})$$
$$\text{and} \quad h(t) = 1 \quad \text{for} \quad t > 0. \qquad (2.1\text{-}14\text{b})$$

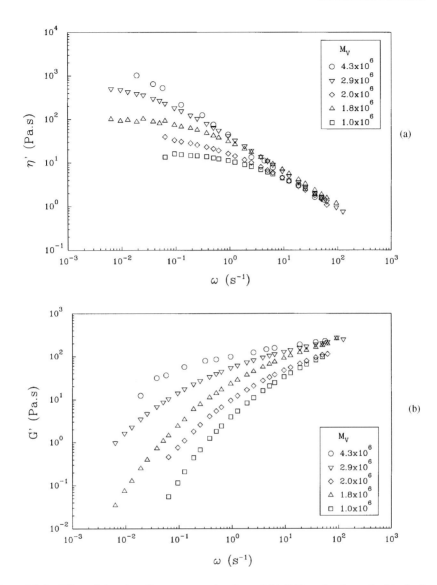

Figure 2.1-8 Effect of the viscosity average molecular weight (M_v) on the dynamic data for 3 mass % PEO solutions in water and glycerine:
(a) η' data
(b) G' data
(Data from Ortiz, 1992, see Ortiz et al., 1994)

28   Material Functions and Generalized Newtonian Fluid

We define the time-dependent shear stress and normal stress coefficients as follows:

$$\eta^+(t, \dot{\gamma}_\infty) = -\frac{\sigma_{yx}(t)}{\dot{\gamma}_\infty}, \tag{2.1-15}$$

$$\psi_1^+(t, \dot{\gamma}_\infty) = -\frac{[\sigma_{xx}(t) - \sigma_{yy}(t)]}{\dot{\gamma}_\infty^2}, \tag{2.1-16}$$

and $\quad \psi_2^+(t, \dot{\gamma}_\infty) = -\frac{[\sigma_{yy}(t) - \sigma_{zz}(t)]}{\dot{\gamma}_\infty^2}. \tag{2.1-17}$

Figure 2.1-9 illustrates this experiment schematically. The lower part of the figure shows the effect of the imposed shear rate $\dot{\gamma}_\infty$ on the reduced shear stress growth function. $\eta(\dot{\gamma}_\infty)$ is the steady shear viscosity value. At low $\dot{\gamma}_\infty$ the function increases monotonously. At higher values of $\dot{\gamma}_\infty$ stress overshoot occurs. The time at which the maximum overshoot occurs decreases with increasing shear rate. The magnitude of the overshoot increases with increasing shear rate. The higher the shear rate, the sooner steady state is attained. The response of a Newtonian fluid (η^+/η), in the absence of inertial effects, is given by the unit step function.

Similar behavior would be observed for the reduced normal stress growth function. The normal stress growth process evolves over a longer time period than the shear stress growth process. This behavior is typical for viscoelastic solutions. Figure 2.1-10 illustrates the stress growth function $\eta^+(\dot{\gamma}_\infty, t)$ for a 20 mass % polystyrene (PS) of molecular weight 1.3×10^6 kg/kmol in dibutylphthalate. For imposed shear rates exceeding 1 s^{-1}, overshoots are observed. A single monotonously increasing function of time is obtained for small values of time and/or small values of shear rate, as predicted by linear viscoelastic behavior (see Chapter 5).

Figure 2.1-11 shows the normal stress growth function $\psi_1^+(t, \dot{\gamma}_\infty)$ for a 25 mass % PS of molecular weight 1.3×10^6 kg/kmol in dibutylphthalate. Here too, we observe an overshoot for $\dot{\gamma}_\infty$ exceeding 1 s^{-1}. As in the case of stress growth, linear viscoelastic behavior is evident for short times even at higher shear rates.

Reduced normal stress growth functions for the 6.0 mass % PIB solution referred to earlier (see Fig. 2.1-3) are shown in Fig. 2.1-12. The observed undershoot at large time for

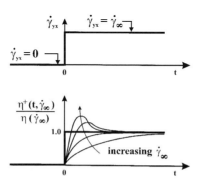

Figure 2.1-9   Stress growth experiment

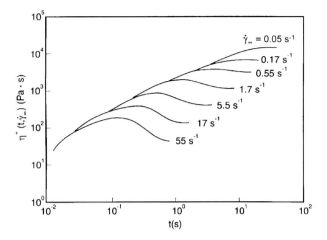

Figure 2.1-10 Shear stress growth function for a 20 mass % PS solution (Data from Attané, 1984, with permission)

$\dot{\gamma}_\infty = 4.34 \text{ s}^{-1}$ is probably due to the interaction of the sample with the measuring instrument. This type of coupling effect is discussed in Section 3.3-4.

2.1-3.2 Stress Relaxation Following Steady-Shear Flow

Similarly, for stress relaxation after cessation of steady simple shear (the lower plate in Fig. 2.1-1 is suddenly stopped), the shear rate is given by

$$\dot{\gamma}(t) = \dot{\gamma}_0[1 - h(t)]. \tag{2.1-18}$$

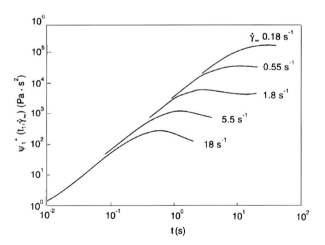

Figure 2.1-11 Normal stress growth function for a 25 mass % PS solution (Data from Attané, 1984, with permission)

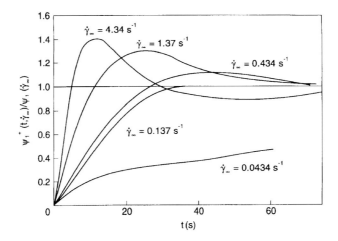

Figure 2.1-12 Normal stress growth function for a 6.0 mass % PIB solution in Primol (Data from DeKee, 1977)

Here $\dot{\gamma}_0$ is the initial constant shear rate, and the transient shear stress and normal stress coefficients are defined as

$$\eta^-(t, \dot{\gamma}_0) = -\frac{\sigma_{yx}(t)}{\dot{\gamma}_0}, \tag{2.1-19}$$

$$\psi_1^-(t, \dot{\gamma}_0) = -\frac{[\sigma_{xx}(t) - \sigma_{yy}(t)]}{\dot{\gamma}_0^2}, \tag{2.1-20}$$

and

$$\psi_2^-(t, \dot{\gamma}_0) = -\frac{[\sigma_{yy}(t) - \sigma_{zz}(t)]}{\dot{\gamma}_0^2}. \tag{2.1-21}$$

Figure 2.1-13 illustrates this experiment schematically. The lower part of the figure shows the effect of the imposed shear rate on the reduced shear stress relaxation function. $\eta(\dot{\gamma}_0)$ is the steady shear viscosity value.

A more rapid decrease in shear stress is observed with increasing initial shear rate $\dot{\gamma}_0$. For a Newtonian fluid, the stress relaxes instantaneously to zero. The normal stress relaxation function $\psi_1^-(t, \dot{\gamma}_0)$ follows a similar pattern, but evolves over a longer time period.

Figure 2.1-14 shows the stress relaxation of the 20 mass % PS solution of Fig. 2.1-10. For very small values of shear rate ($\dot{\gamma}_0 \leqslant 0.015 \text{ s}^{-1}$), the function is independent of shear rate, as predicted by linear viscoelasticity (see Chapter 5).

Figure 2.1-15 illustrates the normal stress relaxation for the 25 mass % PS solution of Fig. 2.1-11.

No experiment at low enough values of initial shear rate $\dot{\gamma}_0$ could be performed to verify the linear viscoelastic behavior. Linear viscoelastic behavior would be characterized

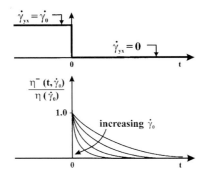

Figure 2.1-13 Stress relaxation following steady shear flow

by a single response curve, independent of $\dot{\gamma}_0$. Reduced normal stress relaxation for the 2.0 mass % PAA solution referred to earlier (see Fig. 2.1-3) is shown in Fig. 2.1-16.

2.1-3.3 Stress Relaxation Following a Sudden Deformation

The stress relaxation after a sudden deformation is defined as

$$\sigma_{yx}(t) = -G(t, \gamma_0)\gamma_0, \tag{2.1-22}$$

where $G(t, \gamma_0)$ is the relaxation modulus and γ_0 is the magnitude of the applied shear strain. The experiment and typical qualitative results are illustrated in Fig. 2.1-17. For small enough applied deformation $\gamma_0 \ll 1$, the relaxation modulus is a unique monotonously decreasing function of time. For large values of the deformation, $\gamma_0 > 1$, the modulus is also a decreasing function of the applied deformation, and the data frequently depict parallel curves on log-log plots, as shown in the Fig. 2.1-17.

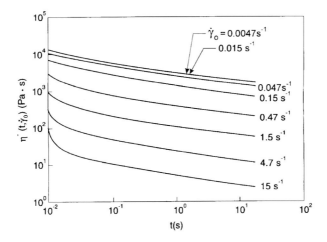

Figure 2.1-14 Shear stress relaxation for the 20 mass % PS solution of Fig. 2.1-10 (Data from Attané, 1984, with permission)

32   Material Functions and Generalized Newtonian Fluid

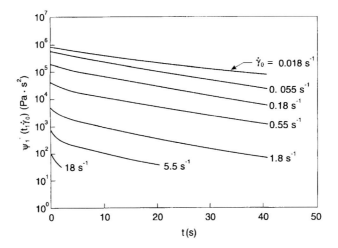

Figure 2.1-15   Normal stress relaxation for the 25 mass % PS solution of Fig. 2.1-11 (Data from Attané, 1984, with permission)

2.1-4   Elongational Flow

2.1-4.1   Uniaxial Elongation

In the case of simple (uniaxial) elongation at constant volume, the flow is non-viscometric, and the velocity profile is given by

$$V_z = \dot{\varepsilon} z, \tag{2.1-23}$$

$$V_x = -\frac{\dot{\varepsilon}}{2} x, \tag{2.1-24}$$

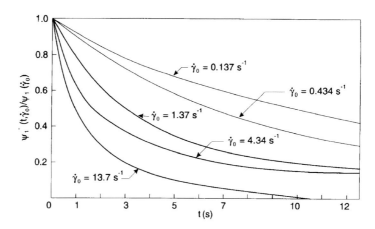

Figure 2.1-16   Normal stress relaxation for a 2.0 mass % solution of polyacrylamide in a 50 mass % mixture of water and glycerine (Data from De Kee, 1977)

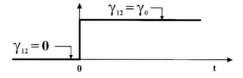

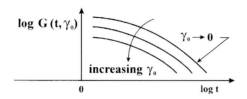

Figure 2.1-17 Shear stress relaxation following a sudden deformation

and

$$V_y = -\frac{\dot{\varepsilon}}{2}y, \qquad (2.1\text{-}25)$$

where $\dot{\varepsilon}$ is the elongational rate, which is constant at steady state. The elongational viscosity is defined as

$$\eta_E(\dot{\varepsilon}) = -\frac{(\sigma_{zz} + P)}{\dot{\varepsilon}} = -\frac{(\sigma_{zz} - \sigma_{xx})}{\dot{\varepsilon}}. \qquad (2.1\text{-}26)$$

Note that $\sigma_{zz} + P = \pi_{zz}$, the total normal stress component in the flow direction. The isotropic term P is equal to $-\sigma_{xx}$ or $-\sigma_{yy}$ for uniaxial elongational flow.

Furthermore, we can also define an elongational stress growth function as

$$\eta_E^+(t, \dot{\varepsilon}_\infty) = -\frac{[\sigma_{zz}(t) - \sigma_{xx}(t)]}{\dot{\varepsilon}_\infty}, \qquad (2.1\text{-}27)$$

by analogy to the definition in equation 2.1-15. It is experimentally extremely difficult to generate steady elongational viscosity at high elongational rates. Most authors report elongational growth functions as illustrated in Fig. 2.1-18 for a molten polymethyl methacrylate (PMMA) using an extensiometer in which the sample is clamped at both ends, with controlled elongational velocity (Froelich et al., 1986). For short times, the data describe a unique and monotonously increasing function of time. A plateau indicative of steady state is attained for the smallest value of the elongational rate, $\dot{\varepsilon} = 0.002$ s^{-1}. For higher elongational rate, the elongational growth function increases with the rate at large times, and no steady values are attained, mostly because of instrument limitations. The figure also shows that the classical Trouton relation, valid for Newtonian fluids, is verified here at low elongational rates under steady-state conditions:

$$\eta_E = 3\eta_0. \qquad (2.1\text{-}28)$$

For measuring the elongational properties of low viscosity fluids, many techniques have been developed. So far, these techniques generate results which can differ by several orders

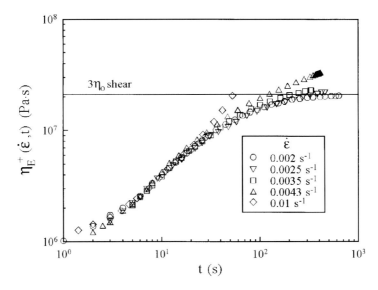

Figure 2.1-18 Uniaxial elongational properties of PMMA at 170 °C
The polymer is a commercial sample supplied by Norsolor, France, under the trade name of Altulite 2773. The average molecular weight is $M_w \approx 130\,000$ kg/kmol with a polydispersity, M_w/M_n, of 1.9 (Data of Bousmina, 1992, with permission)

of magnitude (Hudson and Jones, 1993). One of the more promising techniques has been developed by Tirtaatmadja and Sridhar (1993). Their technique involves a controlled exponential separation of two disks, which hold the fluid sample to be tested. Figure 2.1-19 illustrates the extensional viscosity growth function of a polyisobutylene (PIB) solution for different elongational rates. Note that steady-state conditions appear to be reached for only a few values of the elongational rate ($\dot{\varepsilon} = 2.0$ and 2.7 s^{-1}).

At very short times, the Trouton ratio is less than 3, increasing exponentially with time as predicted by linear viscoelasticity. Then, for times up to two seconds, a value of 3 is observed for the data at low $\dot{\varepsilon}$. The increase of η_E^+ with time is more pronounced as a larger value of the elongational rate is imposed. This behavior is referred to as strain hardening in extensional flow, in contrast to the shear-thinning behavior usually observed in simple shear flow. For flexible high molecular weight polymers in solution, the magnitude of the elongational viscosity, as shown in Fig. 2.1-19, can be two to four decades higher than the shear viscosity. This PIB solution is a so-called Boger fluid, and its recipe is very similar to the M1 fluid discussed in Section 2.1-1 (see Fig. 2.1-6).

2.1-4.2 Biaxial Elongation

Biaxial extension or elongation is defined by the following velocity components:

$$V_1 = \dot{\varepsilon}_B x_1, \quad (2.1\text{-}29)$$

$$V_2 = \dot{\varepsilon}_B x_2, \quad (2.1\text{-}30)$$

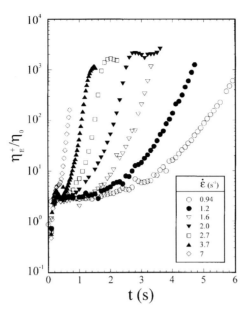

Figure 2.1-19   Elongational properties of 0.31% PIB in a mixture of polybutene and tetradecane at 19.5 °C The zero shear viscosity at 20 °C is 21.63 Pa·s (Data from Tirtaatmadja, 1993, with permission)

and

$$V_3 = -2\dot{\varepsilon}_B x_3, \tag{2.1-31}$$

where $\dot{\varepsilon}_B$ is a positive extension rate. The biaxial extensional viscosity is defined as

$$\eta_B(\dot{\varepsilon}_B) = -\frac{(\sigma_{11} - \sigma_{33})}{\dot{\varepsilon}_B}. \tag{2.1-32}$$

Biaxial extensional properties play an essential role in many polymer processing situations, such as film blowing. More general definitions for extensional flow can be found in Dealy and Wissbrun (1990).

These relations define only a fraction of the material functions which are currently in use. Several other tests have been developed that give rise to material response functions that we did not mention. Nevertheless, the definitions presented here will suffice for our purposes.

2.2   Generalized Newtonian Models

Fluids such as gases, water, and organic solvents are made up of relatively small molecules, and their viscosity (μ or η_0) at a given pressure and temperature is constant. Such fluids are

36 Material Functions and Generalized Newtonian Fluid

Newtonian. That is, μ is not a function of shear rate $\dot{\gamma}_{21}$ or of shear stress σ_{21}. When larger molecules are involved, such as polymer molecules, the viscosity is no longer constant under flow.

2.2-1 Generalized Newtonian Fluid

A generalized Newtonian fluid (GNF) is defined as a purely viscous fluid. That is its viscosity depends only on the shear rate or shear stress. In many flow situations, viscoelasticity will not play an important role, and the non-Newtonian viscosity (η) is sufficient to describe the rheology of the material (see criteria defined in Section 4.1). Newton's law of viscosity is generalized for incompressible fluids by

$$\underline{\underline{\sigma}} = -\eta \underline{\underline{\dot{\gamma}}}, \qquad (2.2\text{-}1)$$

where η is the non-Newtonian viscosity, usually expressed as a function of shear rate.

As illustrated in Figs. 2.1-3 and 2.1-5, viscosity vs shear rate data are usually represented on log-log plots, as shown schematically in Fig. 2.2-1. The simplest viscosity model, the Newtonian (one-parameter) model, would be represented by a horizontal line in this figure. That is, the viscosity η is given by the constant value μ (or η_0). In SI units, the viscosity is expressed in Pa·s. The shear rate $\dot{\gamma}$ is expressed in s^{-1}, independent of the system of units used. Clearly, the typical behavior depicted in the figure suggests the need for a more realistic viscosity model. Quite a variety of non-Newtonian, (more than one parameter) viscosity models have been proposed in the literature. Here, we introduce a few examples of two-, three-, four-, and five-parameter models.

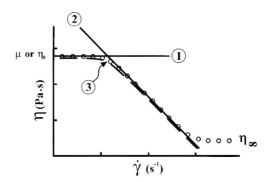

Figure 2.2-1   Sketch of viscosity-shear rate data (log-log plot) as well as predictions of some viscosity models

2.2-2 The Power-Law Model (Ostwald, 1925)

For engineering applications, the power-law or Ostwald-De Waele model is the most frequently used two-parameter GNF model. It is given by

$$\eta = m|\dot\gamma|^{n-1}. \tag{2.2-2}$$

The parameters are m and n. On the usual log-log coordinates, this equation results in a linear relation between η and $\dot\gamma$. Line (2) in Fig. 2.2-1 illustrates the model capability. The power-law model represents the data in the so-called power-law region.

The parameter n is less than one for shear-thinning or pseudoplastic materials and is greater than one for shear-thickening or dilatant materials. Note that in the special case of $n = 1$, the model reduces to the one-parameter Newtonian case, and the parameter m represents μ or η_0, a constant viscosity. Although the power-law model, due to its simplicity, is used in a vast majority of industrial situations, we want to draw the reader's attention to the following problems. The units of the parameter m are $Pa \cdot s^n$, meaning that they depend on the value of the parameter n. Consequently, we cannot construct a parameter with units of time based on the model parameters m and n. As will be outlined later, a characteristic time is necessary to describe viscoelastic behavior. Finally, for shear-thinning fluids, the model predicts a viscosity that goes to infinity as the shear rate approaches zero. This could lead to drastic viscosity overpredictions for flow situations or flow regions where the rate of deformation is very small.

2.2-3 The Ellis Model (Bird, Armstrong, and Hassager, 1977)

This is an example of a three-parameter model, where the viscosity is given as a function of shear stress, as follows:

$$\eta = \frac{\eta_0}{1 + \left|\frac{\sigma_{yx}}{\sigma_{\frac{1}{2}}}\right|^{\alpha-1}}. \tag{2.2-3}$$

The parameters are η_0, $\sigma_{1/2}$, and α. Line (3) in Fig. 2.2-1 is representative of the predictive capability of this model. At low values of shear rate, the model predicts a zero shear rate viscosity η_0. The units of η_0 are $Pa \cdot s$. At higher values of $\dot\gamma$, the model predicts a shear-thinning (power-law) behavior, and at very high shear rates, the model underpredicts the data. The parameter $\sigma_{1/2}$ represents the shear stress value for which the viscosity is $\eta_0/2$. The dimensionless parameter α is related to the slope in the power-law region, given by $(1 - \alpha)/\alpha$. Note that equation 2.2-3 approaches the Newtonian fluid behavior as $\sigma_{1/2} \to \infty$.

2.2-4 The Carreau Model (1972)

This widely cited three-parameter model is given by

$$\eta = \frac{\eta_0}{[1 + (t_1\dot{\gamma})^2]^{\frac{(1-n)}{2}}}. \qquad (2.2\text{-}4)$$

The parameters are η_0, t_1, and n. Line (3) in Fig. 2.2-1 also represents the predictive capabilities of this model. Frequently, the three-parameter Ellis and Carreau models represent the data equally well (Carreau et al., 1979b). The characteristic time t_1 has units of time and the parameter n is dimensionless. For $\dot{\gamma} \to 0$, the model also predicts a zero shear rate viscosity η_0, and the slope in the power-law region is given by $n - 1$. Figure 2.2-2 compares the model predictions to viscosity data for two polyethylene oxide solutions. Also shown are the predictions of the Cross-Williamson model. The model parameters obained by nonlinear regression are reported in Table 2.2-1.

The three-parameter Carreau model fits the data of the aqueous solution very well. For the PEO solution in the mixture of water and glycerine, the model predicts a too sharp transition from the low shear plateau to the power-law behavior.

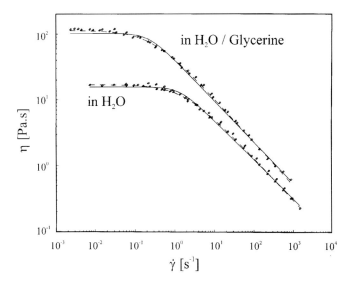

Figure 2.2-2 Viscosity model predictions compared to data of 3 mass % PEO ($M_v = 1.8 \times 10^6$ kg/kmol) in water and in a mixture (50/50) of water and glycerine (Data from Ortiz, 1992)
———— equation 2.2-4
- - - - equation 2.2-5

Table 2.2-1 Model Parameters for the Viscosity Data of Polyethylene Oxide Solutions of Fig. 2.2-2

Solution	3 parameter Carreau equation 2.2-4			Cross-Williamson equation 2.2-5			Carreau-Yasuda equation 2.2-8*			
	η_0 Pa·s	t_1 s	n —	η_0 Pa·s	t_1 s	n —	η_0 Pa·s	t_1 s	a —	n —
3% in H_2O	15.6	0.721	0.408	16.9	0.578	0.367	16.9	0.571	0.983	0.365
3% in 50/50 H_2O/glycerine	102.0	4.36	0.375	113.5	3.78	0.342	120.1	3.42	0.812	0.323

* The parameter η_∞ is set equal to zero

2.2-5 The Cross-Williamson Model (1965)

This is another three-parameter model given by Cross (1965):

$$\eta = \frac{\eta_0}{1 + |t_1 \dot\gamma|^{1-n}} \; . \tag{2.2-5}$$

The parameters are η_0, t_1, and n. Again, this model is capable of predictions comparable to those obtained with other three parameter models. In general, the Carreau three-parameter model tends to slightly underpredict the zero shear rate viscosity, whereas the Cross-Williamson model tends to overpredict the η_0 value. Model predictions are compared to PEO solutions data in Fig. 2.2-2, and the model parameters are reported in Table 2.2-1. The Cross-Williamson model predicts smoother or longer transitions from the zero shear to the power-law behavior. Here, the fit obtained by using the Cross-Williamson model is better than that with the three-parameter Carreau model, mainly for the PEO solution in the mixture of water and glycerine.

It is clear that in order to adequately predict the viscosity behavior at high shear rate ($\dot\gamma \to \infty$), we must introduce a fourth parameter. A four-parameter model should not only describe the power-law region but should also reduce to the observed viscosity values of η_0 and η_∞ at low and high shear rates respectively. Two such models are introduced next.

2.2-6 The Four-Parameter Carreau Model (Carreau et al., 1979b)

This model extends equation 2.2-4 by introducing a parameter η_∞ as follows:

$$\frac{\eta - \eta_\infty}{\eta_0 - \eta_\infty} = \frac{1}{[1 + (t_1 \dot\gamma)^2]^{\frac{(1-n)}{2}}} \; . \tag{2.2-6}$$

Equation 2.2-6 is capable of predicting the complete viscosity range of Fig. 2.2-1 from η_0 to η_∞. The corresponding extension for the Cross-Williamson model can be readily made.

2.2-7 The De Kee Model (1977)

This four-parameter model is obtained as a special case of a series expansion and can be written as

$$\eta = \eta_1 e^{-t_1|\dot\gamma|} + \eta_2 e^{-(0.1)t_1|\dot\gamma|} + \eta_\infty. \tag{2.2-7}$$

The parameters are η_1, η_2, η_∞, and t_1. The first three parameters have units of viscosity, and the parameter t_1 has units of time. Note that as $\dot\gamma \to \infty$, $\eta \to \eta_\infty$ and as $\dot\gamma \to 0$, $\eta \to \eta_1 + \eta_2 + \eta_\infty$ ($=\eta_0$). Because of its exponential formulation, it is easier to treat the data in a semi-logarithmic fashion.

Figure 2.2-3 shows such a semi-log plot of data for two polystyrene solutions. The upper curve is the prediction of equation 2.2-7. The lower curve represents the predictions of equation 2.2-7 without η_∞. These are the predictions of a three-parameter model. Note also that if the data follow a straight line on such a semi-log representation, they will be

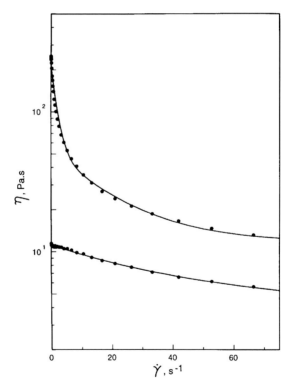

Figure 2.2-3 Semi-logarithmic viscosity-shear rate plots for narrow molecular weight polystyrene solutions in Aroclor 1248 (Data from Ashare, 1968)
The upper curve is for a 4 mass % solution of $M_w = 1.8 \times 10^6$ kg/kmol. The model parameters of equation 2.2-7 are $\eta_1 = 166.7$ Pa·s, $\eta_2 = 39.6$ Pa·s, $\eta_\infty = 11.7$ Pa·s, and $t_1 = 0.526$ s. The lower curve predicts the data of a 2.5 mass % solution of $M_w = 8.6 \times 10^5$ kg/kmol. The three parameters are $\eta_1 = 4.71$ Pa·s, $\eta_2 = 6.47$ Pa·s, and $t_1 = 0.035$ s

represented by a two-parameter version of equation 2.2-7, given by the first term on the right side. This would be the case, for example, of a low molecular weight small concentration polystyrene solution.

If four parameters are sufficient to describe a "complete" viscosity curve from η_0 up to and including η_∞, we would expect models containing more parameters to perform at least equally well. While such models exist, keeping the number of parameters as low as possible is the main concern.

2.2-8 The Carreau-Yasuda Model (Yasuda, 1979)

The Carreau-Yasuda model is an example of a highly successful five-parameter model, given by

$$\frac{\eta - \eta_\infty}{\eta_0 - \eta_\infty} = \frac{1}{[1 + |t_1 \dot\gamma|^a]^{\frac{1-n}{a}}}. \tag{2.2-8}$$

The parameters are: η_0, η_∞, t_1, a, and n. Note that $a = 2$ gives the Carreau four-parameter model. The additional (fifth) parameter improves the description of the transition zone between the zero shear rate region and the power-law region. Taking η_∞ equal to zero, the PEO solution data of Fig. 2.2-2 were used to assess the Carreau-Yasuda model. The best fit showed very little difference compared to the Cross-Williamson model predictions (dashed lines in the figure). The model parameters are listed in Table 2.2-1. Note the close agreement for the values of η_0, t_1, and n for the Cross-Williamson and Carreau-Yasuda models. In general, the Carreau-Yasuda model will yield improved fits, and it is believed to be the most reliable model for estimating the zero shear viscosity of polymers.

Some systems, such as foodstuffs, blood, and drilling muds contain more than one phase. A three-dimensional structure can develop, and the sample exhibits a yield stress σ_0. The behavior of such complex systems is usually not represented by a straight line on a shear stress–shear rate plot, and the curve does not pass through the origin. The intercept on the shear stress axis is a dynamic measure of a yield stress σ_0. Next, we introduce a few models that take the existence of a yield stress into account.

2.2-9 The Bingham Model (1922)

The Bingham model is a straightforward extension of Newton's law of viscosity, and is given by

$$\begin{cases} \sigma_{yx} = -\eta_0 \dot\gamma_{yx} \pm \sigma_0 & \text{if } |\sigma_{yx}| > |\sigma_0| \\ \dot\gamma_{yx} = 0 & \text{if } |\sigma_{yx}| \leqslant |\sigma_0| \end{cases}. \tag{2.2-9}$$

The model parameters are η_0 and σ_0. The model indicates that the sample behaves like a Newtonian material when the shear stress exceeds the yield stress σ_0. For $|\sigma_{yx}| \leqslant |\sigma_0|$, the material does not deform.

2.2-10 The Casson Model (1959)

The Casson equation is another two-parameter model, popular in the area of biorheology. It is given by (for $|\sigma_{yx}| > \sigma_0$)

$$\sqrt{|\sigma_{yx}|} = \sqrt{\sigma_0} + \sqrt{\eta_0 |\dot{\gamma}_{yx}|}. \tag{2.2-10}$$

The two parameters are η_0 and σ_0, and a plot of $\sqrt{|\sigma_{yx}|}$ vs $\sqrt{|\dot{\gamma}_{yx}|}$ produces $\sqrt{\sigma_0}$ as the intercept.

2.2-11 The Herschel-Bulkley Model (1926)

This is a three-parameter model that includes a yield stress. It is an extension of the power-law model and is given by (for $|\sigma_{yx}| > \sigma_0$)

$$\eta = \frac{\sigma_0}{|\dot{\gamma}|} + m|\dot{\gamma}|^{n-1}. \tag{2.2-11}$$

The parameters are σ_0, m, and n. Note that materials that exhibit a yield stress produce viscosity data somewhat like line (2) in Fig. 2.2-1. That is, the viscosity tends to infinity (solid-like behavior) as $\dot{\gamma}$ tends to zero. The η_0 plateau is not there. For this reason, the power-law model can just as well describe the data, and the third (extra) parameter σ_0, associated with the Herschel-Bulkley model is really just a mathematical fitting parameter, which frequently cannot be associated with the physics of a yield stress.

2.2-12 The De Kee-Turcotte Model (1980)

This is also a three-parameter model involving a yield stress, inspired by equation 2.2-7. It is written as

$$\eta = \frac{\sigma_0}{|\dot{\gamma}|} + \eta_1 e^{-t_1|\dot{\gamma}|}. \tag{2.2-12}$$

The parameters are σ_0, η_1, and t_1. A semi-logarithmic representation of η vs $\dot{\gamma}$ data results in a straight line at high shear rates (the second term on the right side of the equation). The first term adds contributions to this straight line which intercepts the η-axis at η_1. The magnitude of these contributions increases as $\dot{\gamma}$ decreases. Indeed the first term on the right side involves a constant yield stress σ_0. In this case, the parameter σ_0 is more closely related to the physical picture conjured up by a microstructure, responsible for a yield stress. The merits of equation 2.2-12 over equation 2.2-11 have been discussed in detail by De Kee and Turcotte (1980).

Figure 2.2-4 illustrates the predictive capabilities of equations 2.2-11 and 2.2-12. The data are for yogurt, and the figure clearly shows the sensitivity to σ_0 of equation 2.2-12, as well as the lack of sensitivity to σ_0 of equation 2.2-11. For all practical purposes, the data

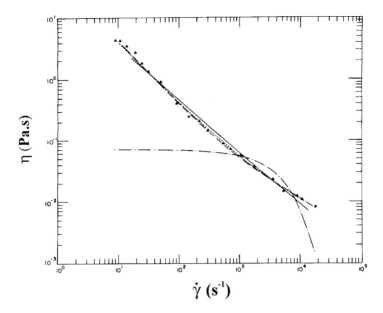

Figure 2.2-4 Viscosity as a function of shear rate for yogurt. (Data from Turcotte, 1980)
- - - -   Herschel-Bulkley model with $m = 1.87 \times 10^{-1}$ Pa·s^n, $n = 0.644$, and $\sigma_0 = 39.5$ Pa
- - -   De Kee-Turcotte model with $\eta_1 = 1.15 \times 10^{-2}$ Pa·s, $t_1 = 4.52 \times 10^{-5}$ s, and $\sigma_0 = 41.7$ Pa
———   Power-law model (Hershel-Bulkley model) with $\sigma_0 = 0$, $m = 25.03$ Pa·s^n and $n = 0.14$
-·-·   De Kee-Turcotte model with $\sigma_0 = 0$

are equally well predicted by equations 2.2-11 and 2.2-12. Setting $\sigma_0 = 0$ has little effect on the predictions of the Herschel-Bulkley model. It reduces to the power-law model, which also describes the data rather well. Setting $\sigma_0 = 0$ in a relation of the form of equation 2.2-12 has a disastrous effect. As mentioned earlier, σ_0 has no physical meaning in the context of equation 2.2-11. This is not the case for equation 2.2-12.

The yield stress values obtained via models such as those given by equations 2.2-9 to 2.2-12 should be considered model parameters and not real material properties.

2.2-13 Viscosity Models for Complex Flow Situations

All the viscosity models presented so far are correctly defined for simple shear. For more complex flow situations, the rate-of-deformation tensor contains more than one independent component, and we have to express the viscosity as functions of combinations of components of the rate-of-deformation tensor (or stress tensor). Such functions have to be independent of the frame of reference. Appendix A identifies certain combinations of these components, which do not depend on the coordinate system and are therefore invariants.

We obtain three independent combinations (invariants) of the tensor $\underline{\underline{\dot{\gamma}}}$, denoted by $I_{\dot{\gamma}}$, $II_{\dot{\gamma}}$, and $III_{\dot{\gamma}}$, defined by

$$I_{\dot{\gamma}} = \text{tr}\, \underline{\underline{\dot{\gamma}}} = \underline{\underline{\dot{\gamma}}} : \underline{\underline{\delta}}, \tag{2.2-13}$$

where

$$\text{tr}\, \underline{\underline{\dot{\gamma}}} = \sum_i \dot{\gamma}_{ii} \quad \text{(see Appendix A)}, \tag{2.2-14}$$

$$II_{\dot{\gamma}} = \text{tr}\, \underline{\underline{\dot{\gamma}}}^2 = \underline{\underline{\dot{\gamma}}} : \underline{\underline{\dot{\gamma}}}, \tag{2.2-15}$$

and

$$III_{\dot{\gamma}} = \det \underline{\underline{\dot{\gamma}}}. \tag{2.2-16}$$

Other invariants can be obtained via various combinations of $I_{\dot{\gamma}}$, $II_{\dot{\gamma}}$, and $III_{\dot{\gamma}}$ (see equations A.5-27 and A.5-28), and it is possible to combine them, since adding invariants results in another invariant. The first invariant ($I_{\dot{\gamma}}$) is equal to zero for incompressible fluids, and the third invariant of the rate-of-deformation tensor ($III_{\dot{\gamma}}$) is zero for simple shear. The viscosity equations can be written as functions of $II_{\dot{\gamma}}$, where the magnitude of the rate-of-deformation tensor can be defined as

$$\bar{\dot{\gamma}} = \sqrt{\frac{1}{2} II_{\dot{\gamma}}}. \tag{2.2-17}$$

In simple shear, $II_{\dot{\gamma}}$ is $2\dot{\gamma}^2$. In a more general way, the power-law model given by equation 2.2-1 can be written as

$$\eta = m\left(\sqrt{\frac{II_{\dot{\gamma}}}{2}}\right)^{n-1} = m\bar{\dot{\gamma}}^{n-1}, \tag{2.2-18}$$

where $\bar{\dot{\gamma}}$ is the rate-of-deformation tensor, which reduces to the shear rate in simple shear. The same principle applies for the other viscosity models.

In the case of an elongational flow of an incompressible fluid, only the first invariant would vanish, and the viscosity could be expressed as a function of $II_{\dot{\gamma}}$ and $III_{\dot{\gamma}}$.

2.3 Thixotropy, Rheopexy, and Hysteresis

Thixotropy and rheopexy or anti-thixotropy are time effects, associated with non-Newtonian, structured materials. In the case of thixotropy, we observe a decrease of viscosity with time, when the sample is subjected to a constant rate of shear. This decrease in viscosity is associated with a change in structure, as a result of the applied shear rate. A time-independent, steady shear viscosity is eventually reached. De Kee et al. (1983) observed such time effects to last well over ten minutes in the case of several foodstuffs.

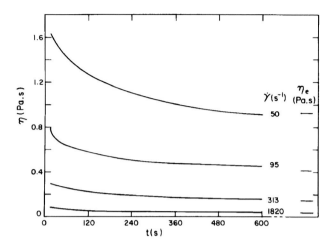

Figure 2.3-1 Viscosity decay for yogurt
η_e is the equilibrium viscosity (Data from Turcotte, 1980)

Figure 2.3-1 shows the thixotropic effect for the yogurt data of Fig. 2.2-4. The viscosity values decrease with time, at a given shear rate, to eventually reach the steady-state equilibrium value η_e, which is the viscosity reported in Fig. 2.2-4.

Thixotropic behavior has been observed for many materials. Figure 2.3-2 illustrates the structure recovery for a kaolin suspension following the application of a shear rate for a sufficiently long time to reach a steady shear viscosity. The figure shows four material functions (η, η', G', and G''). The complex viscosity data for small amplitude oscillatory

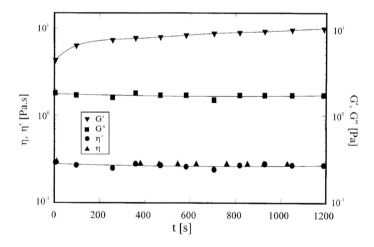

Figure 2.3-2 Rheological properties as a function of time for a 44 vol % kaolin suspension at $\omega = 6$ s^{-1}, following a steady shear experiment at $\dot{\gamma} = 6$ s^{-1} (Data from Carreau and Lavoie, 1993)

flow were obtained at an angular frequency of 6 s^{-1}, corresponding to the value of the previously imposed shear rate, $\dot{\gamma} = 6$ s^{-1}.

Only the elastic modulus (G') increases markedly with time. Steady state is still not obtained after 1200 s. That is, the structure recovery in a thixotropic system is a slow process, which can be probed via dynamic measurements. Note that for such a kaolin suspension, the elastic modulus is much larger than the loss modulus. That is, the sample behaves more like a solid under small-amplitude oscillatory flow. Note also that the dynamic viscosity (η') coincides with the steady shear viscosity (η). The reason for this analogy is not clear.

Anti-thixotropy or rheopexy refers to the case where the viscosity increases with time at a constant shear rate. This increase of viscosity with time does not involve an overshoot. This is a characteristic of viscoelasticity, as will be discussed in Chapter 5. An aqueous suspension of titanium dioxide is an example of a rheopectic material, as illustrated in Fig. 2.3-3. The viscosity increases with time due to a structure buildup.

We close this section by presenting a recently developed model for such structured materials, introduced by De Kee and Chan Man Fong (1994). The model is given by

$$\eta = \frac{\eta_0 k_c}{\alpha_0}[1 + (bf_1 - cf_2)e^{-\alpha_0 t}], \qquad (2.3\text{-}1)$$

where

$$\alpha_0 = k_c(1 + bf_1 - cf_2) \qquad (2.3\text{-}2)$$

The model predicts shear thinning, shear thickening, thixotropy, and anti-thixotropy in different ranges of shear rate. The parameters are η_0, k_c, b, and c. f_1 and f_2 are functions of

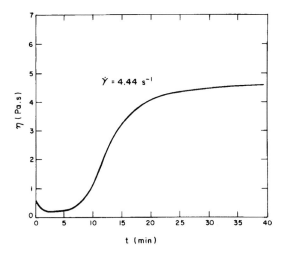

Figure 2.3-3 Rheopectic behavior for a 20 vol % aqueous suspension of titanium dioxide stabilized with sodium pyrophosphate at a concentration of 5 mg/g TiO$_2$
The suspension was previously deflocculated at a shear rate of 178 s^{-1} (Data from Umeya and Kanno, 1979)

shear rate. For $(bf_1 - cf_2) > 0$, the model predicts thixotropic behavior. For $(bf_1 - cf_2) < 0$, the model predicts anti-thixotropic or rheopectic behavior. Shear thinning is predicted if $(bf_1' - cf_2') > 0$. Shear thickening is predicted if $(bf_1' - cf_2') < 0$. The ' denotes the derivative with respect to γ. The first term on the right side of equation 2.3-1 represents the steady shear viscosity. The model reduces to the three-parameter equation of Liu et al. (1983), if only shear-thinning behavior is portrayed. This equation is given by

$$\eta = \frac{\eta_0}{1 + b\dot{\gamma}^{1-n}}, \qquad (2.3\text{-}3)$$

which is equivalent to the Cross-Williamson equation (2.2-5).

This model also allows for the prediction of the hysteresis phenomenon frequently associated with structured materials. In a hysteresis experiment, the shear rate $\dot{\gamma}$ is linearly increased from zero to a maximum value $\dot{\gamma}_m$, and is then linearly decreased from $\dot{\gamma}_m$ to zero in a time of $2t_0$, as illustrated in Fig. 2.3-4. The corresponding shear stress σ_{yx} is measured. Figure 2.3-5 illustrates some of the predictive capabilities of the model for the case where the shear rate is programmed to increase at a constant rate $\ddot{\gamma}$, and $f_1 = f_2 = \dot{\gamma}^m$. Figure 2.3-5 shows computed hysteresis loops for one value of k_c and various values of $\ddot{\gamma}$. Note that as $\ddot{\gamma}$ increases, the area enclosed by the loops decreases and the maximum value of σ_{yx} increases. For low values of $\ddot{\gamma}$, the curves cross once, and for high values of $\ddot{\gamma}$ they cross more than once. The lower the value of $\ddot{\gamma}$, the lower the value of $\dot{\gamma}$ at which the first crossing occurs. The magnitude of the area of the loop is indicative of the amount of structural change in the sample. These predictions are in qualitative agreement with experimental observations on materials such as tomato juice.

Finally, we note that Quemada (1991) introduced a structural viscosity model assuming that flow-induced structural changes can be described by a viscosity functional of the shear rate and of the volume fraction of particles. The Quemada model can be related in a nontrivial way to the De Kee-Chan Man Fong equation 2.3-1 for the special case where $f_1 = \dot{\gamma}$ and $f_2 = 0$. At steady state, the Quemada model reduces to

$$\eta = \frac{\eta_P}{\left(1 - \frac{1}{2}kH\right)^2}, \qquad (2.3\text{-}4)$$

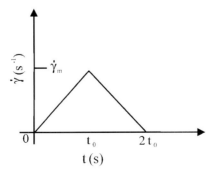

Figure 2.3-4 Schematic representation of a hysteresis experiment

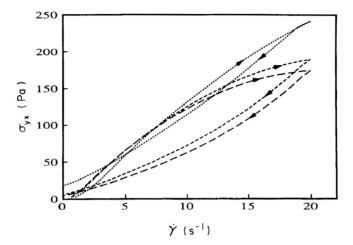

Figure 2.3-5 Hysteresis curves for $k_c = 10^{-3}$ s^{-1} and various values of $\ddot{\gamma}$
--- $\ddot{\gamma} = 0.67$ s^{-2}, $t_0 = 30$ s
— — — $\ddot{\gamma} = 0.80$ s^{-2}, $t_0 = 25$ s
····· $\ddot{\gamma} = 2.0$ s^{-2}, $t_0 = 10$ s

where

$$k = \frac{k_0 + k_\infty \sqrt{\dfrac{\dot{\gamma}}{\dot{\gamma}_c}}}{1 + \sqrt{\dfrac{\dot{\gamma}}{\dot{\gamma}_c}}}. \tag{2.3-5}$$

The model parameters are H, η_p, k_0, k_∞, and $\dot{\gamma}_c$.

2.4 Relations Between Material Functions

In this chapter, we have introduced a variety of material functions, such as the non-Newtonian viscosity, the dynamic viscosity, and the primary normal stress coefficient. A number of experiments conducted on a given material may generate valuable information concerning that material. This information appears in the form of material functions. It is therefore reasonable to look for possible relations between material functions. It would, for example, be very rewarding if information concerning the material elasticity could be deduced from viscosity measurements only. Indeed, viscosity is more easily measured, and viscosity measurements require less expensive instruments than those needed to measure, for example, normal stress differences. While establishing relations between material functions is very advantageous, in practice, most of these relations are empirical and valid

2.4 Relations Between Material Functions

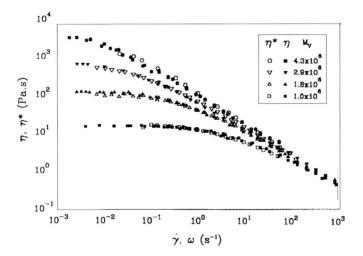

Figure 2.4-1 Cox-Merz relation for 3 mass % solutions of PEO of different molecular weights in a mixture containing 50 mass % of glycerine and 50 mass % water (Data from Ortiz, 1992)

Table 2.4-1 Relations Between Material Functions

Relation	Defining equation	Comments
Bird, Hassager, Abdel-Khalik (1974)	$\psi_1(\dot{\gamma}) = \dfrac{4\beta}{\pi}\int_0^\infty \dfrac{\eta(x)-\eta(\dot{\gamma})}{\dot{\gamma}^2-x^2}dx$	$\beta=2$ for solutions and $\beta=3$ for melts. Generates ψ_1 values which are usually lower than those obtained experimentally. (see Ait-Kadi et al., 1989)
Gleissle (1980)	$\eta(\dot{\gamma}) = \eta^+(t,\dot{\gamma})\vert_{t=\frac{1}{\dot{\gamma}}}$ $\psi_1(\dot{\gamma}) = \psi_1^+(t,\dot{\gamma}_\infty)\vert_{t=\frac{k}{\dot{\gamma}}}$	These so-called mirror relations relate the viscosity to the stress growth and the primary normal stress coefficient to its transient: $2.5<k<3$. These relations have been successfully tested with LDPE.
Laun (1986)	$\psi_1(\dot{\gamma}) = \dfrac{2\eta''(\omega)}{\omega}\left[1+\left(\dfrac{\eta''}{\eta'}\right)^2\right]^{0.7}\bigg\vert_{\omega=\dot{\gamma}}$	Predicts the primary normal stress coefficient from more readily available linear viscoelastic data.
Wagner (1977)	$\psi_1^+(t,\dot{\gamma}_\infty) = -\dfrac{1}{n}\dfrac{\partial \eta^+(t,\dot{\gamma}_\infty)}{\partial \dot{\gamma}}$ $\psi_1(\dot{\gamma}_\infty) = -\dfrac{1}{n}\dfrac{d\eta(\dot{\gamma})}{d\dot{\gamma}}$ (at steady-state)	Relates the primary normal stress functions to the derivative of the viscosity; n is a shear-thinning parameter: $0.10 \leqslant n \leqslant 0.20$. Works well for polymer melts.

only for a limited class of materials and/or in limited ranges of conditions. Therefore they should be used with extreme caution.

A most useful empiricism is known as the Cox-Merz (1958) relation. It is given by

$$\eta(\dot{\gamma})|_{\dot{\gamma}=\omega} = |\eta^*(\omega)| = \sqrt{\eta'^2(\omega) + \eta''^2(\omega)}. \qquad (2.4\text{-}1)$$

For highly elastic fluids and polymer melts, it is very difficult to measure the steady shear viscosity at high shear rates using a rotational device. In this case, the shear viscosity can be estimated from the Cox-Merz relation. This empiricism seems to work well for homogeneous polymer solutions and melts, as shown in Fig. 2.4-1 for polyethylene oxide solutions. Note that it was possible to cover a larger range for the shear rate than for the frequency, by using a variety of rheometers (Ortiz et al., 1994). For polymer melts, it is virtually impossible to measure the steady shear viscosity for shear rates much larger than one, with basic flow geometries such as cone-and-plate and parallel plate arrangements (see Chapter 3). High frequency small amplitude oscillatory data are more easily accessible, and the steady shear viscosity of polymer melts can be estimated via the Cox-Merz relation.

Other possibly useful relations can be found in Table 2.4-1. These relations should be used with caution, and for polymeric systems similar to those for which the relations have been verified.

2.5   Temperature, Pressure, and Molecular Weight Effects

This section is largely taken from Agassant et al. (1991).

2.5-1   Effect of Temperature on Viscosity

The viscosity of polymer solutions usually varies with temperature, according to an exponential Arrhenius type relation given by

$$\eta = A e^{\frac{B}{T}}. \qquad (2.5\text{-}1)$$

A is a pre-exponential factor with units of viscosity, and $B (= E/R)$ is associated with the activation energy E for viscous flow and with the gas constant R. In general, A and B depend on the solvent, on the concentration, and on the molecular weight distribution of the polymer (Shah and Parsania, 1984).

The activation energy for the viscosity of low molecular weight liquids is of the order of 10^4 J/mol. The activation energy for polymers is much larger and increases with both the chain rigidity and the size of the lateral branches. Typical values are 2.9×10^4 J/mol for polyethylene and 9.3×10^4 J/mol for polystyrene, both at 100 K above the glass transition temperature.

The viscosity of polymer melts changes drastically with temperature. The variation is given by

$$\frac{\Delta \eta}{\eta} = -\frac{E}{RT^2} \Delta T, \qquad (2.5\text{-}2)$$

where E is 10^5 J/mol and $T=400$ K, $E/RT=30$, and $\Delta\eta/\eta = 30\, \Delta T/T$. Therefore, a 1% error in the temperature measurement could lead to a 30% error in the viscosity!

The shape of viscosity-shear rate plots is hardly affected by temperature. Higher temperatures yield lower viscosities. The similarity of curves suggests a possibility of combining data, obtained at different temperatures, into a single master curve at an arbitrary reference temperature. Usually, this reference temperature is the glass transition temperature T_g. Glass transition can experimentally be observed by a net change in a narrow range of temperature of the thermal expansion coefficient, of the specific heat, or of some mechanical properties. This phenomenon is not specific to polymers. The master curve is obtained by the following procedure:

(i) Shift the curve at temperature T vertically upward by an amount of $\eta_0|_{T_g}/\eta_0|_T$.
(ii) Shift the resulting curve horizontally by an amount $\log a_T$, so that the overlapping regions of the curve at T_g and the shifted curve at T superpose.

Williams, Landel, and Ferry (1955) found, empirically, that for a wide variety of polymer melts, the shift factor a_T is given by

$$\log a_T = \frac{-17.44(T - T_g)}{51.6 + (T - T_g)}, \qquad (2.5\text{-}3)$$

for $T_g < T < (T_g + 100\text{ K})$, and the viscosity can be expressed as

$$\eta = \eta_g a_T \frac{T\rho}{T_g \rho_g}, \qquad (2.5\text{-}4)$$

where η_g and ρ_g are respectively the viscosity and the density at T_g. The factor a_T renders varying temperature at a fixed shear rate equivalent to varying shear rate at a fixed temperature. That is, measuring the viscosity over two or three decades of shear rate at several temperatures and combining the results into a master curve yields a viscosity-shear rate plot which could span between five and ten decades of shear rate. This is equivalent to the time-temperature superposition principle discussed in Chapter 5.

Doolittle (1951) showed that the viscosity could be better correlated with the free volume than with the temperature. The relative free volume is defined by

$$\frac{V_f(T)}{V_0} = \frac{V(T) - V_0}{V_0}, \qquad (2.5\text{-}5)$$

where V is the volume at temperature T and V_0 is the volume extrapolated to 0 K without phase changes. Cohen and Turnbull (1959) proposed the following expression, based on the free volume:

$$\eta = A' \exp\left[B' \frac{V_0}{V_f(T)} \right]. \qquad (2.5\text{-}6)$$

Sridhar et al. (1978) confirmed that the shear viscosity of polymer solutions is inversely proportional to the relative free volume available.

2.5-2 Effect of Pressure on Viscosity

Viscosity also increases exponentially with pressure. Using a pressure dependence coefficient defined by

$$\Gamma = \frac{d \ln \eta}{dP}, \quad (2.5\text{-}7)$$

we can express the viscosity as

$$\eta(T, P) = A e^{\frac{B}{T}} e^{\Gamma P}. \quad (2.5\text{-}8)$$

Typical values for Γ are between 2×10^{-8} and 6×10^{-8} Pa^{-1} (Gupta et al., 1969).

Example 2.5-1 Viscosity Variations with Pressure
Determine the variation of the viscosity with pressure for a thermoplastic material for which the pressure dependence coefficient Γ is 3.3×10^{-8} Pa^{-1}.

Solution
The viscosity can be expressed as

$$\eta = \eta_0 e^{\frac{P}{30}} \quad (\text{with } P[=] \text{ MPa})$$

$$\approx \eta_0 e^{\frac{P}{300}} \quad (\text{with } P[=] \text{ atm}),$$

where η_0 is the viscosity at very low pressure. Table 2.5-1 shows how drastically the viscosity increases with pressure for values larger than 100 atm.

2.5-3 Effect of Molecular Weight on Viscosity

The viscosity of polymer melts and polymer solutions is considerably larger than that of low molecular weight liquids. Experimental observations show that the viscosity of polymers is dependent on molecular weight as follows:

- For polymers of molecular weight less than a critical value, M_c, the viscosity increases linearly with molecular weight. Moreover, these low molecular weight polymers (or oligomers) obey (or do not deviate much from) Newton's law of viscosity.
- Above M_c, the viscosity increases much more rapidly with molecular weight. Linear polymers with narrow molecular weight distribution follow generally the well known relation (Ferry, 1970).

$$\eta_0 = K M^{3.4}, \quad (2.5\text{-}9)$$

where K is a constant for a given polymer family.

2.5 Temperature, Pressure, and Molecular Weight Effects

Table 2.5-1 Variation of the Viscosity of Polymers with Pressure for an Average Value of the Pressure Dependence Coefficient Equal to 0.0033 atm^{-1} (From Agassant et al., 1991)

P atm	η/η_0
30	1.11
100	1.39
300	2.70
500	5.29
1000	27.9
3000	22026

High molecular weight polymer melts and solutions are generally non-Newtonian in simple shear flow. The viscosity-molecular weight relation is sketched in Fig. 2.5-1. Most commercial polymers have a large distribution of molecular weight (they are polydisperse) and the viscosity is a more complex function of the average molecular weight and of the distribution.

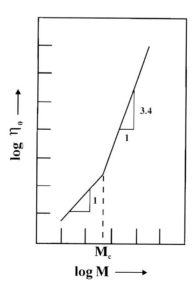

Figure 2.5-1 Molecular weight dependence of viscosity

2.5-3.1 Low Molecular Weight Polymers

For polymers of molecular weight less than M_c, or dilute polymer solutions, the macromolecules take a random configuration (in the amorphous state) known as a statistical coil for flexible enough chains. The radius, R, of the coil is approximately equal to $a\sqrt{x}$, where a is the length of a chain segment and x is the number of segments. Figure 2.5-2 illustrates this concept schematically.

Let us consider a macromolecule under a shear flow of magnitude $\dot{\gamma}$ (see Fig. 2.5-3). If the liquid has a low molecular weight, the molecules have a low but uniform relative velocity of the order of $\dot{\gamma}d/2$, where d is the diameter of a typical molecule. If we consider a polymeric liquid, each macromolecule is subjected to an average rotational speed of $\dot{\gamma}/2$, and the relative velocity of the periphery of the macromolecule, of the order of $\dot{\gamma}R/2$ ($\approx \dot{\gamma}d\sqrt{x}/2$), increases with x (or molecular weight).

The molecular structure implies that the relative velocities between adjacent molecules (or chain segments) could be quite large and, if one assumes that the energy dissipated is proportional to the square of the velocity (R^2 or x), the viscosity then increases linearly with molecular weight, as shown by the following reasoning. The rate of energy dissipated per unit volume is expressed by

$$\dot{W} = n \int_{\text{molecule}} \zeta_0 (V_{\text{molecule}} - V_{\text{average}})^2 dV, \tag{2.5-10}$$

where n is the concentration of molecules, and ζ_0 is the drag coefficient of a chain segment defined such that the force necessary to bring a segment to the relative velocity V with respect to the surrounding segments is

$$F = \zeta_0 V. \tag{2.5-11}$$

Then, from the Debye calculation, and by using equation 2.5-10 and the fact that n is inversely proportional to xd^3, we obtain

$$\dot{W} = \frac{1}{36}\frac{x}{d}\zeta_0 \dot{\gamma}^2. \tag{2.5-12}$$

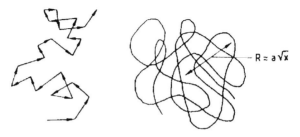

Figure 2.5-2 Coil conformation of a macromolecule (From Agassant et al., 1991)

2.5 Temperature, Pressure, and Molecular Weight Effects

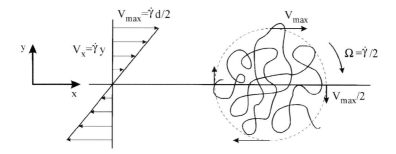

Figure 2.5-3 Deformation of a macromolecule under shear flow (Adapted from Agassant et al., 1991)

It is known that the rate of energy dissipated by viscous forces is proportional to the product of the viscosity and the square of the shear rate (i.e., $\dot{W} = \eta\dot{\gamma}^2$). From equation 2.5-12 we have

$$\eta = \frac{1}{36}\frac{x}{d}\zeta_0. \tag{2.5-13}$$

Using Stokes' law to express ζ_0,

$$F = \zeta_0 V = 6\pi\eta_1 dV, \tag{2.5-14}$$

where η_1 is the viscosity of the liquid monomer, we obtain

$$\eta \approx x\eta_1. \tag{2.5-15}$$

Hence, the viscosity is proportional to the number of segments and the molecular weight. However, we would expect ζ_0 to increase slightly with x, thus explaining why η increases in a nonlinear fashion with x. This analysis is valid for $x < x_c$, that is, the macromolecules are small enough so that there are no chain entanglements.

2.5-3.2 High Molecular Weight Polymers

In concentrated polymer solutions or melts, the coils of adjacent molecules interpenetrate. The following reasoning of Bueche (1952) explains why polymeric liquids exhibit such large viscosities for values larger than M_c. His idea, based on chain entanglements, is now largely accepted. The force necessary to bring a macromolecule containing x segments to the velocity V is

$$F = x\zeta_0 V = \zeta V, \tag{2.5-16}$$

where $\zeta = x\zeta_0$ is the drag coefficient for the whole macromolecule. If the macromolecule is large enough, it is reasonable to assume that it drags along other macromolecules entangled at random positions, due to the Brownian motion. The number of entangled molecules is approximately proportional to the number of chain segments x:

$$n_1 = Kx. \tag{2.5-17}$$

The dragged molecules slip slightly with respect to the first one at the entanglement points. Their average velocity is then less than V. That is,

$$V_1 = sV \quad \text{(where } 0 < s < 1\text{)}. \tag{2.5-18}$$

The net drag force is now given by the following expression:

$$F_1 = \zeta V + n_1 \zeta s V = \zeta_0 V x (1 + Ksx). \tag{2.5-19}$$

These molecules drag other molecules in a similar fashion, with their number given by

$$n_2 = C_2 n_1 K x, \tag{2.5-20}$$

where $C_2 < 1$, since the entanglements with the first molecule should be included. Their velocity V_2 is given by

$$V_2 = sV_1 = s^2 V. \tag{2.5-21}$$

In this pattern, illustrated in Fig. 2.5-4, each dragged molecule is surrounded by a cluster of other molecules dragged at a lower velocity. The intensity is decreasing, since they are not directly entangled with the first molecule. The total drag force is then given by

$$F = \zeta_0 x V [1 + Ksx + C_2(Ksx)^2 + C_3(Ksx)^3 + \cdots], \tag{2.5-22}$$

where $1 > C_2 > C_3 \ldots$ is a rapidly converging series.

If we assume that the transition occurs at $x_c \approx 500$, then $K \approx 1/500$, and for $x < x_c$, Ksx is small and $F \approx \zeta_0 xV$. For $x > x_c$, Bueche (1952) computed the function F corresponding to various values of s (from 0.1 to 0.5). He showed that, for $s \approx 0.3$, F is given by

$$F = A x^{2.5}. \tag{2.5-23}$$

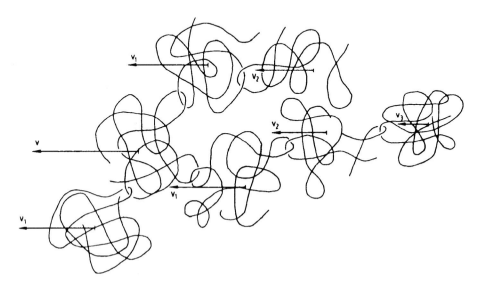

Figure 2.5-4 Deformation of entangled macromolecules (From Agassant et al., 1991)

Via equations 2.5-13 and 2.5-14, we can compute a correlation for the viscosity of the type

$$\eta = Bx^{3.5}, \qquad (2.5\text{-}24)$$

which predicts that the viscosity increases with the molecular weight at the power 3.5, which is close to the experimentally determined value of 3.4 for linear polymers of narrow molecular weight distribution.

2.6 Problems*

2.6-1$_a$ Viscosity Data of a PIB Solution

The following viscosity-shear rate data have been obtained for a 2 mass % solution of polyisobutylene ($M_W = 15 \times 10^6$ kg/kmol) in primol 355, at 25 °C:

Table 2.6-1

η Pa·s	$\dot{\gamma}$ s^{-1}	η Pa·s	$\dot{\gamma}$ s^{-1}
6560	0.00173	490	0.274
5770	0.00274	348	0.434
4990	0.00434	223	0.866
4200	0.00866	163	1.37
3220	0.0137	104	2.74
2190	0.0274	76.7	4.34
1640	0.0434	68.1	5.46
1050	0.0866	58.2	6.88
766	0.137		

Plot these data on log-log paper, determine the parameters for the Carreau and Cross models of equations 2.2-4 and 2.2-5, and compare the model predictions with the data. Are the model predictions satisfactory?

2.6-2$_a$ Viscosity Data of a CMC Solution

The following viscosity-shear rate data were measured for a 1 mass % aqueous solution of carboxy methyl cellulose, at 25 °C:

---

* Problems are labelled by a subscript a or b indicating the level of difficulty.

58   Material Functions and Generalized Newtonian Fluid

Table 2.6-2

η Pa·s	$\dot\gamma$ s^{-1}	η Pa·s	$\dot\gamma$ s^{-1}	η Pa·s	$\dot\gamma$ s^{-1}
1.503	0.435	1.428	0.687	1.344	1.095
1.202	1.727	1.076	2.736	0.94	4.348
0.823	6.872	0.732	10.95	0.62	17.27
0.511	27.36				

Plot these data on log-log paper and determine the parameters of the Ellis model. Compare the predictions of the model with the data.

2.6-3$_a$ The Ellis Model

Show that the following representation of the Ellis model is equivalent to the one given by equation 2.2-3:

$$\sigma_{yx} = \frac{-\dot\gamma_{yx}}{\phi_0 + \phi_1 |\sigma_{yx}|^{\alpha-1}}. \tag{2.6-1}$$

Verify that the slope in the power-law region of η vs $\dot\gamma$, in a log-log representation, is given by $(1-\alpha)/\alpha$.

2.6-4$_b$ Viscosity Data for a PS Solution

The following viscosity-shear rate data were measured for a 1.0 mass % solution of polystyrene ($M_W = 8.6 \times 10^5$ kg/kmol) in Aroclor 1248, at 25 °C:

Table 2.6-3

η Pa·s	$\dot\gamma$ s^{-1}
2.40	0.053
2.04	0.53
1.98	5.28
1.94	13.27
1.88	21.02
1.74	41.95
1.62	66.48

Plot these data on semi-log paper and determine the parameters for the De Kee model. Hint: the model (equation 2.2-7) may have to be simplified.

2.6-5$_b$ Rheological Behavior of Drilling Muds

The rheological behavior of drilling muds can be represented by a model such as

$$\sigma_{yx} = -\eta_\infty \dot{\gamma}_{yx}\left(1 + \frac{\sigma_0}{m\eta_\infty |\dot{\gamma}_{yx}|}\right)^m, \qquad (2.6\text{-}2)$$

with parameters η_∞, σ_0, and m, where $0 < m \leqslant 1$.

(a) Determine the units of the model parameters.
(b) Show that for $m = 1$, the Bingham model is obtained, and that Newtonian behavior is predicted for large shear rate values.
(c) Show that power-law behavior is predicted for lower shear rate values.
(d) Obtain the model parameters, using the data shown in Fig. 2.6-1.

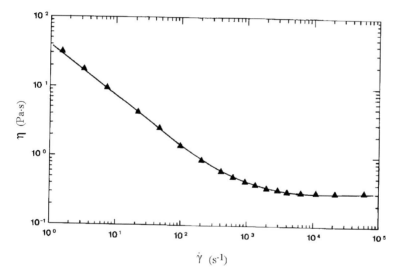

Figure 2.6-1 Viscosity-shear rate data for a drilling mud

2.6-6$_b$ The Cross-Williamson Model

(a) Show that the Cross-Williamson equation, given by

$$\frac{\eta - \eta_\infty}{\eta_0 - \eta_\infty} = \frac{1}{1 + (t_1 \dot{\gamma})^{1-n}}, \qquad (2.6\text{-}3)$$

cannot simultaneously predict Newtonian behavior at low shear rates, as well as shear-thinning behavior at higher shear rates.

Assume: (i) $\eta_\infty \ll \eta_0$
(ii) Newtonian behavior is associated with $\lim_{\dot{\gamma} \to 0}(d\eta/d\dot{\gamma}) = 0$.

(b) Give an example of a simple model that obeys assumption (ii), and predicts acceptable behavior at high shear rates as well.

2.6-7$_b$ Viscosity-Molecular Weight Relationship

Figure 2.6-3 illustrates the viscosity-molecular weight relationship for a typical polymer.

(a) Discuss the behavior depicted in the figure.
(b) Explain the importance of the point $\dot{\gamma}_1 M_1$.

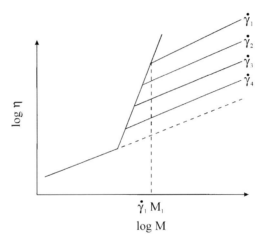

Figure 2.6-2   Viscosity-molecular weight relation for a typical polymer

3 Rheometry

3.1 Capillary Rheometry. 62
   3.1-1 Rabinowitsch Analysis 64
   3.1-2 End Effects or Bagley Correction 68
      3.1-2.1 Fluid Elasticity from End Corrections 71
   3.1-3 Mooney Correction 72
   3.1-4 Intrinsic Viscosity Measurements 73
      3.1-4.1 Comments. 75
3.2 Coaxial-Cylinder Rheometers 76
   3.2-1 Calculation of Viscosity 77
      3.2-1.1 Non-Newtonian Viscosity 79
      3.2-1.2 Comments. 80
   3.2-2 End Effect Corrections. 81
   3.2-3 Normal Stress Determination 82
3.3 Cone-and-Plate Geometry. 84
   3.3-1 Viscosity Determination 86
   3.3-2 Normal Stress Determination 88
   3.3-3 Inertial Effects. 90
      3.3-3.1 Torque Correction 91
      3.3-3.2 Normal Force Corrections 92
   3.3-4 Criteria for Transient Experiments 94
3.4 Concentric Disk Geometry 98
   3.4-1 Viscosity Determination 99
   3.4-2 Normal Stress Difference Determination 101
3.5 Yield Stress Measurements. 103
3.6 Problems. 106
   3.6-1_a Rabinowitsch-Type Analysis. 106
   3.6-2_b Rabinowitsch Analysis for a Yield Stress Fluid 107
   3.6-3_a Viscosity of a High Density Polyethylene 107
   3.6-4_b Cone-and-Plate Flow 107
   3.6-5_b Parallel Plate Rheometer 108
   3.6-6_b Falling Cylinder Viscometer 108
   3.6-7_a Weissenberg Effect 109
   3.6-8_a Normal Stress Measurements 109
   3.6-9_b Normal Stress Determination Via Exit Pressure 109
   3.6-10_a Maxwell Extruder. 110
   3.6-11_b Yield Stress Determination 111

62   Rheometry

The major methods for measuring rheological properties are discussed in many textbooks, including those of Walters (1975) and Dealy (1982), as well as the book edited by Collyer and Clegg (1988). In this chapter we review the principles on which measurements using various flow geometries are based. We analyze the flow in capillaries, concentric cylinders (Couette), cone-and-plate, and concentric disk geometries. These are the basic viscometric flows. Other indirect techniques will not be discussed in this chapter. We close this chapter with some remarks on yield stress measurements. Small amplitude oscillatory shear, as well as creep and recovery experiments, will be discussed in Chapter 5.

3.1   Capillary Rheometry

The capillary rheometer is the simplest and most popular apparatus used to measure the viscosity of fluids. Capillary rheometers were the first viscometers to be used. The measurement of a Newtonian fluid viscosity, based on the relationship between pressure drop and flow rate, was developed independently by Hagen (1839) and Poiseuille (1841). One of the simplest viscometers is the Ostwald glass viscometer illustrated in Fig. 3.1-1a. The pressure drop is given by the hydrostatic head of the liquid and the flow rate is determined by measuring the time required for the level of liquid to fall from the upper index to the lower one, given the volume of the bulb. Obviously, such a viscometer is to be used for low viscosity liquids. Improved versions of the Ostwald viscometer, such as the Ubbelohde

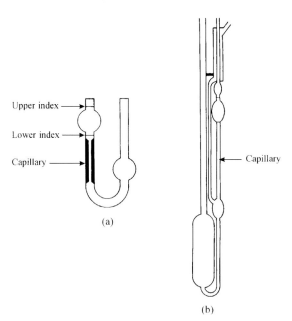

Figure 3.1-1   Ostwald viscometer (a) and improved Ubbelohde viscometer (b)

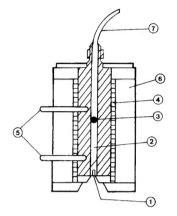

1. Die
2. Polymer reservoir
3. Sphere
4. Heating band
5. Thermocouples
6. Insulation
7. Flexible hose

Figure 3.1-2   Pressure-type capillary rheometer

viscometer illustrated in Fig. 3.1-1b, are currently used to measure the intrinsic viscosity of polymer solutions, as discussed in Section 3.1-4.

For the measurement of the properties of viscous liquids, essentially two types of forced flow capillary rheometers have been developed: the pressure type as illustrated in Fig. 3.1-2, and the piston-driven type as shown in Fig. 3.1-3.

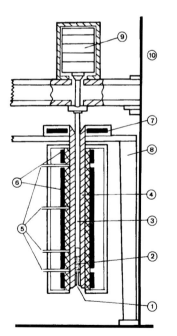

1. Die
2. Polymer
3. Piston
4. Barrel
5. Thermocouples
6. Heating elements
7. Heating disk
8. Frame
9. Load
10. Machine frame

Figure 3.1-3   Piston-driven capillary rheometer

In the pressure type rheometer, the viscous fluid, contained in reservoir (2), is pushed through a capillary die (1) by applying a gas pressure (usually nitrogen) fed from a flexible connector (7). A liquid-gas divider such as a metallic sphere is sometimes used to avoid extrusion of gas along with the liquid. Electrical heating (4) is normally used and the temperature is measured via thermocouples (5). Proper insulation (6) and a good temperature controller are required, especially for measuring polymer melt viscosities. The pressure is measured with a pressure gauge and the flow rate is determined by collecting the extrudate during a given time interval. These rheometers can be operated under a wide range of pressure and flow rates, and are thus quite flexible. However, they cannot be easily automated, and a complete rheological characterization requires a long and tedious procedure.

For other capillary rheometers, the liquid in the reservoir is pushed through the capillary die with the help of a piston driven at constant speed, as illustrated in Fig. 3.1-3. A clear advantage of this type over the pressure-driven one is that the flow rate is readily obtained from the piston speed. The pressure is easily measured by a load cell mounted on the piston (Instron rheometer) or pressure transducers flush-mounted on the reservoir wall (Gottfert rheometer). These instruments can be automated with a data acquisition system for a less tedious rheological characterization. They are mostly used for polymer melts.

The reader should note that the Melt Indexer, well known in the plastics industry, is a simplified version of the piston-driven rheometer. Standard weights are used for driving the piston, and standard, very short capillary dies are used (ASTM D 1238). The melt index, MI, used by the polyethylene industry is the number of grams extruded in 10 min through a capillary die with L/D (length to diameter) approximately equal to 4, using a weight of 2.16 kg with a reservoir temperature of 190 °C. As we will make clear in the remainder of this chapter, such a specific measurement is of very limited use.

3.1-1   Rabinowitsch Analysis

Capillary viscosity measurements are based on the relationships between the pressure drop across the capillary and the flow rate (Fig. 3.1-4). This is not a direct measurement and a

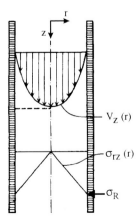

Figure 3.1-4   Flow in a capillary

constitutive relation for the fluid (Newtonian behavior, power-law, Ellis model) must first be chosen. For Newtonian fluids, the viscosity is readily obtained through the use of the Hagen-Poiseuille relation (Bird, Stewart, and Lightfoot, 1960). For non-Newtonian fluids, the complication of choosing an appropriate viscosity model can be avoided by following the Rabinowitsch (1929) procedure (wrongly called the Rabinowitsch correction in the literature).

This approach assumes the following:

- laminar flow: that is, the Reynolds number, Re, is smaller than 2000;
- steady state: $\dfrac{\partial}{\partial t} = 0$;
- fully developed unidirectional flow:

$$V_r = V_\theta = 0$$

and

$$V_z = V_z(r);$$

- No slip at the capillary walls:

$$V_z(r = R) = 0$$

and

$$\frac{dV_z}{dr} = \frac{dV_z}{dr}(\sigma_{rz})$$

or

$$\eta = \eta(\dot{\gamma}_{rz}),$$

that is, the viscosity is a unique function of the shear rate or shear stress.

The volumetric flow rate is expressed by

$$Q = 2\pi \int_0^R V_z r \, dr, \tag{3.1-1}$$

and integrating by parts yields

$$Q = \pi(V_z r^2)\Big|_0^R - \int_0^R \pi \left(\frac{dV_z}{dr}\right) r^2 \, dr. \tag{3.1-2}$$

The no-slip condition at the capillary wall requires the integrated term to be equal to zero. On the other hand, a momentum balance on a differential fluid element $2\pi r \Delta r L$ yields

$$\sigma_{rz} = \frac{r}{R}\sigma_R, \tag{3.1-3}$$

where $\sigma_R \ (= (\Delta P) R / 2L)$ is the shear stress at the wall, and ΔP the total pressure drop. Equation 3.1-2 becomes, after the change of variables, $r \to \sigma_{rz}$,

$$\frac{Q}{\pi R^3} = \frac{1}{\sigma_R^3} \int_0^{\sigma_R} \sigma_{rz}^2 \left(-\frac{dV_z}{dr}\right) d\sigma_{rz}. \tag{3.1-4}$$

Taking the derivative of this expression with respect to σ_R and using Leibnitz's rule, we have

$$\frac{d}{d\sigma_R}\left(\frac{\sigma_R^3 Q}{\pi R^3}\right) = \frac{d}{d\sigma_R}\int_0^{\sigma_R}\sigma_{rz}^2\left(-\frac{dV_z}{dr}\right)d\sigma_{rz} \qquad (3.1\text{-}5a)$$

$$= \left\{\sigma_{rz}^2\left(-\frac{dV_z}{dr}\right)\right\}_{|\sigma_R}. \qquad (3.1\text{-}5b)$$

Hence the shear rate evaluated at the wall is expressed by

$$-\dot{\gamma}_R = \left(-\frac{dV_z}{dr}\right)_{|r=R} = \frac{1}{\pi R^3 \sigma_R^2}\cdot\frac{d}{d\sigma_R}(\sigma_R^3 Q). \qquad (3.1\text{-}6)$$

This result can now be used to determine the viscosity from basic measurements of Q and ΔP. For example, at σ_{R1}, the slope of $(\sigma_R^3 Q)$ vs σ_R is p_1, hence,

$$\left(-\frac{dV_z}{dr}\right)_{1|r=R} = -\dot{\gamma}_{R_1} = \frac{p_1}{\pi R^3 \sigma_{R_1}^2},$$

and the corresponding viscosity is

$$\eta(\dot{\gamma}_{R_1}) = -\frac{\sigma_{R_1}}{\dot{\gamma}_{R_1}} = \frac{\pi R^3 \sigma_{R_1}^3}{p_1}.$$

In practice, it is not easy to determine with accuracy a slope from graphical data. It is preferable to fit a monotonous simple polynomial expression to equation 3.1-6, rewritten in the following form:

$$-\dot{\gamma}_R = \frac{1}{4\sigma_R^2}\frac{d}{d\sigma_R}\left(\sigma_R^3 \frac{4Q}{\pi R^3}\right)$$

$$= \frac{3}{4}\left(\frac{4Q}{\pi R^3}\right) + \frac{1}{4}\left(\frac{4Q}{\pi R^3}\right)\frac{d\ln\left(\frac{4Q}{\pi R^3}\right)}{d\ln\sigma_R}. \qquad (3.1\text{-}7)$$

Defining

$$n' = \frac{d\ln\sigma_R}{d\ln\left(\frac{4Q}{\pi R^3}\right)}, \qquad (3.1\text{-}8)$$

equation 3.1-7 can be written as

$$-\dot{\gamma}_R = \frac{3n'+1}{4n'}\left(\frac{4Q}{\pi R^3}\right). \qquad (3.1\text{-}9)$$

The term $(4Q/\pi R^3)$ is the so-called apparent shear rate, $\dot{\gamma}_a$, which is the correct expression for the wall shear rate of a Newtonian fluid.

The wall shear rate can be obtained indirectly from a log-log plot of σ_R vs $4Q/\pi R^3$. The curve is frequently close to a straight line, and the slope, n', is then a constant equal to the power-law index. Otherwise, the curve can be linearized part by part or a polynomial can be used to fit the data. An alternative approach is to calculate the apparent viscosity from

$$\eta_a = -\frac{\sigma_R}{\dot{\gamma}_a} = -\left(\frac{\left[\frac{(\Delta P)R}{2L}\right]}{\left[\frac{4Q}{\pi R^3}\right]}\right). \quad (3.1\text{-}10)$$

This is the correct expression for the viscosity of a Newtonian fluid. A log-log plot of the apparent viscosity for a high density polyethylene (HDPE 16A of DuPont Canada) at 180 °C is shown in Fig. 3.1-5 (the data shown are not corrected). The slope is equal to $n' - 1$, but it is not constant over the whole shear rate range. In the power-law region, $\dot{\gamma}_a > 100$ s^{-1}, the slope is constant, and n' is equal to n. Knowing n', the true shear rate at the wall can now be calculated from equation 3.1-9, and the non-Newtonian viscosity at the wall shear rate is by definition expressed by

$$\eta(\dot{\gamma}_R) = -\frac{\sigma_R}{\dot{\gamma}_R}. \quad (3.1\text{-}11)$$

We note that the corrected viscosity is slightly lower than the apparent viscosity. The Rabinowitsch "correction" is, however, not negligible. Note that we are dealing with a log-log representation.

Example 3.1-1   Rabinowitsch Analysis for Power-Law Fluids
Show how the power-law parameters, n and m, can be obtained from a log-log plot of σ_R vs $4Q/\pi R^3$.

Solution
For a power-law fluid, we have

$$\sigma_R = m|\dot{\gamma}_R|^n, \quad (3.1\text{-}12)$$

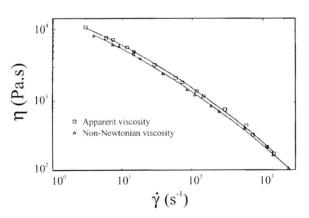

Figure 3.1-5   Apparent and non-Newtonian viscosities of a high density polyethylene melt (HDPE 16A of DuPont) at 180 °C

or

$$\ln \sigma_R = \ln m + n \ln |\dot{\gamma}_R|. \quad (3.1\text{-}13)$$

Combining equations 3.1-9 and 3.1-13 we have

$$\ln \sigma_R = \ln m + n \ln\left(\frac{3n'+1}{4n'}\right) + n \ln\left(\frac{4Q}{\pi R^3}\right). \quad (3.1\text{-}14)$$

Hence

$$\frac{d \ln \sigma_R}{d \ln\left(\frac{4Q}{\pi R^3}\right)} = n' = n. \quad (3.1\text{-}15)$$

That is, the slope is equal to the power-law index n, and the intercept of $\ln \sigma_R$ vs $\ln(4Q/\pi R^3)$ at $4Q/\pi R^3$ equal 1 is

$$\ln K' = \ln m + n \ln\left(\frac{3n+1}{4n}\right). \quad (3.1\text{-}16)$$

In summary, the power-law parameters are

$$n = \frac{d \ln \sigma_R}{d \ln\left(\frac{4Q}{\pi R^3}\right)} \quad (3.1\text{-}15)$$

and

$$m = K'\left(\frac{4n}{3n+1}\right)^n. \quad (3.1\text{-}17)$$

3.1-2 End Effects or Bagley Correction

Unless a very long capillary is used ($L/D > 100$), end effects (entrance and exit excess pressure drops) may considerably affect the accuracy of the determined viscosity. The empirical method proposed by Bagley (1957) can be used to correct for end effects. It can also be used to estimate the fluid's elasticity. The method consists of determining the excess pressure drops at the entrance and exit of the capillary. A typical wall pressure profile is illustrated in Fig. 3.1-6, showing the contribution of the normal stress, σ_{22}, developed for polymeric liquids under shear at the capillary wall.

The shear stress at the wall is corrected by

$$\sigma_R = \frac{(\Delta P)R}{2(L + e_0 R)}. \quad (3.1\text{-}18)$$

This concept is useful provided that the end correction factor, e_0, is independent of the capillary geometry. That is, the capillary has to be long enough for fully-developed flow to be attained before the fluid exits the capillary. The parameter e_0 is determined by constructing curves of ΔP vs L/R at constant shear rate or apparent shear rate ($4Q/\pi R^3$). The

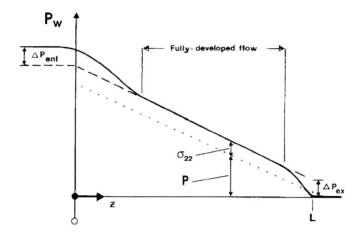

Figure 3.1-6 Pressure profile along the capillary axis

procedure is illustrated in Fig. 3.1-7 for HDPE data obtained with four different capillaries on an Instron rheometer. For four different shear rates, the applied load or pressure in the reservoir is plotted as a function of L/R. Straight lines are obtained, and the values of e_0 are determined by extrapolation to zero pressure. These values vary from 4 to 13. Equation 3.1-18 can then be used to calculate a corrected shear stress, and the viscosity is recalculated using equation 3.1-11. These corrections are rather important, as shown in Fig. 3.1-8 for two sets of data. Clearly, the uncorrected viscosity data obtained with the capillary of L/D equal to 5 show very large departures from the corrected data.

In Fig. 3.1-9, we report the corrected viscosity data for HDPE 16A obtained at four different temperatures and using three different rheometers. The low shear rate data were obtained on a constant stress rheometer (Rheometrics RSR). The medium range shear rate data were obtained from a gas driven capillary viscometer, and the high shear rate data

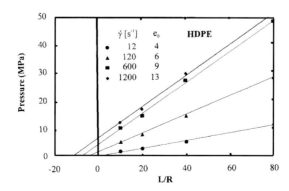

Figure 3.1-7 Bagley corrections for a high density polyethylene at 180 °C (DuPont 16A)

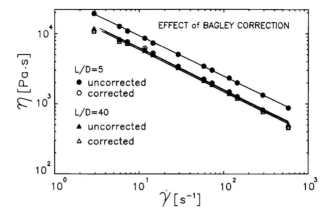

Figure 3.1-8 Uncorrected and corrected viscosities of HDPE 16A of DuPont at 180 °C

from an Instron capillary rheometer. Although there is virtually no overlap, the continuity between the three sets of data is good. Notice that more than six decades of shear rate are covered by these data.

Finally, in Fig. 3.1-10, we present Bagley plots for a molten polypropylene filled with 20 mass % of glass fibers. The applied load (or reservoir pressure) is not a linear function of L/D at most values of the piston speed. Many effects, such as fiber orientation or migration, viscous dissipation, and pressure effects on viscosity, may be responsible for the nonlinearity. The required extrapolation to obtain e_0 is not valid. The Bagley correction for this filled polymeric system is meaningless, and uncorrected data should be reported for the viscosity, which is clearly dependent on the hydrodynamics and on the geometry of the capillary system. Similar results have been obtained for other glass fiber reinforced polymeric systems.

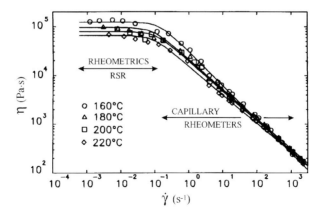

Figure 3.1-9 Shear viscosity as a function of shear rate for HDPE 16A at different temperatures

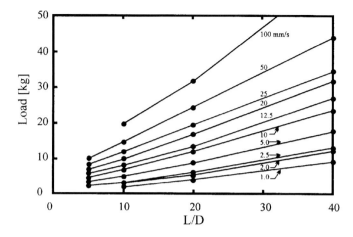

Figure 3.1-10 Bagley plots for a polypropylene filled with 20 mass % of glass fibers at 180 °C, using an Instron capillary rheometer. The piston speed is indicated on the curves

3.1-2.1 Fluid Elasticity from End Corrections

Over the last three decades, there have been numerous attempts to explain the high values of the end corrections for viscoelastic fluids, and to distinguish between the various contributions of viscous dissipation, kinetic energy, and stored elastic energy. From a force balance at the entrance, Philippoff and Gaskins (1958) proposed the following relation:

$$e_0 = e_c + S_R, \tag{3.1-19}$$

where S_R is the recoverable shear defined by

$$S_R = \frac{N_{1w}}{2\sigma_w}, \tag{3.1-20}$$

where N_{1w} is the primary normal stress difference introduced in Section 2.1, evaluated at the wall shear rate; and σ_w is the wall shear stress. This is the expression of recoil after cessation of steady shear flow for a simple Maxwell fluid (see Example 5.2-3), e_c is known as the Couette correction, which for purely viscous fluids is of the order of 0.75. Hence, large values for end corrections can be correlated with the recoverable shear, S_R, and the primary normal stress difference can be determined from equations 3.1-19 and 3.1-20. A corresponding elastic time for the fluid, evaluated at the capillary wall shear rate, is obtained from the following definition

$$\lambda(\dot{\gamma}_w) = \frac{N_{1w}}{2\sigma_w \dot{\gamma}_w}. \tag{3.1-21}$$

These results, however, should be used with caution. First, it is difficult to obtain accurate values for the end correction factor e_0 using a Bagley plot. Moreover, as pointed out by Choplin and Carreau (1981, 1986), the Bagley correction includes both entrance and exit effects. Using a macroscopic mechanical energy balance, they have shown that e_0 is the contribution of the viscous dissipation due to flow rearrangement at the entrance as well as at the exit. Viscoelasticity may enhance the viscous dissipation, and if elastic energy is stored at the entrance, it is dissipated at the outlet. Finally, we notice that the results expressed by equations 3.1-19 and 3.1-20 are independent of the entrance geometry (the primary normal stress differences and the shear stress are evaluated at the wall shear rate for fully-developed flow), whereas e_0 is known to be affected by the entrance (flat or tapered) geometry.

Exit effects (exit pressure) have been proposed (Han, 1976) for a measurement of primary normal stress differences. As discussed here, this technique presents major difficulties, and these are thoroughly reviewed by Boger and Denn (1980) and by Carreau et al. (1985).

3.1-3 Mooney Correction

The Rabinowitsch method allows for the determination of the shear rate at the wall, provided the rheology of the fluid is homogeneous. That is, the viscosity is a unique function of the shear rate or shear stress. This is not necessarily the case with multiphase systems. For example, due to the finite dimensions of the filler particles, special wall effects resulting in an apparent slip may be observed. Other possible effects include migration and orientation of particles, coalescence or phase inversion in blends, or viscous heating effects. These effects may be detected and eventually corrected using the following method.

Suppose that any wall effects can be modeled by a slip velocity at the wall of the capillary. This velocity can be positive (equivalent to a dilution effect) or negative (equivalent to an adsorption or concentration effect). Equation 3.1-4 becomes

$$\frac{Q}{\pi R^3} = \frac{V_s}{R} + \frac{1}{\sigma_R^3}\int_0^{\sigma_R} \sigma_{rz}^2\left(-\frac{dV_z}{dr}\right)d\sigma_{rz}, \qquad (3.1\text{-}22)$$

where V_s is the apparent slip velocity. This result can be rewritten as

$$\frac{Q}{\pi R^3 \sigma_R} = \frac{\beta}{R} + \frac{1}{\sigma_R^4}\int_0^{\sigma_R} \sigma_{rz}^2\left(-\frac{dV_z}{dr}\right)d\sigma_{rz}. \qquad (3.1\text{-}23)$$

In equation 3.1-23, the slip is expressed in terms of a coefficient β ($=V_s/\sigma_R$). We notice that the integral on the right side of this equation is independent of the capillary radius. It is a unique function of the shear stress (or shear rate) evaluated at the wall. Hence, to ascertain if there are wall effects, the Bagley correction is first determined using a set of different L/R capillaries of the same radius. Then another series of experiments using capillaries of different radii is conducted and plots of $(Q/\pi R^3 \sigma_R)$ vs σ_R (corrected values) are made for each R. In the absence of wall effects, β is equal to zero, and a single curve should be obtained. If a set of different curves is obtained, the value of β can be determined by plotting $(Q/\pi R^3 \sigma_R)$ vs $1/R$ at a given shear stress.

In some cases, the Mooney correction may lead to results which have no physical meaning. Mourniac et al. (1992) reported negative shear rates at the wall for a molten rubber compound. That is, the slip velocity was found to be larger than the experimentally determined average velocity. The Mooney method should therefore be used with extreme caution, possibly for determining non-uniformity and wall effects rather than for a quantitative determination of slip velocities. For slip determination we recommend the use of rotational rheometers equipped with cones and plates of varying surface roughness.

In studying the rheology of multiphase systems with capillary rheometers, we should consider other possible effects, which may considerably limit the usefulness of the data. The most important other sources of difficulty are the heat generated by viscous dissipation and the influence of pressure on viscosity. For polymer melts, these effects may be very important, and render the measurements useless.

3.1-4   Intrinsic Viscosity Measurements

The Ostwald or the improved Ubbelohde capillary viscometer is commonly used to determine the intrinsic viscosity of polymer solutions. For a very dilute polymer solution, there is no interaction between polymer molecules, and the viscosity of the solution is that of the solvent plus the contribution of the individual polymer molecules. The intrinsic viscosity is in fact a measure of the hydrodynamic volume of a polymer molecule in a given solvent. This is obtained from the measurement of the relative viscosity of the solution with respect to that of the solvent.

The density of a very dilute solution is approximately that of the solvent ($\rho_s \approx \rho$), and the relative viscosity can be taken as the ratio of the times required for a given volume of fluid to flow through a bulb of an Ostwald or Ubbelohde viscometer (see Fig. 3.1-1). The specific viscosity is then given by

$$\eta_{sp} = \frac{\eta - \eta_s}{\eta_s} = \eta_r - 1 = \frac{t\rho_s}{t_s\rho} - 1 \approx \frac{t}{t_s} - 1, \qquad (3.1\text{-}24)$$

where the subscript s stands for solvent. The intrinsic viscosity $[\eta]$ is defined as the zero concentration limit of η_{sp}/c and is determined by extrapolating η_{sp}/c or $\ln\eta_r/c$ to zero concentration. The relation between η_{sp}/c and η is given by the Huggins (1942) equation:

$$\frac{\eta_{sp}}{c} = [\eta] + k'[\eta]^2 c, \qquad (3.1\text{-}25)$$

where k' is Huggins' constant. This result is valid up to second order in concentration. The polymer solvent interaction is related to $k'[\eta]^2$, the slope of η_{sp}/c vs the polymer concentration. For flexible polymer chains in a θ solvent (see Larson, 1988), the interaction parameter k' has been found by Sakai (1968) to vary from 0.4 to 1.0 depending on the theory used. The corresponding relation involving the relative viscosity is given by

$$\frac{\ln \eta_r}{c} = [\eta] + k''[\eta]^2 c, \qquad (3.1\text{-}26)$$

74   Rheometry

where k'' is Kramer's constant and $k' - k'' = 0.5$. The relation between the intrinsic viscosity and the molecular weight is expressed via the Mark-Houwink equation as

$$[\eta] = KM_v^\alpha. \tag{3.1-27}$$

Values of the parameters K and α can be found in the literature for different polymer-solvent systems at a given temperature (see for example Flory, 1953). M_v is the viscosity average molecular weight in kg/kmol.

The intrinsic viscosity is related to the molecular structure (Flory and Fox, 1951) by

$$[\eta] = \phi' R_g^3 M^{-1}, \tag{3.1-28}$$

where R_g and M are the radius of gyration and the molecular weight of the polymer molecule respectively; and ϕ' is a universal constant, independent of temperature and polymer-solvent system, and is equal to 2.1×10^{26} kmol^{-1}.

Example 3.1-2   Intrinsic Viscosity of PEO

Figure 3.1-11 reports on the measured values of η_{sp}/c and of $\ln \eta_r/c$ vs concentration for polyethylene oxide (PEO) solutions ranging in concentration from 10^2 to 10^3 ppm in water and in a 50 mass % mixture of water and glycerine. The data were obtained by Ortiz (1992), using Ubbelohde viscometers with four bulbs and capillary diameters of 0.4 mm (#50) and 0.6 mm (#100). The reported values for $[\eta]$ are the average values over the four bulbs. All solutions were observed to be Newtonian, that is, their viscosity, calculated from the value of the flow rate and the capillary diameter, was independent of the shear rate. The flow times, t_s, for the solvent (water) varied from 220 s to 250 s for the different bulbs, and were reproducible to within 2%.

Using the data of Fig. 3.1-11, determine the intrinsic viscosity and the molecular weight of this PEO, and its radius of gyration.

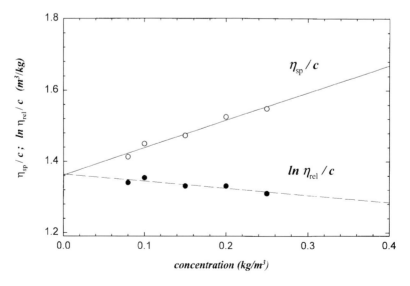

Figure 3.1-11   Determination of the intrinsic viscosity for aqueous PEO solutions

Solution
Straight lines are obtained and both lead to the same intercept, which is the value of the intrinsic viscosity for this polymer-solvent pair with $[\eta]$ equal to 1.37 m^3/kg. The molecular weight is computed from equation 3.1-27 with $K = 1.25 \times 10^{-5}$ m^3/kg and $\alpha = 0.78$, as recommended by Bailey and Koleske (1976):

$$M_v = \left(\frac{1.37}{1.25 \times 10^{-5}}\right)^{\frac{1}{0.78}} = 2.89 \times 10^6 \text{ kg/kmol}.$$

The value of the radius of gyration R_g can now be computed via equation 3.1-28. For this polymer-solvent pair, R_g is then

$$R_g = \left(\left[\frac{1.37}{2.1 \times 10^{26}}\right](2.89 \times 10^6)\right)^{\frac{1}{3}}$$

$$= 2.66 \times 10^{-7} = 266 \text{ nm}.$$

The hydrodynamic radius of this macromolecule is of the dimension of a colloidal particle.

3.1-4.1 Comments

We observe a decrease in the values of $[\eta]$ and R_g when using a water–glycerine mixture as the solvent, indicating that water is a better solvent. That is, the polymer-solvent interactions are favored if water is used as the solvent. In a poorer solvent such as a water–glycerine mixture, the polymer molecules contract and R_g decreases. For the same PEO, in a 50 mass % mixture of water and glycerine, $[\eta]$ was found to be equal to 0.99 m^3/kg with a corresponding R_g of 243 nm. Polymer-solvent interactions also affect the variation of η_{sp}/c vs c. The Huggins constant, k', in equation 3.1-25, changes according to the nature of the solvent.

The critical concentration c^* is the concentration for which interactions between polymer molecules become important. Following de Gennes (1979), the critical concentration is obtained from

$$[\eta]c^* \cong 1. \tag{3.1-29}$$

The value of c^* depends on the solvent for a polymer of a given molecular weight. In a good solvent, the value c^* is less than that in a poor solvent. This is to be expected, considering that R_g increases as the solvent improves. For a given solvent, we observe that c^* decreases with increasing molecular weight.

The interest in the intrinsic viscosity and c^* for correlating rheological properties of concentrated solutions is illustrated in Fig. 3.1-12, where we have plotted the specific viscosity versus the product $[\eta]c$ for a series of PEO solutions in water. The lower part of the figure for the dilute solutions is a straight line of slope equal to 1 following the definition of the intrinsic viscosity (equation 3.1-25). For the concentrated solutions, the polymer concentration was varied from 0.5 to 7.0 mass % and η_0 is the zero shear viscosity determined from cone-and-plate rheometry. It is interesting to note that all data for the concentrated solutions fall approximately on a single straight line of slope equal to 4.5 (the data for concentrated solutions in glycerine–water (not shown) would fall very closely to

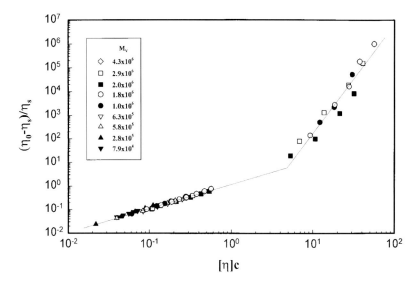

Figure 3.1-12 Specific viscosity versus $[\eta]c$ for dilute and concentrated solutions of different molecular weight PEO in water (Data from Ortiz, 1992)

those shown in the figure). Obviously, these data reflect very strong polymer–polymer interactions (entanglements). The critical polymer concentration given by equation 3.1-29 is somewhat conservative. The rheological properties of the concentrated PEO solutions are presented in Figs. 2.2-2 and 2.4-1.

3.2 Coaxial-Cylinder Rheometers

The majority of modern rheometers are rotating devices, and permit rapid determination of the viscosity curve as well as other important rheological functions. Commercial rheometers are now available with computer control and software packages for data analysis. The measurements are made using three possible geometries: coaxial cylinders (Couette geometry), cone-and-plate geometry, and concentric disks. In this section, the flow in coaxial cylinders is analyzed. The reader will find the other two geometries discussed in Sections 3.3 and 3.4. Advantages and drawbacks of each geometry are also briefly discussed.

Coaxial cylinders were probably the first rotating devices used to measure viscosity. The rheometer is named after Maurice Couette (1890), who invented the concentric cylinder viscometer. Various designs have been proposed. Figure 3.2-1 illustrates an improved version of the Stormer viscometer, used by Dexter (1954). The rotation is obtained by a weight (3) transmitting a torque to an inside cylinder (5) via a string (2) and pulley (1). The outside cylindrical frame is the reservoir for the fluid. Enough fluid is

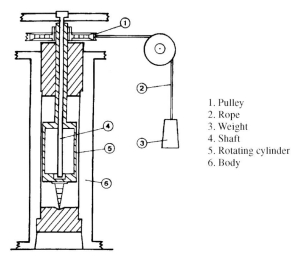

Figure 3.2-1   Dexter's viscometer

initially poured into the reservoir so that, when the inside assembly is put in place, the fluid will occupy the gap between the two cylinders, with some of the fluid covering the top of the inner cylinder. It is assumed that the resistance to flow is due to the flow in the small gap between the two cylinders. The rotation of the inner cylinder is obtained from the velocity of the falling weight. Temperature control can be added by jacketing the outside cylinder. This equipment is really the precursor of modern stress rheometers, such as those commercialized by Bohlin and Carrimed, in which a torque is imposed via an electrical field to a motor. Angular displacement is measured using a transducer.

3.2-1   Calculation of Viscosity

For the flow in a Couette geometry, we follow the analysis of Bird, Stewart, and Lightfoot (1960, Example 3.5-1). Let us consider two coaxial cylinders of length L and of radii R and KR, as shown in Fig. 3.2-2. The inner cylinder is stationary and the outer one is rotating at angular velocity Ω. We assume steady-state, isothermal flow and neglect end effects: that is, we assume that the only non-vanishing velocity component is the tangential component, which is taken to be a unique function of the radial position:

$$V_\theta = V_\theta(r)$$
$$\text{and} \quad V_r = V_z = 0. \tag{3.2-1}$$

The θ-component of the equation of motion reduces to

$$0 = -\frac{1}{r^2}\frac{d}{dr}(r^2 \sigma_{r\theta}). \tag{3.2-2}$$

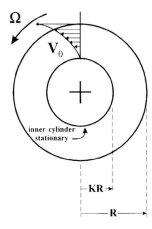

Figure 3.2-2 Flow in coaxial cylinders

Integrating, we have

$$\sigma_{r\theta} = \frac{C_1}{r^2}. \tag{3.2-3}$$

To evaluate the constant C_1, we must make a further assumption or use an appropriate rheological model. Contrary to the flow in a capillary for which the shear rate varies from zero at the center line to a maximum value at the wall, the shear rate variation across the gap is quite small in properly designed coaxial cylinder viscometers. Hence, to assume *power-law behavior* is quite sufficient for most applications. This can be written in the form

$$\sigma_{r\theta} = -m\left[r\frac{d}{dr}\left(\frac{V_\theta}{r}\right)\right]^n, \tag{3.2-4a}$$

where the shear rate in cylindrical coordinates is given by

$$\dot{\gamma}_{r\theta} = r\frac{d}{dr}\left(\frac{V_\theta}{r}\right). \tag{3.2-4b}$$

This is the $r\theta$-component of the rate-of-strain tensor. Combining equations 3.2-3 and 3.2-4, we obtain, after integration and rearrangement,

$$\frac{V_\theta}{r} = \frac{n}{2}\left|-\frac{C_1}{mR^2}\right|^{\frac{1}{n}}\left[\left(\frac{R}{r}\right)^{\frac{2}{n}} + C_2\right]. \tag{3.2-5}$$

The two constants, C_1 and C_2, are evaluated using the no-slip conditions at the wall of the cylinders:

$$\begin{aligned}&\text{B.C. 1:} \quad \text{at} \quad r = KR, \quad V_\theta = 0\\&\text{and}\quad \text{B.C. 2:} \quad \text{at} \quad r = R, \quad V_\theta = \Omega R.\end{aligned} \tag{3.2-6}$$

The velocity profile is then given by the following expression:

$$\frac{V_\theta}{r} = \frac{\Omega}{1 - \left(\frac{1}{K}\right)^{\frac{2}{n}}} \left[\left(\frac{R}{r}\right)^{\frac{2}{n}} - \left(\frac{1}{K}\right)^{\frac{2}{n}}\right]. \qquad (3.2\text{-}7)$$

The expression for the shear rate, the $r\theta$-component of the rate-of-strain tensor, is obtained from the derivative of equation 3.2-7:

$$\dot{\gamma}_{r\theta} = \left(\frac{R}{r}\right)^{\frac{2}{n}} \left[\frac{2\Omega}{n\left[\left(\frac{1}{K}\right)^{\frac{2}{n}} - 1\right]}\right]. \qquad (3.2\text{-}8)$$

We note that for a small gap, $\dot{\gamma}_{r\theta}$ does not vary much and is almost constant. In the limit as $K \to 1$, the shear rate is constant and given by

$$\dot{\gamma}_{r\theta} = \left[\frac{\Omega}{1-K}\right]. \qquad (3.2\text{-}9)$$

The shear stress is obtained from equations 3.2-4 and 3.2-8:

$$\sigma_{r\theta} = -m\left(\frac{R}{r}\right)^2 \left[\frac{2\Omega}{n\left[\left(\frac{1}{K}\right)^{\frac{2}{n}} - 1\right]}\right]^n, \qquad (3.2\text{-}10)$$

and the torque exerted on the outer cylinder is calculated from

$$T = (2\pi RL)(-\sigma_{r\theta})\big|_{r=R} R, \qquad (3.2\text{-}11)$$

which, using equation 3.2-10, becomes

$$T = 2\pi R^2 L m \left[\frac{2\Omega}{n\left[\left(\frac{1}{K}\right)^{\frac{2}{n}} - 1\right]}\right]^n. \qquad (3.2\text{-}12)$$

Obviously, this result reduces to the classical relation for Newtonian fluids (equation 3.5-13 of Bird, Stewart, and Lightfoot (1960)). For non-Newtonian fluids, the power-law index n is obtained from the slope of the log-log plot of T vs Ω. The parameter m is obtained from the intercept. We note that for steady-state conditions the torque at the inner cylinder is identical to the torque exerted on the outer cylinder, and is given by expression 3.2-12.

3.2-1.1 Non-Newtonian Viscosity

Although the power-law expression gives a poor description of the viscosity of polymeric fluids in the low shear rate region, the previous analysis is in general valid for a small gap, in which case the shear rate does not vary much across the gap. Note that the situation is quite different in Poiseuille flow, where the shear rate goes to zero at the center of the tube. From the definition of the non-Newtonian viscosity, at the *inner cylinder*, we obtain

$$\eta(\dot{\gamma}_{r\theta}(KR)) = -\frac{\sigma_{r\theta}(KR)}{\dot{\gamma}_{r\theta}(KR)}, \qquad (3.2\text{-}13)$$

where

$$-\sigma_{r\theta}(KR) = \frac{T}{2\pi K^2 R^2 L} \qquad (3.2\text{-}14)$$

and

$$\dot{\gamma}_{r\theta}(KR) = \frac{2\Omega}{n\left[1 - K^{\frac{2}{n}}\right]}. \qquad (3.2\text{-}15)$$

Also, at the *outer cylinder*,

$$\eta(\dot{\gamma}_{r\theta}(R)) = -\frac{\sigma_{r\theta}(R)}{\dot{\gamma}_{r\theta}(R)}, \qquad (3.2\text{-}16)$$

where

$$-\sigma_{r\theta}(R) = \frac{T}{2\pi R^2 L} \qquad (3.2\text{-}17)$$

and

$$\dot{\gamma}_{r\theta}(R) = \frac{2\Omega}{n\left[\left(\frac{1}{K}\right)^{\frac{2}{n}} - 1\right]}. \qquad (3.2\text{-}18)$$

Note that these results remain valid if the inner cylinder is rotated instead of the outer one. Only the signs of $\sigma_{r\theta}$ and $\dot{\gamma}_{r\theta}$ are interchanged. These results depend on the power-law index, which is a priori unknown. The procedure suggested above is to plot $\log T$ vs $\log \Omega$ and determine the local slope n. If the fluid does not obey the power law, the value of n will vary with Ω, but the locally determined value of n should be used in calculating the viscosity using equation 3.2-13 or 3.2-16. The determination of a minimum of two data points is required to calculate the viscosity of a non-Newtonian fluid from Couette rheometer measurements. Most software packages for commercial rheometers use only one point, based on the assumption that the gap is small enough and that the shear rate across the gap is constant. This leads to slightly incorrect results.

3.2-1.2 Comments

The analysis presented above is based on the assumption that the flow is laminar, that is, there are no secondary flows or vortices induced by centrifugal forces. At high rotational velocities, circulation cells called Taylor vortices are generated, and torques larger than those predicted by equation 3.2-12 are measured. For most applications, the fluids are viscous enough and we can neglect inertial effects even at shear rates larger than 10^4 s^{-1}. Taylor vortices appear at lower values of Ω when the inner cylinder is rotated and the outer cylinder is held stationary. Stability criteria in terms of Reynolds numbers for Newtonian fluids have been proposed (Schlichting, 1968; Bird, Stewart, and Lightfoot, 1960, pp. 95–96). The stability of inelastic non-Newtonian fluids has been analyzed by Lockett et al. (1992) using a finite-element numerical method.

Krieger and Maron (1954) have derived expressions for the shear rate at the inner cylinder of the Couette geometry using a method similar to that of Rabinowistch. Calderbank and Moo-Young (1959) have extended the Krieger-Maron analysis to large gaps.

Obviously, the Couette geometry with a small gap generates a flow which approximates simple shear flow. The disadvantages encountered with the early designs—inadequate temperature control, large sample volumes, and end effects—have been overcome by manufacturers. Rheometers with Couette geometries designed to accurately measure the viscosity of fluids in a wide range of shear rate and shear stress are now commercially available. The Couette geometry is the most suitable geometry for the viscosity determination of low and moderately high viscosity homogeneous fluids. It is not appropriate for polymer melts, pastes, and suspensions containing large size particles. (See Section 3.4 for particle size to gap ratio.) As an example of an appropriate design for low viscosity fluids, the Couette geometry proposed by Bohlin for the VOR rheometer is shown in Fig. 3.2-3. The system also includes a low friction cover which minimizes solvent evaporation during the measurement. End effects are not too important with this design, but these effects can still be estimated using the procedure outlined in the following section.

3.2-2   End Effect Corrections

End effects can be minimized by using a small gap geometry or a more sophisticated geometry. If not, the following simple method can be applied to detect and correct for end effects, which are mostly due to the additional resistance of the cup (outer cylinder) bottom. Using standards or Newtonian fluids of known but comparable viscosities, experiments can be conducted with a series of constant diameter cylinders of different length (height of immersion in the fluid), L. Then, for a constant gap, we can plot the torque as a function of the length for various values of the rotational speed, as illustrated in

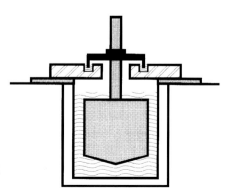

Figure 3.2-3 Couette geometry for the Bohlin VOR rheometer, with evaporation control

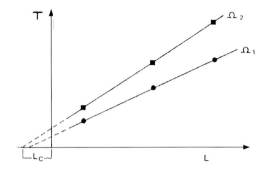

Figure 3.2-4   End effect determination for Couette viscometry

Fig. 3.2-4. The intercept on the x-axis gives a correction term, L_c, which is an equivalent additional length due to the end effects. Obviously this method is empirical: L_c may depend on the fluid properties and on the rotational speed. It may not be valid for viscoelastic fluids, and in all cases, it should not be used if L_c is larger than 10% of the magnitude of L.

The shear stress is calculated, accounting for L_c, as follows:

$$-\sigma_{r\theta}(KR) = \frac{T}{2\pi K^2 R^2 (L + L_c)}. \tag{3.2-19}$$

3.2-3   Normal Stress Determination

In principle, the Couette geometry can be used to measure normal stress differences. This follows from the assumption that the flow approximates simple shearing. We refer to Fig. 3.2-5 and, in addition to the hypotheses used in Section 3.2-1, we further assume that the only nonzero stress tensor components are $\sigma_{r\theta} = \sigma_{\theta r}(r)$, $\sigma_{rr}(r)$, $\sigma_{\theta\theta}(r)$, and $\sigma_{zz}(r)$. It is quite reasonable to assume that the other shear components vanish.

The r-component of the equation of motion yields

$$-\rho \frac{V_\theta^2}{r} = -\frac{\partial P}{\partial r} - \left(\frac{1}{r} \frac{d}{dr}(r\sigma_{rr}) - \frac{\sigma_{\theta\theta}}{r} \right), \tag{3.2-20}$$

which can be written for any position z as

$$\frac{d}{dr}(P + \sigma_{rr}) = \frac{d\pi_{rr}}{dr} = \rho \frac{V_\theta^2}{r} + \frac{\sigma_{\theta\theta} - \sigma_{rr}}{r}. \tag{3.2-21}$$

Integrating with respect to r from $r=KR$ to $r=R$, we get

$$\pi_{rr}(R) - \pi_{rr}(KR) = \int_{KR}^{R} \left(\rho \frac{V_\theta^2}{r} + \frac{\sigma_{\theta\theta} - \sigma_{rr}}{r} \right) dr, \tag{3.2-22}$$

3.2 Coaxial-Cylinder Rheometers

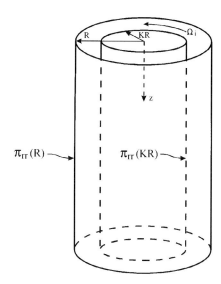

Figure 3.2-5 Normal stress determination in Couette geometry

where $\pi_{rr}(R)$ and $\pi_{rr}(KR)$ are the total pressures measured at the outer and inner cylinder walls respectively, as shown in Fig. 3.2-5. Following the definition of the primary normal stress difference, this result can be written as

$$\pi_{rr}(R) - \pi_{rr}(KR) = \int_{KR}^{R} \left(\rho \frac{V_\theta^2}{r} - \frac{N_1(r)}{r} \right) dr. \tag{3.2-23}$$

This result is not very useful, since the dependence of N_1 on the radius is unknown. However, in the limit of a small gap, that is, for $K \to 1$, we obtain the following useful results. In the case of a *small gap*, the expression for the shear rate (equation 3.2-15 or 3.2-18) reduces to

$$\dot{\gamma}_{r\theta} \approx -\frac{\Omega_i KR}{R - KR} \approx -\frac{\Omega_i}{1 - K}, \tag{3.2-34}$$

which is a constant (simple shear flow), and the velocity profile is

$$V_\theta \approx \frac{\Omega_i KR}{1 - K}\left(1 - \frac{r}{R}\right). \tag{3.2-25}$$

Equation 3.2-23 can easily be integrated to obtain

$$\pi_{rr}(R) - \pi_{rr}(KR) = \rho \frac{\Omega_i^2 (KR)^2}{(1 - K)^2}\left[\frac{1}{2}(1 - K^2) - 2(1 - K) + \ln\left(\frac{1}{K}\right)\right] - N_1 \ln\left(\frac{1}{K}\right). \tag{3.2-26}$$

This result can be further simplified by noting that

$$\ln\left(\frac{1}{K}\right) \approx 1 - K,$$

and

$$1 - K^2 = (1 + K)(1 - K) \approx 2(1 - K).$$

Hence, for a very small gap the inertial term is negligible and we obtain

$$\pi_{rr}(R) - \pi_{rr}(KR) \approx (K-1)N_1 = (K-1)\psi_1 \dot{\gamma}_{r\theta}^2, \quad (3.2\text{-}27)$$

where ψ_1 is the primary normal stress coefficient defined by equation 2.1-7.

The results of equations 3.2-23 and 3.2-27 explain the Weissenberg effect presented in Chapter 1. For a Newtonian fluid, N_1 is zero and the pressure at the outer wall is greater than that at the inner wall, due to inertial forces that create a typical vortex at the fluid-free surface. For polymeric fluids, N_1 is positive, and if the rotational speed is not too large or if a very small gap is used, the integral is negative. That is, the pressure at the inner wall is larger than the pressure at the outer wall. Hence, the primary normal stress difference forces the fluid to climb the inner cylinder wall (Weissenberg effect).

As far as we know, no commercial rheometer offers the option of measuring the primary normal stress difference with the Couette geometry. Technically, it is very difficult to accurately measure normal stresses or pressures at a curved wall surface of a rotating cylinder. However, with recent sophisticated and highly compact electronic sensors, such an option may become commercially available in the not too distant future.

3.3   Cone-and-Plate Geometry

The most popular geometry for rheological measurements of viscoelastic fluids is the cone-and-plate geometry, illustrated in Fig. 3.3-1. A small fluid sample (usually less than 0.5 mL) is placed in the space between a plate of radius R and a cone of the same radius and a very small angle ($\theta_0 < 3°$). The sample should have a free spherical shaped surface at the outer edge, $r = R$. For viscous fluids, the cone can be positioned below the plate, and either the cone or the plate can be rotated. For a very small angle, the velocity profile is linear with respect to the θ position in the gap (see Example 3.5-3 of Bird, Stewart, and

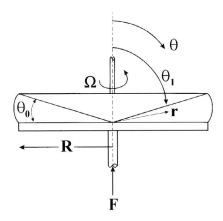

Figure 3.3-1   Cone-and-plate geometry

Lightfoot, 1960) and is expressed as

$$\frac{V_\phi}{r} = \Omega \left[\frac{\left(\frac{\pi}{2}\right) - \theta}{\theta_0} \right], \qquad (3.3\text{-}1)$$

where Ω is the imposed angular rotational speed of the cone (or plate).

The shear rate is the $\theta\phi$-component of the rate-of-strain tensor, and is expressed by

$$\dot{\gamma} = \dot{\gamma}_{\theta\phi} = \frac{\sin\theta}{r} \left[\frac{\partial}{\partial\theta}\left(\frac{V_\phi}{\sin\theta}\right) \right] \approx -\frac{\Omega}{\theta_0}, \qquad (3.3\text{-}2)$$

since $\sin\theta \approx 1$. Hence, for a small angle cone, the shear rate is constant, and the flow corresponds to that of a simple shear situation. This result was obtained for Newtonian fluids, but we will assume that it remains valid for viscoelastic fluids. It is recommended to use a cone with an angle of less than or equal to 3 degrees. However, Attané et al. (1980) used a cone of 8° with very little apparent effect on the measurements.

The cone-and-plate geometry appears to be the ideal geometry. The main advantages are

(i) The shear rate is constant and the determination of rheological properties does not require any assumption about the flow kinematics. Rheological models are not required.
(ii) Very small samples are required, and this is of particular importance for scarce fluids such as biological fluids and polymers synthesized in small quantities in research laboratories.
(iii) The system allows for very good heat transfer and temperature control.
(iv) End effects are negligible, at least for low rotational speeds when using the appropriate quantity of fluid in the gap.

The disadvantages are

(i) The system is limited to very low shear rates, mainly for polymer melts which will not stay in the gap at large rotational speeds. For low viscosity and slightly elastic fluids, a cup as shown in Fig. 3.3-2 can be used to obtain data at large shear rates.

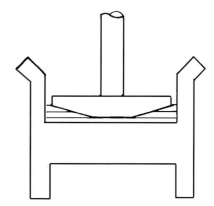

Figure 3.3-2 Cone-and-cup arrangement

(ii) It is difficult to eliminate evaporation and free-boundary effects for solutions made with volatile solvents. Techniques to minimize these effects include coating the outer edge with a non-volatile fluid such as a silicone oil or glycerine. Precautions have to be taken to ensure that this extra coating does not generate appreciable stresses at the boundary.
(iii) Highly erroneous results could be obtained for multiphase systems, such as suspensions of solids and polymer blends where the particles or domain sizes are of the same order of magnitude as the gap size. The preferred geometry for multiphase systems is the concentric disk geometry discussed in Section 3.4.

Most manufacturers propose truncated cones as shown in Fig. 3.3-3. With a truncated cone, it is much easier experimentally to set the correct gap as required by the geometry. The truncation is restricted to a small central portion so that it does not significantly affect the torque and normal force. The use of a large angle cone should be avoided in order to retain a homogeneous shear field. However, the use of a small angle leads to experimental difficulties. For a 1° cone, the gap is of the order of 40 μm, and the measurements become quite sensitive to gap and alignment settings. Figure 3.3-3 also shows vortices or secondary flows. These will be discussed in Section 3.3-3.

3.3-1 Viscosity Determination

As the shear rate is constant across the gap, the viscosity is readily determined using the torque, which is expressed by

$$T = 2\pi \int_0^R \sigma_{\theta\phi} r^2 dr = \frac{2}{3}\pi R^3 \sigma_{\theta\phi}, \qquad (3.3\text{-}3)$$

since the shear stress is also constant. The non-Newtonian viscosity is obtained by combining equations 3.3-2 and 3.3-3 with

$$\sigma_{\theta\phi} = -\eta \dot{\gamma}_{\theta\phi} = -\eta \dot{\gamma}, \qquad (3.3\text{-}4)$$

to obtain

$$\eta = \frac{3\theta_0 T}{2\pi R^3 \Omega}. \qquad (3.3\text{-}5)$$

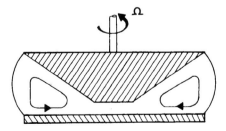

Figure 3.3-3 Truncated cone and secondary flows

That is, the viscosity is directly proportional to the torque T and inversely proportional to the rotational speed, Ω, expressed here in rad/s. The simplicity of this result explains why the cone-and-plate geometry is so popular. The result obtained here for the shear viscosity is also valid for transient shear flows, provided that inertia is not important. This is discussed below for stress growth and relaxation experiments and in Chapter 5 for the complex viscosity.

Figure 3.3-4 presents steady shear viscosity data for a series of 3 mass % polyethylene oxide (PEO) solutions in a 50 mass % mixture of water and glycerine. The data were obtained using different rheometers, but all with the cone-and-plate geometry. The effect of molecular weight on η is striking: we notice that the viscosity, especially in the low shear rate region, increases rapidly with the polymer molecular weight, and the shear thinning becomes more pronounced and appears at a lower shear rate. For the highest molecular weight solution, viscosity data could be obtained at a shear rate as low as 10^{-4} s^{-1}. This is about the lowest shear rate that can be attained using a Rheometrics RSR in the creep mode (see Chapter 5). Although the low shear rate viscosity appears to be constant, the data would show a definite negative slope on a regular scale graph.

Figure 3.3-5 illustrates how the viscosity of 3 mass % PEO solutions in water/glycerine changes with molecular weight for different shear rates (η_s is the viscosity of the solvent, which is equal to 5.24 mPa·s). At high shear rates, the reduced viscosity increases slowly with increasing molecular weight. At low shear rates, the effect of M_v is as expected, and a power-law relation emerges as $\dot{\gamma} \to 0$ and $(\eta_0 - \eta_s) \sim M_v^{3.4}$.

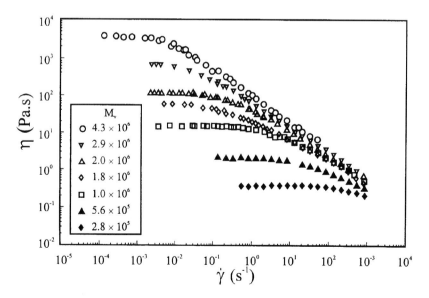

Figure 3.3-4 Viscosity (η) vs shear rate for 3 mass % PEO solutions in water and glycerine, as a function of molecular weight M_v at $T = 25$ °C (Data from Ortiz, 1992)

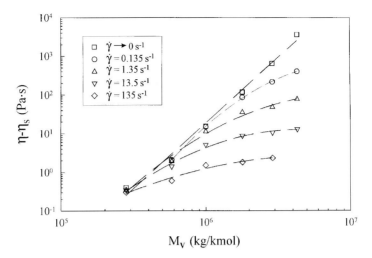

Figure 3.3-5 Effect of molecular weight (M_v) on viscosity ($\eta - \eta_s$) as a function of shear rate ($\dot{\gamma}$) for a 3 mass % PEO solution in a 50 mass % mixture of water and glycerine at $T = 25\ °C$ (Data from Ortiz, 1992); the solvent viscosity is 5.24 mPa·s

3.3-2 Normal Stress Determination

As the shear rate is constant in a cone-and-plate instrument, the geometry is ideal for measuring normal stress differences. Neglecting inertial forces (assuming $\rho V_\theta^2/r \to 0$), the r-component of the equation of motion simplifies to

$$0 = -\frac{\partial P}{\partial r} - \frac{1}{r^2}\frac{\partial}{\partial r}(r^2 \sigma_{rr}) - \frac{1}{r \sin\theta}\frac{\partial}{\partial \theta}(\sigma_{r\theta} \sin\theta) - \frac{1}{r \sin\theta}\frac{\partial \sigma_{r\phi}}{\partial \phi} + \frac{\sigma_{\theta\theta} + \sigma_{\phi\phi}}{r}. \qquad (3.3\text{-}6)$$

The $r\theta$- and $r\phi$-components of the stress tensor are equal to zero. There is no shear force in the r-direction, and the flow is symmetrical with respect to ϕ. We can define the total stress components by

$$\pi_{rr} = P + \sigma_{rr}, \quad \pi_{\theta\theta} = P + \sigma_{\theta\theta}, \quad \text{and} \quad \pi_{\phi\phi} = P + \sigma_{\phi\phi}, \qquad (3.3\text{-}7)$$

since the total stress tensor is

$$\underline{\underline{\pi}} = P\underline{\underline{\delta}} + \underline{\underline{\sigma}}. \qquad (3.3\text{-}8)$$

Equation 3.3-6 can then simply be written as

$$0 = -\frac{1}{r^2}\frac{\partial}{\partial r}(r^2 \pi_{rr}) + \frac{\pi_{\theta\theta} + \pi_{\phi\phi}}{r}, \qquad (3.3\text{-}9)$$

or as

$$\frac{\partial \pi_{rr}}{\partial \ln r} = \pi_{\theta\theta} + \pi_{\phi\phi} - 2\pi_{rr}. \qquad (3.3\text{-}10)$$

3.3 Cone-and-Plate Geometry

However,
$$\pi_{rr} - \pi_{\theta\theta} = \sigma_{rr} - \sigma_{\theta\theta}, \tag{3.3-11}$$

which is a unique function of the shear rate. The normal stress difference, which is a unique function of the shear rate, is also a constant. Hence,

$$\frac{\partial \pi_{rr}}{\partial \ln r} = \frac{\partial \pi_{\theta\theta}}{\partial \ln r} = \frac{\partial (P + \sigma_{\theta\theta})}{\partial \ln r}, \tag{3.3-12}$$

and equation 3.3-10 can be written in terms of normal stress differences as

$$\frac{\partial}{\partial \ln r}(P + \sigma_{\theta\theta}) = (\sigma_{\phi\phi} - \sigma_{\theta\theta}) + 2(\sigma_{\theta\theta} - \sigma_{rr}) = \text{constant}. \tag{3.3-13}$$

This result can be readily integrated from $r = r$ to $r = R$ to obtain

$$\pi_{\theta\theta}(r) = (P + \sigma_{\theta\theta}) = \pi_{\theta\theta}(R) + [(\sigma_{\phi\phi} - \sigma_{\theta\theta}) + 2(\sigma_{\theta\theta} - \sigma_{rr})]\ln\frac{r}{R}. \tag{3.3-14}$$

Equation 3.3-14 is the expression of the pressure profile that we can, in principle, measure on the upper cone or lower plate. As illustrated in Fig. 3.3-6, the wall pressure as a function of $-\ln(r/R)$ should be a straight line of slope equal to

$$-[(\sigma_{\phi\phi} - \sigma_{\theta\theta}) + 2(\sigma_{\theta\theta} - \sigma_{rr})] = \psi_1 \dot{\gamma}^2 + 2\psi_2 \dot{\gamma}^2.$$

The slope is a combination of the primary and secondary normal stress differences.

The measurement of the pressure profile on a plate of radius equal to 25 mm or at most 50 mm is a very difficult task. Because of the hole pressure error, as briefly discussed in Chapter 1, tiny and very sensitive flush-mounted pressure transducers must be used. Very few researchers have succeeded in doing so. (See for example the work of Christiansen and Leppard (1974).) However, the primary normal stress difference can be easily determined from the axial force exerted on the plate or cone.

The net force acting on the plate is given by

$$F = 2\pi \int_0^R \pi_{\theta\theta}|_{\theta = \frac{\pi}{2}} r\, dr - \pi R^2 P_a, \tag{3.3-15}$$

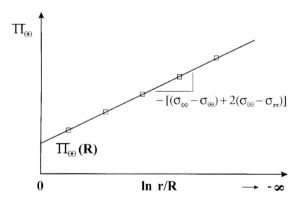

Figure 3.3-6 Radial pressure profile in a cone-and-plate geometry

where P_a is the atmospheric pressure. Combining equations 3.3-14 and 3.3-15, and noting that $\pi_{\theta\theta}$ is constant, we obtain

$$F = \pi R^2 \pi_{\theta\theta}(R) - \tfrac{1}{2}\pi R^2 (\pi_{\phi\phi} + \pi_{\theta\theta} - 2\pi_{rr}) - \pi R^2 P_a. \tag{3.3-16}$$

We now make use of the boundary condition that at the free surface ($r = R$), the sample presents a spherical shape with no effect of surface tension. The internal pressure (total normal stress) is then equal to the atmospheric pressure

$$\pi_{rr}(R) = P_a. \tag{3.3-17}$$

Simplifying equation 3.3-16 we get

$$F = \pi R^2 [\pi_{\theta\theta}(R) - \pi_{rr}(R)] - \tfrac{1}{2}\pi R^2 [(\pi_{\theta\theta} - \pi_{rr}) + (\pi_{\phi\phi} - \pi_{rr})]. \tag{3.3-18}$$

We note that

$$\pi_{\theta\theta}(R) - \pi_{rr}(R) = \pi_{\theta\theta} - \pi_{rr} = \sigma_{\theta\theta} - \sigma_{rr}$$

and

$$\pi_{\phi\phi} - \pi_{rr} = (\sigma_{\phi\phi} - \sigma_{\theta\theta}) + (\sigma_{\theta\theta} - \sigma_{rr}).$$

The normal force expression reduces to

$$F = -\tfrac{1}{2}\pi R^2 (\sigma_{\phi\phi} - \sigma_{\theta\theta}) = -\tfrac{1}{2}\pi R^2 (\sigma_{11} - \sigma_{22}). \tag{3.3-19}$$

The primary normal stress difference is then calculated from

$$N_1 = -(\sigma_{11} - \sigma_{22}) = \frac{2F}{\pi R^2}. \tag{3.3-20}$$

Figure 3.3-7 shows the primary normal stress differences (N_1) data for the same series of 3 mass % polyethylene oxide solutions in a 50 mass % mixture of water and glycerine, for which the viscosity data are reported in Figs. 3.3-4 and 3.3-5. The normal stress data were obtained on two different Weissenberg rheogoniometers (models R-18 and R-20), with a cone-and-plate geometry (the data for both rheometers superimpose). The effect of molecular weight on N_1 is quite important: the magnitude of N_1 increases rapidly with the polymer molecular weight, and for the higher molecular weights, data could be obtained at lower shear rates, but still considerably above the so-called zero shear rate or quadratic region. The normal stress differences in the water–glycerine mixture are much higher than those for the comparable solutions in water. The use of a more viscous solvent (viscosity equal to 5.24 mPa·s) allowed us to obtain reliable normal stress data for relatively dilute solutions or for lower molecular weights.

3.3-3 Inertial Effects

The linear velocity profile expression 3.3-1 obtained for small cone angle will not satisfy the r- and θ-components of the equation of motion, except in the limit of a negligible inertial force. For low viscosity fluids, the centrifugal force becomes important at high rotational speeds, and vortices or secondary flows will develop as illustrated in Fig. 3.3-3.

3.3 Cone-and-Plate Geometry

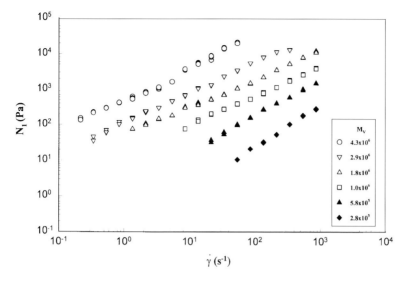

Figure 3.3-7 Primary normal stress difference N_1 for 3 mass % PEO solutions in water and glycerine, as a function of molecular weight M_v at $T = 25$ °C (Data from Ortiz, 1992)

Clearly this is no longer a simple shear situation, and the torque and normal force readings are affected by inertia.

The centrifugal force is responsible for the secondary flows, with circulation of fluid toward the center at the fixed plate and away from the center at the rotating cone. This results in higher torque values and negative normal forces for Newtonian fluids. The importance of the inertial effects can be assessed by using the following methods of correction.

3.3-3.1 Torque Correction

Cheng (1968) has experimentally measured the torque for low viscosity Newtonian fluids at high rotational speeds and compared his data to the theoretical expression developed by Walters and Waters (1968). The comparison is shown in Fig. 3.3-8 as the ratio of the actual torque to the theoretical torque that would be obtained in the absence of inertial effects (equation 3.3-3) as a function of the Reynolds number. Three fluids were used, water, n-hexane, and a 20 mass % aqueous solution of saccharose, using three cone angles, 1°, 2°, and 4°. The experimental data can be expressed by the following correlation (Cheng, 1968):

$$\frac{T}{T_0} = 1 + B(\theta_0)\left(\frac{4\rho R^2 \Omega}{\mu}\right)^2, \qquad (3.3\text{-}21)$$

92   Rheometry

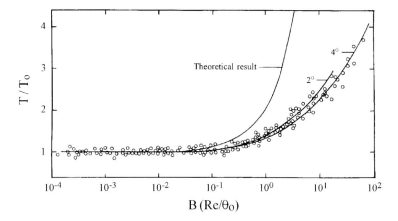

Figure 3.3-8 Effect of inertia on torque measurements for Newtonian fluids (From Cheng, 1968)

where $B(\theta_0)$ is equal to 2.505×10^{-10} for a 2° cone. A similar expression has been proposed by Turian (1972). We note that the inertial effects are strongly dependent on the Reynolds number defined as

$$Re = \frac{4\rho R^2 \Omega \theta_0}{\mu}. \qquad (3.3\text{-}22)$$

The theoretical result of Walters and Waters (1968) predicts correctly the onset of inertial effects, but overestimates the torque. This is because terms of order higher than $(Re/\theta_0)^2$ are neglected in their analysis.

These results can be used with caution to estimate inertial effects for inelastic shear-thinning fluids, replacing the Newtonian viscosity μ by the non-Newtonian viscosity η. For viscoelastic fluids, the effect of secondary flows on the viscosity measurements is still a question that needs to be resolved. We could limit the measurements to low rotational speeds, or still better conduct experiments using different geometries.

3.3-3.2   Normal Force Corrections

The following approximate solution obtained for Newtonian fluids can be used to correct the normal force for inertial effects. For inelastic fluids, the r-component of the equation of motion with a linear velocity profile (equation 3.3-1) becomes

$$\frac{1}{\rho}\frac{\partial P}{\partial r} = \frac{V_\phi^2}{r} = r\Omega^2 \left(\frac{\theta_0 - \theta}{\theta_0}\right)^2. \qquad (3.3\text{-}23)$$

Integrating with respect to r and noting that $P = P_a$ at $r = R$, we obtain

$$P - P_a = \frac{1}{2}\rho\Omega^2 R^2 \left[\left(\frac{r}{R}\right)^2 - 1\right] \left(\frac{\theta_0 - \theta}{\theta_0}\right)^2. \qquad (3.3\text{-}24)$$

This result incorrectly predicts a strong dependence of the pressure on the angle θ. This is because the velocity profile is no longer linear. Nevertheless, as an approximate solution, we take the average pressure obtained by integrating equation 3.3-24 with respect to θ. The result is

$$\langle P \rangle - P_a = \frac{\frac{1}{2}\rho\Omega^2 R^2 \left[\left(\frac{r}{R}\right)^2 - 1\right]}{\theta_0} \int_0^{\theta_0} \left(\frac{\theta_0 - \theta}{\theta_0}\right)^2 d\theta \quad (3.3\text{-}25a)$$

$$= \frac{1}{6}\rho\Omega^2 R^2 \left[\left(\frac{r}{R}\right)^2 - 1\right]. \quad (3.3\text{-}25b)$$

The normal force due to inertial effects is then given by

$$F_{\text{inertia}} = 2\pi \int_0^R (\langle P \rangle - P_a) r \, dr \quad (3.3\text{-}26a)$$

$$= -\frac{1}{12}\pi\rho\Omega^2 R^4. \quad (3.3\text{-}26b)$$

We note that the inertial effects result in a negative force in contrast to the viscoelastic effects. Figure 3.3-9 reports measurements of the normal force for a Newtonian glycerine using two different cone angles on a Weissenberg rheogoniometer. The solid lines are the predictions of equation 3.3-26b. Obviously this approximate solution is quite satisfactory.

It is not at all obvious that equation 3.3-26b can be used as a inertial correction term for viscoelastic fluids, but in the absence of other experimental evidence or viscoelastic calculations, it can be used to estimate corrections to add (opposite signs) to the normal force measurements. Figure 3.3-10 reports on the primary normal stress differences for a 3 mass % solution of polyisobutylene in decalin at 25 °C, obtained on a Weissenberg rheogoniometer using two different cones. The uncorrected normal stress data clearly show

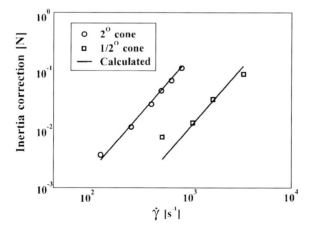

Figure 3.3-9 Normal force in cone-and-plate geometries for a Newtonian fluid, glycerine, at 25 °C The solid lines are the theoretical results of equation 3.3-26b

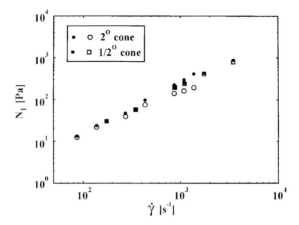

Figure 3.3-10 Primary normal stress difference data for a 3 mass % polyisobutylene in decalin at 25 °C
Open symbols: uncorrected data
Closed symbols: corrected data

differences, mainly at high shear rates (high rotational speeds). Both sets of corrected (closed symbols) and uncorrected (open symbols) data coincide. Note that the corrections for the 1/2° cone are negligible even at high shear rates. However, as mentioned before, the problems of gap setting and alignment are substantial with such a small cone angle. On the other hand, the use of a standard 2° cone can lead to large errors at high shear rates. This is an area where more research is needed.

3.3-4 Criteria for Transient Experiments

Experimental complications associated with the use of a rotational rheometer with a cone-and-plate geometry to determine transient viscometric functions were extensively studied during the last decades by Chang et al. (1975), Nazem and Hansen (1976), Crawley and Graessley (1977), and Attané et al. (1981), following the pioneering work of Meissner (1972). The major problems encountered in measuring transient properties can be summarized as follows:

(i) It is not possible to generate true step changes for the speed of the rotating assembly of the rheometer. The time t_s required to obtain the desired change in velocity depends upon the maximum acceleration rate of the rheometer, which is of the order of 1000 rad/s^2 for the best commercial rheometers.
(ii) The existence of a lag time or retardation time (t_r) between the instant of the command and the effective initiation of the start-up or cessation of flow is a problem easy to account for, but not to be ignored.
(iii) Major problems arise from the lack of torsional and axial rigidities of the rheometer. The measuring device consisting of a torsion or flexion bar and the fluid in the gap of

the cone and plate are in fact a viscoelastic body. There is a coupling between the instrument and the fluid properties. According to the analysis of Nazem and Hansen (1976), the torsional and axial response times for a cone-and-plate rheometer containing a Newtonian fluid of viscosity η_0 are respectively

$$t_t = \frac{20\pi\eta_0 R^3}{3K_t\theta_0} \qquad (3.3-27)$$

and

$$t_a = \frac{6\pi\eta_0 R}{K_a\theta_0^3}, \qquad (3.3-28)$$

where K_t and K_a are the torsional and axial compliances respectively, θ_0 is the cone angle, and R is the platen radius. Thus, if τ_n and τ_t are the characteristic times of the normal thrust response of the fluid and of the torque response respectively during a transient test, we should have

$$\begin{aligned} \tau_n &\gg t_s, \\ \tau_t &\gg t_s, \\ \tau_n &\gg t_a, \\ \text{and} \quad \tau_t &\gg t_t. \end{aligned} \qquad (3.3-29)$$

More useful forms of these requirements have been introduced by Vrentas and Graessley (1981). We can define a mean relaxation time for the material as

$$\lambda_w = J_e^0 \eta_0 = \frac{\psi_{10}}{2\eta_0}, \qquad (3.3-30)$$

where J_e^0 is the steady-state recoverable compliance and ψ_{10} is the zero shear first normal stress coefficient. If τ_n and τ_t are assumed to be approximately equal to λ_w, then equations 3.3-27 to 3.3-29 lead to

$$J_e^0 \eta_0 \gg t_s, \qquad (3.3\text{-}31\text{a})$$

$$J_e^0 \gg \frac{20\pi R^3}{3K_t\theta_0}, \qquad (3.3\text{-}31\text{b})$$

$$\text{and} \quad J_e^0 \gg \frac{6\pi R}{K_a\theta_0^3}. \qquad (3.3\text{-}31\text{c})$$

We should stress that criteria 3.3-31 are not on the conservative side, as τ_n and τ_t are of the order of λ_w only in the linear range (in the zero shear viscosity region), and could be considerably lower at higher shear rates. For example, results on monodisperse polystyrene solutions obtained by Attané et al. (1985a, b) yield the following empirical correlations:

$$\tau_t \approx \frac{\lambda_w \eta(\dot\gamma)}{\eta_0} \qquad (3.3\text{-}32\text{a})$$

$$\text{and} \quad \tau_n = \lambda_w \left[\frac{\eta(\dot\gamma)}{\eta_0}\right]^{0.75}. \qquad (3.3\text{-}32\text{b})$$

However, as t_a and t_t can be taken to be proportional to the steady-state viscosity $\eta(\dot\gamma)$, equations 3.3-31b and c remain valid for the whole range of shear rate. In contrast, equation 3.3-31a should be replaced by

$$J_e^0 \eta(\dot\gamma) \gg t_s. \qquad (3.3\text{-}33)$$

This is the criterion for the upper shear rate limit for transient experiments. (Bear in mind that t_s is a function of the angular velocity $\dot\gamma\theta_0$.) Equations 3.3-31b and c define the experimental window of a given rheometer in terms of linear properties of the fluid to be tested. We note also that the use of the mean relaxation time "in number", λ_n, instead of λ_w, would lead to slightly different relations:

$$\lambda_n = \frac{\eta_0}{G_N^0}, \qquad (3.3\text{-}34)$$

$$\frac{\eta(\dot\gamma)}{G_N^0} \gg t_s, \qquad (3.3\text{-}35a)$$

$$\frac{1}{G_N^0} \gg \frac{20\pi R^3}{3 K_t \theta_0}, \qquad (3.3\text{-}35b)$$

and $\quad \dfrac{1}{G_N^0} \gg \dfrac{6\pi R}{K_a \theta_0^3}, \qquad (3.3\text{-}35c)$

where G_N^0 is the plateau modulus. For solutions of monodisperse polystyrene, $G_N^0 J_e^0$ was found to be roughly equal to 2 (Attané et al., 1985a, b), and thus, equations 3.3-35 are equivalent to equations 3.3-33 and 3.3-31b and c. For polydisperse polymers, however, criteria 3.3-35 can be more restrictive and should be used.

Obviously the design of rheometers for transient experiments has to be a compromise between sensitivity and rigidity. Simple tests with Newtonian fluids of different viscosity should be conducted to evaluate the performance of the equipment. As an example of such tests, Fig. 3.3-11 shows the relaxation of the shear stress reported by Attané et al. (1980) for a Newtonian mineral oil of viscosity equal to 4.2 Pa·s, following a sudden cessation of a steady shear rate. They used a modified Weissenberg rheogoniometer, with an improved braking clutch and a much more rigid torque sensor. The results shown were obtained for an initial shear rate of 560 s^{-1} with a 0.07 rad and 100 mm diameter cone. Since a Newtonian fluid should relax instantaneously to zero, the response shown is that of the instrument. Correcting for the initial delay time, we observe that the response time of the equipment is approximately equal to 16 ms. Thus, even this improved rheometer cannot correctly measure the relaxation of viscoelastic fluids with characteristic times less than 0.1 s.

The rheometer response associated with the shear stress relaxation of a typical viscoelastic fluid (NBS standard) is shown in Fig. 3.3-12. Different cone angles were used (1° and 4°) with an almost constant shear rate for response times less than 20 ms. Two different curves were obtained. For the 1° cone, the response time, t_s, of the instrument was 6 ms, compared with 13 ms for the 4° cone. Hence, curve A is a better representation of the transient behavior of the NBS standard.

3.3 Cone-and-Plate Geometry

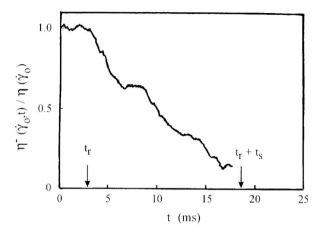

Figure 3.3-11 Shear stress relaxation following cessation of steady shear rate (equal to 560 s^{-1}) of a Newtonian oil for a 4° cone of 100 mm diameter
The y-axis is the reduced shear function: the transient viscosity divided by the steady shear viscosity (Data from Attané, 1984)

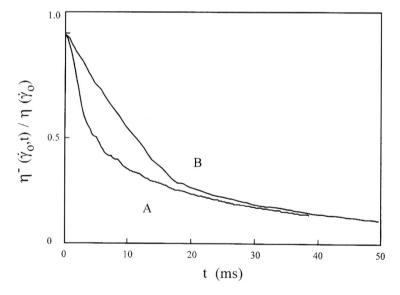

Figure 3.3-12 Reduced shear relaxation function for a NBS standard
Curve A was obtained for a 1° cone after an initial shear rate of 360 s^{-1}; the corresponding viscosity is 14.9 Pa · s. Curve B was obtained for a 4° cone at a shear rate equal to 354 s^{-1} (Data from Attané, 1984)

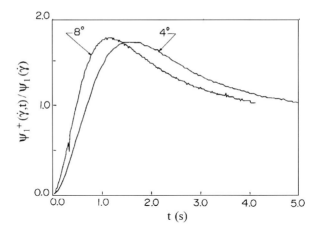

Figure 3.3-13 Reduced primary normal stress growth function of a 25 mass % polystyrene solution for two cone angles at an applied shear rate of 5.5 s^{-1} (Data from Attané, 1984)

For times in excess of $2t_s$, the relaxation properties should be correctly determined, provided that the criteria defined above are respected. That is, the torque and axial force measuring devices should be stiff enough. For other transient experiments, such as stress growth and stress relaxation following a small deformation, the same principles can be applied. More information can be obtained from Attané et al. (1985a and b, 1988), and from Vrentas and Graessley (1981). Because the response time is shorter for smaller angles, there is a temptation to use the smallest cone angle to conduct transient experiments. However, the characteristic time of the normal force measurement device, and hence the coupling effects, increases rapidly with decreasing cone angle (see equation 3.3-28). Figure 3.3-13 reports two transient normal stress growth measurements made by Attané (1984) on a 25 mass % polystyrene ($M_w = 1.3 \times 10^6$ kg/kmol) solution in dibutylphthalate at 20 °C, using two cone angles at the same applied shear rate of 5.5 s^{-1}. Although, as mentioned earlier, a very stiff system was used, the coupling effects with a cone angle of 4° are quite important, and Attané had to use an 8° cone to obtain reliable stress growth data for this very viscous polymer solution. Obviously the overshoot, in particular the time at which the maximum is observed, is considerably affected by coupling effects. We should, however, stress that with the use of a larger cone angle, the shear rate is no longer homogeneous in the gap. This effect is expected to be small on the steady-state viscosity and primary normal stress difference.

3.4 Concentric Disk Geometry

This parallel plate geometry is frequently used for measuring rheological properties of polymer melts and of multiphase polymer systems (composites and blends). The geometry is illustrated in Fig. 3.4-1.

3.4 Concentric Disk Geometry

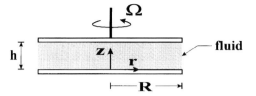

Figure 3.4-1 Concentric disk geometry

This geometry consists of two parallel concentric disks of radius R with a constant separation or gap h. Either the upper or lower plate rotates and the torque and normal thrust can be measured at either one. The edge represents a free boundary with the surrounding atmosphere. The effects of the capillary pressure and of stresses at the free boundary on the measured torque and axial force are usually negligible. This geometry is advantageous for high temperature measurements and in particular when multiphase systems are studied. The gap can be easily varied: a value between 1 mm and 2 mm is frequently used with disks of 25 mm diameter, but if necessary, larger gaps can be used. For high temperature, thermal expansion effects are minimized. Errors in gap setting are not important, and with multiphase systems, the gap can be much larger than the sizes of the domains or particles. The usual criterion is

$$\frac{d_p}{h} \ll 1, \tag{3.4-1}$$

where d_p is the particle or domain diameter. Also, with a large gap, capillary effects at the free boundary are negligible. The major disadvantage of this geometry is that the flow is not homogeneous within the gap. That is, the shear rate varies linearly with the radial position. Also, as in the case of the cone-and-plate geometry, viscoelastic materials will not stay in the gap at high shear rates.

For a small enough gap ($h/R \ll 1$), or for low rotational speed, inertia can be neglected, and the velocity profile for steady-state conditions is given by

$$V_\theta = \Omega r \left(1 - \frac{z}{h}\right), \tag{3.4-2}$$

and the shear rate is expressed by

$$\dot{\gamma} = \dot{\gamma}_{z\theta} = \Omega \frac{r}{h}. \tag{3.4-3}$$

3.4-1 Viscosity Determination

For non-Newtonian fluids, because the shear rate varies with the radial position, the viscosity is no longer proportional to the torque. Hence, a Rabinowitsch-type development has to be performed. The torque is given by

$$T = 2\pi \int_0^R -\sigma_{z\theta}(r) r^2 \, dr \qquad (3.4\text{-}4a)$$

$$= 2\pi \int_0^R \frac{\eta(r) \Omega r^3}{h} \, dr. \qquad (3.4\text{-}4b)$$

Making a change of variables $r \to \dot{\gamma}$ ($=\Omega r/h$) in equation 3.4-4b, we obtain

$$T = 2\pi \left(\frac{h}{\Omega}\right)^3 \int_0^{\dot{\gamma}_R} \eta(\dot{\gamma}) \dot{\gamma}^3 \, d\dot{\gamma}. \qquad (3.4\text{-}5)$$

Combining with equation 3.4-3, this result can be written as

$$T = 2\pi \left(\frac{R}{\dot{\gamma}_R}\right)^3 \int_0^{\dot{\gamma}_R} \eta(\dot{\gamma}) \dot{\gamma}^3 \, d\dot{\gamma}. \qquad (3.4\text{-}6)$$

Taking the derivative with respect to $\dot{\gamma}_R$ and using Leibnitz's rule, we obtain

$$\frac{d\left(\frac{T}{2\pi R^3}\right)}{d\dot{\gamma}_R} = \eta(\dot{\gamma}_R) - 3\dot{\gamma}_R^{-4} \int_0^{\dot{\gamma}_R} \eta(\dot{\gamma}) \dot{\gamma}^3 \, d\dot{\gamma}. \qquad (3.4\text{-}7)$$

Using equation 3.4-6 and rearranging we obtain the final expression for the viscosity:

$$\eta(\dot{\gamma}_R) = \frac{T}{2\pi R^3 \dot{\gamma}_R} \left[3 + \frac{d \ln\left(\frac{T}{2\pi R^3}\right)}{d \ln \dot{\gamma}_R} \right]. \qquad (3.4\text{-}8)$$

To obtain the viscosity of a non-Newtonian fluid we must first plot $\ln T$ vs $\ln \dot{\gamma}_R$. Then the viscosity is calculated using equation 3.4-8 with the local value of the slope. For a *power-law fluid*, the torque is expressed by

$$T = 2\pi m \int_0^R (\dot{\gamma}_{z\theta})^n r^2 \, dr, \qquad (3.4\text{-}9a)$$

and

$$\ln T \sim n \ln \dot{\gamma}_R. \qquad (3.4\text{-}9b)$$

The viscosity is then given by the following simplified expression:

$$\eta(\dot{\gamma}_R) = \frac{T}{2\pi R^3 \dot{\gamma}_R} [3 + n]. \qquad (3.4\text{-}10)$$

In most commercially available software packages, the viscosity is calculated from a single torque value, assuming Newtonian behavior (that is, $n = 1$ in equation 3.4-10). This is obviously incorrect, except when the viscosity is relatively constant, for example in the plateau region at low shear rate. The error depends on the value of the power-law index. For example, in the case of a typical polyethylene melt, n is of the order of 0.3 and the error in using the simplified Newtonian result is $((4 - 3.3)/3.3) \times 100 \approx 22\%$, which is not negligible.

3.4-2 Normal Stress Difference Determination

Under the assumption that the inertial forces are negligible, the shear rate is given by equation 3.4-3. We may further assume that the only non-vanishing shear stress components are $\sigma_{z\theta}(r) = \sigma_{\theta z}(r)$. The r-component of the equation of motion for steady-state conditions reduces to

$$0 = -\frac{\partial P}{\partial r} - \left(\frac{1}{r}\frac{\partial}{\partial r}(r\sigma_{rr}) - \frac{\sigma_{\theta\theta}}{r}\right). \tag{3.4-11}$$

This equation can be written as

$$\frac{\partial}{\partial r}(P + \sigma_{rr}) = -\frac{(\sigma_{rr} - \sigma_{\theta\theta})}{r} = \frac{(\sigma_{\theta\theta} - \sigma_{zz})}{r} + \frac{(\sigma_{zz} - \sigma_{rr})}{r}$$
$$= -\frac{(\psi_1 + \psi_2)}{r}\dot{\gamma}^2, \tag{3.4-12}$$

where ψ_1 and ψ_2 are the primary and secondary normal stress coefficients respectively. Because it is not possible to measure pressures and normal stresses in the radial direction, the rr-component is rewritten from the definition of the secondary normal stress coefficient as

$$\sigma_{rr} = \psi_2 \dot{\gamma}^2 + \sigma_{zz}. \tag{3.4-13}$$

Equation 3.4-12 can be integrated with respect to r to obtain

$$(P + \psi_2 \dot{\gamma}^2 + \sigma_{zz})\Big|_0^r = -\int_0^r \frac{\psi_1 + \psi_2}{r}\dot{\gamma}^2 dr, \tag{3.4-14}$$

or, in terms of the pressure at the disk, this result can be written as

$$\pi_{zz}(r) - \pi_{zz}(0) = -\psi_2 \dot{\gamma}^2 - \int_0^r \frac{\psi_1 + \psi_2}{r}\dot{\gamma}^2 dr. \tag{3.4-15}$$

Obviously this result is quite complex and of little practical use. To obtain a useful expression in terms of the measurable normal force, we proceed with the following development. First, we write from equation 3.4-15,

$$\pi_{zz}(r) - \pi_{zz}(R) = -\psi_2 \dot{\gamma}^2(r) + \psi_2 \dot{\gamma}^2(R) - \int_R^r \frac{\psi_1 + \psi_2}{r}\dot{\gamma}^2 dr. \tag{3.4-16}$$

From the definition of the secondary normal stress coefficient, we can write

$$\psi_2 \dot{\gamma}^2(R) = -[\pi_{zz}(R) - \pi_{rr}(R)], \tag{3.4-17}$$

and at the boundary, the internal pressure is equal to the outside pressure, that is,

$$\pi_{rr}(R) = P_a, \tag{3.4-18}$$

where P_a is the surrounding (atmospheric) pressure. Combining equations 3.4-16, 3.4-17, and 3.4-18, we obtain

$$\pi_{zz}(r) = -\psi_2 \dot{\gamma}^2(r) + P_a - \int_R^r \frac{\psi_1 + \psi_2}{r}\dot{\gamma}^2 dr. \tag{3.4-19}$$

Next we make the change of variables $r \to \dot{\gamma}$ ($=\Omega r/h$), and equation 3.4-19 can be rewritten as

$$\pi_{zz}(r) = -\psi_2 \dot{\gamma}^2(r) + P_a + \int_{\dot{\gamma}}^{\dot{\gamma}_R} (\psi_1 + \psi_2) \dot{\gamma} d\dot{\gamma}. \tag{3.4-20}$$

The net axial force acting on the plate is expressed by

$$\begin{aligned} F &= 2\pi \int_0^R [\pi_{zz}(r) - P_a] r\,dr \\ &= -2\pi \int_0^R \psi_2 \dot{\gamma}^2(r)\,r\,dr + 2\pi \int_0^R \int_{\dot{\gamma}}^{\dot{\gamma}_R} (\psi_1 + \psi_2) \dot{\gamma}\,d\dot{\gamma}\,r\,dr. \end{aligned} \tag{3.4-21}$$

We again make the change of variables $r \to \dot{\gamma}$ ($=\Omega r/h$) to obtain

$$F = \frac{2\pi R^2}{\dot{\gamma}_R^2} \left(-\int_0^{\dot{\gamma}_R} \psi_2 \dot{\gamma}^3 d\dot{\gamma} + \int_0^{\dot{\gamma}_R} \int_{\dot{\gamma}}^{\dot{\gamma}_R} (\psi_1 + \psi_2)(\xi) \xi\,d\xi\,\dot{\gamma}\,d\dot{\gamma} \right), \tag{3.4-22}$$

and changing the order of integration (in the double integral) we get

$$\begin{aligned} F &= \frac{2\pi R^2}{\dot{\gamma}_R^2} \left(-\int_0^{\dot{\gamma}_R} \psi_2 \dot{\gamma}^3 d\dot{\gamma} + \int_0^{\dot{\gamma}_R} \int_0^{\xi} (\psi_1 + \psi_2)(\xi) \dot{\gamma}\,d\dot{\gamma}\,\xi\,d\xi \right) \\ &= \frac{2\pi R^2}{\dot{\gamma}_R^2} \left(-\int_0^{\dot{\gamma}_R} \psi_2 \dot{\gamma}^3 d\dot{\gamma} + \frac{1}{2}\int_0^{\dot{\gamma}_R} (\psi_1 + \psi_2) \dot{\gamma}^3 d\dot{\gamma} \right) \\ &= \frac{\pi R^2}{\dot{\gamma}_R^2} \left(\int_0^{\dot{\gamma}_R} (\psi_1 - \psi_2) \dot{\gamma}^3 d\dot{\gamma} \right). \end{aligned} \tag{3.4-23a}$$
$$\tag{3.4-23b}$$

We then take the derivative with respect to $\dot{\gamma}_R$:

$$\begin{aligned} \frac{d}{d\dot{\gamma}_R}(F\dot{\gamma}_R^2) &= \pi R^2 (\psi_1 - \psi_2) \dot{\gamma}_R^3 \\ &= \dot{\gamma}_R F \frac{d\ln F}{d\ln \dot{\gamma}_R} + 2\dot{\gamma}_R F \\ &= \pi R^2 (N_1 - N_2) \dot{\gamma}_R. \end{aligned} \tag{3.4-24}$$

Finally, the difference between the normal stresses is given by the following expression:

$$N_1(\dot{\gamma}_R) - N_2(\dot{\gamma}_R) = \frac{2F}{\pi R^2} \left(1 + \frac{1}{2} \frac{d\ln F}{d\ln \dot{\gamma}_R} \right). \tag{3.4-25}$$

In general, the secondary normal stress difference, N_2, is negligible compared to the primary normal difference N_1 ($N_2 = 0$ is the *Weissenberg hypothesis*). The primary normal stress difference is calculated using the approximate result

$$N_1(\dot{\gamma}_R) = \psi_1(\dot{\gamma}_R)\dot{\gamma}_R^2 \approx \frac{2F}{\pi R^2} \left(1 + \frac{1}{2} \frac{d\ln F}{d\ln \dot{\gamma}_R} \right), \tag{3.4-26}$$

in which the derivative is determined from the slope of $\ln F$ vs $\ln \dot{\gamma}_R (\sim \Omega)$.

3.5 Yield Stress Measurements

Materials such as suspensions, coatings, foodstuffs ... possess a three-dimensional microstructure, which imparts solid-like properties, but is susceptible to breakdown under applied forces. These are referred to as viscoplastic materials.

Consider the shear stress (σ_{21})-shear rate ($\dot{\gamma}$) relation derived from steady viscometry. For Newtonian fluids, a plot of σ_{21} vs $\dot{\gamma}$ is represented by a straight line passing through the origin. The constant slope of this line is the viscosity μ. The behavior of more complex fluids, possessing microstructure, is usually not represented by a straight line, and often the σ_{21} vs $\dot{\gamma}$ curve does not pass through the origin as $\dot{\gamma}$ is decreased, as shown in Fig. 3.5-1. The intercept on the shear stress axis is a dynamic measure of a yield stress, σ_0. In practice, we must extrapolate the σ_{21} vs $\dot{\gamma}$ data to obtain σ_0. As indicated in Fig. 3.5-1, extrapolation can result in a variety of values for σ_0, depending on the distance from the shear stress axis experimentally accessible. This depends on the ability of the instrument used to generate accurate σ_{21} vs $\dot{\gamma}$ data in the low $\dot{\gamma}$ region. The vast majority of yield stress data reported results from such extrapolations, making most values in the literature instrument-dependent. A yield stress so obtained should be referred to as a dynamic yield stress, since it reflects a limiting microstructure under dynamic condition of shear flow at the lowest shear rate accessible.

On the other hand, we can refer to a yield stress, measured directly without extrapolation and without disturbing the microstructure during the measurement, as a static yield stress. Such measurements can be made by experimentally controlling the applied shear stress rather than the shear rate as in steady viscometry. That is, we can perform shear creep measurements. If we apply a shear stress $\sigma_{21} < \sigma_0$ (the situation represented by A in Fig. 3.5-1), no flow occurs, and the sample deforms like a solid. A fragile three-dimensional microstructure (e.g., interconnected fibers or particles) makes this solid-like behavior possible. The application of a shear stress exceeding the yield stress σ_0 puts us on the flow

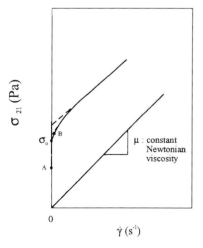

Figure 3.5-1 Yield stress determination from shear stress-shear rate behavior

curve after a steady state is achieved (say at point B in Fig. 3.5-1). Here, the structure breaks down as the material flows, eventually reaching a structure consistent with the flow curve. By carrying out a series of shear creep measurements with increasing applied stress we can discern the minimum value needed to induce flow, and thereby determine a static yield σ_0. Such measurements are extremely tedious and most suffer from the ambiguity of judging when flow is perceptible (Cheng, 1986). In practice, both steady viscometry and creep type measurements give yield stresses that depend on the time scale of observation. From steady viscometry, σ_0 depends on the minimum shear rate used in extrapolation $\dot{\gamma}_{min}$ (i.e., on the time scale $\dot{\gamma}_{min}^{-1}$). From creep type measurements, σ_0 depends on the time scale allowed for the perception of the flow, σ_{obs}. Time scale effects may enter in an even more complex way if the microstructural response to shear flow is time-dependent (e.g., thixotropy).

Despite the pragmatic difficulties, we can operationally define a yield stress that is reproducible, but that does not necessarily represent a true material property. The magnitude of an observed yield stress depends on those factors which affect the microstructure. For example, in suspensions, it depends on the particle concentration, particle shape, and any other factors contributing to the nature of particle interactions. Typical reported yield stress values range from 10^{-3} Pa (for blood) to 40 kPa (for propellant doughs).

If the yield stress is going to be used as an engineering quality control tool, the choice of observation time, or minimum shear rate, should be related to a characteristic time for the process under consideration. For example, in a coating process, the characteristic time could be chosen as the time for which unevenness becomes too difficult to measure. There exists a variety of techniques used to determine the yield stress. There are dynamic methods and static methods, and a number of these have recently been reviewed by De Kee and Durning (1990) and by Nguyen and Boger (1992). Static methods are preferable, as they do not rely on disturbing the three-dimensional structure of the sample prior to the measurement. We close this chapter with a brief discussion of a static technique proposed by De Kee et al. (1986a).

The technique makes use of an automatic surface tension analyzer, where a Wilhelmy plate is kept completely immersed in the sample at all times during an experiment. Figure 3.5-2 depicts the instrument used for the yield stress determination. Transducer signals can be fed to a recorder. Prior to an experiment, the stirrup pan and a clean plate are suspended in air from the balance beam. The recorder is then zeroed at a convenient level, and this point is taken as a reference point. Calibration is achieved by placing analytical masses on the stirrup pan. The transducer produces a signal that is proportional to the weight of the mass placed on the stirrup pan. To measure the yield stress, the plate is first immersed in the sample in the container placed on the platform. When thermal equilibrium is reached, the platform is raised so that the plate can be reattached to the stirrup pan. At this stage the weight of the plate can often be completely supported by the sample, and zero load is registered. (Note that this observation strongly supports the existence of a yield stress of the sample, as buoyancy alone does not allow the plate to be suspended indefinitely.) The experiment is started by lowering the platform at a controlled and constant rate.

As the platform is lowered, the sample in the vessel exerts a force on the plate. Since the plate is attached to the cantilever beam, the plate, in turn, exerts a force on the beam. It is this net force (corrected by buoyancy) that is recorded. Due to the force on the plate, the

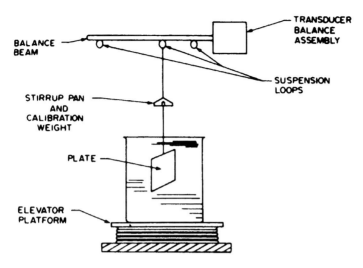

Figure 3.5-2 Plate method used in yield stress measurements

plate moves down with the sample until flow commences. A typical output is shown in Fig. 3.5-3. When the acting force exceeds the sample yield stress, flow begins. That is, a relative motion between two adjacent layers of sample close to the plate is induced, implying that the structure of the sample is beginning to disintegrate. This is characterized by a change in slope in the recorded trace, at $t=t_r$. As the platform continues to descend, the recorded force increases to a steady value at a given rate of descent.

The yield stress σ_0 is given by

$$\sigma_0 = \frac{F_r}{S}, \qquad (3.5\text{-}1)$$

where F_r is the recorded force at $t=t_r$ and S is the area of the plate,

$$S = 2LB. \qquad (3.5\text{-}2)$$

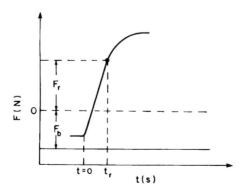

Figure 3.5-3 Response curve obtained during the measurement of the yield stress F_b is the buoyancy force on the plate. F_r marks the sudden transition of the force trace

B and L are the width and the length of the plate respectively. Since the experiment is done by immersing the plate in the sample, a correction factor, the buoyant force, should be taken into account in evaluating the yield stress. The buoyant force F_b is given by

$$F_b = \rho V g, \qquad (3.5\text{-}3)$$

where ρ is the density of the sample and V is the volume of the displaced sample. This technique also offers a simple method for evaluating the elastic modulus of viscoplastic materials (De Kee and Chhabra, 1994).

3.6 Problems*

3.6-1$_a$ Rabinowitsch-Type Analysis

Perform a Rabinowitsch-type analysis for the pressure flow between two plates of width $2W$, separated by a distance $2H$, where $W \gg H$.

(a) Obtain the expression for the shear rate at the wall ($x = \pm H$).
(b) Obtain the expression of the wall shear rate for a power-law fluid, in terms of the overall flow rate.

Table 3.6-1 Capillary Data for a HDPE Melt

	ΔP total MPa	$4Q/\pi R^3$ s^{-1}
L/D = 4	2.556	93.5
	3.172	189.2
	4.066	357.7
	4.683	497.2
L/D = 8	3.656	82.7
	4.335	148.2
	6.221	359.3
	6.771	457.1
L/D = 12	5.155	93
	6.652	181.8
	8.391	361.4
	8.899	423.4
L/D = 16	6.728	84.9
	8.571	182.6
	10.59	356.2

* Problems are labelled by a subscript a or b indicating the level of difficulty.

3.6-2b Rabinowitsch Analysis for a Yield Stress Fluid

A sample obeys the following equations:

$$\eta = \infty \quad \text{for} \quad |\sigma_{21}| < \sigma_0.$$
$$\text{and} \quad \eta = \eta(\sigma_{21}) \quad \text{for} \quad |\sigma_{21}| > \sigma_0. \tag{3.6-1}$$

(a) Show that a Rabinowitsch analysis for capillary flow results in the following expression:

$$\frac{Q}{\pi R^3} = -\frac{1}{\sigma_R^3} \int_{\sigma_0}^{\sigma_R} \sigma_{rz}^2 \left(\frac{dV_z}{dr}\right) d\sigma_{rz}. \tag{3.6-2}$$

(b) Obtain the expression for the shear rate at the wall. Does this result allow for the determination of σ_0?

3.6-3a Viscosity of a High Density Polyethylene

The data of Table 3.6-1 for a high density polyethylene (HDPE) were obtained with a capillary viscometer of diameter 3.175 mm. The polymer density at 180 °C is 745 kg/m^3.

(a) Compute the Bagley correction factor.
(b) Is this factor a function of the shear rate?
(c) Generate the corrected viscosity curve. Show that all the corrected viscosity data approximately reduce to a single curve.

3.6-4b Cone-and-Plate Flow

Bird, Stewart, and Lightfoot (1960) obtained the following velocity profile for the flow of a Newtonian fluid in the gap of a cone-and-plate viscometer (see Fig. 3.3-1):

$$\frac{V_\phi}{r} = \Omega \sin \theta_1 \left[\frac{\cot \theta + \frac{1}{2}\left(\ln \frac{1+\cos\theta}{1-\cos\theta}\right)\sin\theta}{\cot \theta_1 + \frac{1}{2}\left(\ln \frac{1+\cos\theta_1}{1-\cos\theta_1}\right)\sin\theta_1} \right]. \tag{3.6-3}$$

(a) Show that the shear rate is essentially constant across the gap, provided that the cone angle is small.
(b) Discuss the validity of this result for non-Newtonian fluids in general.

3.6-5b Parallel Plate Rheometer

A parallel plate arrangement as shown in Fig. 3.4-1 can be used to determine the normal stresses. The gap h is such that $\dot{\gamma}_{21} = \Omega r/h$. The velocity profile is given by

$$V_\theta = \Omega r \left(1 - \frac{z}{h}\right).$$

(a) Show that the θ-component of the equation of motion is satisfied.
(b) Calculate the relation between the torque and the rotational velocity for a non-Newtonian fluid.
(c) Using a Rabinowitsch analysis, compute the relation between the viscosity, the torque, and the rotational velocity.
(d) Verify equations 3.4-15 to 3.4-25.

3.6-6b Falling Cylinder Viscometer

Consider a cylinder of length L and radius KR falling through a power-law fluid as illustrated in Fig. 3.6-1. The fluid is contained in a beaker of radius R. Assume $K \to 1$. We wish to obtain an equation relating the velocity (V_0) and the mass of the cylinder (M) to the viscosity of the fluid.

(a) Simplify the equations of motion, assuming steady-state conditions.
(b) Identify the boundary conditions.
(c) Obtain the expression for the velocity profile across the gap, in terms of the pressure drop $(P_0 - P_L)$.
(d) Via a force balance, eliminate the pressure drop to generate a relation between the cylinder velocity and the rheological properties of the fluid.

Hint: Note that the velocity gradient changes sign across the gap. Therefore, the power-law expression must be used with care.

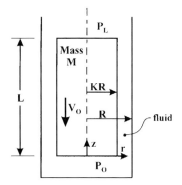

Figure 3.6-1 Falling cylinder viscometer

3.6-7$_a$ Weissenberg Effect

Consider the Couette flow illustrated in Fig. 1.2-1 or Fig. 3.2-5.

(a) Explain in a few words, using the equation of motion, why a viscoelastic fluid climbs the rotating cylinder.
(b) Under which conditions will the free surface not be deformed? Explain in terms of appropriate equations.

3.6-8$_a$ Normal Stress Measurements

Consider the fully developed flow situation between two coaxial cylinders as shown in Fig. 3.6-2.
 Show how pressure measurements at positions A and B can be used to determine the secondary normal stress difference N_2.

3.6-9$_b$ Normal Stress Determination Via Exit Pressure

Consider the exit flow illustrated in Fig. 3.6-3. Han (1976) suggests that the primary normal stress difference can be obtained from the so-called exit pressure by extrapolating the pressure profile along a capillary (see Fig. 3.1-6).
 To obtain the expression relating the exit pressure P_c^0 ($=\Delta P_{ex}$ in Fig. 3.1-6) to the primary normal stress difference, we suggest the procedure followed by Boger and Denn (1980). That is, assume fully-developed flow in section I, plug flow in section II, negligible inertial effects, and no external forces acting on the emerging jet.

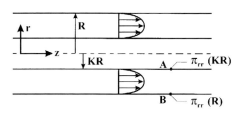

Figure 3.6-2 Flow between coaxial cylinders

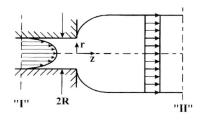

Figure 3.6-3 Sketch of the exit flow from a capillary

(a) Verify that a macroscopic momentum balance over the control volume shown in Fig. 3.6-3 yields

$$\int_0^{\sigma_w} \left[N_1(\sigma_{21}) + \frac{1}{2} N_2(\sigma_{21}) \right] \sigma_{21} d\sigma_{21} = \frac{\sigma_w^2}{2} P_c^0 + \frac{L}{R} \sigma_w^3 \left[1 - \int_0^1 f(\zeta; \sigma_w, Re) d\zeta \right], \quad (3.6\text{-}4)$$

where f describes the effect of the velocity rearrangement near the exit.

(b) Take the derivative of equation 3.6-4 with respect to σ_w. Assume that

$$N_2 = -\epsilon N_1 = -\epsilon m' |\sigma_{21}|^{n'}, \quad (3.6\text{-}5)$$

and show that

$$N_{1w} = \frac{n' + 2}{1 - \frac{\epsilon}{2}} \left\{ \frac{P_c^0}{2} + \frac{L}{R} \sigma_w \left[1 - \int_0^1 f(\zeta; \sigma_w, Re) d\zeta \right] \right\}. \quad (3.6\text{-}6)$$

Note that the factor 2 dividing P_c^0 is missing in equation 25 of Boger and Denn (1980).

(c) Han (1976) implicitly assumed the function f to be equal to one. Discuss the physical implications of this assumption.

Further discussion concerning the validity of normal stress determinations via exit pressure measurements can be found in Boger and Denn (1980) and in Carreau et al. (1985).

3.6-10$_a$ Maxwell Extruder

Professor B. Maxwell proposed to exploit the Weissenberg effect to pump viscoelastic fluids. Figure 3.6-4 shows the suggested arrangement consisting of two parallel disks, with the lower one rotating at an angular velocity Ω.

Neglecting inertial effects and assuming that the flow component in the radial direction is small compared to the rotational component, we can approximate the shear rate by $\dot{\gamma} = \Omega r / h$.

Assuming $N_2 = 0$ and $N_1 = m' |\dot{\gamma}|^{n'}$,

(a) Obtain the expression of the net force acting on the disks.
(b) Discuss the applicability of such a device to pump viscoelastic fluids.

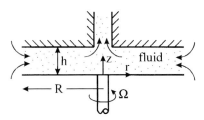

Figure 3.6-4   The Maxwell extruder

3.6-11$_b$ Yield Stress Determination

(a) Determine the yield stress of a bentonite suspension, via the following data which pertain to the plate method described in Section 3.5. The plate area is 0.00237 m² and the net force corresponds to a weight of 50 mg.

  Answer: $\sigma_0 = 0.207$ Pa

(b) Explain how a trace such as the one shown in Fig. 3.5-3 can be used to determine the elastic modulus of a suspension.

4 Transport Phenomena in Simple Flows

4.1 Criteria for Using Purely Viscous Models 113
4.2 Isothermal Flow in Simple Geometries. 114
    4.2-1 Flow of a Shear-Thinning Fluid in a Circular Tube. 114
    4.2-2 Film Thickness for the Flow on an Inclined Plane 116
    4.2-3 Flow in a Thin Slit 118
    4.2-4 Helical Flow in an Annular Section 119
    4.2-5 Flow in a Disk-Shaped Mold 122
        4.2-5.1 Velocity Profile. 124
        4.2-5.2 Pressure Profile. 124
4.3 Heat Transfer to Non-Newtonian Fluids 126
    4.3-1 Convective Heat Transfer in Poiseuille Flow. 126
        4.3-1.1 Lévêque Analysis 127
            4.3-1.1.1 Comments 132
        4.3-1.2 Corrections for Temperature Effects on the Viscosity. 133
    4.3-2 Heat Generation in Poiseuille Flow 134
        4.3-2.1 Equilibrium Regime. 135
        4.3-2.2 Transition Regime (Approximate Solution) 136
4.4 Mass Transfer to Non-Newtonian Fluids 138
    4.4-1 Mass Transfer to a Power-Law Fluid Flowing on an Inclined Plate 138
    4.4-2 Mass Transfer to a Power-Law Fluid in Poiseuille Flow. 141
4.5 Boundary Layer Flows. 144
    4.5-1 Laminar Boundary Layer Flow of Power-Law Fluids over a Plate. 144
    4.5-2 Laminar Thermal Boundary Layer Flow over a Plate 149
4.6 Problems 152
    4.6-1$_a$ Pressure Drop in a Tube 152
    4.6-2$_a$ Generalized Reynolds Number for Poiseuille Flow 152
    4.6-3$_a$ Flow Characteristics of a Suspension. 153
    4.6-4$_b$ Generalized Non-Newtonian Poiseuille Flow 153
    4.6-5$_b$ Tolerance in Machining an Extrusion Die 154
    4.6-6$_b$ Wire Coating 154
    4.6-7$_b$ Axial Flow Between Two Concentric Cylinders 155
    4.6-8$_b$ Generalized Couette Flow 156
    4.6-9$_b$ Velocity Controller. 157
    4.6-10$_b$ Drainage of a Power-Law Fluid 157
    4.6-11$_b$ Heat Transfer by Convection in a Slit 158
    4.6-12$_b$ Heat Transfer to a Falling Film 159
    4.6-13$_b$ Mass Transfer to a Falling Film 160
    4.6-14$_b$ Heat and Mass Transfer in Boundary Layers 160

In this chapter, we present a series of transport phenomena problems of interest for various applications with non-Newtonian fluids. This chapter is restricted to purely viscous flow situations, that is, we will assume that elastic effects play a negligible role. Viscoelastic effects are covered mainly in Chapters 5 and 6. The criteria for using purely viscous rheological models are discussed first, then momentum, heat, and mass transfer problems in simple geometries are presented. We complete the chapter with a few boundary layer flow problems.

4.1 Criteria for Using Purely Viscous Models

As illustrated in Chapter 1, most non-Newtonian fluids exhibit time-dependent behavior and elastic (viscoelastic) effects. For a variety of fluids of industrial interest, for example suspensions of minerals, pulp, foodstuffs, ..., elastic properties play a negligible role. The rheological properties of such materials can be adequately described by a purely viscous model, that is, a generalized non-Newtonian fluid (GNF), as described in Chapter 2. In a more formal way, we will state that viscoelastic fluids can be described by a GNF model under fully-developed, steady-state, simple, one-dimensional flows. For example, the flow rate-pressure drop relation for the steady fully-developed flow of a viscoelastic fluid in a tube is independent of the generated normal stresses. In more complex and unsteady flow situations, the Deborah number can be used as an appropriate criterion to assess the importance of elasticity in a given flow situation. The Deborah number is defined by

$$De = \frac{\lambda}{t_r}, \quad (4.1\text{-}1)$$

where λ is a characteristic time associated with the fluid's elasticity. Such a characteristic time can be obtained, for example, from normal stress or storage modulus measurements. t_r is the residence (characteristic) time of the flow, or the contact time.

Elastic effects are negligible if

$$De \ll 1. \quad (4.1\text{-}2)$$

An alternative to the Deborah number is the Weissenberg number, which can be defined in general as

$$Wi = \frac{\lambda}{t_{\dot\gamma}}, \quad (4.1\text{-}3)$$

where $t_{\dot\gamma}$ is a characteristic time of the fluid deformation rate. It is common to use the inverse of the shear rate as $t_{\dot\gamma}$. For the flow in a tube, the Weissenberg number can be taken as

$$Wi = \lambda \dot\gamma_W, \quad (4.1\text{-}4)$$

where $\dot\gamma_W$ is the shear rate at the tube wall, while the Deborah number for tube flow becomes

$$De = \lambda \frac{\langle V \rangle}{L}, \quad (4.1\text{-}5)$$

where $\langle V \rangle$ is the average velocity and L the tube length. Obviously, for a given flow geometry, both dimensionless numbers are related. For example, for tube flow, the wall shear rate is proportional to $\langle V \rangle/D$ and

$$De \sim Wi\left(\frac{D}{L}\right). \tag{4.1-6}$$

Therefore, for a given flow geometry, the Weissenberg number can be used as an alternative to assess the importance of elastic properties. If $Wi \ll 1$, then the elastic forces are negligible compared to the viscous forces and a GNF model can be used to solve the momentum, heat, and mass transfer equations.

4.2 Isothermal Flow in Simple Geometries

In this section, we examine the solutions of a variety of one-dimensional flow problems. Other situations of interest will be found in the Problems at the end of the chapter.

4.2-1 Flow of a Shear-Thinning Fluid in a Circular Tube

The problem is illustrated in Fig. 4.2-1. We consider the laminar and isothermal flow of an incompressible fluid in a tube, under steady state, with an established velocity profile. We postulate no slip at the wall, and

$$V_z = V_z(r)$$
$$\text{and} \quad V_r = V_\theta = 0. \tag{4.2-1}$$

The z-component of the equation of motion reduces to

$$\sigma_{rz} = \frac{P_0 - P_L}{2L} r = \sigma_R\left(\frac{r}{R}\right), \tag{4.2-2}$$

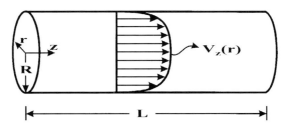

Figure 4.2-1 Flow in a circular tube

4.2 Isothermal Flow in Simple Geometries

where σ_R is the shear stress evaluated at the wall and $P_0 - P_L$ is the pressure drop over the length L, including the possible gravitational term.

For a power-law fluid, the rheological behavior is given by

$$\sigma_{rz} = -m\left|\frac{dV_z}{dr}\right|^{n-1}\left(\frac{dV_z}{dr}\right) = +m\left(-\frac{dV_z}{dr}\right)^n, \tag{4.2-3}$$

since the velocity gradient dV_z/dr is negative. Combining equations 4.2-2 and 4.2-3, we obtain

$$m\left(-\frac{dV_z}{dr}\right)^n = \sigma_R \frac{r}{R}. \tag{4.2-4}$$

Taking the nth root and integrating, we obtain

$$V_z = -\left(\frac{\sigma_r}{mR}\right)^{\frac{1}{n}} \frac{r^{(\frac{1}{n}+1)}}{\frac{1}{n}+1} + C_1. \tag{4.2-5}$$

The constant C_1 is evaluated with the help of the no-slip boundary condition: at $r = R$, $V_z = 0$:

$$C_1 = \left(\frac{\sigma_R}{mR}\right)^{\frac{1}{n}} \frac{R^{(\frac{1}{n}+1)}}{\frac{1}{n}+1}. \tag{4.2-6}$$

The velocity profile is then given by

$$V_z = \left(\frac{\sigma_R}{m}\right)^{\frac{1}{n}} \frac{R}{\frac{1}{n}+1}\left[1 - \left(\frac{r}{R}\right)^{\frac{1}{n}+1}\right]. \tag{4.2-7}$$

This velocity profile is illustrated in Fig. 4.2-2 for various values of n.

We notice that, for $n = 1$, we recover the Poiseuille parabolic velocity profile. For shear-thinning fluids, the velocity profiles become more blunt as n decreases. The flow rate Q is obtained by integrating the velocity profile over the tube cross section, that is,

$$Q = \pi R^2 \langle V \rangle = \int_0^{2\pi}\int_0^R V_z r\, dr\, d\theta$$

$$= \frac{\pi R^3}{\frac{1}{n}+3}\left(\frac{\sigma_R}{m}\right)^{\frac{1}{n}}, \tag{4.2-8}$$

where $\langle V \rangle$ is the average fluid velocity. The pressure drop for the flow of a power-law fluid in a tube may be calculated from equations 4.2-2 and 4.2-8 to be

$$P_0 - P_L = \frac{2mL}{R}\left[\frac{Q(1/n+3)}{\pi R^3}\right]^n. \tag{4.2-9}$$

This equation overestimates the pressure drop, since the viscosity is overestimated by the power-law model in the low shear rate region (i.e., center of pipe). For design purposes, the error introduced by using the power-law model is generally not very important. If a

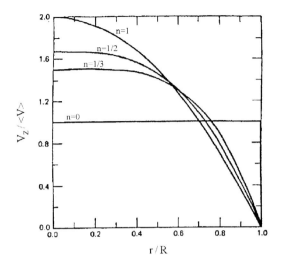

Figure 4.2-2   Tube flow velocity for a power-law fluid

better prediction is required, we may use the more adequate three-parameter *Ellis model* and rework this problem to obtain the corresponding expression for flowrate as

$$Q = \frac{\pi R^3 \sigma_R}{4\eta_0}\left[1 + \frac{4}{\alpha+3}\left(\frac{\sigma_R}{\sigma_{\frac{1}{2}}}\right)^{\alpha-1}\right], \quad (4.2\text{-}10)$$

where η_0, α, and $\sigma_{1/2}$ are the three parameters of the model defined by equation 2.2-3.

For any GNF, the flow rate can be expressed as

$$Q = \frac{\pi R^3 \dot{\gamma}_R}{3} - \frac{\pi}{3}\left(\frac{R}{\sigma_R}\right)^3 \int_0^{\dot{\gamma}_R} \eta^3 \dot{\gamma}^3 d\dot{\gamma}, \quad (4.2\text{-}11)$$

where $\dot{\gamma}_R$ is the wall shear rate and η is expressed as a function of $\dot{\gamma}$ (see Chapter 2 for a variety of models). In general, a numerical integration of the second term of equation 4.2-11 has to be performed to compute Q.

4.2-2   Film Thickness for the Flow on an Inclined Plane

Consider the isothermal flow of an incompressible fluid on an inclined plane as illustrated in Fig. 4.2-3.

For steady fully-developed flow, we make the following hypotheses:

$$V_z = V_z(x)$$
$$\text{and} \quad V_x = V_y = 0. \quad (4.2\text{-}12)$$

4.2 Isothermal Flow in Simple Geometries

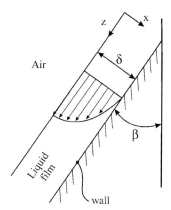

Figure 4.2-3 Flow on an inclined plane

The fluid obeys the Ellis model of equation 2.2-3, which we write in the following form:

$$\sigma_{xz} = \frac{-\eta_0}{1 + \left|\frac{\sigma_{xz}}{\sigma_{\frac{1}{2}}}\right|^{\alpha-1}} \left(\frac{dV_z}{dx}\right). \qquad (4.2\text{-}13)$$

From the z-component of the equation of motion we get

$$0 = -\frac{d\sigma_{xz}}{dx} + \rho g \cos \beta. \qquad (4.2\text{-}14)$$

Integrating and using the condition of no air resistance on the liquid film, that is, at $x=0$, $\sigma_{xz}=0$, we obtain

$$\sigma_{xz} = \rho g x \cos \beta$$
$$= \sigma_\delta \left(\frac{x}{\delta}\right), \qquad (4.2\text{-}15)$$

where $\sigma_\delta = \rho g \delta \cos \beta$ is the shear stress at the wall. Substituting the Ellis model for σ_{xz} in equation 4.2-15 yields

$$V_z = -\frac{\sigma_\delta \delta}{2\eta_0}\left[\left(\frac{x}{\delta}\right)^2 + \frac{2}{\alpha+1}\left|\frac{\sigma_\delta}{\sigma_{\frac{1}{2}}}\right|^{\alpha-1}\left(\frac{x}{\delta}\right)^{\alpha+1}\right] + C_1, \qquad (4.2\text{-}16)$$

where the constant C_1 is evaluated using the no-slip boundary condition $V_z=0$ at $x=\delta$. Hence, the velocity profile is given by

$$V_z = \frac{\sigma_\delta \delta}{2\eta_0}\left[1 - \left(\frac{x}{\delta}\right)^2 + \frac{2}{\alpha+1}\left|\frac{\sigma_\delta}{\sigma_{\frac{1}{2}}}\right|^{\alpha-1}\left(1 - \left(\frac{x}{\delta}\right)^{\alpha+1}\right)\right]. \qquad (4.2\text{-}17)$$

As before, the flow rate is obtained by integrating this expression over the flow cross section, that is,

$$Q = \int_0^W \int_0^\delta V_z dx dy = W \int_0^\delta V_z dx, \qquad (4.2\text{-}18)$$

$$= \frac{W \sigma_\delta \delta^2}{\eta_0} \left[\frac{1}{3} + \frac{1}{\alpha+2} \left(\frac{\sigma_\delta}{\sigma_{1/2}} \right)^{\alpha-1} \right], \qquad (4.2\text{-}19)$$

where W is the width of the inclined plane.

The equivalent expression for a power-law fluid is

$$Q = \frac{W \delta^2}{\frac{1}{n}+2} \left(\frac{\sigma_\delta}{m} \right)^{\frac{1}{n}}. \qquad (4.2\text{-}20)$$

Example 4.2-1   Paint Film Thickness

Estimate the maximum film thickness of a latex paint that can be applied on a vertical wall without dripping. The paint rheology is described by a Bingham model with a yield stress σ_0 of 10 Pa.

Solution

For non-dripping (no-flow) conditions,

$$\frac{dV_z}{dx} = 0, \qquad (4.2\text{-}21)$$

and $\sigma_{xz} \leq \sigma_0$.

The maximum film thickness is obtained from equation 4.2-15 with $\beta = 0$:

$$\sigma_0 = \sigma_\delta = \rho g \delta_{\max}. \qquad (4.2\text{-}22)$$

Taking $\rho = 1000$ kg/m³, $\delta_{\max} = 10/(1000)(9.8) = 1.02 \times 10^{-3}$ m. Hence, a 1 mm paint coat can be applied. Note that this result is independent of the fluid viscosity.

4.2-3   Flow in a Thin Slit

Let us consider the flow in a thin rectangular slit as illustrated in Fig. 4.2-4. The origin of the coordinates is at the center of the flow cross section. This flow geometry is of considerable importance in polymer processing, because various extrusion dies and injection molds can be approximated by thin slits.

For $H \ll W \ll L$, we neglect side wall effects and assume a fully-developed flow situation. That is,

$$V_z = V_z(y)$$

and $V_x = V_y = 0$. $\qquad (4.2\text{-}23)$

4.2 Isothermal Flow in Simple Geometries

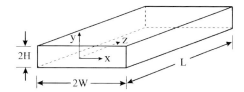

Figure 4.2-4   Thin slit geometry

Then the z-component of the equation of motion for steady laminar flow of an incompressible fluid yields

$$\sigma_{yz} = \frac{P_0 - P_L}{L} y = \sigma_H \left(\frac{y}{H}\right), \tag{4.2-24}$$

where σ_H is the shear stress at the wall at $y = \pm H$. Substituting the power-law expression for σ_{yx} and proceeding as in the case of a tube, we get the following equation for the flow rate:

$$Q = \frac{4WH^2}{\frac{1}{n}+2} \left(\frac{\sigma_H}{m}\right)^{\frac{1}{n}}. \tag{4.2-25}$$

The corresponding expression for the *Ellis model* is

$$Q = \frac{4WH^2}{3\eta_0} \sigma_H \left[1 + \frac{3}{\alpha+2}\left(\frac{\sigma_H}{\sigma_{\frac{1}{2}}}\right)^{\alpha-1}\right]. \tag{4.2-26}$$

4.2-4   Helical Flow in an Annular Section

This flow geometry is illustrated in Fig. 4.2-5. It consists of two coaxial cylinders of length L and of radii KR and R. The inner cylinder is rotating at a constant angular speed Ω, and a pressure difference, $P_L - P_0$, is applied to generate axial flow. This flow situation is encountered in oil fields where drilling mud is used to remove rock fragments from the well. It is also encountered in polymer extrusion. This problem was first analyzed by Dierckes and Schowalter (1966).

We will assume that the fluid is inelastic and obeys a power-law expression, which for a two-directional flow is given by

$$\eta = m\left[\left(\frac{1}{2}\Pi_{\dot\gamma}\right)^{\frac{1}{2}}\right]^{n-1}, \tag{4.2-27}$$

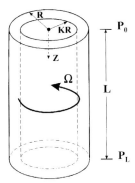

Figure 4.2-5 Helical flow in an annular flow section

where $\Pi_{\dot{\gamma}}$ is the second invariant of the rate-of-deformation tensor. We will further assume that inertial effects are negligible and that the fluid is incompressible. Hence, for steady-state conditions, we have

$$\frac{\partial(\ )}{\partial t} = 0,$$
$$V_r = 0,$$
$$V_\theta = V_\theta(r), \quad (4.2\text{-}28)$$
$$V_z = V_z(r),$$
$$\text{and} \quad P = P(z).$$

The rate-of-deformation tensor is given by

$$\underline{\underline{\dot{\gamma}}} = \nabla \underline{V} + \nabla \underline{V}^+ = \begin{pmatrix} 0 & \dot{\gamma}_{r\theta} & \dot{\gamma}_{rz} \\ \dot{\gamma}_{\theta r} & 0 & 0 \\ \dot{\gamma}_{zr} & 0 & 0 \end{pmatrix}, \quad (4.2\text{-}29)$$

with

$$\dot{\gamma}_{zr} = \dot{\gamma}_{rz} = \frac{\partial V_z}{\partial r} + \frac{\partial V_r}{\partial z} = V_z', \quad (4.2\text{-}30)$$

since

$$V_r = 0$$

and

$$\dot{\gamma}_{\theta r} = \dot{\gamma}_{r\theta} = r\frac{\partial}{\partial r}\left(\frac{V_\theta}{r}\right) + \frac{1}{r}\frac{\partial V_r}{\partial \theta} = r\omega', \quad (4.2\text{-}31)$$

where

$$\omega' = \frac{\partial}{\partial r}\left(\frac{V_\theta}{r}\right).$$

4.2 Isothermal Flow in Simple Geometries

Hence,

$$\Pi_{\dot\gamma} = \dot{\underline{\underline{\gamma}}} : \dot{\underline{\underline{\gamma}}} = tr\begin{pmatrix} 0 & \dot\gamma_{r\theta} & \dot\gamma_{rz} \\ \dot\gamma_{r\theta} & 0 & 0 \\ \dot\gamma_{rz} & 0 & 0 \end{pmatrix}\begin{pmatrix} 0 & \dot\gamma_{r\theta} & \dot\gamma_{rz} \\ \dot\gamma_{r\theta} & 0 & 0 \\ \dot\gamma_{rz} & 0 & 0 \end{pmatrix} = 2V_z'^2 + 2r^2\omega'^2, \quad (4.2\text{-}32)$$

and

$$\eta = m(V_z'^2 + r^2\omega'^2)^{\frac{n-1}{2}}. \quad (4.2\text{-}33)$$

Because the fluid is taken to be inelastic, we will assume that the only nonzero stress components are

$$\sigma_{rz}(r) = \sigma_{zr}(r)$$
$$\text{and} \quad \sigma_{r\theta}(r) = \sigma_{\theta r}(r). \quad (4.2\text{-}34)$$

The θ- and z-components of the equation of motion respectively reduce to

$$0 = -\frac{1}{r^2}\frac{d}{dr}(r^2\sigma_{r\theta}) \quad (4.2\text{-}35)$$

$$\text{and} \quad 0 = -\frac{\partial P}{\partial z} - \frac{1}{r}\frac{d}{dr}(r\sigma_{rz}). \quad (4.2\text{-}36)$$

Since $P(z)$ is a unique function of z, these two equations yield, after integration,

$$\sigma_{r\theta} = \frac{C_1}{r^2} \quad (4.2\text{-}37)$$

$$\text{and} \quad \sigma_{rz} = -\frac{(\Delta P)r}{2L} + \frac{C_2}{r}, \quad (4.2\text{-}38)$$

where $\Delta P = P_L - P_0$. The stress components can be replaced, using the viscosity expression, by

$$\sigma_{rz} = -\eta\dot\gamma_{rz} \quad (4.2\text{-}39)$$
$$\text{and} \quad \sigma_{r\theta} = -\eta\dot\gamma_{r\theta}, \quad (4.2\text{-}40)$$

where η is given by equation 4.2-33, and the following two equations are obtained:

$$-m(V_z'^2 + r^2\omega'^2)^{\frac{n-1}{2}}r\omega' = \frac{C_1}{r^2} \quad (4.2\text{-}41)$$

and

$$-m(V_z'^2 + r^2\omega'^2)^{\frac{n-1}{2}}V_z' = -\frac{\Delta P}{2L}r + \frac{C_2}{r}. \quad (4.2\text{-}42)$$

These equations are solved with the following four boundary conditions:

B.C. 1: at $r = KR$, $V_\theta = \Omega KR$,
B.C. 2: at $r = KR$, $\omega = \Omega$,
B.C. 3: at $r = R$, $V_z = 0$,
and B.C. 4: at $r = R$, $V_\theta = 0$.

Equations 4.2-41 and 4.2-42 can be uncoupled and integrated to obtain, via boundary conditions 3 and 4,

$$\omega = \frac{V_\theta}{r} = \int_R^r \left(-\frac{C_1}{r^3 m}\right)^{\frac{1}{n}} \left[\frac{1}{C_1^2}\left(-\frac{\Delta P}{2L}r^4 + C_2 r^2\right)^2 + r^2\right]^{\frac{1-n}{2n}} dr, \quad (4.2\text{-}43)$$

and

$$V_z = \int_R^r \left[-\frac{1}{m}\left(-\frac{\Delta P r}{2L} + \frac{C_2}{r}\right)\right]^{\frac{1}{n}} \left[1 + \frac{C_1^2}{\left(-\frac{\Delta P r^3}{2L} + C_2 r\right)^2}\right]^{\frac{1-n}{2n}} dr. \quad (4.2\text{-}44)$$

For a given ΔP, the constants C_1 and C_2 have to be determined numerically using boundary conditions (1) and (2). The constants can also be determined from the torque and the axial force:

$$T = \sigma_{r\theta}|_{KR} 2\pi (KR)^2 L \quad (4.2\text{-}45)$$

$$\text{and} \quad F_z = -2\pi KRL \sigma_{rz}|_{KR}. \quad (4.2\text{-}46)$$

Then, from equations 4.2-37 and 4.2-38, we get

$$C_1 = \frac{T}{2\pi L} \quad (4.2\text{-}47)$$

$$\text{and} \quad C_2 = \frac{-F_z - \pi K^2 R^2 \Delta P}{2\pi L}. \quad (4.2\text{-}48)$$

4.2-5 Flow in a Disk-Shaped Mold

Injection molding is an important operation in the plastics industry. It involves, in many practical situations, radial flow between parallel disks as illustrated in Fig. 4.2-6.

We are interested in obtaining the pressure profile as a function of time, and the time required for filling the disk-shaped cavity. This flow geometry is also encountered in other applications, such as lubrication.

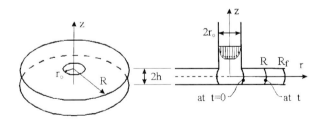

Figure 4.2-6 Radial flow between two parallel disks

Let us assume that at time $t=0$ the fluid starts to flow in the mold, as illustrated in Fig. 4.2-6 with a constant flow rate, Q. We will examine how the injection pressure varies with time for an incompressible inelastic shear-thinning fluid under isothermal conditions. At some reasonable distance from the injection point, we will assume that the flow is axisymmetrical and purely radial. The velocity components in cylindrical coordinates are

$$V_r = V_r(r,z), \quad \text{and} \quad V_\theta = V_z = 0, \tag{4.2-49}$$

and the rate-of-deformation tensor is given by

$$\underline{\dot{\gamma}} = \begin{pmatrix} 2\dfrac{\partial V_r}{\partial r} & 0 & \dfrac{\partial V_r}{\partial z} \\ 0 & 2\dfrac{V_r}{r} & 0 \\ \dfrac{\partial V_r}{\partial z} & 0 & 0 \end{pmatrix}. \tag{4.2-50}$$

The continuity equation for incompressible fluids is given by $1/r\, \partial(rV_r)/\partial r\ (rV_r)=0$, hence, $V_r = A(z)/r$, and the rate-of-strain tensor can be written as

$$\underline{\dot{\gamma}} = \begin{pmatrix} \dfrac{-2}{r^2}A(z) & 0 & \dfrac{1}{r}\dfrac{dA}{dz} \\ 0 & \dfrac{2}{r^2}A(z) & 0 \\ \dfrac{1}{r}\dfrac{dA}{dz} & 0 & 0 \end{pmatrix}. \tag{4.2-51}$$

The second invariant of the rate-of-strain tensor is

$$\Pi_{\dot{\gamma}} = \frac{8}{r^4}A^2(z) + \frac{2}{r^2}\left(\frac{dA}{dz}\right)^2. \tag{4.2-52}$$

Hence, for a power-law fluid, the stress tensor can be expressed as

$$\underline{\pi} = P\underline{\delta} + \underline{\sigma} = \begin{pmatrix} P + \dfrac{2m}{r^2}A(z)\bar{\dot{\gamma}}^{n-1} & 0 & -\dfrac{m}{r}\dfrac{dA}{dz}\bar{\dot{\gamma}}^{n-1} \\ 0 & P - \dfrac{2m}{r^2}A(z)\bar{\dot{\gamma}}^{n-1} & 0 \\ -\dfrac{m}{r}\dfrac{dA}{dz}\bar{\dot{\gamma}}^{n-1} & 0 & P \end{pmatrix}, \tag{4.2-53}$$

where the rate of deformation is defined by

$$\bar{\dot{\gamma}} = \sqrt{\frac{1}{2}\Pi_{\dot{\gamma}}} = \frac{1}{r}\sqrt{4\frac{A^2}{r^2} + \left(\frac{dA}{dz}\right)^2}. \tag{4.2-54}$$

For viscous (creeping) flow of inelastic fluids, the term A/r is negligible with respect to the term dA/dz, and (A/r^2) can be neglected compared with $(dA/dz)^2$. (This is equivalent to

124   Transport Phenomena in Simple Flows

assuming that the elongational contribution to the flow is weak with respect to the shear contribution.) The *r*-component of the equation of motion reduces to

$$\frac{r^n}{m}\frac{dP}{dr} = -\frac{d}{dz}\left(-\frac{dA}{dz}\right)^n = C. \tag{4.2-55}$$

Both members of the equation can be integrated separately, and using no-slip conditions at $z = \pm h$, we obtain the expressions for the velocity and pressure profiles.

4.2-5.1 Velocity Profile

$$V_r(r,z) = \frac{A(z)}{r} = -\frac{n}{n+1}\frac{(-C)^{\frac{1}{n}}}{r}(z^{1+\frac{1}{n}} - h^{1+\frac{1}{n}}), \ 0 \le z \le h. \tag{4.2-56}$$

The flow rate is given by

$$Q = 4\pi r \int_0^h V_r(r,z)dz$$
$$= \frac{4\pi n}{2n+1}(-C)^{\frac{1}{n}}h^{2+\frac{1}{n}}. \tag{4.2-57}$$

Eliminating C allows us to write the velocity profile as

$$V_r(r,z) = -\frac{2n+1}{n+1}\frac{Q}{4\pi r h^{2+\frac{1}{n}}}(z^{1+\frac{1}{n}} - h^{1+\frac{1}{n}}), \ 0 \le z \le h. \tag{4.2-58}$$

4.2-5.2 Pressure Profile

The integration of the left side of equation 4.2-55 leads to

$$P(r) = -\frac{Cm}{1-n}(r^{1-n} - R^{1-n}). \tag{4.2-59}$$

Eliminating C, with the help of result 4.2-57, we get

$$P(r) = \frac{m}{h(1-n)}\left[\frac{(2n+1)Q}{4\pi nh^2}\right]^n (R^{1-n} - r^{1-n}), \tag{4.2-60}$$

and, as a particular case, the injection pressure P_I is (taking $P_R = 0$)

$$P_I = \frac{m}{h(1-n)}\left[\frac{(2n+1)Q}{4\pi nh^2}\right]^n (R^{1-n} - r_0^{1-n}). \tag{4.2-61}$$

The radial position of the melt front is obtained, for a constant injection rate, from an overall mass balance. That is

$$R^2 - r_0^2 = \frac{Qt}{2\pi h} \tag{4.2-62}$$

Example 4.2-2 Pressure Profile in a Disk-Shaped Mold

Determine the injection pressure and the pressure profile for the injection in a disk-shaped mold for the following conditions:

$$R_f = 0.15 \text{ m}$$
$$r_0 = 5 \text{ mm}$$
$$h = 1 \text{ mm}$$
$$Q = 50 \text{ mL/s}$$
$$\text{power-law parameters}: \quad m = 6.4 \text{ kPa} \cdot \text{s}^n$$
$$n = 0.39.$$

Solution

Results for the injection pressure as a function of the melt front position are compared with those for the Newtonian case in Fig. 4.2-7. As expected, the required pressure is considerably lower for the shear-thinning polymer. This is a well-known result, but the comparison is misleading, since the Newtonian fluid ($n = 1$ and $\mu = 6.4$ kPa·s) is much more viscous than the shear-thinning fluid in the shear rate range occurring in the mold. Note that the inclusion of temperature effects results in a slightly greater injection pressure.

Pressure profiles for three different times are shown in Fig. 4.2-8. For highly shear-thinning fluids, the profiles are considerably more linear than we would observe with Newtonian fluids.

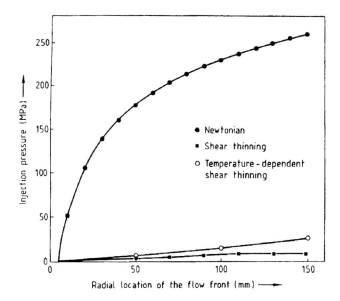

Figure 4.2-7 Injection pressure as a function of melt front position in a disk-shaped mold (From Agassant et al., 1991)

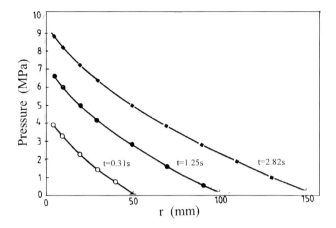

Figure 4.2-8 Radial pressure profile in a disk-shaped mold—Power-law fluid ($n = 0.39$) (From Agassant et al., 1991)

4.3 Heat Transfer to Non-Newtonian Fluids

Heat transfer to non-Newtonian fluids is a complex subject. Some analytical solutions obtained for Newtonian fluids can be generalized to shear-thinning inelastic fluids. However, many difficulties are encountered, which limit the use of these results. Some problems encountered are due to viscous heating effects, temperature dependence of physical properties, combined effects of convection (free and forced), and radiative heat transfer. It is beyond the scope of this chapter to cover all these topics. We will restrict ourselves to a few examples which will illustrate the importance of shear-thinning effects in non-isothermal flow problems.

In the following examples, we will assume that the fluid is incompressible, inelastic, and obeys a power-law expression, and that the physical properties are independent of the temperature. The following simplified energy equation will be retained:

$$\rho \hat{C}_p \frac{DT}{Dt} = k\nabla^2 T + \frac{1}{2}\underline{\underline{\eta\dot{\gamma}}} : \underline{\underline{\dot{\gamma}}}, \tag{4.3-1}$$

where the left side is the rate of accumulation of internal energy, the first term on the right side is the rate of heat addition by conduction, and the second term is the rate of energy generated by viscous forces. All rates are per unit volume.

4.3-1 Convective Heat Transfer in Poiseuille Flow

In the first example, we analyze the heat transfer for the laminar flow of a shear-thinning fluid in a circular tube. We will assume that the heat generated by viscous dissipation can be neglected, and that the heat conduction in the axial (flow) direction is negligible with respect to the convective term. This problem was initially analyzed by Graetz (1885), who obtained an analytical solution for a plug flow situation (flat velocity profile) and isothermal wall conditions. Lévêque (1928) proposed an analytical solution for short contact time, that is, for the case where the heat transfer is confined to the region near the tube wall and for which the velocity profile of the fluid is assumed to be linear. Here, we will examine extensions of the Lévêque solution to power-law fluids.

The flow and heat transfer situation is illustrated in Fig. 4.3-1. The general boundary conditions are

B.C. 1 : at $z = 0$, $T = T_{b_1}$,

B.C. 2 : at $r = 0$, T is finite $\left(\text{or } \dfrac{\partial T}{\partial r} = 0\right)$,

and B.C. 3a : at $r = R$, $-k\dfrac{\partial T}{\partial r} = q_1$

or B.C. 3b : at $r = R$, $T = T_0$,

where T_{b_1} and T_{b_2} are the bulk temperatures of the fluid at the inlet and outlet of the tube respectively. For the constant heat flux case, boundary condition (3a) will be used; for the isothermal wall case, boundary condition (3b) will be taken. We consider steady state and we neglect viscous dissipation as well as axial heat conduction. Equation 4.3-1 reduces to

$$\rho \hat{C}_p V_z \dfrac{\partial T}{\partial z} = k \left[\dfrac{1}{r}\dfrac{\partial}{\partial r}\left(r \dfrac{\partial T}{\partial r}\right)\right]. \qquad (4.3\text{-}2)$$

4.3-1.1 Lévêque Analysis

The following analysis is valid for a very short contact time or equivalently for a large Graetz number, defined by

$$Gz = \dfrac{\pi R^2 \langle V \rangle \rho \hat{C}_p}{kL}, \qquad (4.3\text{-}3)$$

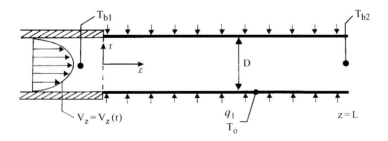

Figure 4.3-1 Heat transfer by convection in Poiseuille flow

where $\langle V \rangle$ is the average velocity of the fluid in the tube. The heat transfer is confined to the tube wall region and we assume that the velocity profile is linear near the wall, that is

$$V_z \approx -\left(\frac{dV_z}{dr}\right)_{r=R}(R-r) = -\dot{\gamma}_R y, \qquad (4.3\text{-}4)$$

where y is the distance from the wall and $\dot{\gamma}_R$ is the shear rate evaluated at the wall. For a power-law fluid,

$$\dot{\gamma}_R = -\left|\frac{\sigma_R}{m}\right|^{\frac{1}{n}}, \qquad (4.3\text{-}5)$$

and the velocity profile becomes

$$V_z = \left|\frac{\sigma_R}{m}\right|^{\frac{1}{n}} y = V_0\left(\frac{y}{R}\right), \qquad (4.3\text{-}6)$$

where

$$V_0 = \left|\frac{\sigma_R}{m}\right|^{\frac{1}{n}} R = -\dot{\gamma}_R R. \qquad (4.3\text{-}7)$$

Since the heat transfer is occurring within a short distance of the wall, curvature effects can also be neglected, and we write

$$\frac{1}{r}\frac{\partial}{\partial r}\left(r\frac{\partial T}{\partial r}\right) = \frac{\partial^2 T}{\partial r^2} + \frac{1}{r}\frac{\partial T}{\partial r} \approx \frac{\partial^2 T}{\partial r^2} = \frac{\partial^2 T}{\partial y^2}, \qquad (4.3\text{-}8)$$

and equation 4.3-2 reduces to

$$V_0 \frac{y}{R}\frac{\partial T}{\partial z} = \alpha \frac{\partial^2 T}{\partial y^2}, \qquad (4.3\text{-}9)$$

where $\alpha \; (= k/\rho \hat{C}_p)$ is the thermal diffusivity.

Example 4.3-1 Constant Heat Flux

In the case of constant heat flux q_1 at the tube wall, we define the following dimensionless variables:

$$\psi = \frac{q_y}{q_1},$$

$$\xi = \frac{y}{R}, \qquad (4.3\text{-}10)$$

$$\text{and} \quad \zeta = \frac{\alpha z}{V_0 R^2},$$

and equation 4.3-9 is rewritten in terms of $q_y = -k\partial T/\partial y$ by taking the derivative of each term with respect to y:

$$\frac{V_0}{R}\frac{\partial q_y}{\partial z} = \alpha \frac{\partial}{\partial y}\left(\frac{1}{y}\frac{\partial q_y}{\partial y}\right), \qquad (4.3\text{-}11)$$

4.3 Heat Transfer to Non-Newtonian Fluids

which becomes in dimensionless form

$$\frac{\partial \psi}{\partial \zeta} = \frac{\partial}{\partial \xi}\left(\frac{1}{\xi}\frac{\partial \psi}{\partial \xi}\right). \qquad (4.3\text{-}12)$$

This equation is to be solved with the following boundary conditions expressed in dimensionless form as

$$\begin{aligned}\text{B.C. 1:} &\quad \text{at} \quad \zeta = 0, \quad \psi = 0,\\ \text{B.C. 2:} &\quad \text{at} \quad \xi = 0, \quad \psi = 1,\\ \text{and} \quad \text{B.C. 3:} &\quad \text{at} \quad \xi = \infty, \quad \psi = 0.\end{aligned}$$

Boundary condition (3) is a mathematical convenience and is physically valid, since the transfer is confined to a short distance (small ξ values) from the wall. Boundary condition (1) and (3) suggest a solution by combining independent variables as follows:

$$\psi = \psi(\chi), \qquad (4.3\text{-}13)$$

where

$$\chi = \frac{\xi}{(9\zeta)^{\frac{1}{3}}}. \qquad (4.3\text{-}14)$$

Substituting these new variables in equation 4.3-12, we obtain the following ordinary differential equation:

$$\chi \frac{d^2\psi}{d\chi^2} + (3\chi^3 - 1)\frac{d\psi}{d\chi} = 0, \qquad (4.3\text{-}15)$$

to be solved with the two boundary conditions

$$\begin{aligned}\text{B.C. 1} &\quad \text{at} \quad \chi = 0, \quad \psi = 1\\ \text{and} \quad \text{B.C. 2} &\quad \text{at} \quad \chi = \infty, \quad \psi = 0.\end{aligned}$$

Equation 4.3-15 can be readily integrated in terms of $p\ (= d\psi/d\chi)$ to obtain

$$\psi = -C_1 \int_\chi^\infty \chi e^{-\chi^3} d\chi + C_2. \qquad (4.3\text{-}16)$$

From boundary condition (2), we deduce $C_2 = 0$. We use boundary condition (1) to evaluate the constant C_1 and the solution is

$$\psi = \frac{q_y}{q_1} = \frac{\int_\chi^\infty \chi e^{-\chi^3} d\chi}{\int_0^\infty \chi e^{-\chi^3} d\chi} = \frac{3}{\Gamma\left(\frac{2}{3}\right)} \int_\chi^\infty \chi e^{-\chi^3} d\chi \qquad (4.3\text{-}17)$$

where $\Gamma(\ )$ is the Gamma function.

$$\int_T^{T_{b_1}} dT = T_{b_1} - T = -\frac{1}{k}\int_y^\infty q_y dy, \qquad (4.3\text{-}18)$$

130   Transport Phenomena in Simple Flows

which follows from Fourier's law of heat conduction. In dimensionless form, equation 4.3-18 becomes

$$\theta(\xi, \zeta) = \frac{T - T_{b_1}}{q_1 \frac{R}{k}} = (9\zeta)^{\frac{1}{3}} \int_\chi^\infty \psi d\chi. \tag{4.3-19}$$

Integration by parts yields

$$\theta = \frac{3(9\zeta)^{\frac{1}{3}}}{\Gamma(\frac{2}{3})} \left[\frac{e^{-\chi^3}}{3} - \chi \int_\chi^\infty \chi e^{-\chi^3} d\chi \right], \tag{4.3-20}$$

where $\int_\chi^\infty \chi e^{-\chi^3} d\chi$ is the incomplete gamma function.

Heat transfer is usually expressed in terms of the Nusselt number. The local Nusselt number is defined by

$$Nu_{loc} = \frac{2 h_{loc} R}{k} = \frac{2R}{k} \left(\frac{q_1}{T|_{y=0} - T_{b_1}} \right) = \frac{2}{\theta(0, \zeta)}. \tag{4.3-21}$$

From equation 4.3-20, we have

$$\theta(0, \zeta) = \frac{(9\zeta)^{\frac{1}{3}}}{\Gamma(\frac{2}{3})} \tag{4.3-22}$$

Hence,

$$Nu_{loc} = \frac{2\Gamma(\frac{2}{3})}{(9\zeta)^{\frac{1}{3}}} = \frac{2\Gamma(\frac{2}{3})}{\left(\frac{9\alpha z}{V_0 R^2} \right)^{\frac{1}{3}}}. \tag{4.3-23}$$

For a power-law fluid, the wall shear rate is obtained from equation 3.1-9 with $n' = n$. It follows that

$$V_0 = -\dot{\gamma}_R R = \frac{3n+1}{4n} \left(\frac{8\langle V \rangle}{2R} \right) R = \frac{3n+1}{n} \langle V \rangle, \tag{4.3-24}$$

and the final expression for the Nusselt number becomes

$$Nu_{loc} = \frac{2\Gamma(\frac{2}{3})}{\left[\frac{9\alpha z}{\left(3 + \frac{1}{n} \right) \langle V \rangle R^2} \right]^{\frac{1}{3}}}. \tag{4.3-25}$$

4.3 Heat Transfer to Non-Newtonian Fluids

The expression for the average Nusselt number can be easily obtained by integration over the entire tube:

$$Nu_a = \frac{1}{L}\int_0^L Nu_{loc}\, dz = \frac{3\Gamma(\frac{2}{3})}{\left[\dfrac{9\alpha L}{\left(3+\dfrac{1}{n}\right)\langle V\rangle R^2}\right]^{\frac{1}{3}}}. \tag{4.3-26}$$

This result is valid for Graetz numbers larger than 30. We note that the heat transfer decreases with increasing tube length and that shear thinning enhances heat transfer. For a power-law fluid of $n=0.30$, the heat transfer rate is 17% larger than for the comparable Newtonian case.

Example 4.3-2 Isothermal Wall

For the isothermal wall case (wall maintained at T_0), we define the following dimensionless variables:

$$\theta = \frac{T - T_0}{T_{b_1} - T_0},$$
$$\xi = \frac{y}{R}, \tag{4.3-27}$$
$$\text{and}\quad \zeta = \frac{\alpha z}{\langle V\rangle R^2}.$$

For a power-law fluid, $V_0 = (3 + 1/n)\langle V\rangle$, and equation 4.3-9 in dimensionless form is given by

$$\left(3+\frac{1}{n}\right)\xi\frac{\partial\theta}{\partial\zeta} = \frac{\partial^2\theta}{\partial\xi^2}, \tag{4.3-28}$$

to be solved with the three boundary conditions:

$$\begin{aligned}
\text{B.C. 1:} &\quad \text{at}\ \zeta = 0,\ \theta = 1,\\
\text{B.C. 2:} &\quad \text{at}\ \xi = 0,\ \theta = 0,\\
\text{and}\quad \text{B.C. 3:} &\quad \text{at}\ \xi = \infty,\ \theta = 1.
\end{aligned}$$

Here again the solution is obtaining by combining the variables as follows:

$$\theta = \theta(\chi), \tag{4.3-29}$$

with

$$\chi = \frac{\xi}{\left[\dfrac{9\zeta}{3+\dfrac{1}{n}}\right]^{\frac{1}{3}}}. \tag{4.3-30}$$

The solution is obtained by following the same procedure as described for Example 4.3-1. The dimensionless temperature is given by

$$\theta = \frac{1}{\Gamma(\frac{4}{3})}\int_0^\chi e^{-\chi^3}\,d\chi. \tag{4.3-31}$$

Here, the local Nusselt number is expressed by

$$Nu_{loc} = 2\left(\frac{\partial\theta}{\partial\xi}\right)_{\xi=0} = 2\left(\frac{\partial\theta}{\partial\chi}\right)_{\chi=0}\left(\frac{\partial\chi}{\partial\xi}\right)_\zeta$$

$$= \frac{2}{\Gamma(\frac{4}{3})}\left[\frac{3+\frac{1}{n}}{9\zeta}\right]^{\frac{1}{3}}$$

$$= \frac{2}{\Gamma(\frac{4}{3})}\left[\frac{\left(3+\frac{1}{n}\right)\langle V\rangle R^2}{9\alpha z}\right]^{\frac{1}{3}}. \tag{4.3-32}$$

The numerical value for $\Gamma(4/3) \approx 0.893$, and this result in terms of the Graetz number is

$$Nu_{loc} = 1.167\left[\frac{3+\frac{1}{n}}{4}\right]^{\frac{1}{3}} Gz^{\frac{1}{3}}, \tag{4.3-33}$$

where Gz is the local Graetz number defined as

$$Gz = \frac{\pi R^2 \langle V\rangle}{\alpha z}. \tag{4.3-34}$$

The average Nusselt number is obtained by integrating equation 4.3-33 over the entire length:

$$Nu_a = 1.75\left[\frac{3+\frac{1}{n}}{4}\right]^{\frac{1}{3}} Gz^{\frac{1}{3}}, \tag{4.3-35}$$

with Gz evaluated at $z = L$.

4.3-1.1.1 Comments
(i) For a Newtonian fluid ($n=1$), this result approaches the expression proposed by Sieder and Tate (1936) for laminar flow (see Bird, Stewart, and Lightfoot, 1960).
(ii) Note that equation 4.3-35 is similar to the previous result of equation 4.3-26, which can be written in terms of the Graetz number as

$$Nu_a = 2.12\left[\frac{3+\frac{1}{n}}{4}\right]^{\frac{1}{3}} Gz^{\frac{1}{3}}. \tag{4.3-36}$$

As expected, constant heat flux heat transfer is more efficient.

(iii) In all cases, shear thinning enhances the heat transfer rate.
(iv) For very long contact times ($Gz \to 0$), the fluid bulk temperature reaches the wall temperature T_0. An overall energy balance then yields

$$-q_r|_{r=R} 2\pi RL = \langle V \rangle \pi R^2 \hat{C}_p \rho (T_{b_2} - T_{b_1})$$
$$= \langle V \rangle \pi R^2 \hat{C}_p \rho (T_0 - T_{b_1}). \qquad (4.3\text{-}37)$$

Using an average heat transfer coefficient defined by

$$h_a = -\frac{q_r|_{r=R}}{T_0 - T_a}, \qquad (4.3\text{-}38a)$$

where

$$T_a = \frac{(T_{b_1} + T_{b_2})}{2}, \qquad (4.3\text{-}38b)$$

we obtain the following simple expression for the average Nusselt number:

$$Nu_a = h_a 2\frac{R}{k} = \frac{2\langle V \rangle R^2}{\alpha L} = \frac{2}{\pi} Gz. \qquad (4.3\text{-}39)$$

This asymptotic solution, as well as equation 4.3-35 are compared to three experimental data in Fig. 4.3-2. The data obtained by Griskey and Wiehe (1966) for a polymer melt of flow index n equal to 0.7 are in reasonable agreement with the theoretical results.

4.3-1.2 Corrections for Temperature Effects on the Viscosity

The results presented in this section are valid provided that the viscosity as well as the other properties do not vary much with temperature. In many practical situations, the temperature gradients are quite large, and the variations of the viscosity with temperature cannot be

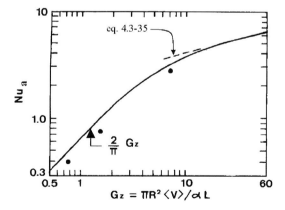

Figure 4.3-2 Average Nusselt number as a function of the Graetz number for Poiseuille flow with isothermal wall • (Data of Griskey and Wiehe (1966) for a polymer melt ($n = 0.7$))

ignored. One useful empiricism is that proposed by Metzner et al. (1957), which is a generalization of the Sieder and Tate (1936) idea. Equation 4.3-35 is corrected in the following way:

$$Nu_a = 1.75 \left(\frac{3n+1}{4n}\right)^{\frac{1}{n}} Gz^{\frac{1}{3}} \left(\frac{K_b}{K_w}\right)^{0.14} \qquad (4.3\text{-}40)$$

where $K = 8^{n-1} m$ is a parameter related to the non-Newtonian viscosity. The subscripts b and w stand for evaluation at the bulk and wall temperature respectively.

Equation 4.3-40 should be used with caution, and the correction term $(K_b/K_w)^{0.14}$ should be taken as an indicator of possible temperature effects. If this term is large, we should rely on numerical solutions of the energy equation with temperature-dependent parameters, and eventually account for viscous dissipation effects. The interested reader should refer to the abundant literature on the subject. (See for example Christiansen and Craig (1962), Christiansen et al. (1966), and Mahalingam et al. (1975a and b).)

4.3-2 Heat Generation in Poiseuille Flow

For viscous fluids, the heat generated by the viscous forces may represent an important contribution to the overall heat balance. For simplified situations, the expression for the temperature profile can be obtained analytically, as illustrated here.

For Poiseuille flow of a power-law fluid, the expression for the velocity profile can be written by combining equations 4.2-7 and 4.2-8 to yield

$$V_z(r) = \left(\frac{3n+1}{n+1}\right) \langle V \rangle \left[1 - \left(\frac{r}{R}\right)^{1+\frac{1}{n}}\right], \qquad (4.3\text{-}41)$$

where the average velocity $\langle V \rangle$ is equal to $Q/\pi R^2$. The expression for the shear rate is obtained by taking the derivative of $V_z(r)$ with respect to r:

$$\dot{\gamma}(r) = \frac{dV_z}{dr} = -\left(\frac{3n+1}{n}\right) \frac{\langle V \rangle}{R} \left(\frac{r}{R}\right)^{\frac{1}{n}}. \qquad (4.3\text{-}42)$$

The heat generation rate per unit volume is then

$$\frac{1}{2} \underline{\underline{\eta}} \underline{\underline{\dot{\gamma}}} : \underline{\underline{\dot{\gamma}}} = m|\dot{\gamma}|^{n+1} = m \left(\frac{3n+1}{n}\langle V \rangle\right)^{n+1} \frac{r^{1+\frac{1}{n}}}{R^{\frac{(n+1)^2}{n}}}. \qquad (4.3\text{-}43)$$

The energy equation 4.3-2 is now modified to the following form:

$$\frac{3n+1}{n+1} \rho \hat{C}_p \langle V \rangle \left[1 - \left(\frac{r}{R}\right)^{1+\frac{1}{n}}\right] \frac{\partial T}{\partial z}$$

$$= k \frac{1}{r}\left[\frac{\partial}{\partial r}\left(r \frac{\partial T}{\partial r}\right)\right] + m \left(\frac{3n+1}{n}\langle V \rangle\right)^{n+1} \frac{r^{1+\frac{1}{n}}}{R^{\frac{(n+1)^2}{n}}}. \qquad (4.3\text{-}44)$$

4.3-2.1 Equilibrium Regime

In the equilibrium regime, the temperature reaches a constant value with respect to z and $T = T(r)$. The convection term becomes equal to zero and equation 4.3-44 reduces to

$$\frac{d}{dr}\left(r\frac{dT}{dr}\right) = -\frac{m}{k}\left[\frac{3n+1}{n}\langle V \rangle\right]^{n+1} \frac{r^{\frac{(2n+1)}{n}}}{R^{\frac{(n+1)^2}{n}}}, \qquad (4.3\text{-}45)$$

which can be integrated to obtain the temperature profile

$$T(r) = T_0 + \frac{m}{k}\left(\frac{n}{3n+1}\right)^{1-n} \langle V \rangle^{n+1} \frac{R^{3+\frac{1}{n}} - r^{3+\frac{1}{n}}}{R^{\frac{(n+1)^2}{n}}}. \qquad (4.3\text{-}46)$$

The radial temperature profile is illustrated in Fig. 4.3-3 for different values of the power-law index. As n decreases, the velocity profile becomes more blunt. The shear rate is more important near the tube wall, and hence the temperature profile also becomes more blunt, because the heat source by viscous forces is more and more confined to the region near the tube wall.

The maximum temperature is at the tube center, and the maximum difference is expressed by

$$\Delta T_{max} = T(0) - T_0 = \frac{m}{k}\left(\frac{n}{3n+1}\right)^{1-n} \frac{\langle V \rangle^{n+1}}{R^{n-1}}. \qquad (4.3\text{-}47)$$

This expression reduces correctly to $\Delta T_{max} = \eta V^2/k$ for a Newtonian fluid ($m = \eta$ and $n = 1$). It can also be expressed in terms of the pressure drop (using the relationship between the flow rate and the pressure drop):

$$\Delta T_{max} = \frac{m}{k}\left(\frac{n}{3n+1}\right)^2 \left|\frac{1}{2m}\frac{\Delta P}{L}\right|^{1+\frac{1}{n}} R^{3+\frac{1}{n}}. \qquad (4.3\text{-}48)$$

Figure 4.3-4 shows the value of ΔT_{max} as a function of n for a polymer of consistency index m equal to 10^4 Pa·s^n flowing at an average velocity of 0.2 m/s in a capillary of 5 mm radius. The thermal conductivity is equal to 0.2 W/(m·°C). Obviously, as n decreases, the

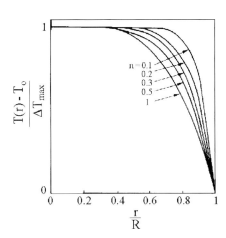

Figure 4.3-3 Radial temperature profiles in Poiseuille flow for different values of the power-law index. Thermal equilibrium regime (Adapted from Agassant et al., 1991)

136   Transport Phenomena in Simple Flows

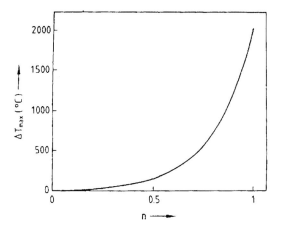

Figure 4.3-4 Viscous heating as a function of the power-law index in the thermal equilibrium regime Flow conditions and polymer properties are reported in the text (From Agassant et al., 1991)

polymer at the same flow rate is much less viscous, and viscous heating is reduced considerably. The extremely large values of ΔT_{max} for n close to 1 are unrealistic.

4.3-2.2 Transition Regime (Approximate Solution)

In the general case, the temperature varies along the flow axis, and it is not possible to obtain an analytical solution of $T(r, z)$ from equation 4.3-44. We look for a solution for the bulk temperature $T_b(z)$. From an energy balance over a differential volume $\pi R^2 dz$, we obtain

$$\rho \hat{C}_p \pi R^2 \langle V \rangle dT_b(z) = -2\pi R dz q + \left[\int_0^R \frac{1}{2} \eta(\underline{\underline{\dot{\gamma}}} : \underline{\underline{\dot{\gamma}}}) 2\pi r dr \right] dz. \tag{4.3-49}$$

As for the Newtonian case, the heat flux at the wall is approximated from the equilibrium temperature profile (equation 4.3-46). The mean bulk temperature at equilibrium is obtained from the definition for the bulk temperature (see Bird, Stewart, and Lightfoot, 1960):

$$T_b - T_0 = \frac{\int_0^R (T(r) - T_0) V_z(r) r dr}{\int_0^R V_z(r) r dr}. \tag{4.3-50}$$

Using equations 4.3-46 and 4.3-41, we obtain

$$T_b - T_0 = \left(\frac{4n+1}{5n+1} \right) \frac{m}{k} \left(\frac{n}{3n+1} \right)^{1-n} \left(\frac{\langle V \rangle}{R} \right)^{n+1} R^2. \tag{4.3-51}$$

Taking the derivative of equation 4.3-46 with respect to r, we get an approximate expression for the heat flux at the tube wall, that is

$$q = -k\left.\frac{dT}{dr}\right|_{r=R} \approx \left(\frac{3n+1}{n}\right)\left(\frac{5n+1}{4n+1}\right)k\left(\frac{T_b - T_0}{R}\right). \tag{4.3-52}$$

However, for any axial distance z,

$$2\pi \int_0^R \frac{1}{2}\eta(\underline{\underline{\dot{\gamma}}} : \underline{\underline{\dot{\gamma}}}) r\, dr\, dz = \pi R^2 \langle V \rangle \frac{|\Delta P|}{L} dz, \tag{4.3-53}$$

that is, the rate of energy production by viscous dissipation is equal to the rate of work done by pressure forces. Using equations 4.3-52 and 4.3-53, equation 4.3-49 can be written approximately as

$$\frac{dT_b}{dz} \approx -\frac{2\pi}{Gz}\left(\frac{3n+1}{4n+1}\right)\left(\frac{5n+1}{n}\right)\left(\frac{T_b - T_0}{L}\right) + \frac{1}{\rho \hat{C}_p}\frac{|\Delta P|}{L}, \tag{4.3-54}$$

where Gz is the Graetz number defined by equation 4.3-3. This result is readily integrated to obtain, for the case of $T_b(0) = T_0$,

$$T_b(z) - T_0 = \left(\frac{4n+1}{5n+1}\right)\Delta T_{\max}\left[1 - \exp\left\{-\frac{2\pi}{Gz}\left(\frac{5n+1}{n}\right)\left(\frac{3n+1}{4n+1}\right)\left(\frac{z}{L}\right)\right\}\right], \tag{4.3-55}$$

where $\Delta T_{\max}$ is expressed by equation 4.3-47 or 4.3-48. The magnitude of the Graetz number allows us to distinguish between the adiabatic, the transition, and the equilibrium regimes.

Example 4.3-3 Viscous Heating in Poiseuille Flow

We examine the flow of a polymer in a tube of radius equal to 5 mm and length equal to 1 m. The average velocity is 0.2 m/s, and the polymer properties are $m = 10^4$ Pa·s^n, $k = 0.2$ W/m·°C, and $\alpha = 10^{-7}$ m^2/s. The Graetz number for these conditions is equal to 157, hence the flow is in the transition regime.

The solutions for the temperature increase predicted by equation 4.3-55 are presented in Fig. 4.3-5 for different values of the power-law index, n.

It is interesting to note that the equilibrium state is somewhat affected by the shear-thinning behavior of the polymer. For $n \geq 0.5$, we are near the adiabatic regime even at the tube exit, whereas for $n = 0.2$, we are in the transition regime at $z/L = 1$. It is clear that equilibrium is attained more rapidly, since the factor in the exponential term of equation 4.3-55 increases with increasing shear-thinning properties (smaller values of the parameter n).

We stress that these results are approximate and should eventually be compared with results of numerical solutions of equation 4.3-49.

For very large values of the Graetz number, equation 4.3-54 reduces to

$$\frac{dT_b}{dz} = \frac{1}{\rho \hat{C}_p}\frac{|\Delta P|}{L}, \tag{4.3-56}$$

138 Transport Phenomena in Simple Flows

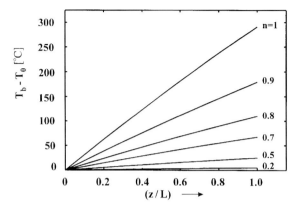

Figure 4.3-5 Influence of the power-law index parameter on the polymer temperature increase
The Graetz number is equal to 157

which yields a linear temperature profile given by

$$T_b(z) - T_b(0) = \frac{|\Delta P|}{\rho \hat{C}_p}\left(\frac{z}{L}\right). \qquad (4.3\text{-}57)$$

This is the *adiabatic regime* (also valid for any Graetz number in the case of a perfectly insulated tube wall).

4.4 Mass Transfer to Non-Newtonian Fluids

Following the philosophy used in the previous section on heat transfer, we present here a few analytical solutions of mass transfer problems involving power-law fluids in simple flow geometries.

4.4-1 Mass Transfer to a Power-Law Fluid Flowing on an Inclined Plate

In this example, we examine how the mass transfer coefficient is affected by the shear rate for the flow of a shear-thinning fluid on an inclined surface. The fluid is assumed to obey the power-law expression. The mass transfer rate is small and no chemical reaction takes place.

Consider the flow configuration shown in Fig. 4.4-1. Let us assume that downstream of a position $z=0$, where the flow is hydrodynamically fully-developed, the solid surface

4.4 Mass Transfer to Non-Newtonian Fluids

consists of a soluble material of length L. The mass transfer is governed by the differential equation

$$D_A \frac{\partial^2 c}{\partial x^2} = V_z(x) \frac{\partial c}{\partial z}, \tag{4.4-1}$$

where D_A is the diffusivity of the solute in the liquid. Note that the diffusive contribution in the z-direction has been neglected. The boundary conditions for this process are

B.C. 1: at $z = 0$, $c = c_0$ (inlet concentration),

B.C. 2: at $x = 0$, $c = c^*$ (solubility),

and B.C. 3: at $x = \delta$, $\dfrac{\partial c}{\partial x} = 0$ (no mass transfer).

The shear stress distribution in the film is given by

$$\sigma_{xz} = \rho g (x - \delta) \cos \beta, \tag{4.4-2}$$

which linearly decreases, eventually becoming zero at the free surface.

We further assume that the mass transfer is confined to a region near the solid surface. (This is strictly valid for short contact times, as discussed in Section 17.5 of Bird, Stewart, and Lightfoot, 1960.) The wall shear stress can be approximated by

$$\sigma_{xz} \approx -\rho g \delta \cos \beta. \tag{4.4-3}$$

Also, boundary condition (3) can be replaced by

B.C. 3': at $x \to \infty$, $c = c_0$.

For power-law fluids, equation 4.4-3 can be written as

$$m \left| \frac{dV_z}{dx} \right|^n \approx \rho g \delta \cos \beta \tag{4.4-4}$$

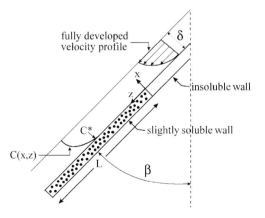

Figure 4.4-1 Mass transfer to a falling film with fully-developed velocity profile

or

$$V_z = \left(\frac{\rho g \delta \cos \beta}{m}\right)^{\frac{1}{n}} x. \tag{4.4-5}$$

The constant of integration is zero for the no-slip boundary condition. Note that the shear rate at $x = 0$, that is, at the plane surface, is

$$\dot{\gamma}_w = \left(\frac{\rho g \delta \cos \beta}{m}\right)^{\frac{1}{n}}. \tag{4.4-6}$$

With the help of equation 4.4-5, we can rewrite the diffusion equation as

$$D_A \frac{\partial^2 c}{\partial x^2} = \dot{\gamma}_w x \frac{\partial c}{\partial z}. \tag{4.4-7}$$

The solution can be obtained by combining variables as done in Section 17.5 of Bird, Stewart, and Lightfoot (1960) or by using Laplace transforms, as done here. We introduce the Laplace transformed concentration $\bar{c}$ as

$$\bar{c}(s) = \int_0^\infty \exp(-sz)(c - c_0) dz, \tag{4.4-8}$$

and equation 4.4-7 becomes

$$D_A \frac{\partial^2 \bar{c}(s)}{\partial x^2} = \dot{\gamma}_w x s \bar{c}(s). \tag{4.4-9}$$

This is a form of the Bessel equation, which, combined with boundary conditions (2) and (3), leads to the following result:

$$\bar{c}(s) = (c_0^* - c_0)(3^{-\frac{1}{3}})\Gamma\left(\frac{2}{3}\right)\left(\frac{D_A}{\dot{\gamma}_w s}\right)^{\frac{1}{6}} \sqrt{x}\left[I_{-\frac{1}{3}}(t) - I_{\frac{1}{3}}(t)\right], \tag{4.4-10}$$

where

$$t = \left(\frac{2}{3}\right)\left(\frac{\dot{\gamma}_w s}{D_A}\right)^{\frac{1}{2}} x^{\frac{1}{6}}. \tag{4.4-11}$$

$I_{-1/3}$ and $I_{1/3}$ are the modified Bessel functions of the first kind of order $-1/3$ and $1/3$ respectively, and $\Gamma()$ is the gamma function.

In order to calculate the average mass transfer rate, the concentration gradient $(\partial c/\partial x)_{x=0}$ is evaluated from $(\partial \bar{c}(s)/\partial x)_{x=0}$:

$$\left.\frac{\partial \bar{c}}{\partial x}\right|_{x=0} = -(c^* - c_0)(3)^{-\frac{1}{6}} \frac{\Gamma\left(\frac{2}{3}\right)}{\Gamma\left(\frac{4}{3}\right)} \left(\frac{\dot{\gamma}_w s}{D_A}\right)^{\frac{1}{3}}. \tag{4.4-12}$$

This result can be inverted to the real concentration c as follows:

$$\left.\frac{\partial c}{\partial x}\right|_{x=0} = -(c^* - c_0)\frac{3^{\frac{1}{3}}}{\Gamma\left(\frac{1}{3}\right)} \left(\frac{\dot{\gamma}_w}{D_A z}\right)^{\frac{1}{3}}. \tag{4.4-13}$$

The average (over the length L) mass transfer coefficient, k_a, is evaluated as

$$k_a = \frac{1}{L} \int_0^L \frac{-D_A \frac{\partial c}{\partial x}\big|_{x=0}}{(c^* - c_0)} dz \qquad (4.4\text{-}14)$$

$$= \frac{3^{\frac{4}{3}} D_A^{\frac{2}{3}}}{2\Gamma\left(\frac{1}{3}\right)} \left(\frac{\dot{\gamma}_w}{L}\right)^{\frac{1}{3}},$$

where $\dot{\gamma}_w$ is given by equation 4.4-6.

Thus, equation 4.4-14 enables us to elucidate effects of wall shear rate on mass transfer through this simple experiment. This development is based on the work of Astarita (1966), who employed this method to determine the diffusion coefficient of benzoic acid in carboxy methyl cellulose solutions. Unfortunately, apparent wall-slip effects are known to be significant in this flow configuration, and must be accounted for in the interpretation of mass transfer data (Astarita et al., 1964; Astarita, 1966; Carreau et al., 1979a).

4.4-2 Mass Transfer to a Power-Law Fluid in Poiseuille Flow

As a second example, we discuss the mass transfer associated with the flow of a power-law fluid in a tube. The flow configuration is illustrated in Fig. 4.3-1. We consider that the flow is fully-developed, and that from $z=0$ to $z=L$, the tube inner wall is coated with a solute that is slightly soluble in the flowing fluid. Thus, for low transfer rates and in the absence of chemical reactions, the problem is completely analogous to that of convective heat transfer in Poiseuille flow. We will restrict the problem to short contact times, and use the Lévêque (1928) approach, assuming a linear velocity profile near the tube wall. By analogy to the heat transfer problem, the governing equation for the solute concentration (Fick's second law of diffusion) is expressed as

$$V_0 \frac{y}{R} \frac{\partial c}{\partial z} = D_A \frac{\partial^2 c}{\partial y^2}, \qquad (4.4\text{-}15)$$

where $c = c(y, z)$ is the solute concentration, D_A is the diffusivity of the solute in the fluid, y is the distance from the tube wall, and V_0 is a characteristic value of the velocity profile given by equation 4.3-7. Obviously, in equation 4.4-15, the tube curvature has been neglected and the effect of molecular diffusion in the axial direction is assumed negligible compared to the convective term (left side of equation 4.4-15). Equation 4.4-15 must be solved with the following conditions:

B.C. 1: at $z = 0$, $c = c_0$ (inlet concentration),
B.C. 2: at $y = 0$, $c = c^*$ (solubility),
and B.C. 3: at $y = \infty$, $c = c_0$.

Boundary condition (3) is valid only for the case of short contact times for which the mass transfer, or diffusion, proceeds in the vicinity of the wall.

We introduce the following dimensionless quantities:

$$C = \frac{c^* - c}{c^* - c_0}, \tag{4.4-16}$$

$$\xi = \frac{y}{R}, \tag{4.4-17}$$

and

$$\zeta = \frac{D_A z}{\langle V \rangle R^2}. \tag{4.4-18}$$

The velocity profile for power-law fluids may be expressed as

$$\frac{V_z}{\langle V \rangle} = \frac{V_z}{\frac{Q}{\pi R^2}} = \frac{3 + \frac{1}{n}}{1 + \frac{1}{n}}[1 - (1-\xi)^{1+\frac{1}{n}}]. \tag{4.4-19}$$

The wall shear rate is then

$$\dot{\gamma}_R = -\left(\frac{dV_z}{dy}\right)_{y=0} = -\left(3 + \frac{1}{n}\right)\frac{Q}{\pi R^3}, \tag{4.4-20}$$

and

$$V_0 = -\dot{\gamma}_R R = \left(3 + \frac{1}{n}\right)\frac{Q}{\pi R^2}. \tag{4.4-21}$$

In dimensionless variables, equation 4.4-15 reduces to

$$\left(3 + \frac{1}{n}\right)\xi\frac{\partial C}{\partial \zeta} = \frac{\partial^2 C}{\partial \xi^2}, \tag{4.4-22}$$

which is to be solved with the following boundary conditions:

B.C. 1: at $\zeta = 0$, $C = 1$,
B.C. 2: at $\xi = 0$, $C = 0$,
and B.C. 3: at $\xi = \infty$, $C = 1$.

Boundary conditions (1) and (3) suggest a solution by combination of the independent variables

$$C = C(\chi), \tag{4.4-23}$$

where

$$\chi = \frac{\xi}{\sqrt[3]{\frac{9\zeta}{\left(3 + \frac{1}{n}\right)}}}. \tag{4.4-24}$$

Substituting the new variable and using the chain rule, we obtain the ordinary differential equation

$$\frac{d^2 C}{d\chi^2} + 3\chi^2 \frac{dC}{d\chi} = 0, \tag{4.4-25}$$

with $C(0) = 0$ and $C(\infty) = 1$. The solution for the dimensionless concentration profile is

$$C = \frac{1}{\Gamma(\frac{4}{3})} \int_0^\chi e^{-\chi^3} d\chi. \qquad (4.4\text{-}26)$$

The local mass transfer coefficient is defined by

$$-D_A \frac{\partial c}{\partial y}\bigg|_{y=0} = k_{loc}(c_b - c^*), \qquad (4.4\text{-}27)$$

where c_b is the bulk solute concentration in the fluid. For short contact time, $c_b \approx c_0$ and the local Sherwood number, Sh_{loc} (or Nusselt number for mass transfer) is given by

$$Sh_{loc} = \frac{2k_{loc}R}{D_A} = \frac{2R\left(\frac{\partial c}{\partial y}\right)\big|_{y=0}}{(c_0 - c^*)} = \frac{2\partial C}{\partial \xi}\bigg|_{\xi=0}. \qquad (4.4\text{-}28)$$

Using result 4.4-26 to evaluate the derivative in equation 4.4-28, we obtain

$$Sh_{loc} = 2\frac{dC}{d\chi}\bigg|_{\chi=0}\left(\frac{\partial \chi}{\partial \xi}\right) = \frac{2}{\Gamma(\frac{4}{3})}\sqrt[3]{\frac{\left(3 + \frac{1}{n}\right)\langle V \rangle R^2}{9D_A z}}. \qquad (4.4\text{-}29)$$

It is more practical to use an average mass transfer coefficient or average Sherwood number. This is readily obtained from equation 4.4-29 by integrating over the entire tube length:

$$Sh_a = \frac{2k_a R}{D_A} = \frac{1}{L}\int_0^L Sh_{loc} dz$$

$$= \frac{3}{\Gamma(\frac{4}{3})}\sqrt[3]{\frac{\left(3 + \frac{1}{n}\right)\langle V \rangle R^2}{9D_A L}}. \qquad (4.4\text{-}30)$$

This result can be expressed in terms of the Graetz number (noting that $\Gamma(4/3) \approx 0.893$):

$$Sh_a = 1.75\left[\frac{3 + \frac{1}{n}}{4}\right]^{\frac{1}{3}} Gz^{\frac{1}{3}}, \qquad (4.4\text{-}31)$$

where Gz is the Graetz number defined for mass transfer as

$$Gz = \frac{\pi R^2 \langle V \rangle}{D_A L}. \qquad (4.4\text{-}32)$$

Note that equation 4.4-31 is identical to result 4.3-35 obtained for heat transfer, except for the definition of the Graetz number. This result can be written in terms of other dimensionless numbers by noting that

$$Gz = \frac{\pi}{2}\left(\frac{2R\langle V\rangle\rho}{\eta}\right)\left(\frac{\eta}{\rho D_A}\right)\left(\frac{R}{L}\right)$$

$$= \frac{\pi}{2} ReSc\left(\frac{R}{L}\right), \qquad (4.4\text{-}33)$$

where $Sc\ (=\eta/\rho D_A)$ is the Schmidt number.

As observed in heat transfer, shear thinning (lower values of the power-law index n) enhances the mass transfer rate. The analogy shown here between heat and mass transfers is also applicable to other flow situations. However, it should be used with caution. It is valid only for low mass transfer rates in the absence of a source term (e.g., negligible viscous dissipation, no chemical reaction). The analogy will also fail if the physical properties exhibit important variations with temperature and concentration.

4.5 Boundary Layer Flows

We conclude this chapter by presenting extensions of two boundary layer problems to power-law fluids.

4.5-1 Laminar Boundary Layer Flow of Power-Law Fluids over a Plate

Consider the flow shown schematically in Fig. 4.5-1. The exact boundary layer equations require a numerical solution due to the nonlinear power-law constitutive relation. However, reasonably accurate predictions of the drag on a plate can be obtained via the integral momentum approach due to Von Kármán.

Consider the control volume ABCD shown in Fig. 4.5-1. The velocity components normal to the wall are assumed to be small compared to those in the x-direction, and are neglected. From the equation of continuity, we can easily show that the mass flow rate entering the control volume through the boundary CD is given by

$$\dot{m}_{CD} = \frac{d}{dx}\left\{\int_0^H \rho V_x dy\right\}dx. \qquad (4.5\text{-}1)$$

Applying the momentum balance to the control volume in the x-direction yields

$$\sum F_x = +\sigma_W dx - \frac{dP}{dx}Hdx = \frac{d}{dx}\left(\int_0^H \rho V_x^2 dy\right)dx - V_0\frac{d}{dx}\left(\int_0^H \rho V_x dy\right)dx. \qquad (4.5\text{-}2)$$

In equation 4.5-2, the first term on the right side arises simply as a difference in momentum leaving and entering the control volume through the planes BC and AD respectively,

4.5 Boundary Layer Flows

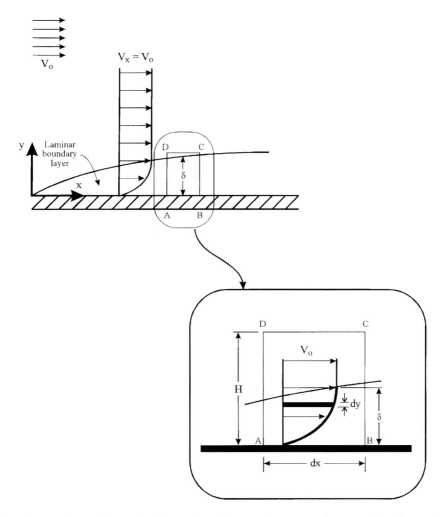

Figure 4.5-1 Boundary layer formation on a flat plate oriented at zero incidence angle to the undisturbed stream

whereas the second term represents the momentum entering the control volume via plane CD. The latter can be slightly rearranged to rewrite equation 4.5-2 as follows

$$-\sigma_w - \frac{dP}{dx} H = -\rho \frac{d}{dx} \left\{ \int_0^H (V_0 - V_x) V_x \, dy \right\}. \tag{4.5-3}$$

Note that in arriving at equation 4.5-3, use has been made of the fact that $dV_0/dx = 0$, which in turn, coupled with the Bernoulli equation, yields that the pressure is effectively constant, that is, $dP/dx = 0$. Equation 4.5-3 further simplifies to

$$-\sigma_w = \rho \frac{d}{dx}\left\{\int_0^\delta (V_0 - V_x)V_x dy\right\}. \tag{4.5-4}$$

The upper limit of integration has been changed to δ, since the integrand is zero for $\delta \leq y \leq H$.

The thickness δ of the boundary layer at any value of x may be determined by assuming a plausible expression for the velocity profile, which satisfies the following boundary conditions:

B.C. 1 : at $y = 0$, $V_x = 0$ (no slip), (4.5-5a)

B.C. 2 : at $y = \delta$, $V_x = V_0$ (free stream velocity), (4.5-5b)

and B.C. 3 : at $y = \delta$, $\dfrac{\partial V_x}{\partial y} = 0$ (no momentum transfer). (4.5-5c)

Skelland (1967) suggested that the velocity profiles, which have proved to be quite successful for Newtonian fluids, also provide adequate results for power-law fluids. We begin by choosing a third degree polynomial expression

$$V_x = a + by + cy^3. \tag{4.5-6}$$

Using the three boundary conditions outlined in equation 4.5-5, we can rewrite the velocity profile as

$$\frac{V_x}{V_0} = \frac{3}{2}\left(\frac{y}{\delta}\right) - \frac{1}{2}\left(\frac{y}{\delta}\right)^3. \tag{4.5-7}$$

For a power-law fluid,

$$-\sigma_w = m\left(\frac{\partial V_x}{\partial y}\right)^n\bigg|_{y=0}. \tag{4.5-8}$$

Combining equations 4.5-4, 4.5-7, and 4.5-8 and integrating, we obtain

$$m\left(\frac{3\,V_0}{2\,\delta}\right)^n = \frac{39}{280}\rho V_0^2 \frac{d\delta}{dx}, \tag{4.5-9}$$

which can be further integrated, with the boundary condition $\delta = 0$ at $x = 0$, to obtain the following expression for δ:

$$\delta = \left(\frac{3}{2}\right)^{\frac{n}{n+1}}\left(\frac{280(n+1)m}{39}\right)^{\frac{1}{n+1}}\left(\frac{x}{\rho V_0^{2-n}}\right)^{\frac{1}{n+1}}. \tag{4.5-10}$$

This can be written in a more familiar form as

$$\left(\frac{\delta}{x}\right) = \left[\left(\frac{3}{2}\right)^n\left(\frac{280}{39}\right)(n+1)\right]^{\frac{1}{n+1}}(Re_x)^{-\frac{1}{n+1}}, \tag{4.5-11}$$

where Re_x is the Reynolds number defined for power-law fluids by

$$Re_x = \frac{\rho V_0^{2-n} x^n}{m}. \tag{4.5-12}$$

4.5 Boundary Layer Flows

The shear stress on the plate, σ_w, is given by

$$-\sigma_w = \rho V_0^2 \left[\left(\frac{1.5}{n+1}\right)\left(\frac{39}{280}\right) \right]^{\frac{n}{n+1}} Re_x^{-\frac{1}{n+1}}. \qquad (4.5\text{-}13)$$

The total frictional drag, F_D, on one side of the plate of length L and width W is expressed by

$$\begin{aligned}
F_D &= W \int_0^L -\sigma_w \, dx \\
&= W\rho V_0^2 \left[\left(\frac{1.5}{n+1}\right)\left(\frac{39}{280}\right) \right]^{\frac{n}{n+1}} \int_0^L Re_x^{-\frac{1}{n+1}} dx \qquad (4.5\text{-}14) \\
&= WL\rho V_0^2(n+1) \left[\left(\frac{1.5}{n+1}\right)\left(\frac{39}{280}\right) \right]^{\frac{n}{n+1}} Re_L^{-\frac{1}{n+1}},
\end{aligned}$$

where $Re_L = Re_x|_{x=L} = \rho V_0^{2-n} L^n / m$.

It is customary to introduce a drag coefficient, C_f, as

$$C_f = \frac{F_D}{WL\left(\frac{1}{2}\rho V_0^2\right)}. \qquad (4.5\text{-}15)$$

Combining equations 4.5-14 and 4.5-15, we obtain

$$C_f = \frac{2(n+1)c(n)}{Re_L^{\frac{1}{n+1}}}, \qquad (4.5\text{-}16)$$

where

$$c(n) = \left[\left(\frac{1.5}{n+1}\right)\left(\frac{39}{280}\right) \right]^{\frac{n}{n+1}}. \qquad (4.5\text{-}17)$$

Note that for $n=1$, equation 4.5-16 correctly reduces to the Newtonian result

$$C_f = \frac{1.293}{\sqrt{Re_L}}. \qquad (4.5\text{-}18)$$

Other velocity expressions which have been quite successful are of the following form (Schlichting, 1968):

$$\frac{V_x}{V_0} = 2\left(\frac{y}{\delta}\right) - 2\left(\frac{y}{\delta}\right)^3 + \left(\frac{y}{\delta}\right)^4 \qquad (4.5\text{-}19)$$

and

$$\frac{V_x}{V_0} = \sin\left(\frac{\pi}{2}\frac{y}{\delta}\right). \qquad (4.5\text{-}20)$$

Following an identical procedure, the resulting expressions for the drag coefficient corresponding to the fourth degree polynomial and the sinusoidal profiles are given by equation 4.5-16 with $c(n)$ expressed respectively by

$$c(n) = \left[\frac{74}{315(n+1)}\right]^{\frac{n}{n+1}} \qquad (4.5\text{-}21)$$

and

$$c(n) = \left[\frac{0.136}{n+1}\left(\frac{\pi}{2}\right)\right]^{\frac{n}{n+1}}. \tag{4.5-22}$$

Despite the differences inherent in the three aforementioned velocity expressions, the corresponding drag predictions are extremely close to each other as well as to the more rigorous numerical analysis of Acrivos et al. (1960). A comparison for power-law fluids is shown in Fig. 4.5-2. Good correspondence exists under most conditions of practical interest. For $n \leq 1$, the fourth degree polynomial velocity profile yields drag predictions close to the numerical solution. In general, the lower the value of n, the larger the deviation. On the other hand, for $1 < n \leq 3$, the third degree polynominal velocity profile yields the best estimate of the friction.

The extensions to two- and three-dimensional boundary layer flows of power-law fluids have been discussed by Bizzell and Slattery (1962) and Schowalter (1978). Lee and Ames (1966) have attempted similarity solutions. The approximate analysis for turbulent boundary layers has been discussed by Skelland (1967).

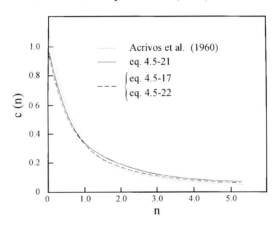

Figure 4.5-2 Comparison between approximate and numerical solutions for laminar boundary layer flow over a plate

Example 4.5-1 Drag Force on a Plate

A thin plate (0.1 m × 0.1 m) is immersed in a power-law fluid ($m = 0.15$ Pa·s$^{0.4}$; $n = 0.4$; $\rho = 1000$ kg/m^3). The free stream velocity is 2 m/s. Estimate the total drag force (both sides) on the plate, the boundary layer thickness, and the value of the shear stress at $x = 0.05$ m from the leading edge.

Solution

Since $n < 1$, we shall be using the expressions based on the fourth degree polynomial velocity profile.

$$Re_L = \frac{\rho V_0^{2-n} L^n}{m} = \frac{(1000)(2)^{2-0.4}(0.1)^{0.4}}{0.15} = 8050$$

The flow is laminar. Now from equation 4.5-16,

$$C_f = 2(n+1)\left\{\frac{74}{315(n+1)}\right\}^{\frac{n}{n+1}}(8050)^{-\frac{1}{n+1}} = 0.00273$$

and

$$F_D = 2\left(\frac{1}{2}\rho V_0^2\right)WLC_f$$

$$= 2\left(\frac{1}{2}\right)(1000)(2^2)(0.1)(0.1)(0.00273)$$

$$= 0.11 \text{ Pa}.$$

The corresponding expressions for δ/x and σ_w are

$$\frac{\delta}{x} = \left\{\frac{315}{37}(2)^n(n+1)\right\}^{\frac{1}{n+1}} Re_x^{-\frac{1}{n+1}}$$

and

$$-\sigma_w = \rho V_0^2 \left\{\frac{74}{315(n+1)}\right\}^{\frac{n}{n+1}} Re_x^{-\frac{1}{n+1}}.$$

At $x = 0.05$ m,

$$Re_x = \frac{(1000)(2)^{2-0.4}(0.05)^{0.4}}{0.15}$$

$$= 6100,$$

$$\delta = \left(\frac{315}{37}(2)^{0.4}(0.4+1)\right)^{\frac{1}{1.4}}(6100)^{-\frac{1}{1.4}}(0.05)$$

$$= 0.707 \text{ mm},$$

and

$$-\sigma_w = 1000 \times 2^2 \left[\frac{74}{315 \times 1.4}\right]^{\frac{0.4}{1.4}}(6100)^{-\frac{1}{1.4}}$$

$$= 4.76 \text{ Pa}.$$

4.5-2 Laminar Thermal Boundary Layer Flow over a Plate

Consider an incompressible power-law fluid at a temperature T_0 flowing over an immersed plate at temperature T_w, as shown schematically in Fig. 4.5-3. The thermal boundary layer thickness δ_t is usually different from the momentum boundary layer thickness δ. It is assumed here that the heating begins at $x = 0$. The simplified continuity, momentum, and energy equations for this flow configuration are written as

$$\frac{\partial V_x}{\partial x} + \frac{\partial V_y}{\partial y} = 0, \quad (4.5\text{-}23)$$

$$V_x \frac{\partial V_x}{\partial x} + V_y \frac{\partial V_y}{\partial y} = -\frac{1}{\rho}\frac{\partial \sigma_{yx}}{\partial y}, \quad (4.5\text{-}24)$$

and

$$V_x \frac{\partial T}{\partial x} + V_y \frac{\partial T}{\partial y} = \frac{k}{\rho \hat{C}_p}\frac{\partial^2 T}{\partial y^2}. \quad (4.5\text{-}25)$$

The assumptions of constant physical properties, no viscous dissipation, constant pressure, and negligible heat conduction in the x-direction are inherent in equations 4.5-23 to 4.5-25.

The pertinent boundary conditions are:

B.C. 1 : at $y = 0$, $V_x = V_y = 0$, $T = T_w$, (4.5-26a)

B.C. 2 : at $y = \infty$, $V_x = V_0$, $T = T_0$, (4.5-26b)

and

B.C. 3 : at $x = 0$, $V_x = V_0$, $T = T_0$. (4.5-26c)

For small temperature differences $(T_w - T_0)$, the assumption of constant physical properties is reasonable. Acrivos et al. (1960) obtained the following expressions for the local heat transfer coefficient for an isothermal plate. For $x > x^*(n < 1)$,

$$Nu_x = \frac{xh_x}{k} = 1.12 \left[\frac{c(n)^{\frac{1}{n}}}{9} - \frac{(2n+1)}{(2n+2)} \right]^{\frac{1}{3}} Re_L^{\frac{1}{n+1}} Pr^{\frac{1}{3}} \left(\frac{x}{L} \right)^{\frac{2n+1}{3(n+1)}}. \quad (4.5\text{-}27)$$

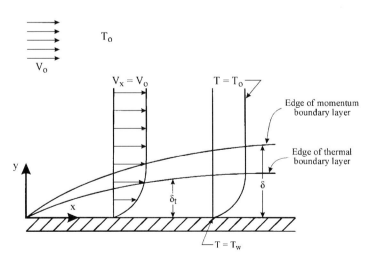

Figure 4.5-3 Thermal boundary layer over a plate

For $0 < x < x^*(n < 1)$,

$$Nu_x = \left(\frac{x}{\pi L}\right)^{\frac{1}{2}} Re_L^{\frac{1}{n+1}} Pr^{\frac{1}{2}}, \qquad (4.5\text{-}28)$$

where x^* is defined by

$$\left(\frac{x^*}{L}\right)^{\frac{n-1}{6(n+1)}} = 1.316 Pr^{\frac{1}{6}} \left[\frac{2n+2}{2n+1}\left(\frac{1}{c(n)}\right)^{\frac{1}{n}}\right]^{\frac{1}{3}}, \qquad (4.5\text{-}29)$$

and the Prandtl number is expressed by

$$Pr = \frac{\hat{C}_p V_0 \rho L}{k(Re_L)^{\frac{2}{1+n}}}. \qquad (4.5\text{-}30)$$

The numerical value of $c(n)$ is shown in Fig. 4.5-2.

For shear-thinning fluids, the local Nusselt number is given by equation 4.5-27 for $x > x^*$, and by equation 4.5-28 for $0 \leq x \leq x^*$. On the other hand, for $n > 1$, we must use equation 4.5-27 for $0 \leq x \leq x^*$ and equation 4.5-28 for $x \geq x^*$. We can thus define the mean heat transfer coefficient, h_a, as

$$h_a = \frac{1}{L}\left\{\int_0^{x^*} h_x dx + \int_{x^*}^L h_x dx\right\}, \qquad (4.5\text{-}31)$$

where h_x is evaluated using either equation 4.5-27 or 4.5-28 depending upon the values of n and x^*. For instance, for $n < 1$, we obtain

$$Nu_a = \frac{h_a L}{k} = \left\{\frac{2}{\sqrt{\pi}}\left(\frac{x^*}{L}\right)^{\frac{1}{2}} Pr^{\frac{1}{6}} + \frac{3(n+1)}{0.893(2n+1)}\left[\frac{C(n)^{\frac{1}{n}}}{9}\left(\frac{2n+1}{2n+2}\right)\right]^{\frac{1}{3}}\left[1 - \left(\frac{x^*}{L}\right)^{\frac{2n+1}{3(n+1)}}\right]\right\} Re_L^{\frac{1}{n+1}} Pr^{\frac{1}{3}}.$$

$$(4.5\text{-}32)$$

Similar analyses to predict the local rate of heat transfer from a cylinder and a sphere at constant wall temperature and wall flux conditions have been carried out by Shah et al. (1962). Boundary layer concepts have also been applied to investigate the laminar natural convective heat transfer to power-law fluids from isothermal surfaces of various geometries (Acrivos, 1960; Stewart, 1971). The preliminary comparisons between predictions and experimental results on free convective heat and mass transfer for isothermal spheres and plates are acceptable (Amato and Tien, 1976; Liew and Adelman, 1975; Lee and Donatelli, 1989).

The combined free and forced thermal convection from vertical and horizontal plates to power-law fluids has been studied by Lin and Shih (1980) and Gorla (1986). The influence of temperature-dependent physical properties on heat transfer from Nylon-6 melts to spheres has been studied by Westerberg and Finlayson (1990). The extensive literature on convective heat and mass transport from immersed bodies to rheologically complex media has been reviewed recently by Chhabra (1993b) and Ghosh et al. (1994).

152   Transport Phenomena in Simple Flows

4.6   Problems*

4.6-1$_a$ Pressure Drop in a Tube

A non-Newtonian solution of aluminium laurate is to be pumped through a tube that is 20 mm in diameter and 100 m long. For flow rates of 1, 10, and 100 L/h, obtain the pressure drop assuming

(a) power-law behavior, and
(b) Ellis model behavior.

Compare the results and briefly discuss the differences.
    The fluid properties at 25 °C are

$$\rho = 940 \text{ kg/m}^3,$$
$$m = 62 \text{ Pa} \cdot \text{s}^n,$$
$$n = 0.228,$$
$$\eta_0 = 89.8 \text{ Pa} \cdot \text{s},$$
$$\alpha = 5.08,$$
$$\text{and} \quad \sigma_{1/2} = 63.8 \text{ Pa}.$$

4.6-2$_a$ Generalized Reynolds Number for Poiseuille Flow

The friction factor for Poiseuille flow is given by equation 1.2-1

$$f = \frac{1}{4}\left(\frac{D}{L}\right)\frac{\Delta P}{\frac{1}{2}\rho\langle V\rangle^2}. \tag{4.6-1}$$

Show that for the laminar flow of a power-law fluid, this definition leads to

$$f = \frac{16}{Re_g}, \tag{4.6-2}$$

where the generalized Reynolds number is defined by

$$Re_g = \frac{D^n \langle V\rangle^{2-n}\rho}{8^{n-1} m \left(\frac{3n+1}{4n}\right)^n}. \tag{4.6-3}$$

m and n are the power-law parameters. Consider the flow to be laminar if

$$Re_g < 2100. \tag{4.6-4}$$

* Problems are labelled by a subscript a or b indicating the level of difficulty.

4.6-3ₐ Flow Characteristics of a Suspension

The following viscosity data at 25 °C have been determined for a mineral suspension ($\rho = 1500$ kg/m^3).
Using these data and an appropriate viscosity model:

(a) Determine the pressure drop in a circular conduit 50 m long and 50 mm in diameter for an average velocity of 0.46 m/s.
(b) What would be the effect of doubling the average velocity?
(c) Verify if the flow is laminar in all cases and briefly discuss the validity of the viscosity model used.

Table 4.6-1 Rheological Data of a Mineral Suspension

$\dot{\gamma}$ s^{-1}	σ_{21} Pa
5	95.7
20	133.5
50	166.5
100	192.2
500	289.5
1000	403.4

4.6-4ᵦ Generalized Non-Newtonian Poiseuille Flow

A GNF model may in general be written in the following form for tubular flow:

$$\sigma_{rz} = -\eta_0 f(\sigma_{rz}) \left(\frac{dV_z}{dr}\right). \qquad (4.6\text{-}5)$$

(a) Show that the mass flow rate for the fully-developed steady-state flow in a tube can be expressed as

$$w = \frac{\rho \pi R^3}{\eta_0 \sigma_R^3} \int_0^{\sigma_R} \frac{\sigma_{rz}^3}{f(\sigma_{rz})} d\sigma_{rz}. \qquad (4.6\text{-}6)$$

(b) Using this result, obtain the mass flow rate expression for an Ellis fluid.

4.6-5ᵦ Tolerance in Machining an Extrusion Die

An extrusion die for polyethylene consists of ten slit orifices of cross section as shown in Fig. 4.6-1.
The relevant properties for the polyethylene at 200 °C are

$$\rho = 925 \text{ kg/m}^3,$$
$$m = 4.10 \text{ kPa} \cdot \text{s}^n,$$
$$\text{and} \quad n = 0.50.$$

Neglecting entrance and lateral wall effects, calculate the construction tolerances (ΔH) on H so that the flow rate will be within 5% of the design flow rate of 3.0 kg/h per orifice.

Answer: $\Delta H = \pm 3.125 \ \mu m$

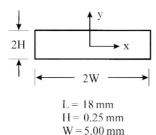

L = 18 mm
H = 0.25 mm
W = 5.00 mm

Figure 4.6-1 Slit die cross-section

4.6-6ᵦ Wire Coating

A metallic wire moves at velocity V through an extrusion die as illustrated in Fig. 4.6-2. As a first approximation, we may assume that the pressure inside the die, P_0, is equal to the atmospheric pressure; the only resistance to flow is the small orifice of radius R and length L; and entrance effects are negligible.

(a) Show that the z-component of the equation of motion reduces to

$$\frac{d}{dr}(r\sigma_{rz}) = 0. \tag{4.6-7}$$

(b) Integrate this equation to obtain the velocity profile for a power-law fluid:

$$\frac{V_z(r)}{V} = \frac{\xi^{1-\frac{1}{n}} - 1}{k^{1-\frac{1}{n}} - 1}, \tag{4.6-8}$$

where $\xi = r/R$.

(c) Obtain the corresponding expression for the flow rate, and the expression for the coating thickness, δ, on the wire. (Assume ρ to be constant.)
(d) Obtain the expression for the axial force exerted on the wire.

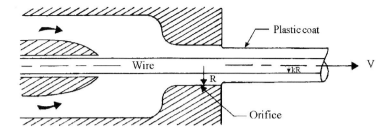

Figure 4.6-2 Wire coating die

(e) Calculate the values of the force and the coating thickness for the following conditions:

$$\begin{aligned}
\text{wire speed} &: 10 \text{ m/s}, \\
\text{wire diameter} &: 0.5 \text{ mm}, \\
\text{orifice diameter} &: 1.1 \text{ mm}, \\
\text{orifice length} &: 10 \text{ mm}, \\
\text{power-law parameters} &: n = 0.3, \\
m &= 1 \text{ kPa} \cdot \text{s}^{0.3}, \\
\text{and} \quad \rho_{\text{(polymer)}} &: 890 \text{ kg/m}^3.
\end{aligned}$$

(f) Rework this problem (very difficult) and make all calculations for a die pressure equal to 1 MPa, assuming $L/R = 20$.

4.6-7$_b$ Axial Flow Between Two Concentric Cylinders

Consider the flow geometry of Section 4.2-5 for the case of stationary cylinders.

(a) Show that the velocity profile is given by

$$V_z(r) = R\left[\frac{(P_0 - P_L)}{2mL} R\right]^{\frac{1}{n}} \int_K^{\frac{r}{R}} \left(\frac{B^2}{\xi} - \xi\right)^{\frac{1}{n}} d\xi, \quad K \le \xi \le B, \qquad (4.6\text{-}9)$$

$$V_z(r) = R\left[\frac{(P_0 - P_L)}{2mL} R\right]^{\frac{1}{n}} \int_{\frac{r}{R}}^{1} \left(\xi - \frac{B^2}{\xi}\right)^{\frac{1}{n}} d\xi, \quad B \le \xi \le 1, \qquad (4.6\text{-}10)$$

where B is the dimensionless radius at which the velocity profile goes through a maximum. It can be obtained from equating the velocity in expressions 4.6-9 and 4.6-10, that is,

$$\int_K^B \left(\frac{B^2}{\xi} - \xi\right)^{\frac{1}{n}} d\xi = \int_B^1 \left(\xi - \frac{B^2}{\xi}\right)^{1/n} d\xi. \qquad (4.6\text{-}11)$$

Values for B as a function of n have been calculated by Hanks and Larsen (1979). Also see Bird et al. (1987).

(b) Obtain the expression for the flow rate in terms of B.
(c) Rework the problem for the Ellis model.

4.6-8$_b$ Generalized Couette Flow

In the generalized plane Couette flow, a pressure gradient is superimposed on the drag action of the plate displaced at constant velocity V. The flow is illustrated in Fig. 4.6-3, for a positive pressure gradient. This flow situation is of practical importance in extrusion, gear pumping, and lubrication.

In the most general case, the velocity gradient can change sign (negative velocity region) across the gap. We will restrict the problem to the case where the velocity gradient does not change sign.

(a) For a power-law fluid, show that the steady-state velocity profile is given by

$$\frac{V_x}{V} = \frac{(C_1 - \xi)^{1+\frac{1}{n}} - (C_1 - 1)^{1+\frac{1}{n}}}{C_1^{1+\frac{1}{n}} - (C_1 - 1)^{1+\frac{1}{n}}}, \qquad (4.6\text{-}12)$$

where $\xi = y/H$ and C_1 is determined from

$$V = \left[\frac{(P_L - P_0)}{mL} H\right]^{\frac{1}{n}} \left(\frac{nH}{n+1}\right) [C_1^{1+\frac{1}{n}} - (C_1 - 1)^{1+\frac{1}{n}}]. \qquad (4.6\text{-}13)$$

(b) Obtain the flow rate expression

$$\frac{Q}{WHV} = \frac{-(C_1 - 1)^{1+\frac{1}{n}} + \dfrac{n}{2n+1}(C_1^{2+\frac{1}{n}} - (C_1 - 1)^{2+\frac{1}{n}})}{C_1^{1+\frac{1}{n}} - (C_1 - 1)^{1+\frac{1}{n}}}, \qquad (4.6\text{-}14)$$

where W is the width of the plate.

(c) Rework the problem for the cases where the shear rate changes sign across the fluid gap.

This problem has been analyzed by Flumerfelt et al. (1969); see also Bird et al. (1987).

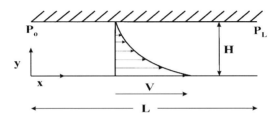

Figure 4.6-3 Generalized plane Couette flow

4.6-9$_b$ Velocity Controller

A velocity controller consists of a disk rotating in a cylindrical box containing a non-Newtonian fluid, as illustrated in Fig. 4.6-4.

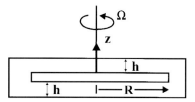

Figure 4.6-4 Speed controller

The disk is concentric with respect to the sides of the box and the gap h is very small, so that the velocity profile may be assumed to be linear with respect to the axial distance, z, from the disk.

(a) Obtain the expression for the torque exerted on the disk as a function of the angular velocity Ω for a power-law fluid.
(b) For the following conditions:

$$h = 2.5 \text{ mm},$$
$$R = 120 \text{ mm},$$
$$m = 60 \text{ Pa} \cdot \text{s}^n,$$
$$\text{and} \quad n = 0.80,$$

obtain the value of the torque (N·m) for $\Omega = 2$ rad/s.

4.6-10$_b$ Drainage of a Power-Law Fluid

A power-law fluid is being drained from a large reservoir, and we want to determine how much liquid sticks to the inner wall. The situation is illustrated in Fig. 4.6-5. This problem was originally analyzed by Fredrickson and Bird (1958), and has been extended by Tallmadge (1971) and by De Kee et al. (1988).

(a) For an incompressible fluid, show from a mass balance that

$$\frac{\partial}{\partial z}(\langle V \rangle \delta) = -\frac{\partial \delta}{\partial t}. \tag{4.6-15}$$

(b) Using the result for the steady-state flow of a power-law fluid on an inclined plane, show that equation 4.6-15 can be written as

$$\frac{\partial \delta}{\partial t} + \left[\frac{\rho g}{m}\right]^{\frac{1}{n}} \delta^{\frac{1}{n}+1} \frac{\partial \delta}{\partial z} = 0. \tag{4.6-16}$$

Indicate for which conditions this result is valid.

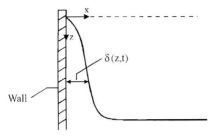

Figure 4.6-5 Drainage of a fluid on a vertical wall

(c) Obtain a solution of the form

$$\delta = \phi(\zeta),$$
$$\text{where} \quad \zeta = \frac{az^b}{t^c}. \tag{4.6-17}$$

(d) Show that in the limit $n=1$, the solution reduces to

$$\delta = \sqrt{\frac{\eta}{\rho g}\frac{z}{t}}. \tag{4.6-18}$$

4.6-11$_b$ Heat Transfer by Convection in a Slit

Consider a slit die of rectangular cross section as illustrated in Fig. 4.6-6.

(a) For the case of $W \gg H$, rework the corresponding problem of Section 4.3-1 and show that the local Nusselt number for short contact times (large Graetz number) in the case of a power-law fluid are expressed as

Isothermal walls:

$$Nu_{loc} = \frac{4hH}{k} = \frac{4}{9^{\frac{1}{3}}\Gamma(\frac{4}{3})}\left[\frac{H^2 \langle V \rangle \left(2+\frac{1}{n}\right)}{\alpha z}\right]^{\frac{1}{3}}. \tag{4.6-19}$$

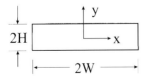

Figure 4.6-6 Rectangular slit

Constant wall heat flux:

$$Nu_{loc} = \frac{4\Gamma(\frac{2}{3})}{9^{\frac{1}{3}}} \left[\frac{H^2 \langle V \rangle \left(2 + \frac{1}{n}\right)}{\alpha z} \right]^{\frac{1}{3}} \qquad (4.6\text{-}20)$$

(b) For a slit length L, obtain the corresponding average Nusselt numbers and compare the results with the predictions of equations 4.3-33 and 4.3-35.

4.6-12$_b$ Heat Transfer to a Falling Film

A liquid film of thickness δ is flowing along a vertical wall as illustrated in Fig. 4.6-7. The fluid is initially at T_0 and the wall is maintained at T_w. Assuming that the velocity profile is fully developed, obtain the expression for the heat transfer rate from the wall to the liquid film for short contact times.

(a) In the case of a power-law fluid, show that the energy equation reduces to

$$\frac{\partial T}{\partial z} \left(\frac{\rho g \delta}{m}\right)^{\frac{1}{n}} y \approx \alpha \frac{\partial^2 T}{\partial y^2}, \qquad (4.6\text{-}21)$$

where α is the thermal diffusivity.

(b) Defining the dimensionless variables

$$\theta = \frac{T - T_0}{T_w - T_0} \qquad (4.6\text{-}22)$$

and

$$\eta = \frac{y}{(9Bz)^{\frac{1}{3}}}, \qquad (4.6\text{-}23)$$

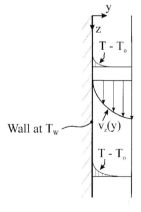

Figure 4.6-7 Heat transfer to a falling liquid film

where

$$B = \alpha\left(\frac{m}{\rho g \delta}\right)^{\frac{1}{n}}, \quad (4.6\text{-}24)$$

show that the reduced temperature profile is given by

$$\theta = \frac{1}{\Gamma(\frac{4}{3})}\int_\eta^\infty e^{-\eta^3}d\eta. \quad (4.6\text{-}25)$$

(c) Obtain the corresponding expression for the average Nusselt number.

4.6-13$_b$ Mass Transfer to a Falling Film

Consider the mass transfer problem discussed in Section 4.4-1.

(a) Solve the differential equation 4.4-7 by combining variables along the technique suggested in Section 4.4-2. Express the wall shear rate in terms of the average fluid velocity and the power-law index n. Obtain an expression for the mass transfer coefficient and show that this is equivalent to the result given by equation 4.4-14.

(b) Express the previous result in terms of the average Sherwood number as a function of the Graetz number. Compare this result with the analogous heat transfer result of Problem 4.6-12.

4.6-14$_b$ Heat and Mass Transfer in Boundary Layers

(a) Use the integral heat balance together with the results derived for the boundary layer thickness to show that the local Nusselt number for the laminar boundary layer flow over a flat plate maintained at constant temperature is given by

$$Nu_x = \frac{h_{eoc}x}{k} = \frac{3}{2}\left[\frac{30c(n)(n+1)}{(2n+1)}\right]^{-\frac{1}{3}} Pr_x^{\frac{1}{3}} Re_x^{\frac{n+2}{3(n+1)}}, \quad (4.6\text{-}26)$$

where

$$Re_x = \frac{\rho V_0^{2-n} x^n}{m}, \quad (4.6\text{-}27)$$

and

$$Pr_x = \frac{\hat{C}_p}{k} m \left(\frac{V_0}{x}\right)^{n-1}, \quad (4.6\text{-}28)$$

where $c(n)$ is the function given by equation 4.5-17. Assume that the temperature can be approximated by a third order polynomial in y. Compare these predictions with the numerical results of Acrivos et al. (1960) presented in Fig. 4.5-2.

(b) Show that these results also apply to mass transfer in laminar boundary layer flow of power-law fluids over a plate by establishing the following equivalence:

$$Nu_x \Rightarrow Sh_x = \frac{k_{loc} x}{D_A} \tag{4.6-29}$$

and

$$Pr_x \Rightarrow Sc_x = \frac{m}{\rho D_A} \left(\frac{V_0}{x}\right)^{n-1}. \tag{4.6-30}$$

where $\Rightarrow$ means : is replaced by.

(c) Verify whether the Chilton-Colburn analogy between heat, mass, and momentum transfer is obeyed.

5 Linear Viscoelasticity

5.1 Importance and Definitions 162
5.2 Linear Viscoelastic Models 163
    5.2-1 The Maxwell Model 164
    5.2-2 Generalized Maxwell Model 170
        5.2-2.1 Unspecified Forms 173
    5.2-3 The Jeffreys Model 178
    5.2-4 The Voigt-Kelvin Model 180
    5.2-5 Other Linear Models 182
5.3 Relaxation Spectrum 184
5.4 Time-Temperature Superposition 186
5.5 Problems 189
    5.5-1_a Rheological Model with Friction 189
    5.5-2_a Maxwell Model 189
    5.5-3_a Stress Relaxation for a Maxwell Fluid 189
    5.5-4_b Complex Viscosity of a Generalized Maxwell Fluid 190
    5.5-5_b The Jeffreys Model 190
    5.5-6_b Maxwell and Voigt-Kelvin Elements 191
    5.5-7_a Storage and Loss Moduli of a Voigt-Kelvin Material 191
    5.5-8_b Complex Compliance 192
    5.5-9_b Relaxation Modulus 193

A viscoelastic material was defined in Chapter 1 as a body for which the deformation or the rate of deformation does not respond instantaneously to an applied stress or vice versa. The rheology of a viscoelastic material is intermediate between that of a purely viscous fluid and that of a perfectly elastic solid. We devote this entire chapter to the basic concepts of linear viscoelasticity.

5.1 Importance and Definitions

Most phenomena discussed in Chapter 1 are due to viscoelasticity. Polymeric materials, such as polymer solutions, polymer melts, rubber, and plastics usually possess viscoelastic characteristics. The viscoelastic nature of solid polymers can rarely be ignored. Similarly, the viscoelastic properties of polymeric fluids are of importance in the following three practical situations:

(i) time-dependent flows, for example start-up flows;
(ii) developing (accelerated or decelerated) flows, such as the flow in porous media; and
(iii) complex flow situations, for example 3-D flows encountered in a mechanically stirred polymerization reactor.

Even in these situations the fluid's elasticity can be ignored if the Deborah or Weissenberg number is much smaller than 1 (see Section 4.1).

Linear viscoelasticity is the field of rheology devoted to the study of viscoelastic materials under very small strain or deformation. In principle, the strain has to be small enough so that the structure of the material, for example the configuration of the macromolecules, remains unperturbed by the flow history. In shear experiments, the usual criterion for linearity is

$$\gamma_{21} = \gamma_{yx} \leq 1. \tag{5.1-1}$$

The maximum value allowed for γ_{21} depends on the type of experiment and on the material investigated. In some cases, a strain of 0.5 is acceptable, whereas in other situations, the strain must be under 0.1 or even 0.01 to obtain linear behavior. It is believed that the magnitude of the rate of strain is irrelevant provided the strain remains very small. This is still an open question and more research is needed to determine the effect of the rate of strain on linear viscoelasticity.

The goals of the study of linear viscoelasticity are twofold. First, we must elucidate the equilibrium structure of polymeric materials and determine its main characteristics in terms of molecular weight and molecular weight distribution, chain flexibility, formulation, etc. Second, we need to determine qualitative effects of elasticity and assess their impact on flow behavior. For example, the complex viscosity data can be used to obtain an elastic characteristic time defined by

$$\lambda = \frac{G'}{G''\omega} = \frac{\eta''}{\eta'\omega}, \tag{5.1-2}$$

where G' and G'' are the storage and loss modulus respectivley. η' and η'' are the dynamic viscosity and dynamic rigidity respectively, as defined in Section 2.1-2. This elastic time constant can be used to compute the Deborah or the Weissenberg number and to qualitatively assess the effects of elasticity in a given process. It is recommended to use the frequency (rad/s) corresponding to the characteristic shear rate of the actual process. However, there is no guarantee that this empirical method, based on the Cox-Merz analogy, will work in all cases. In many complex flow situations, the deformation is highly extensional (for example the entry flow to an extrusion die), and the shear rate may not be an appropriate parameter or characteristic of the flow process.

5.2 Linear Viscoelastic Models

Linear viscoelastic models are useful tools for analyzing experimental data obtained in small deformation experiments, and for interpreting important viscoelastic phenomena, at least qualitatively. In this section we will restrict ourselves to the well-known linear models, and we will show the important features of these models for characterization purposes.

5.2-1 The Maxwell Model

The simplest and best known Maxwell model is obtained by a series arrangement of a purely viscous element and a perfectly elastic body. We can derive the resulting differential equation by solving the force and displacement response of the mechanical element illustrated in Fig. 5.2-1. In this dashpot and spring element, we will neglect inertia and assume that the velocity of the piston in the dashpot is proportional to the applied force, that is

$$F_1 = K_1 V_1, \tag{5.2-1}$$

whereas for the spring, the displacement is proportional to the applied force:

$$F_2 = K_2 \Delta x. \tag{5.2-2}$$

Because the two components are in series, the force on both components is the same, but the velocity of the entire element is the sum of the piston velocity and that of the spring, $V_2 = d(\Delta x)/dt$. The response of this mechanical Maxwell element is then given by the following simple differential equation ($F_1 = F_2 = F$):

$$V = V_1 + V_2 = \frac{F}{K_1} + \frac{1}{K_2}\frac{dF}{dt}, \tag{5.2-3a}$$

which can be written as

$$F + \frac{K_1}{K_2}\frac{dF}{dt} = K_1 V. \tag{5.2-3b}$$

For a fluid element under shear flow, we must replace the force by the shear stress, and the velocity by the shear component of the rate-of-strain tensor (or simply the shear rate). The resulting equation is the simple Maxwell model, which we will call *Maxwell model-A*, given by

$$\sigma_{yx} + \lambda_0 \frac{\partial}{\partial t}\sigma_{yx} = -\eta_0 \dot{\gamma}_{yx}. \tag{5.2-4}$$

The model contains two parameters:

η_0, the constant viscosity term corresponding to K_1, with units of Pa·s; and
$\lambda_0 = \eta_0/G_0$, the characteristic time, which is the ratio of the viscosity and the elastic modulus G_0, corresponding to K_2, with units of s.

In equation 5.2-4, we have used the partial derivative with respect to t, since in general the stress and the shear rate are functions of the spatial position. For homogeneous flows,

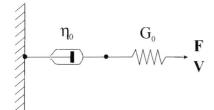

Figure 5.2-1 Simple mechanical Maxwell element

σ_{yx} is a unique function of time and the partial time derivative becomes the total time derivative.

Example 5.2-1 Integrated Forms for the Maxwell Model

The Maxwell model-A is a differential type model. Obtain the corresponding integral models based on the rate-of-deformation and deformation history.

Solution

Equation 5.2-4 is an ordinary linear differential equation of first order which can be integrated using the integration factor e^{-t/λ_0} to obtain

$$\sigma_{yx} = -e^{-\frac{t}{\lambda_0}} \left(\int_{-\infty}^{t} \frac{\eta_0}{\lambda_0} e^{\frac{t}{\lambda_0}} \dot{\gamma}_{yx}(t) dt + C_1 \right). \tag{5.2-5}$$

The shear stress is assumed to remain finite for $t \to -\infty$, hence the constant C_1 has to be equal to zero. Replacing the dummy variable t in the integral by t', which physically refers to the time in the past (t is the current time), we obtain the *Maxwell model-B*:

$$\sigma_{yx} = -\int_{-\infty}^{t} \frac{\eta_0}{\lambda_0} e^{-\frac{(t-t')}{\lambda_0}} \dot{\gamma}_{yx}(t') dt'. \tag{5.2-6}$$

This integral equation expresses the shear stress as a function of the history of the rate of deformation. The exponential term is a factor which gives more weight to recent events (t' close to t) as opposed to rates of strain imposed in a more distant past. This relates to the notion that viscoelastic fluids have a "fading memory," since they are more affected by recent flow kinematics.

An integral model based on the deformation history is obtained by integrating equation 5.2-6 by parts, noting that $\dot{\gamma}_{yx} dt' = d\gamma_{yx}$. The result is

$$\sigma_{yx} = \left[\frac{\eta_0}{\lambda_0^2} e^{-\frac{(t-t')}{\lambda_0}} \gamma_{yx} \right]_{-\infty}^{t} + \int_{-\infty}^{t} \frac{\eta_0}{\lambda_0^2} e^{-\frac{(t-t')}{\lambda_0}} \gamma_{yx}(t') dt'. \tag{5.2-7a}$$

As it is a relative measure, the deformation is arbitrarily set equal to zero at time t, and the first term on the right is equal to zero. The *Maxwell model-C* is then given by

$$\sigma_{yx} = +\int_{-\infty}^{t} \frac{\eta_0}{\lambda_0^2} e^{-\frac{(t-t')}{\lambda_0}} \gamma_{yx}(t') dt'. \tag{5.2-7b}$$

In this integral model the stress is expressed as a function of the strain (or deformation) history. Here, the weight factor in the integral is called a memory function $m(t - t')$. That is,

$$m(t - t') = \frac{\eta_0}{\lambda_0^2} e^{-\frac{(t-t')}{\lambda_0}}. \tag{5.2-8}$$

Obviously, these three results (models A, B, and C) are equivalent, but for solving a given problem, one form may be more appropriate than the other two. Also, generalizations to obtain nonlinear viscoelastic models will lead to different results depending on which equation is used as the starting point.

166   Linear Viscoelasticity

Example 5.2-2   Complex Viscosity of a Maxwell Fluid

Obtain the expressions for the complex viscosity, the dynamic viscosity, and the dynamic rigidity for a Maxwell fluid.

Solution

The material functions for small amplitude oscillatory shear flow are defined in Section 2.1-2. The imposed oscillatory shear rate is given by equation 2.1-10.

In the case of linear viscoelastic behavior, the stress response has to be of the same form as the imposed shear rate. That is, the frequency is the same but the stress signal is in general out of phase. Using complex number notation, we write

$$\sigma_{yx} = \text{Re}[\sigma^0 e^{i\omega t}], \tag{5.2-9}$$

where σ^0 is the complex amplitude of the shear stress response, and Re refers to the real part.

The shear stress response for a Maxwell fluid under oscillatory shear flow is given by (using equations 5.2-4 and 2.1-10)

$$\text{Re}[\sigma^0 e^{i\omega t}] + \lambda_0 \text{Re}[i\omega \sigma^0 e^{i\omega t}] = -\eta_0 \text{Re}[\dot{\gamma}^0 e^{i\omega t}]. \tag{5.2-10}$$

Simplification yields

$$\sigma^0 = -\frac{\eta_0}{1 + i\omega\lambda_0}\dot{\gamma}^0. \tag{5.2-11}$$

From the definition of the complex viscosity (equation 2.1-9), we can write

$$\sigma^0 = -\eta^* \dot{\gamma}^0 = -\eta' \dot{\gamma}^0 + i\eta'' \dot{\gamma}^0. \tag{5.2-12}$$

We rewrite equation 5.2-11 as

$$\sigma^0 = -\frac{\eta_0(1 - i\lambda_0\omega)\dot{\gamma}^0}{(1 + i\lambda_0\omega)(1 - i\lambda_0\omega)}$$

$$= -\frac{\eta_0\dot{\gamma}_0}{1 + \lambda_0^2\omega^2} + i\frac{\eta_0\lambda_0\omega\dot{\gamma}_0}{1 + \lambda_0^2\omega^2}. \tag{5.2-13}$$

Via equation 5.2-12, we obtain the following results:

$$\frac{\eta'}{\eta_0} = \frac{(G''/\omega)}{\eta_0} = \frac{1}{1 + \lambda_0^2\omega^2}, \tag{5.2-14}$$

and

$$\frac{\eta''}{\eta_0} = \frac{(G'/\omega)}{\eta_0} = \frac{\lambda_0\omega}{1 + \lambda_0^2\omega^2}. \tag{5.2-15}$$

As mentioned in Chapter 2, η' is the dynamic viscosity (viscous response) and η'' is the dynamic rigidity; and $G' = \eta''\omega$ is the storage modulus, associated with the elastic response of the material. A clearer physical interpretation is obtained by writing the result in terms of the shear stress response. Starting with equation 5.2-9, noting that

$$e^{i\omega t} = \cos\omega t + i\sin\omega t, \tag{5.2-16}$$

and using equation 5.2-13, we obtain

$$\sigma_{yx} = -\frac{\eta_0 \dot{\gamma}^0}{1 + \lambda_0^2 \omega^2}(\cos \omega t + \lambda_0 \omega \sin \omega t), \qquad (5.2\text{-}17a)$$

or

$$\sigma_{yx} = -\eta' \dot{\gamma}^0 \cos \omega t - \eta'' \dot{\gamma}^0 \sin \omega t. \qquad (5.2\text{-}17b)$$

The stress response is compared to the imposed oscillatory shear rate in Fig. 5.2-2.

The stress signal is shown to lag behind the imposed shear rate by a time (Φ/ω). The second term on the right side of equation 5.2-17a or b is clearly responsible for the lag. For a purely viscous fluid, $\lambda_0 = 0$ and the stress response is in phase with the shear rate. Note that due to the sign convention adopted for the stress tensor, the stress response is always negative compared to the applied shear rate. The phase shift is usually reported in terms of δ, which is the phase shift angle with respect to the response of a perfectly elastic solid. The stress response of a perfectly elastic solid ($\eta' = 0$) would be 90° out of phase with the imposed shear rate, that is, in phase with the imposed shear ($\gamma_{yx} = \gamma^\circ \sin \omega t = \omega \dot{\gamma}^\circ \sin \omega t$).

The *loss tangent* is defined by

$$\tan \delta = \frac{G''}{G'}, \qquad (5.2\text{-}18)$$

which is calculated for a Maxwell fluid using equations 5.2-14 and 5.2-15 to be

$$\tan \delta = \frac{1}{\lambda_0 \omega}. \qquad (5.2\text{-}19)$$

Note that the inverse of $\tan \delta$ can be taken as a Weissenberg number to assess the effect of elasticity, as discussed in Section 4.1.

The results for the Maxwell model in terms of η' and G' are compared in Fig. 5.2-3 with experimental data for a 4% polystyrene solution in Aroclor. The predictions for a single Maxwell element (dashed lines in the figure) are clearly not satisfactory. In the next section, we propose a generalization that is shown to be more adequate (solid lines in the figure).

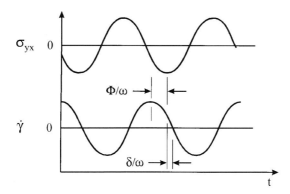

Figure 5.2-2 Imposed shear rate and stress response in oscillatory shear flow

168   Linear Viscoelasticity

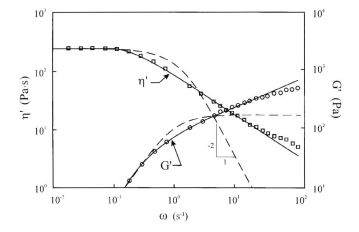

Figure 5.2-3 Complex viscosity data for a 4 mass % polystyrene solution in Aroclor (Data from Ashare, 1968)
– – – single Maxwell element
——— generalized Maxwell model

Example 5.2-3   Recoil of a Maxwell Fluid

Recoil in simple shear is defined in Fig. 5.2-4. For time t prior to t_1, we consider a steady shear flow, with a linear velocity profile and a constant applied shear stress σ_0. At $t = t_1$, the stress is suddenly set equal to zero. We wish to determine the deformation as a function of time for a Maxwell fluid. This experiment is carried out using a constant stress rheometer.

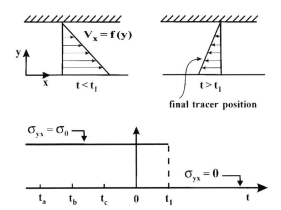

Figure 5.2-4 Recoil in shear flow

Solution

Since the applied stress is described by a step function, it is easier to solve the problem using Laplace transforms. The Laplace transform of the shear stress is defined by

$$\overline{\sigma}_{yx}(s) = \int_0^\infty e^{-st}\sigma_{yx}\,dt. \tag{5.2-20}$$

For the situation considered in Fig. 5.2-4, we obtain

$$\overline{\sigma}_{yx}(s) = \int_0^{t_1} \sigma_0 e^{-st}\,dt$$

$$= \sigma_0\left[\frac{1}{s} - \frac{1}{s}e^{-st_1}\right]. \tag{5.2-21}$$

The Laplace transform for Maxwell model-A is

$$\overline{\sigma}_{yx} + \lambda_0 s\overline{\sigma}_{yx} - \lambda_0\sigma_0 = -\eta_0[s\overline{\gamma}_{yx} - \gamma_{yx}(0)], \tag{5.2-22}$$

where $\gamma_{yx}(0)$ is the deformation at time zero. This result combined with equation 5.2-21 yields

$$\overline{\gamma}_{yx}(s) = \frac{\gamma_{yx}(0)}{s} - \frac{\sigma_0}{\eta_0 s^2}(1 - e^{-st_1}) + \left(\frac{\lambda_0\sigma_0}{\eta_0 s}\right)e^{-st_1}. \tag{5.2-23}$$

The inverse functions are

$$\frac{1}{s} \to 1,$$

$$\frac{1}{s^2} \to t,$$

$$\frac{e^{-st_1}}{s^2} \to \begin{cases} 0, & 0 < t < t_1, \\ t - t_1, & t > t_1, \end{cases} \tag{5.2-24}$$

and

$$\frac{e^{-st_1}}{s} \to \begin{cases} 0, & 0 < t < t_1 \\ 1, & t > t_1. \end{cases}$$

Hence, the deformation is given by

$$\gamma_{yx}(t) = \gamma_{yx}(0) - \left(\frac{\sigma_0}{\eta_0}\right)\left(t - \begin{cases} 0, & 0 < t < t_1 \\ t - t_1, & t > t_1 \end{cases}\right) + \left(\frac{\lambda_0\sigma_0}{\eta_0}\right)\left(\begin{cases} 0, & 0 < t < t_1 \\ 1, & t > t_1 \end{cases}\right). \tag{5.2-25}$$

This result can be simply expressed by the following two equations:

$$\gamma_{yx}(t) = \gamma_{yx}(0) - \frac{\sigma_0}{\eta_0}t, \qquad t < t_1, \tag{5.2-26a}$$

and $\quad\gamma_{yx}(t) = \gamma_{yx}(0) - \dfrac{\sigma_0 t_1}{\eta_0} + \dfrac{\lambda_0\sigma_0}{\eta_0}, \qquad t > t_1. \tag{5.2-26b}$

For $t < t_1$, the deformation is a linearly decreasing function of time. Figure 5.2-5 illustrates this behavior. A tracer mark in the fluid is deformed as shown in the upper part of the figure; the corresponding deformation follows a straight line of negative slope presented in the bottom part of the figure. At $t = t_1$, the tracer snaps back to the vertical position and the

170   Linear Viscoelasticity

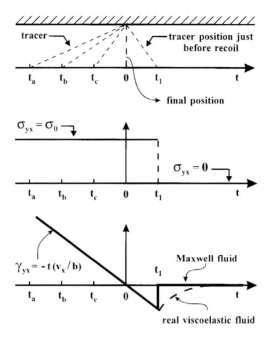

Figure 5.2-5 Deformation and recovery for a Maxwell fluid

deformation becomes equal to zero. Note that the deformation has been arbitrarily set to zero at $t=0$, and t_1 has been chosen so that the deformation, which is constant for $t > t_1$, is also equal to zero. Recoil is defined by

$$\gamma_R = \gamma_{yx}(t_1^+) - \gamma_{yx}(t_1^-). \tag{5.2-27a}$$

Subtracting equation 5.2-26a from equation 5.2-26b we obtain

$$\gamma_R = \frac{\lambda_0 \sigma_0}{\eta_0}. \tag{5.2-27b}$$

This is the expression for the recovery or recoil of a Maxwell fluid. Note that the strain is recovered instantaneously. This is never observed in practice. Real viscoelastic fluids, as in the cases depicted in Figs. 1.2-6 and 1.2-7, will recoil over a finite time. Nevertheless, the simple Maxwell model is capable of qualitatively explaining this very important viscoelastic phenomenon.

5.2-2   Generalized Maxwell Model

The structure of macromolecules in a melt or in solution is more complex than that described by a simple spring and dashpot element. Indeed, we do not expect a polymeric

5.2 Linear Viscoelastic Models

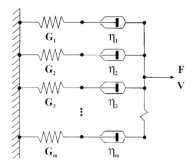

Figure 5.2-6 Parallel arrangement of Maxwell elements

system of a broad molecular weight distribution to be characterized in terms of a single relaxation time λ_0. Various generalized forms of the single Maxwell element have been proposed to more adequately describe the linear properties of polymers (at least from an engineering point of view).

The first generalization is based on the idea that the polymer macromolecules are of different length, and possibly form a network with junctions or interactions of variable resistance. The simplest representation is that of Fig. 5.2-6, which illustrates a number of Maxwell elements placed in parallel. Note that Maxwell elements placed in series cannot be distinguished from a single Maxwell element.

For this parallel arrangement, the overall stress response is the sum of the stress response for each element. That is,

$$\sigma_{yx} = \sum_{p=1}^{M} \sigma_{yx}^{(p)}, \qquad (5.2\text{-}28a)$$

with

$$\sigma_{yx}^{(p)} + \lambda_p \frac{\partial}{\partial t}\sigma_{yx}^{(p)} = -\eta_p \dot{\gamma}_{yx}. \qquad (5.2\text{-}28b)$$

The equivalent integral forms, corresponding to equations 5.2-6 and 5.2-7b, are

$$\sigma_{yx} = -\int_{-\infty}^{t} \left\{\sum_{p=1}^{M} \frac{\eta_p}{\lambda_p} e^{-\frac{(t-t')}{\lambda_p}}\right\} \dot{\gamma}_{yx}(t')dt' \qquad (5.2\text{-}29)$$

and

$$\sigma_{yx} = \int_{-\infty}^{t} \left\{\sum_{p=1}^{M} \frac{\eta_p}{\lambda_p^2} e^{-\frac{(t-t')}{\lambda_p}}\right\} \gamma_{yx}(t')dt'. \qquad (5.2\text{-}30)$$

These equations contain $2p$ parameters (η_p and λ_p ($=G_p/\eta_p$)) to describe the linear properties of a polymer melt or solution. For practical reasons, the number of parameters has to be as small as possible. In some cases, five ($M=5$) elements may be adequate. Another approach is to consider an infinite number of elements and reduce the number of parameters, using the following empirical method. First, we notice that the limiting viscosity at low frequency is the sum of the viscosity of all the elements:

$$\sum_{p=1}^{M} \eta_p = \eta_0. \qquad (5.2\text{-}31)$$

Secondly, inspired by the Rouse (1953) and Zimm (1956) theories (see Chapter 7), we can postulate

$$\eta_p = \frac{\lambda_p}{\sum \lambda_p} \eta_0 \qquad (5.2\text{-}32)$$

and

$$\lambda_p = \frac{\lambda}{p^\alpha}, \qquad (5.2\text{-}33)$$

where λ is the longest relaxation time, and α is a dimensionless parameter. In the Rouse theory, α is equal to 2; it is equal to 1.5 in the Zimm theory. For $M = \infty$, we have an infinite series of η_p and λ_p, but defined in terms of only three parameters, η_0, λ, and α.

Example 5.2-4 Complex Viscosity of a Generalized Maxwell Model

The result for the complex viscosity of a generalized Maxwell model is a direct extension of the result obtained in example 5.2-2. The complex amplitude for the stress is given by

$$\sigma^0 = -\sum_{p=1}^{M} \frac{\eta_p}{1 + i\omega\lambda_p} \dot{\gamma}^0, \qquad (5.2\text{-}34)$$

from which we can extract the expressions for η' and η'', which are given by

$$\eta' = \frac{G''}{\omega} = \sum_{p=1}^{M} \frac{\eta_p}{1 + (\lambda_p \omega)^2} \qquad (5.2\text{-}35)$$

and $\quad \eta'' = \frac{G'}{\omega} = \sum_{p=1}^{M} \frac{\lambda_p \eta_p \omega}{1 + (\lambda_p \omega)^2}. \qquad (5.2\text{-}36)$

Using the reduction method proposed by equations 5.2-32 and 5.2-33 with $M = \infty$, we obtain

$$\frac{\eta'}{\eta_0} = \frac{1}{Z(\alpha)} \sum_{p=1}^{\infty} \frac{p^\alpha}{p^{2\alpha} + \lambda^2 \omega^2} \qquad (5.2\text{-}37)$$

and $\quad \dfrac{\eta''}{\eta_0} = \dfrac{\lambda \omega}{Z(\alpha)} \sum_{p=1}^{\infty} \dfrac{1}{p^{2\alpha} + \lambda^2 \omega^2}, \qquad (5.2\text{-}38)$

where $Z(\alpha)$ is the Riemann zeta function defined by

$$Z(\alpha) = \sum_{p=1}^{\infty} \frac{1}{p^\alpha}. \qquad (5.2\text{-}39)$$

Using equations 5.2-37 and 5.2-38, a very reasonable fit for the complex viscosity data of most polymer melts or solutions can be obtained. An example is shown in Fig. 5.2-3 for a polystyrene solution. We reiterate that the main advantage of this empirical method is the small number of parameters:

5.2 Linear Viscoelastic Models

η_0, the zero shear viscosity, Pa·s,
λ, the longest relaxation time, s,
and α, related to the slope of $\log \eta'$ versus $\log \omega$. The slope is $(1 - \alpha)/\alpha$.

5.2-2.1 Unspecified Forms

The generalized Maxwell model can be written in terms of an unspecified memory or relaxation function as

$$\sigma_{yx} = \int_{-\infty}^{t} m(t - t')\gamma_{yx}(t')dt' \qquad (5.2\text{-}40)$$

or

$$\sigma_{yx} = -\int_{-\infty}^{t} G(t - t')\dot{\gamma}_{yx}(t')dt', \qquad (5.2\text{-}41)$$

where $m(t - t')$ and $G(t - t')$ are respectively the memory and the relaxation functions. The use of the term "relaxation function" is clarified in Example 5.2-5.

These equations are written for a shear flow experiment. For any flow situation including the contribution of the solvent to the stress tensor, equation 5.2-41 can be written in a tensorial form as

$$\underline{\underline{\sigma}} - \underline{\underline{\sigma}}_s = -\int_{-\infty}^{t} G(t - t')\underline{\underline{\dot{\gamma}}}(t')dt'. \qquad (5.2\text{-}42)$$

Example 5.2-5 Relaxation Following a Small Deformation
Stress relaxation following a sudden deformation is defined in Section 2.1-3. This example will clarify the definition. We consider a fluid submitted to a constant shear rate for a very short time, t_1, as illustrated in Fig. 5.2-7. For a very small value of t_1, the imposed deformation, $\dot{\gamma}_0 t_1$, is small, so that the experiment is in the linear viscoelastic domain. The behavior of the fluid is described by a generalized Maxwell model expressed by equation 5.2-41.

To solve the problem, we consider three regions: I: for $t < 0$, II: for $0 < t < t_1$, and III: for $t > t_1$. The shear stress responses for the three regions are given by the following equations:

$$\text{Region I,} \quad t < 0: \qquad \sigma_{yx}(t) = -\int_{-\infty}^{0} G(t - t')0\, dt' = 0, \qquad (5.2\text{-}43)$$

$$\text{Region II,} \quad 0 < t < t_1: \qquad \sigma_{yx}(t) = -\int_{0}^{t} G(t - t')\dot{\gamma}_0\, dt', \qquad (5.2\text{-}44)$$

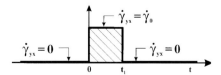

Figure 5.2-7 Small strain experiment

174   Linear Viscoelasticity

and

$$\text{Region III,} \quad t > t_1: \quad \sigma_{yx}(t) = -\int_0^{t_1} G(t-t')\dot{\gamma}_0 dt' - \int_{t_1}^t G(t-t')0 dt'. \tag{5.2-45}$$

For region III, we rewrite the result as

$$\sigma_{yx}(t) = -\frac{\dot{\gamma}_0 t_1}{t_1}\int_0^{t_1} G(t-t')dt' = -\frac{\gamma_0}{t_1}\int_0^{t_1} G(t-t')dt'. \tag{5.2-46}$$

The limit as t_1 tends to zero is

$$\lim_{t_1 \to 0} \sigma_{yx}(t) = -\gamma_0 \frac{\frac{d}{dt_1}\int_0^{t_1} G(t-t')dt'}{\frac{d}{dt_1}(t_1)}$$

$$= -\gamma_0 G(t). \tag{5.2-47}$$

The stress relaxation following a small deformation is then described in terms of the relaxation modulus $G(t)$—hence the term "relaxation function."

Figure 5.2-8 reports the torque measurement for a 6% polyisobutylene solution in Primol 355 corresponding to regions II and III.

For a simple (single element) Maxwell model, the stress response for region III is given by

$$\sigma_{yx}(t) = \gamma_0 \frac{\eta_0}{\lambda_0} e^{-\frac{t}{\lambda_0}} \tag{5.2-48}$$

A few relaxation times are usually required to give a good description of the relaxation function, which in principle contains the same information as the complex modulus G^*. We stress, however, that the relaxation following a small deformation is an experiment that requires a rheometer capable of a very rapid acceleration and deceleration. It also requires a high rigidity of the torque measuring device to reduce coupling effects between the fluid and the instrument (see Section 3.3-4).

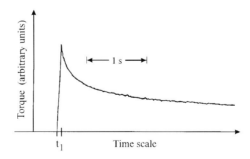

Figure 5.2-8 Typical stress response for the 6% PIB solution: $t_1 = 0.092$ s, $\dot{\gamma}_0 = 4.34$ s^{-1} (Adapted from De Kee and Carreau, 1984)

Example 5.2-6 Creep Experiment

The creep experiment is defined in Fig. 5.2-9. A sudden shear stress is imposed on a fluid initially at rest.

This experiment is carried out using a constant stress rheometer, in which the shear stress is imposed via a specified magnetic field applied to a DC motor. The torque is transmitted to the upper plate (or cone) and the angular rotation of the plate is measured via a transducer.

To obtain the solution, we first note that

$$\gamma_{yx} = \int_0^t \dot{\gamma}_{yx}(t')dt'. \qquad (5.2\text{-}49)$$

Hence,

$$\gamma_0 = \int_0^\infty (\dot{\gamma}_{yx}(t') - \dot{\gamma}_\infty)dt'. \qquad (5.2\text{-}50)$$

Neglecting the contribution of the solvent to the stress and using equation 5.2-41, we obtain

$$\sigma_0 = -\int_0^t G(t-t')\dot{\gamma}_{yx}(t')dt'. \qquad (5.2\text{-}51)$$

For steady-state conditions,

$$\sigma_0 = -\int_0^\infty G(t-t')\dot{\gamma}_\infty dt'. \qquad (5.2\text{-}52)$$

Eliminating σ_0 and using $s = t - t'$, we obtain

$$-\dot{\gamma}_\infty \int_0^\infty G(s)ds = -\int_0^t G(s)\dot{\gamma}_{yx}(t-s)ds, \qquad (5.2\text{-}53)$$

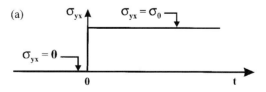

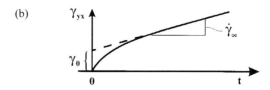

Figure 5.2-9 Creep experiment
(a) imposed shear stress
(b) typical viscoelastic strain response

176 Linear Viscoelasticity

which we can write as

$$\dot{\gamma}_\infty \int_t^\infty G(s)ds = \int_0^t G(s)[\dot{\gamma}_{yx}(t-s) - \dot{\gamma}_\infty]ds. \tag{5.2-54}$$

Integrating and changing the order of integration, we obtain

$$\dot{\gamma}_\infty \int_0^\infty \underbrace{\int_0^s G(s)dt\,ds}_{G(s)s\,ds} = \int_0^\infty \underbrace{\int_s^\infty G(s)[\dot{\gamma}_{yx}(t-s) - \dot{\gamma}_\infty]dt}_{G(s)\gamma_0}\,ds. \tag{5.2-55}$$

Therefore, the result simplifies to

$$\dot{\gamma}_\infty \int_0^\infty G(s)s\,ds = \gamma_0 \int_0^\infty G(s)ds. \tag{5.2-56}$$

Defining the equilibrium compliance by

$$J_e = -\frac{\gamma_0}{\sigma_0}, \tag{5.2-57}$$

we obtain

$$J_e = \frac{1}{\eta_0^2}\int_0^\infty G(s)s\,ds = \frac{\int_0^\infty G(s)s\,ds}{\left[\int_0^\infty G(s)ds\right]^2}, \tag{5.2-58}$$

since

$$\eta_0 = -\frac{\sigma_0}{\dot{\gamma}_\infty} = \int_0^\infty G(s)ds. \tag{5.2-59}$$

For a single Maxwell element, the relaxation function is

$$G(s) = \frac{\eta_0}{\lambda_0}e^{-\frac{s}{\lambda_0}}. \tag{5.2-60}$$

The equilibrium compliance is obtained by solving equation 5.2-58:

$$J_e = \frac{\lambda_0}{\eta_0}. \tag{5.2-61}$$

As illustrated in Fig. 5.2-9, the deformation is instantaneous, with

$$\gamma_0 = -\frac{\lambda_0 \sigma_0}{\eta_0}, \tag{5.2-62}$$

which is the same value as γ_R, the recoil or recoverable strain of a Maxwell fluid obtained in Example 5.2-3 (see eq. 5.2-27b). Finally, the deformation under creep for a Maxwell fluid is given by

$$\gamma_{yx} = \gamma_0 - \frac{\sigma_0 t}{\eta_0}, \tag{5.2-63}$$

which is obtained by solving the problem using the Laplace transform as in Example 5.2-3. The deformation described by equation 5.2-63 is a linear function of time, whereas a real viscoelastic material will describe an exponential creep function as shown in Fig. 5.2-9.

The real behavior is better described by using models that include a second characteristic time such as a retardation time, as in the Jeffreys model discussed in the next section.

The generalized Maxwell model can also be used to describe the relaxation of stresses in solid polymers. For example, the stress response following a small deformation in tension can be obtained by extending the result of equation 5.2-48 to multimodes in tension. The relaxation modulus of a generalized Maxwell fluid is expressed by

$$E_r(t) = -\frac{\sigma_{xx}(t)}{\varepsilon_0} = \sum_{p=1}^{M} E_p e^{-\frac{t}{\lambda_p}}, \tag{5.2-64}$$

where $-\sigma_{xx}(t)$ is the tensile stress, ε_0 is the imposed deformation, and E_p are tensile moduli of the polymer. Equation 5.2-64 can adequately describe the relaxation of stresses in amorphous polymers, as illustrated in Fig. 5.2-10 for a polycarbonate at different temperature levels.

Stress relaxation in a polymer following a sudden small deformation is phenomenologically explained by the internal motion or slip of the polymeric chains. Initially, the short chains or segments between entanglements react to an imposed deformation. With time, the longer chains or segments start reacting, and contribute to the total deformation by relaxing the stress level.

For a solid crystalline or crosslinked polymer, the generalized Maxwell model can be used to describe the relaxation of stresses if one of the relaxation times is taken to be infinitely large. The relaxation modulus is then given by

$$E_r(t) = E_\infty + \sum_{p=1}^{M-1} E_p e^{-\frac{t}{\lambda_p}}, \tag{5.2-65}$$

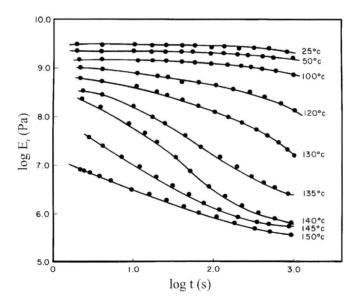

Figure 5.2-10 Relaxation modulus for a polycarbonate, bisphenol A (Adapted from Malik et al., 1989)

178 Linear Viscoelasticity

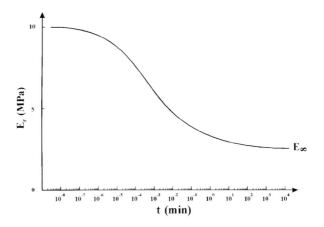

Figure 5.2-11 Relaxation modulus of a natural rubber at 25 °C

where E_∞ is the relaxation plateau for very long times. The behavior of a natural rubber, shown in Fig. 5.2-11, can be described by equation 5.2-65. Notice that natural rubber initially behaves like a very stiff material ($E_r \approx 10$ MPa), but after a minute, the modulus drops by a factor of approximately 5. Vulcanized rubber would have a much larger modulus and would change very little with time.

5.2-3 The Jeffreys Model

We can show that the mechanical model illustrated in Fig. 5.2-12 leads to the so-called Jeffreys model.

The Jeffreys equation in simple shear is

$$\sigma_{yx} + \lambda_1 \frac{\partial}{\partial t}\sigma_{yx} = -\eta_0\left(\dot{\gamma}_{yx} + \lambda_2 \frac{\partial}{\partial t}\dot{\gamma}_{yx}\right). \tag{5.2-66}$$

This model contains two time constants, λ_1 and λ_2, related to the mechanical elements by

$$\lambda_1 = \frac{\eta_1 + \eta_2}{G_3} = \frac{\eta_0}{G_3} \tag{5.2-67}$$

and

$$\lambda_2 = \frac{\eta_2}{G_3}. \tag{5.2-68}$$

λ_1 is a relaxation time, and λ_2 is a retardation time, which will qualitatively describe the correct behavior for a creep experiment. Equation 5.2-66 can be integrated to obtain

$$\sigma_{yx}(t) = -\int_{-\infty}^{t} \frac{\eta_0}{\lambda_1}\left(1 - \frac{\lambda_2}{\lambda_1}\right)e^{-\frac{(t-t')}{\lambda_1}}\dot{\gamma}_{yx}(t')dt' - \eta_0\frac{\lambda_2}{\lambda_1}\dot{\gamma}_{yx}(t). \tag{5.2-69}$$

5.2 Linear Viscoelastic Models

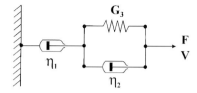

Figure 5.2-12 Mechanical analog of the Jeffreys model

As for the case of the integral Maxwell model, this result is based on the assumption that the stress is finite as $t \to \infty$. Notice that equation 5.2-69 makes sense only if $\lambda_2 \leq \lambda_1$, otherwise the memory function is negative, which is physically unacceptable.

Using the Dirac delta function, $\delta(t-t')$, the integral model can be rewritten as

$$\sigma_{yx}(t) = -\int_{-\infty}^{t} \left\{ \frac{\eta_0}{\lambda_1}\left(1-\frac{\lambda_2}{\lambda_1}\right) e^{-\frac{(t-t')}{\lambda_1}} + 2\eta_0 \frac{\lambda_2}{\lambda_1}\delta(t-t') \right\} \dot{\gamma}_{yx}(t')dt'. \tag{5.2-70}$$

Example 5.2-7 Complex Viscosity of a Jeffreys Fluid

Under small amplitude oscillatory shear flow, the strain rate signal is given by equation 2.1-10. Using the complex notation, the Jeffreys model (equation 5.2-66) can be written as

$$\text{Re}\{\sigma^0(1+i\omega\lambda_1)e^{i\omega t}\} = -\text{Re}\{\eta_0 \dot{\gamma}^0(1+i\omega\lambda_2)e^{i\omega t}\}. \tag{5.2-71}$$

Simplifying,

$$\eta^* = \eta' - i\eta'' = -\frac{\sigma_0}{\dot{\gamma}^0} = \eta_0 \frac{(1+i\omega\lambda_2)}{(1+i\omega\lambda_1)}$$

$$= \eta_0 \frac{(1+\omega^2\lambda_1\lambda_2)}{(1+\omega^2\lambda_1^2)} - i\frac{\eta_0\omega(\lambda_1-\lambda_2)}{(1+\omega^2\lambda_1^2)}. \tag{5.2-72}$$

Hence, the dynamic viscosity and the storage modulus are respectively given by

$$\eta' = \frac{G''}{\omega} = \frac{\eta_0(1+\omega^2\lambda_1\lambda_2)}{(1+\omega^2\lambda_1^2)} \tag{5.2-73}$$

and

$$G' = \eta''\omega = \frac{\eta_0\omega^2(\lambda_1-\lambda_2)}{(1+\omega^2\lambda_1^2)}. \tag{5.2-74}$$

For $\lambda_2 = 0$, we recover the simple Maxwell result. For $\lambda_2 < \lambda_1$, the Jeffreys model predicts a limiting dynamic viscosity at very high frequencies, that is

$$\lim_{\omega \to \infty} \eta' = \eta_0 \frac{\lambda_2}{\lambda_1}. \tag{5.2-75}$$

5.2-4 The Voigt-Kelvin Model

The mechanical analog for the Voigt-Kelvin model is illustrated in Fig. 5.2-13. In its simplest form, it consists of a dashpot and a spring placed in parallel.

The force acting on a Voigt-Kelvin element is the sum of the forces acting on the dashpot and on the spring. The corresponding rheological model in simple shear is given by

$$\sigma_{yx} = -G_0 \gamma_{yx} - \eta_0 \dot{\gamma}_{yx}$$
$$= -G_0(\gamma_{yx} + \lambda_0 \dot{\gamma}_{yx}). \tag{5.2-76}$$

The Voigt-Kelvin model can easily be generalized by considering M elements in parallel as shown in Fig. 5.2-14.

The corresponding rheological equation is

$$\sigma_{yx} = -\sum_{p=1}^{M} G_p(\gamma_{yx} + \lambda_p \dot{\gamma}_{yx}). \tag{5.2-77}$$

Example 5.2-8   Creep Function of a Voigt-Kelvin Solid

Obtain the deformation and compliance in creep for a solid described by the generalized Voigt-Kelvin model in a series arrangement as illustrated by Fig. 5.2-15.

Solution

The experiment is described by Fig. 5.2-9. For Voigt-Kelvin elements in series, the total deformation in creep is the sum of the deformations of each element. Since the shear stress is described by a step function, we solve the problem using the Laplace transform. The Laplace transform of equation 5.2-76, written for any element p, gives

$$\bar{\sigma}_{yx} = -G_p[\bar{\gamma}_{yx} + \lambda_p(s\bar{\gamma}_{yx} - \gamma_{yx}(0))]$$
$$= \int_0^\infty e^{-st} \sigma_0 dt = \frac{\sigma_0}{s}, \tag{5.2-78}$$

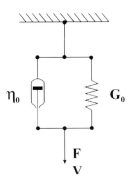

Figure 5.2-13   The simple Voigt-Kelvin element

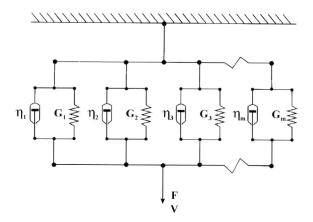

Figure 5.2-14   The generalized Voigt-Kelvin model in a parallel arrangement

where the deformation at $t=0$, $\gamma_{yx}(0)$ is arbitrarily set equal to zero. Simplifying equation 5.2-78, we obtain

$$\bar{\gamma}_{yx} = -\sigma_0 \frac{1}{G_p(s + \lambda_p s^2)}. \tag{5.2-79}$$

Taking the inverse and summing over all elements, we obtain the expression for the deformation under creep, that is,

$$\gamma_{yx}(t) = -\sigma_0 \sum_{p=1}^{M} \frac{1}{G_p}[1 - e^{-\frac{t}{\lambda_p}}], \tag{5.2-80a}$$

and the compliance is given by

$$J_c(t) = \frac{\gamma_{yx}(t)}{-\sigma_0} = \sum_{p=1}^{M} \frac{1}{G_p}[1 - e^{\frac{t}{\lambda_p}}]. \tag{5.2-80b}$$

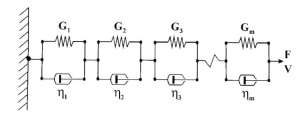

Figure 5.2-15 M Voigt-Kelvin elements in series

5.2-5 Other Linear Models

Various other linear models can be obtained by combining springs and dashpots. One useful combination is obtained by placing a Maxwell and a Voigt-Kelvin element in series as illustrated in Fig. 5.2-16. This can be seen as an extension of the Jeffreys model presented in Section 5.2-3.

The rheological equation corresponding to the mechanical analog of Fig. 5.2-16 is written as

$$\frac{\partial^2 \sigma_{yx}}{\partial t^2} + \left(\frac{G_1}{\eta_1} + \frac{G_1}{\eta_2} + \frac{G_2}{\eta_2}\right)\frac{\partial \sigma_{yx}}{\partial t} + \left(\frac{G_1 G_2}{\eta_1 \eta_2}\right)\sigma_{yx}$$
$$= -G_1 \frac{\partial^2 \gamma_{yx}}{\partial t^2} - \left(\frac{G_1 G_2}{\eta_2}\right)\frac{\partial \gamma_{yx}}{\partial t}. \qquad (5.2\text{-}81)$$

This is a second order differential equation that describes the creep and recovery behavior of viscoelastic materials rather well. The compliance in creep is the sum of the compliances for the Maxwell and the Voigt-Kelvin elements. That is,

$$J_c(t) = \frac{1}{G_1} + \frac{t}{\eta_1} + \frac{1}{G_2}(1 - e^{-\frac{t}{\lambda_2}}), \qquad (5.2\text{-}82)$$

where $\lambda_2 = \eta_2/G_2$ is the elastic characteristic time of the Voigt-Kelvin element. For the recovery, the compliance is expressed by

$$J_r(t) = \frac{t_c}{\eta_1} + \left(J_c(t_c) - \frac{t_c}{\eta_1} - \frac{1}{G_1}\right)e^{\left(\frac{t_c-t}{\lambda_2}\right)}, \qquad (5.2\text{-}83)$$

where t_c is the creep time. Figure 5.2-17 compares the experimental creep and recovery data with the model predictions for a molten polyvinylchloride (PVC) at four different temperatures. The PVC formulation is a typical one for profile extrusion (Huneault et al., 1992).

The creep data for the four different temperatures are very well described by the model. For the recovery, only the data obtained at the higher temperatures are correctly predicted, using the same parameters as determined from the creep data. For the lower temperatures, a non-negligible part of the PVC is not molten and the possible orientation of the non-molten crystallites obtained under creep is not recovered under cessation of the imposed shear stress. The creep and recovery data and the model parameters are reported in Table 5.2-1.

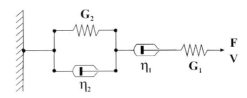

Figure 5.2-16 Combined Maxwell and Voigt-Kelvin elements

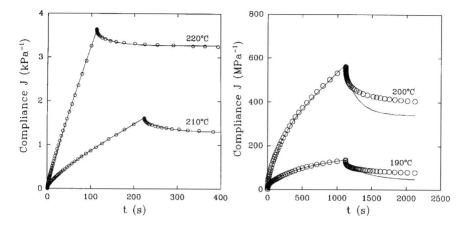

Figure 5.2-17 Creep and recovery data for a molten PVC for an imposed stress of 300 Pa
Creep and recoverable compliances and model parameters are reported in Table 5.2-1 (Data of Huneault, 1992)

We note that the compliance under recovery, J_r, increases with increasing temperature. This correlates fairly well with extrudate swelling of PVC, which was also observed to increase with temperature (Huneault et al., 1992).

Table 5.2-1 Creep and Recovery Data and Model Parameters for the PVC of Figure 5.2-17

Experimental data				Model parameters			
T °C	J_0 MPa^{-1}	J_r MPa^{-1}	η MPa·s	G_1 kPa	η_1 MPa·s	G_2 kPa	λ_2 s
190	78	54	23	50.1	25.0	13.2	340
200	280	156	4.0	41.0	3.1	5.46	146
210	200	380	0.170	13.7	0.173	3.94	51
220	210	410	0.028	9.22	0.034	3.92	38

5.3 Relaxation Spectrum

Many authors prefer to use a continuous spectrum of relaxation times for characterization purposes, instead of a discrete set as in the generalized Maxwell model. The relaxation function of the Maxwell model,

$$G(t - t') = \sum_{p=1}^{M} \frac{\eta_p}{\lambda_p} e^{-\frac{(t-t')}{\lambda_p}}, \quad (5.3\text{-}1)$$

is replaced by

$$G(t - t') = \int_0^\infty \frac{H(\lambda)}{\lambda} e^{-\frac{(t-t')}{\lambda}} d\lambda \quad (5.3\text{-}2a)$$

$$= \int_{-\infty}^{\infty} H(\lambda) e^{-\frac{(t-t')}{\lambda}} d\ln \lambda. \quad (5.3\text{-}2b)$$

We note that $G(s)$ is simply the Laplace transform of $\lambda H(\lambda)$, where $H(\lambda)$ is a relaxation time spectrum.

Complex viscosity data are normally used to determine the relaxation spectrum. First, we note that the storage and loss moduli are related to the relaxation function, $G(t)$, by

$$G'(\omega) = \eta''\omega = \omega \int_0^\infty G(t)\sin(\omega t)dt \quad (5.3\text{-}3)$$

and

$$G''(\omega) = \eta'\omega = \omega \int_0^\infty G(t)\cos(\omega t)dt. \quad (5.3\text{-}4)$$

These equations can be inverted using Fourier transforms (Problem 5.5-4) to obtain

$$G(t) = \frac{2}{\pi}\int_0^\infty \frac{G'(\omega)}{\omega}\sin(\omega t)d\omega \quad (5.3\text{-}5a)$$

or

$$G(t) = \frac{2}{\pi}\int_0^\infty \frac{G''(\omega)}{\omega}\cos(\omega t)d\omega. \quad (5.3\text{-}5b)$$

Equations 5.3-5a and b can be used as a consistency test. This is particularly useful for very low or very high frequency ranges for which one set of data may not be reliable.

Approximate methods have been proposed to estimate $H(\lambda)$. The first method is due to Alfrey and Dotey (1945) and is based on the behavior of typical exponential functions encountered in a relaxation modulus. This is illustrated by Fig. 5.3-1.

The exponential shows a very sudden increase for $\lambda > s$. Neglecting the influence of the shorter relaxation times ($\lambda < s$), we write equation 5.3-2b as

$$G(s) \approx \int_{\ln s}^{\infty} H(\lambda)d\ln \lambda, \quad (5.3\text{-}6)$$

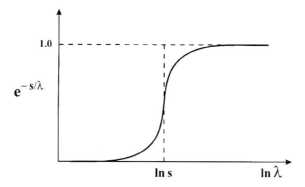

Figure 5.3-1 Typical exponential function in the relaxation modulus

and as a first approximation,

$$H_1(\lambda) \approx -\frac{dG(s)}{d\ln\lambda}\bigg|_{s=\lambda}, \tag{5.3-7}$$

which follows from equation 5.3-6 by taking the derivative. A better approximation has been proposed by Schwarzl and Stavermann (1953). The second approximation due to Schwarzl and Stavermann is

$$H_2(\lambda) \approx H_1(\lambda) + \frac{d^2 G(s)}{d(\ln\lambda)^2}\bigg|_{s=2\lambda}. \tag{5.3-8}$$

Ferry and Williams (1952) proposed yet another method for obtaining the second approximation:

$$H_2(\lambda) \approx \frac{H_1(\lambda)}{\Gamma(m+1)}, \tag{5.3-9}$$

where Γ is the gamma function, and m is the slope of $\log H_1(\lambda)$ vs $\log \lambda$. For $\Gamma(m+1) = 1$, the first approximation is recovered.

Typical relaxation spectra are illustrated in Fig. 5.3-2 for solutions of poly-n-butyl methacrylate in diethyl phthalate, reduced to 100 °C, using the time-temperature superposition principle presented in the next section. The symbols with black tops are H values calculated using G' data, the symbols with black bottoms are associated with G'' data. The dashed lines of slope $-1/2$ are the theoretical predictions from the Rouse theory. Note the shift to the right of the spectrum as the polymer concentration is increased. That is, longer relaxation times play a more important role in the linear viscoelastic properties of the more concentrated solutions. More detailed information on the relaxation spectra of polymeric systems can be found in Ferry's (1980) book.

186   Linear Viscoelasticity

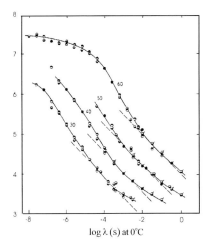

Figure 5.3-2 Relaxation spectra for 30, 40, 50, and 60 mass % of poly-n-butyl methacrylate in diethyl phthalate reduced to 100 °C (Data from Saunders et al., 1959, adapted from Ferry, 1980)

5.4   Time–Temperature Superposition

The time–temperature superposition principle is based on the idea that time and temperature have similar or comparable effects on the relaxation modulus. The principle has been largely verified for amorphous polymers. The principle follows also from molecular theories, such as the Rouse theory (1953) for dilute solutions.

To illustrate the principle, we refer to Fig. 5.4-1, which shows the tensile relaxation modulus of an amorphous polymer as a function of time (right part of the figure). At low temperatures, the modulus is very high, typical of the glassy behavior, and the relaxation time for the polymeric chains is considerably longer than the duration of the experiment (of the order of 1 h). At high temperature, the modulus is that of a rubbery material, with stress relaxation occurring within the experimental time frame. The glass transition is observed in the intermediate temperature range where the relaxation modulus is changing rapidly.

For practical reasons, the measurements are made at constant temperature on a restricted timespan, typical of two to four decades. As shown in the right part of the figure, the different curves of the relaxation modulus can be shifted to obtain a single master curve.

First, a correction to the modulus for the change of density with temperature is made as suggested by molecular and rubber elasticity theories. A reduced modulus is defined:

$$E_r(t) \text{ reduced} = E_r(t)\left(\frac{T_0}{T}\right)\left(\frac{\rho_0}{\rho}\right), \tag{5.4-1}$$

where T_0 is the reference temperature. Then, the horizontal shift factor a_T such that a single master curve is obtained, that is,

$$E_r(t, T_0) = \frac{\rho_0 T_0}{\rho T} E_r(ta_T, T). \tag{5.4-2}$$

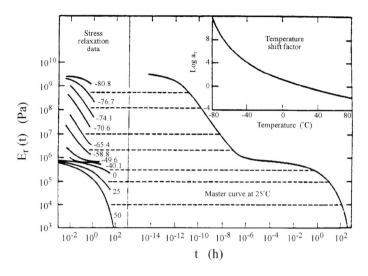

Figure 5.4-1 Time-temperature superposition for the tensile relaxation modulus of a polyisobutylene (Adapted from Williams, 1971)

The shift factor is a unique function of temperature that is equal to 1 at the reference temperature, T_0. The variation of a_T with temperature is shown in the upper right corner of Fig. 5.4-1. It is frequently described by the WLF expression (Williams et al., 1955; Ferry, 1970):

$$\log a_T = -\frac{C_1(T - T_0)}{C_2 + T - T_0}, \quad T_g \leq T \leq T_g + 100 \,°\text{C}, \tag{5.4-3}$$

where C_1 and C_2 depend on the reference temperature and polymer system. If the reference temperature is taken as the glass transition temperature ($T_0 = T_g$), the constants are the same for many amorphous polymers (Ferry, 1970): $C_1 = 17.44$ and $C_2 = 51.6$. For $T_0 = T_g + 45\,°\text{C}$, $C_1 = 8.86$ and $C_2 = 101.6$. As shown in Fig. 5.4-1, the master curve covers over sixteen decades on the time scale. The relaxation process at 25 °C for very short time is that of a glassy material, and becomes that of a liquid at very long times ($>10^2$ h).

Obviously, the time–temperature principle is quite useful for correlating data and expanding the experimental window on the time or frequency scale. This is illustrated in Fig. 5.4-2 in terms of the complex viscosity and moduli of a polystyrene melt. The linear viscoelastic data were obtained at three temperatures (160, 200, and 220 °C) using a Bohlin CSM (constant stress) rheometer. These data are quite accurate in the frequency range covered (10^{-2} to 10^2 rad/s). The superposition of the data is shown in Fig. 5.4-2b. The density correction was assumed to be negligible (i.e., $\rho_0 T_0 \approx \rho T$) and the reference temperature was 200 °C. Note that the frequency range for the master curve is extended by two decades. The high frequency behavior approaches the glassy state, and the lower frequency behavior is also better defined. Since the moduli are shifted horizontally, the complex

Linear Viscoelasticity

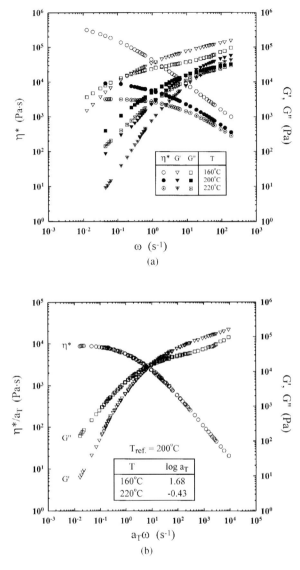

Figure 5.4-2 Complex viscosity and moduli of a polystyrene melt using a Bohlin CSM rheometer (a) Storage and loss moduli at three temperatures (b) master curves for η^*, G', and G'' (Unpublished data)

viscosity has to be shifted vertically. Applying equation 5.4-2 to the complex viscosity yields

$$\eta^*(\omega, T_0) = \frac{\rho_0 T_0}{\rho T} \frac{1}{a_T} \eta^*(\omega a_T, T). \quad (5.4\text{-}4)$$

This procedure is also used for the steady shear viscosity as outlined in Section 2.5-1. The method is strictly speaking valid for amorphous polymers. It has also been applied successfully to semi-crystalline and lightly crosslinked polymers (Tobolsky, 1960).

5.5   Problems*

5.5-1$_a$ Rheological Model with Friction

(a) Derive the rheological model for the mechanical analog shown in Fig. 5.5-1. Neglect inertial effects and assume the mass M exerts a constant drag force on the lower surface.

(b) For the case of $G_0 \to \infty$, the model reduces to a well known model. Which one?

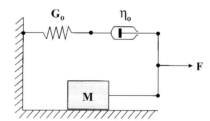

Figure 5.5-1   Mechanical analog of fluid behavior with friction

5.5-2$_a$ Maxwell Model

(a) Compute the stress growth, $\eta^+(t, \dot{\gamma}_\infty)$ and stress relaxation, $\eta^-(t, \dot{\gamma}_0)$, functions for the Maxwell model given by equation 5.2-6.

(b) Sketch the behavior obtained in (a).

5.5-3$_a$ Stress Relaxation for a Maxwell Fluid

(a) Consider the material function $\eta^-(t, \dot{\gamma}_0)$ defined in Section 2.1-3 for stress relaxation following steady shear flow. Obtain its expression for a generalized Maxwell model containing M elements.

---

* Problems are labelled by a subscript a or b indicating the level of difficulty.

(b) Sketch the behavior obtained in (a).
(c) Discuss the advantage of using more than one element.

5.5-4$_b$ Complex Viscosity of a Generalized Maxwell Fluid

Consider the Maxwell fluid described by equation 5.2-42.

(a) Show that the real and imaginary parts of the complex viscosity are given respectively by

$$\eta'(\omega) = \eta_s + \int_0^\infty G(s) \cos \omega s \, ds \qquad (5.5\text{-}1)$$

and

$$\eta''(\omega) = \int_0^\infty G(s) \sin \omega s \, ds, \qquad (5.5\text{-}2)$$

where η_s is the constant solvent viscosity.

(b) Assuming that the solvent viscosity is negligibly small and using the Fourier transform show that the relaxation function can be obtained from the complex viscosity data:

$$G(s) = \frac{2}{\pi} \int_0^\infty \eta'(\omega) \cos \omega s \, d\omega \qquad (5.5\text{-}3)$$

or

$$G(s) = \frac{2}{\pi} \int_0^\infty \eta''(\omega) \sin \omega s \, d\omega. \qquad (5.5\text{-}4)$$

These results are identical to equations 5.3-5a and b.

(c) Using the linear data of Table 5.5-1 for a 3% polyethylene oxide (PEO) solution in a mixture of water and glycerine at 25 °C. (The PEO molecular weight is 1.8×10^6 g/mol.) Obtain the relaxation function $G(s)$ for both G' and G'' data.
(d) Plot the curves for $G(s)$ and discuss the differences observed.
(e) The viscosity of the solvent used is 5.24 mPa·s. Discuss its possible effect on the results obtained for $G(s)$.

5.5-5$_b$ The Jeffreys Model

(a) Show that the mechanical analog, illustrated in Fig. 5.2-12, represents the Jeffreys model given by equation 5.2-66.
(b) Sketch the behavior (on log-log scales) of G', G'', η', and η'' for the case $\lambda_2 = 0.1\lambda_1$.
(c) What are the physical implications if λ_2 is larger than λ_1?
(d) Obtain the expressions of the strain for the creep and recovery experiments.

Table 5.5-1 Linear Viscoelastic Data of a 3% PEO Solution in Water/Glycerine at 25 °C (Data from Ortiz, 1992)

ω s^{-1}	G' Pa	G'' Pa	ω s^{-1}	G' Pa	G'' Pa
0.00628	0.035	0.640	0.94	23.3	29.6
0.00943	0.075	0.876	1.25	28.6	33.7
0.0130	0.128	1.24	1.88	37.6	39.8
0.0190	0.254	1.68	2.51	45.1	44.4
0.0250	0.400	2.61	3.77	57.3	51.3
0.0380	0.732	3.40	5.02	67.0	56.2
0.0500	1.09	4.13	6.28	75.3	60.0
0.0630	1.46	5.86	9.42	91.9	67.1
0.094	2.42	7.14	12.5	105.0	72.0
0.126	3.43	8.68	18.8	125.0	78.9
0.188	5.99	12.0	25.1	140.0	84.0
0.251	7.84	14.6	37.6	164.0	91.0
0.377	11.2	18.4	50.2	183.0	95.8
0.503	14.1	21.5	62.8	199.0	99.9
0.62	17.0	24.2	94.2	262.0	112.0

5.5-6$_b$ Maxwell and Voigt-Kelvin Elements

The mechanical analog for the combined Maxwell and Voigt-Kelvin elements is shown in Fig. 5.2-16. Neglecting inertial effects:

(a) Show that equation 5.2-81 is the corresponding rheological equation in simple shear flow.
(b) Verify expressions 5.2-82 and 5.2-83 for creep and recovery compliance respectively.
(c) The creep and recovery data of Table 5.5-2 have been obtained for a rigid PVC at 190 °C for an imposed stress of 3000 Pa. Plot these data and obtain the values for the steady-state viscosity and equilibrium compliance modulus. Estimate the value of the recovery at equilibrium, determine the parameters for the combined Maxwell-Voigt-Kelvin model, and compare the model predictions to the data.

5.5-7$_a$ Storage and Loss Moduli of a Voigt-Kelvin Material

(a) Obtain the expressions for the storage modulus, loss modulus, and loss tangent of a Voigt-Kelvin fluid.
(b) Illustrate and compare to the corresponding expressions for a single-element Maxwell fluid.

192   Linear Viscoelasticity

Table 5.5-2 Creep and Recovery Data for a Rigid PVC at 190 °C (data of Huneault, 1992)

t s	J MPa^{-1}	t s	J MPa^{-1}	t s	J MPa^{-1}
0.10	6.17	398.1	189	1153	225
1.00	15.5	501.2	206	1161	222
5.01	30.1	794	241	1172	218
10.0	39.9	891	250	1178	216
15.8	48.1	1000	261	1185	214
25.1	57.9	1122	271	1201	210
31.6	63.6	1122	267	1222	206
39.8	70.4	1122	265	1247	202
50.1	78.1	1122	261	1280	197
79.4	95.6	1122	258	1321	193
100.0	105	1123	256	1373	188
125.9	116	1123	255	1438	183
141.3	122	1124	252	1476	181
158.5	129	1125	250	1520	179
177.8	135	1126	247	1623	174
199.5	142	1128	244	1753	170
223.9	149	1130	241	1916	165
251.2	157	1134	238	2122	162
281.8	165	1139	233	2244	160
316.2	173	1144	230	2380	159
354.8	181	1147	229	2534	157

5.5-8$_b$ Complex Compliance

The complex compliance is defined by

$$J^* = \frac{1}{G^*} = J' - iJ'', \tag{5.5-5}$$

where

$$J' = \frac{G'}{G'^2 + G''^2} \tag{5.5-6}$$

and

$$J'' = \frac{G''}{G'^2 + G''^2}. \tag{5.5-7}$$

(a) Show that the energy dissipated per unit volume (that is, the lost work) for a sinusoidal cycle is given by

$$E_v = \pi |\sigma^0|^2 \frac{J''}{2}. \tag{5.5-8}$$

(b) Show that the energy stored during a quarter of a cycle is given by

$$E_p = |\sigma^0|^2 \frac{J'}{2}. \tag{5.5-9}$$

5.5-9$_b$ Relaxation Modulus

The following three parameter model has been suggested for the relaxation modulus $G(s)$ (Bird, Armstrong, and Hassager 1987):

$$G(s) = \frac{\frac{\eta_0}{\lambda}}{\Gamma(1-v)} \left(\frac{\lambda}{s}\right)^v e^{-\frac{s}{\lambda}}, \tag{5.5-10}$$

where Γ is the gamma function and $0 < v < 1$.

(a) Show that this expression generates the following equation for the complex viscosity:

$$\eta^* = \frac{\eta_0}{(1 + i\lambda\omega)^{1-v}}. \tag{5.5-11}$$

(b) Compute the real and imaginary parts for $v = 1/2$.
(c) Discuss the relaxation modulus for the case $v = 1$.

6 Nonlinear Viscoelasticity

6.1 Nonlinear Deformations 195
    6.1-1 Expressions for the Deformation and Deformation Rate 197
    6.1-2 Pure Deformation or Uniaxial Elongation. 200
    6.1-3 Planar Elongation 204
    6.1-4 Expansion or Compression. 205
    6.1-5 Simple Shear 205
        6.1-5.1 Comments 206

6.2 Formulation of Constitutive Equations 208
    6.2-1 Material Objectivity and Formulation of Constitutive Equations 209
    6.2-2 Maxwell Convected Models 210
    6.2-3 Generalized Newtonian Models 215

6.3 Differential Constitutive Equations 220
    6.3-1 The De Witt Model 221
    6.3-2 The Oldroyd Models 222
    6.3-3 The White-Metzner Model. 223
    6.3-4 The Marrucci Model 230
    6.3-5 The Phan-Thien-Tanner Model 232

6.4 Integral Constitutive Equations 234
    6.4-1 The Lodge Model 235
    6.4-2 The Carreau Constitutive Equation 239
        6.4-2.1 Carreau A 241
        6.4-2.2 Carreau B 243
        6.4-2.3 The De Kee Model 247
    6.4-3 The K-BKZ Constitutive Equation 247
    6.4-4 The LeRoy-Pierrard Equation 254

6.5 Concluding Remarks 257

6.6 Problems 258
    6.6-1$_a$ Planar Elongational Flow 258
    6.6-2$_a$ Elongational Viscosity of a Lower-Convected Maxwell Fluid 258
    6.6-3$_b$ Biaxial Elongation 259
    6.6-4$_a$ Admissible Constitutive Equations 259
    6.6-5$_b$ Second-Order Fluid 260
    6.6-6$_b$ Elongational Viscosity of an Oldroyd-B Fluid 260
    6.6-7$_b$ Transient Behavior of a White-Metzner Fluid 260
    6.6-8$_b$ Flow of a White-Metzner Fluid in a Tube Under an Oscillatory Pressure Gradient 260
    6.6-9$_a$ Viscometric Functions for a Marrucci Fluid 261
    6.6-10$_b$ Material Functions for a Carreau Fluid 261
    6.6-11$_b$ Material Functions for a Maxwell Model Involving Slip 261
    6.6-12$_b$ Relations Between Material Functions 262
    6.6-13$_b$ Flow Above an Oscillating Plate 262

In Chapter 5, the basic concepts of linear viscoelasticity were presented. These concepts are useful, but are restricted to flow situations for which the deformation is very small, that is, near equilibrium or non-flow conditions. In most engineering applications, the deformations and rates of deformation are large, and the results of linear viscoelasticity are no longer valid, except for a qualitative evaluation of viscoelastic effects.

Chapter 6 deals with nonlinear strains, and expressions for the deformation under various flow situations are reviewed. Basic principles for the development of constitutive equations are presented and rheological models are assessed with respect to their ability to describe real flow problems. The topic of nonlinear viscoelasticity is a very difficult one and we do not pretend to cover all of the essential elements in this chapter. More comprehensive treatments can be found in Lodge (1964, 1974) for basic principles, and in Bird, Armstrong, and Hassager (1977, 1987) for a more complete description of constitutive equations available in the literature. Another book of considerable interest is that of Tanner (1985), who reviews basic principles and discusses the most important viscoelastic problems encountered in polymer processing. Finally, in Larson (1988), the reader will find a very useful evaluation of constitutive equations derived from continuum mechanics principles as well as from molecular theories.

6.1 Nonlinear Deformations

To describe nonlinear deformations or strains, we make use of embedded or convected coordinates, as suggested by Lodge (1964, 1974) to describe fluid elements. By definition, these coordinates are embedded, and deform and move in space with the fluid elements. That is, the origin moves, the axes deform, and the numerical values of the coordinates of a fluid element remain constant. The notation for covariant and contravariant components of vectors or tensors is used throughout the chapter (see Appendix A).

In fact, the choice of embedded coordinates will ensure that the constitutive equation is valid for all deformations. This is discussed in more detail in Section 6.2. The embedded coordinates are defined by $\hat{x}^i$, in contrast to x^i. The latter are the fluid element coordinates with respect to fixed coordinates. The differences between the two coordinate systems are illustrated in Fig. 6.1-1 for a simple shear flow situation. The lower plate is stationary, whereas the upper plate is displaced at constant velocity V.

At time $t' = t$ (present time), the embedded coordinates are arbitrarily taken to coincide with the fixed coordinates (right side of the figure), that is, $(\hat{x}^1, \hat{x}^2, \hat{x}^3) = (x^1, x^2, x^3)$. However, for past times, $t' < t$, the embedded coordinates do not coincide with the fixed coordinates. For example, the two embedded coordinates of point Q at t' are (8, 6), whereas the fixed coordinates are (4.5, 6). At $t' = t$, both coordinates are (8, 6).

The separations, or distances between two neighboring fluid elements P and Q at times t' and t are expressed respectively by

$$[ds(t')]^2 = g_{ij}(x)dx'^i(t')dx'^j(t') = \hat{g}_{ij}(t')d\hat{x}^i d\hat{x}^j \qquad (6.1\text{-}1)$$

196   Nonlinear Viscoelasticity

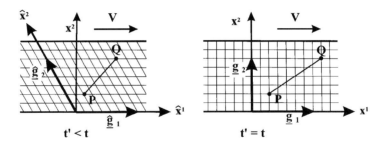

Figure 6.1-1   Embedded and fixed coordinates for a simple shear situation

and

$$[ds(t)]^2 = g_{ij}(x)dx^i(t)dx^j(t) = \hat{g}_{ij}(t)d\hat{x}^i d\hat{x}^j \qquad (6.1\text{-}2)$$

where g_{ij} and $\hat{g}_{ij}$ are the covariant components of the metric tensor (see Appendix A) in fixed and embedded coordinates respectively; and x'^i and x^i denote the fixed coordinates at t' and t respectively. During a flow process, the distance $\sqrt{ds^2}$ between two fluid elements may change, so that ds^2 is a function of time. Note that for a fixed coordinate system, the metric tensor g_{ij} is independent of time. That is, the origin and the axes do not change, and therefore the tangents (defining g_{ij}) do not change with time. Note, however, that the coordinates of the fluid elements change with time. The embedded coordinates, $\hat{x}^i$, are time-independent, but the embedded metric tensor, $\hat{g}_{ij}$, varies with time. Equations 6.1-1 and 6.1-2 identify the dependence of various quantities on time and/or on position. The covariant embedded components of the metric tensor are therefore a complete measure of the deformation of a material element. From results A.2-51,

$$\hat{g}^{ik}\hat{g}_{kj} = \delta^i_j, \qquad (6.1\text{-}3)$$

where δ^i_j is the Kronecker delta. It follows that the contravariant embedded components of the metric tensor, $\hat{g}^{ij}$, are also a complete measure of the deformation. Note, in reference to Fig. 6.1-1, that the covariant 12-component of the metric tensor is nonzero for $t' < t$ and is equal to zero at $t' = t$. That is,

$$\text{at } t' < t, \quad \hat{g}_{12}(t') = \underline{\hat{g}}_1(t') \cdot \underline{\hat{g}}_2(t') \neq 0 \qquad (6.1\text{-}4)$$

and

$$\text{at } t' = t, \quad \hat{g}_{12}(t) = \underline{\hat{g}}_1(t) \cdot \underline{\hat{g}}_2(t)$$
$$= \underline{g}_1(t) \cdot \underline{g}_2(t) = 0, \qquad (6.1\text{-}5)$$

where $\underline{\hat{g}}_1$ and $\underline{\hat{g}}_2$ are the embedded base vectors along coordinates 1 and 2 respectively, which become equal to the fixed base vectors at t ($\underline{g}_1$ and $\underline{g}_2$ or $\underline{\delta}_1$ and $\underline{\delta}_2$ in Cartesian coordinates). In Fig. 6.1-1, we have set the deformation at time t equal to zero. This can be done since a deformation is defined relative to a reference configuration. We choose the configuration at time t to be the reference configuration such that the deformation is zero at time t. In Section 6.1-1 we present various expressions for the deformation and the deformation rate.

6.1-1 Expressions for the Deformation and Deformation Rate

The following expressions can be used to express the deformation and deformation rates in embedded coordinates:

(a) Deformation at time t':

$$\hat{g}^{ij}(\hat{x}, t')$$
$$\hat{g}_{ij}(\hat{x}, t')$$

(b) Quantity related to the separation at time t':

$$\sqrt{\hat{g}^{ij}(\hat{x}, t')}$$
$$\sqrt{\hat{g}_{ij}(\hat{x}, t')}$$

(c) Relative deformation between t' and t:

$$\hat{g}^{ij}(\hat{x}, t) - \hat{g}^{ij}(\hat{x}, t')$$
$$\hat{g}_{ij}(\hat{x}, t') - \hat{g}_{ij}(\hat{x}, t)$$

(d) Rate of deformation:

$$\left(\frac{\partial \hat{g}^{ij}}{\partial t}\right)_{\hat{x}}$$
$$\left(\frac{\partial \hat{g}_{ij}}{\partial t}\right)_{\hat{x}}$$

(e) Acceleration of deformation:

$$\left(\frac{\partial^2 \hat{g}^{ij}}{\partial t^2}\right)_{\hat{x}}$$
$$\left(\frac{\partial^2 \hat{g}_{ij}}{\partial t^2}\right)_{\hat{x}}$$

To obtain the corresponding expressions in fixed coordinates, we use the transformation rules developed in Appendix A, taking the corresponding time at $t' = t$, that is,

$$\text{at } t' = t, \quad \hat{x}^i = x^i \quad (6.1\text{-}6)$$
$$\hat{g}^{ij} = g^{ij}.$$

Following the rules of covariant and contravariant transformations (A.2) from $\hat{x}$-coordinates to x'-coordinates, we get

$$\hat{g}^{ij}(\hat{x}, t') = \left(\frac{\partial \hat{x}^i}{\partial x'^k}\right)\left(\frac{\partial \hat{x}^j}{\partial x'^l}\right) g^{kl}(x'). \quad (6.1\text{-}7)$$

Since $\hat{x}^i = x^i$ at $t' = t$, this result is

$$\hat{g}^{ij}(\hat{x}, t') \overset{t}{\Leftrightarrow} \left(\frac{\partial x^i}{\partial x'^k}\right)\left(\frac{\partial x^j}{\partial x'^l}\right) g^{kl}(x') \equiv C^{-1}{}^{ij}(x', t'), \quad (6.1\text{-}8)$$

where $\overset{t}{\Leftrightarrow}$ means that the equality holds for a correspondence at $t' = t$. $\underline{\underline{C}}^{-1}$ is known as the *Finger tensor*.

Similarly, the covariant components of the metric tensor are transformed to obtain

$$\hat{g}_{ij}(\hat{x}, t') \overset{t}{\Leftrightarrow} \left(\frac{\partial x'^k}{\partial x^i}\right)\left(\frac{\partial x'^l}{\partial x^j}\right) g_{kl}(x') \equiv C_{ij}(x', t') \qquad (6.1\text{-}9)$$

where $\underline{\underline{C}}$ is the *Cauchy-Green tensor*. Note that

$$\underline{\underline{C}} \cdot \underline{\underline{C}}^{-1} = \underline{\underline{\delta}} \qquad (6.1\text{-}10)$$

where $\underline{\underline{\delta}}$ is the unit tensor.

The transformation of the components of the rate-of-deformation tensor yields

$$-\left(\frac{\partial \hat{g}^{ij}}{\partial t'}\right)\bigg|_{\hat{x}|t'=t} = -\frac{\delta}{\delta t} g^{ij} = g^{ir} g^{js}(V_{r,s} + V_{s,r}) \equiv \dot{\gamma}^{ij} \qquad (6.1\text{-}11)$$

and

$$\left(\frac{\partial \hat{g}_{ij}}{\partial t'}\right)\bigg|_{\hat{x}|t'=t} = \frac{\delta}{\delta t} g_{ij} = V_{i,j} + V_{j,i} \equiv \dot{\gamma}_{ij} \qquad (6.1\text{-}12)$$

where $\dot{\gamma}$ $(=\nabla \underline{V} + \nabla \underline{V}^+)$ is the rate-of-deformation tensor, $V_{i,j}$ is the covariant derivative defined in Appendix A, and $\delta/\delta t$ the convected (Oldroyd, 1950) derivative defined by

$$\frac{\delta}{\delta t} A^{ij} = \frac{\partial}{\partial t} A^{ij} + V^k A^{ij}_{,k} - V^i_{,k} A^{kj} - V^j_{,k} A^{ik}$$

$$= \frac{D}{Dt} A^{ij} - V^i_{,k} A^{kj} - V^j_{,k} A^{ik} \qquad (6.1\text{-}13)$$

where D/Dt is the substantial or material derivative (see Bird, Stewart, and Lightfoot, 1960). The derivative defined by equation 6.1-13 is also known as the *upper-convected derivative*. The *lower-convected derivative* operates on the covariant components and is defined by

$$\frac{\delta A_{ij}}{\delta t} = \frac{\partial}{\partial t} A_{ij} + V^k A_{ij,k} + V^k_{,i} A_{kj} + V^k_{,j} A_{ik}. \qquad (6.1\text{-}14)$$

These two convected derivatives are not equivalent, and this explains the difference in sign in the transformations of the rate-of-deformation tensor (equations 6.1-11 and 6.1-12). The fixed components of the metric tensor do not depend on time. Hence,

$$\begin{aligned} \frac{\partial}{\partial t} g^{ij} &= 0, \\ \frac{\partial}{\partial t} g_{ij} &= 0, \\ \frac{D}{Dt} g^{ij} &= 0, \\ \text{and} \quad \frac{D}{Dt} g_{ij} &= 0, \end{aligned} \qquad (6.1\text{-}15)$$

and the upper- and lower-convected derivatives of the metric tensor can be written as

$$\frac{\delta}{\delta t} g^{ij} = -V^i_{,k} g^{kj} - V^j_{,k} g^{ik} = -\dot{\gamma}^{ij} \tag{6.1-16}$$

$$\frac{\delta}{\delta t} g_{ij} = V^k_{,i} g_{kj} + V^k_{,j} g_{ik} = \dot{\gamma}_{ij}. \tag{6.1-17}$$

Note that $V^i_{,k} g^{kj} = V_{r,s} g^{ir} g^{js}$, etc. (see Appendix A).

The convected derivative expresses the rate of change as a fluid element moves and deforms. Both convected derivatives of the metric tensor convey the same physics (the rate of deformation), but the sign is different. Note that for Cartesian coordinates, $g^{ik} = g_{ik} = \delta_{ik}$, and

$$\dot{\gamma}^{ij} = \dot{\gamma}_{ij} = V_{i,j} + V_{j,i}. \tag{6.1-18}$$

Acceleration terms are transformed according to the following rules

$$\left(\frac{\partial^2 \hat{g}_{ij}}{\partial t'^2} \right)\bigg|_{\hat{x}|_{t'=t}} = \frac{\delta^2}{\delta t^2} g_{ij} = \frac{D}{Dt} \dot{\gamma}_{ij} + \dot{\gamma}_{ik} \dot{\gamma}^k_j \tag{6.1-19}$$

and

$$-\left(\frac{\partial^2 \hat{g}^{ij}}{\partial t'^2} \right)\bigg|_{\hat{x}|_{t'=t}} = \frac{\delta^2}{\delta t^2} g^{ij} = \frac{D}{Dt} \dot{\gamma}^{ij} - \dot{\gamma}^{ik} \dot{\gamma}^j_k \tag{6.1-20}$$

where D/Dt is the *Jaumann derivative* defined by:

$$\frac{D}{Dt} A_{ij} = \frac{\partial A_{ij}}{\partial t} + V^k A_{ij,k} + \frac{1}{2} (\omega_{ik} A^k_j - A^k_i \omega_{kj}) \tag{6.1-21}$$

or

$$\frac{D}{Dt} A^{ij} = \frac{\partial A^{ij}}{\partial t} + V^k A^{ij}_{,k} + \frac{1}{2} (\omega^{ik} A^j_k - A^i_k \omega^{kj}) \tag{6.1-22}$$

where $\underline{\underline{\omega}}$ is the vorticity tensor, given by

$$\underline{\underline{\omega}} = \nabla \underline{V} - \nabla \underline{V}^+. \tag{6.1-23}$$

Both expressions for the Jaumann derivative (equations 6.1-21 and 6.1-22) are equivalent, and one can be obtained from the other by raising or lowering indices:

$$g^{ir} g^{js} \frac{D}{Dt} A_{rs} = \frac{D A^{ij}}{Dt}. \tag{6.1-24}$$

Example 6.1-1   Linear Deformation Tensor

Show that for very small deformations, the deformation tensor is expressed by

$$\gamma_{ij} = U_{i,j} + U_{j,i}, \tag{6.1-25}$$

where U^i is the displacement function defined by

$$U^i = x'^i - x^i \tag{6.1-26}$$

and

$$U_i = g_{ik} U^k. \tag{6.1-27}$$

Solution
Assume for the sake of simplicity that the flow kinematics is described by Cartesian coordinates. The Cauchy-Green tensor (equation 6.1-9) is then

$$C_{ij}(x', t') = \left(\frac{\partial x'_k}{\partial x_i}\right)\left(\frac{\partial x'_l}{\partial x_j}\right)\delta_{kl} = \left(\frac{\partial x'_k}{\partial x_i}\right)\left(\frac{\partial x'_k}{\partial x_j}\right). \tag{6.1-28}$$

Using the displacement function (6.1-26):

$$C_{ij}(x', t') = \left(\frac{\partial U_k}{\partial x_i} + \frac{\partial x_k}{\partial x_i}\right)\left(\frac{\partial U_k}{\partial x_j} + \frac{\partial x_k}{\partial x_j}\right)$$

$$= \left(\frac{\partial U_k}{\partial x_i}\right)\left(\frac{\partial U_k}{\partial x_j}\right) + \frac{\partial U_k}{\partial x_j}\delta_{ki} + \frac{\partial U_k}{\partial x_i}\delta_{kj} + \delta_{ki}\delta_{kj}. \tag{6.1-29}$$

For very small deformations, the first term in U_k^2 is negligible and

$$C_{ij}(x', t') \approx \frac{\partial U_i}{\partial x_j} + \frac{\partial U_j}{\partial x_i} + \delta_{ij}$$

$$= \gamma_{ij} + \delta_{ij}. \tag{6.1-30}$$

Note that γ_{ij} is the linear strain used throughout Chapter 5. Hence, for small deformations, the Cauchy-Green tensor is equal to the linear deformation tensor plus the unit tensor. Since the deformation is a relative quantity, the unit tensor can be ignored. Usually we write the relative deformation as $C_{ij}(x', t') - \delta_{ij}$.

Similarly, we can show that the Finger tensor reduces to the linear deformation tensor in the limit of very small deformations. That is, for Cartesian coordinates,

$$-C_{ij}^{-1} \approx U_{i,j} + U_{j,i} \equiv \gamma_{ij} + \delta_{ij} \tag{6.1-31}$$

All these results are summarized in Table 6.1-1. In the following sections, various examples of flow fields are presented.

6.1-2 Pure Deformation or Uniaxial Elongation

Pure deformation or uniaxial elongation is defined in the sketch of Fig. 6.1-2.
The embedded base vectors remain orthogonal: $\hat{g}_1$ is stretched with time whereas the magnitude of the other vectors remains equal ($\hat{g}_2 = \hat{g}_3$), but compressed with time. At some reference time, t_0, the vectors are taken to be orthonormal. The covariant components of the metric tensor are given by

6.1 Nonlinear Deformations

Table 6.1-1 Relations Between Deformation and Rate-of-Deformation Tensors

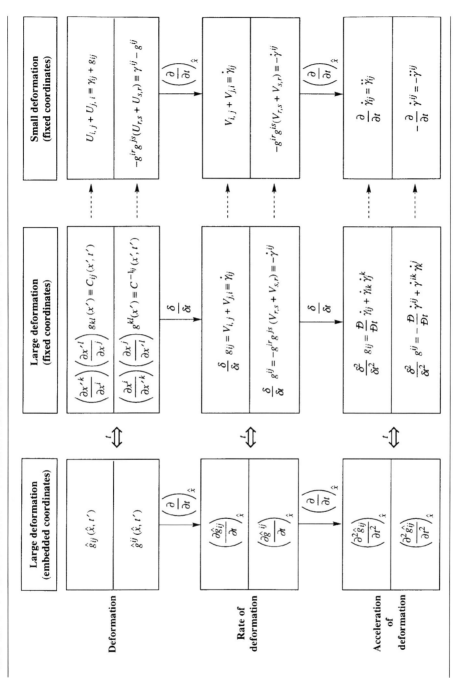

Nonlinear Viscoelasticity

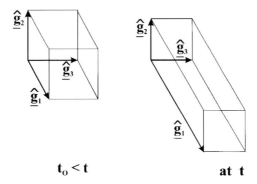

Figure 6.1-2 Pure deformation or uniaxial elongation

(at t_0)

$$\hat{g}_{ij}(t_0) = \underline{\hat{g}}_i \cdot \underline{\hat{g}}_j = \delta_{ij} = \begin{bmatrix} 1 & 0 & 0 \\ 0 & 1 & 0 \\ 0 & 0 & 1 \end{bmatrix}, \quad (6.1\text{-}32)$$

and

(at t)

$$\hat{g}_{ij}(t) = \begin{bmatrix} \hat{g}_1^2 & 0 & 0 \\ 0 & \hat{g}_2^2 & 0 \\ 0 & 0 & \hat{g}_3^2 \end{bmatrix}. \quad (6.1\text{-}33)$$

We note that $\hat{g}_{ij}$ is a direct measure of the elongation, the 11-component increasing as the square of the stretching in the $\hat{x}_1$- direction. To obtain the contravariant components, we note first that the volume of the material element is

$$\begin{aligned} V(t_0) &= 1 \\ \text{and} \quad V(t) &= \hat{g}_1 \hat{g}_2 \hat{g}_3. \end{aligned} \quad (6.1\text{-}34)$$

Hence, the reciprocal base vectors are related to the base vectors by

$$\underline{\hat{g}}^1 = \frac{\underline{\hat{g}}_2 \times \underline{\hat{g}}_3}{\hat{g}_1 \hat{g}_2 \hat{g}_3} = \frac{\underline{\hat{g}}_1}{\hat{g}_1^2}, \quad \text{etc.} \quad (6.1\text{-}35)$$

The contravariant components of the metric tensor are then

$$\hat{g}^{ij}(t) = \underline{\hat{g}}^i \cdot \underline{\hat{g}}^j = \begin{bmatrix} \hat{g}_1^{-2} & 0 & 0 \\ 0 & \hat{g}_2^{-2} & 0 \\ 0 & 0 & \hat{g}_3^{-2} \end{bmatrix}. \quad (6.1\text{-}36)$$

The 11-component of the metric tensor decreases with the square of the stretching ($\hat{g}_1$). The contravariant components of the metric tensor are then representative of an inverse measure of the deformation.

For elongation at constant volume (incompressible material), $V(t) = V(t_0) = 1$, and

$$\hat{g}_2 = \hat{g}_3 = \frac{1}{\hat{g}_1^{1/2}}. \tag{6.1-37}$$

The metric tensor can be expressed by

$$\hat{g}_{ij}(t) = \begin{bmatrix} \hat{g}_1^2 & 0 & 0 \\ 0 & \hat{g}_1^{-1} & 0 \\ 0 & 0 & \hat{g}_1^{-1} \end{bmatrix} \tag{6.1-38}$$

and

$$\hat{g}^{ij}(t) = \begin{bmatrix} \hat{g}_1^{-2} & 0 & 0 \\ 0 & \hat{g}_1 & 0 \\ 0 & 0 & \hat{g}_1 \end{bmatrix}. \tag{6.1-39}$$

For *steady-state uniaxial elongational flow*, the elongation rate defined by

$$\dot{\varepsilon} = \frac{1}{\hat{g}_1} \frac{d\hat{g}_1}{dt} \tag{6.1-40}$$

is constant. Integrating 6.1-40 from t_0 to t, we obtain

$$\hat{g}_1(t) = \exp\{\dot{\varepsilon}(t - t_0)\}. \tag{6.1-41}$$

That is, the stretching or elongation in the $\hat{x}_1$-direction increases exponentially with time. From equation 6.1-37

$$\hat{g}_2 = \hat{g}_3 = \exp\left\{\frac{1}{2}\dot{\varepsilon}(t_0 - t)\right\}, \tag{6.1-42}$$

which is an exponentially decreasing function of time ($t_0 < t$). The covariant and contravariant components of the metric tensor are then given respectively by

$$\hat{g}_{ij}(t) = \begin{bmatrix} \exp\{2\dot{\varepsilon}(t-t_0)\} & 0 & 0 \\ 0 & \exp\{\dot{\varepsilon}(t_0-t)\} & 0 \\ 0 & 0 & \exp\{\dot{\varepsilon}(t_0-t)\} \end{bmatrix} \tag{6.1-43a}$$

and

$$\hat{g}^{ij}(t) = \begin{bmatrix} \exp\{2\dot{\varepsilon}(t_0-t)\} & 0 & 0 \\ 0 & \exp\{\dot{\varepsilon}(t-t_0)\} & 0 \\ 0 & 0 & \exp\{\dot{\varepsilon}(t-t_0)\} \end{bmatrix}. \tag{6.1-43b}$$

To obtain the rate-of-deformation tensor, we must take the derivative of the metric tensor with respect to time and evaluate that derivative at $t=t_0$, which is in this example the reference time (correspondence of the embedded coordinates with the fixed Cartesian ones). That is, the covariant components for the rate-of-deformation tensor are

$$\dot{\gamma}_{ij} = \left.\frac{\partial \hat{g}_{ij}}{\partial t}\right|_{t=t_0} = \begin{bmatrix} 2\dot{\varepsilon} & 0 & 0 \\ 0 & -\dot{\varepsilon} & 0 \\ 0 & 0 & -\dot{\varepsilon} \end{bmatrix}. \tag{6.1-44}$$

6.1-3 Planar Elongation

The planar elongation of an incompressible ($V=1$) fluid is defined by

$$\hat{g}_1 = 1$$
$$\text{and} \quad \hat{g}_2 = \hat{g}_3^{-1}. \tag{6.1-45}$$

This is illustrated in Fig 6.1-3.

This flow situation corresponds to the stretching of a sheet of material at a constant width (in the $\hat{x}_1$-direction). Defining an elongational velocity by

$$\dot{\varepsilon}_p = \frac{1}{\hat{g}_3} \frac{d\hat{g}_3}{dt}, \tag{6.1-46}$$

we obtain for constant $\dot{\varepsilon}_p$

$$\hat{g}_3 = \hat{g}_2^{-1} = \exp\{\dot{\varepsilon}_p(t - t_0)\}. \tag{6.1-47}$$

The covariant and contravariant components of the metric tensor are then

$$\hat{g}_{ij}(t) = \begin{bmatrix} 1 & 0 & 0 \\ 0 & \hat{g}_3^{-2} & 0 \\ 0 & 0 & \hat{g}_3^{2} \end{bmatrix} = \begin{bmatrix} 1 & 0 & 0 \\ 0 & \exp\{2\dot{\varepsilon}_p(t_0 - t)\} & 0 \\ 0 & 0 & \exp\{2\dot{\varepsilon}_p(t - t_0)\} \end{bmatrix} \tag{6.1-48a}$$

and

$$\hat{g}^{ij}(t) = \begin{bmatrix} 1 & 0 & 0 \\ 0 & \hat{g}_3^{2} & 0 \\ 0 & 0 & \hat{g}_3^{-2} \end{bmatrix} = \begin{bmatrix} 1 & 0 & 0 \\ 0 & \exp\{2\dot{\varepsilon}_p(t - t_0)\} & 0 \\ 0 & 0 & \exp\{2\dot{\varepsilon}_p(t_0 - t)\} \end{bmatrix}, \tag{6.1-48b}$$

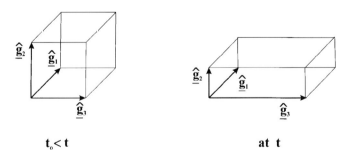

Figure 6.1-3 Planar elongation

and the components of the-rate-of-deformation tensor are

$$\dot{\gamma}_{ij} = \left.\frac{\partial \hat{g}_{ij}}{\partial t}\right|_{t=t_0} = \begin{bmatrix} 0 & 0 & 0 \\ 0 & -2\dot{\varepsilon}_p & 0 \\ 0 & 0 & 2\dot{\varepsilon}_p \end{bmatrix}. \tag{6.1-49}$$

Note that planar elongation is quite different from biaxial elongation defined in Section 2.1-4. For planar elongation, the three invariants of the rate-of-deformation tensor, defined by equations 2.2-13 to 2.2-16, are

$$\begin{aligned} I_{\dot{\gamma}} &= 0, \\ II_{\dot{\gamma}} &= 8\dot{\varepsilon}_p^2, \\ \text{and} \quad III_{\dot{\gamma}} &= 0. \end{aligned} \tag{6.1-50}$$

6.1-4 Expansion or Compression

This type of deformation is defined by

$$\hat{g}_1 = \hat{g}_2 = \hat{g}_3 = V^{\frac{1}{3}}, \tag{6.1-51}$$

and the components of the metric tensor are

$$\hat{g}_{ij}(t) = V^{\frac{2}{3}} \delta_{ij} \tag{6.1-52a}$$

and

$$\hat{g}^{ij}(t) = V^{-\frac{2}{3}} \delta^{ij}. \tag{6.1-52b}$$

This represents approximately the flow situation encountered in bubble expansion and in foaming.

6.1-5 Simple Shear

Simple shear is the most frequently described flow situation in fluid rheology. It is defined by $\hat{g}_3 = $ constant. The variation of $\hat{g}_2$ with time is sketched in Fig. 6.1-4.

The correspondence time is now taken at $t' = t$, and at $t' < t$, the embedded base vector $\hat{\underline{g}}_2$ is at an angle α with respect to its position at time t. Defining the shear strain $\gamma = \tan \alpha$ and noting that

$$\hat{\underline{g}}_1 \cdot \hat{\underline{g}}_2 = \hat{g}_2 \cos\left(\frac{\pi}{2} + \alpha\right) = -\hat{g}_2 \sin \alpha \tag{6.1-53}$$

$$\text{and} \quad \hat{g}_2 = \frac{1}{\cos \alpha}, \tag{6.1-54}$$

206   Nonlinear Viscoelasticity

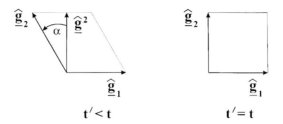

Figure 6.1-4   Simple shearing

we obtain

$$\hat{\underline{g}}_1 \cdot \hat{\underline{g}}_2 = -\tan \alpha = -\gamma \tag{6.1-55}$$

and

$$\hat{\underline{g}}_2 \cdot \hat{\underline{g}}_2 = \frac{1}{\cos^2 \alpha} = 1 + \tan^2 \alpha = 1 + \gamma^2. \tag{6.1-56}$$

The covariant components of the metric tensor are then

$$\hat{g}_{ij}(t') = \begin{bmatrix} 1 & -\gamma & 0 \\ -\gamma & 1+\gamma^2 & 0 \\ 0 & 0 & 1 \end{bmatrix}. \tag{6.1-57}$$

The corresponding contravariant components are obtained by noting that the reciprocal base vectors for incompressible materials are

$$\hat{\underline{g}}^1 = \hat{\underline{g}}_2 \times \hat{\underline{g}}_3, \quad \hat{\underline{g}}^2 = \hat{\underline{g}}_3 \times \hat{\underline{g}}_1, \quad \text{and} \quad \hat{\underline{g}}^3 = \hat{\underline{g}}_1 \times \hat{\underline{g}}_2. \tag{6.1-58}$$

It follows that

$$\hat{g}^{ij}(t') = \begin{bmatrix} 1+\gamma^2 & \gamma & 0 \\ \gamma & 1 & 0 \\ 0 & 0 & 1 \end{bmatrix}. \tag{6.1-59}$$

6.1-5.1   Comments

It is of interest to recall that in the classical theory of elasticity, the shear strain γ is assumed to be very small. Hence, the terms in γ^2 in the metric tensors are negligible and only the shear components have to be considered. The γ^2 are responsible for the normal stresses measured in simple shear for viscoelastic fluids, as discussed in Chapter 2. Normal stresses are clearly phenomena of nonlinear viscoelasticity. For *steady simple shear*,

$$\gamma = \dot{\gamma}(t - t'), \tag{6.1-60}$$

where the shear rate, $\dot{\gamma}$, is a constant. The expressions for the rate-of-deformation tensor are obtained by taking the derivative of the metric tensor with respect to time and making the correspondence at $t' = t$:

$$\dot{\gamma}_{ij} = \left.\frac{d\hat{g}_{ij}}{dt'}\right|_{t'=t} = \begin{bmatrix} 0 & \dot{\gamma} & 0 \\ \dot{\gamma} & 0 & 0 \\ 0 & 0 & 0 \end{bmatrix} \quad (6.1\text{-}61a)$$

$$\text{and} \quad \dot{\gamma}^{ij} = -\left.\frac{d\hat{g}^{ij}}{dt'}\right|_{t'=t} = \begin{bmatrix} 0 & \dot{\gamma} & 0 \\ \dot{\gamma} & 0 & 0 \\ 0 & 0 & 0 \end{bmatrix}. \quad (6.1\text{-}61b)$$

As explained in Section 6.1-1, the derivative of the contravariant components leads to the same numerical value for the rate-of-deformation tensor, but of opposite sign.

Example 6.1-2 Finger and Cauchy-Green Tensors in Simple Shear

Using the general expressions developed in Section 6.1-1 for the Finger and Cauchy-Green tensors (equations 6.1-8 and 6.1-9), obtain their simplified expressions for steady simple shear flow.

Solution

The Cartesian components of $\underline{\underline{C}}^{-1}$ and $\underline{\underline{C}}$ are respectively

$$C_{ij}^{-1} = \left(\frac{\partial x_i}{\partial x'_k}\right)\left(\frac{\partial x_j}{\partial x'_k}\right) \quad (6.1\text{-}62a)$$

$$\text{and} \quad C_{ij} = \left(\frac{\partial x'_k}{\partial x_i}\right)\left(\frac{\partial x'_k}{\partial x_j}\right). \quad (6.1\text{-}62b)$$

For steady simple shear in Cartesian coordinates, the flow kinematics are given by

$$x'_1 = x_1 - \dot{\gamma}(t - t')x_2,$$
$$x'_2 = x_2, \quad (6.1\text{-}63a)$$
$$\text{and} \quad x'_3 = x_3.$$

Hence,

$$\frac{\partial x'_1}{\partial x_1} = \frac{\partial x'_2}{\partial x_2} = \frac{\partial x'_3}{\partial x_3} = 1$$

$$\text{and} \quad \frac{\partial x_1}{\partial x'_2} = -\frac{\partial x'_1}{\partial x_2} = \dot{\gamma}(t - t'). \quad (6.1\text{-}63b)$$

All the other derivatives are zero. (Note that $\partial x_2/\partial x'_1 = 0$ since $x_2 (= x'_2)$ and x'_1 are independent variables.) The components of the Finger tensor are

$$\left.\begin{aligned}
C_{11}^{-1} &= \left(\frac{\partial x_1}{\partial x'_1}\right)\left(\frac{\partial x_1}{\partial x'_1}\right) + \left(\frac{\partial x_1}{\partial x'_2}\right)\left(\frac{\partial x_1}{\partial x'_2}\right) + \left(\frac{\partial x_1}{\partial x'_3}\right)\left(\frac{\partial x_1}{\partial x'_3}\right) = 1 + \dot{\gamma}^2(t-t')^2 \\
C_{12}^{-1} &= \left(\frac{\partial x_1}{\partial x'_1}\right)\left(\frac{\partial x_2}{\partial x'_1}\right) + \left(\frac{\partial x_1}{\partial x'_2}\right)\left(\frac{\partial x_2}{\partial x'_2}\right) + \left(\frac{\partial x_1}{\partial x'_3}\right)\left(\frac{\partial x_2}{\partial x'_3}\right) = C_{21}^{-1} = \dot{\gamma}(t-t') \\
C_{22}^{-1} &= C_{33}^{-1} = 1 \\
C_{23}^{-1} &= C_{32}^{-1} = C_{13}^{-1} = C_{31}^{-1} = 0.
\end{aligned}\right\} \quad (6.1\text{-}64)$$

The Finger tensor is therefore expressed by

$$C_{ij}^{-1} = \begin{bmatrix} 1 + \dot{\gamma}^2(t-t')^2 & \dot{\gamma}(t-t') & 0 \\ \dot{\gamma}(t-t') & 1 & 0 \\ 0 & 0 & 1 \end{bmatrix}, \quad (6.1\text{-}65)$$

which is identical to the result obtained in equation 6.1-59 with $\gamma = \dot{\gamma}(t-t')$. Similarly, the components of the Cauchy-Green tensor are

$$\left.\begin{aligned}
C_{11} &= \left(\frac{\partial x'_1}{\partial x_1}\right)\left(\frac{\partial x'_1}{\partial x_1}\right) + \left(\frac{\partial x'_2}{\partial x_1}\right)\left(\frac{\partial x'_2}{\partial x_1}\right) + \left(\frac{\partial x'_3}{\partial x_1}\right)\left(\frac{\partial x'_3}{\partial x_1}\right) = 1 \\
C_{12} &= \left(\frac{\partial x'_1}{\partial x_1}\right)\left(\frac{\partial x'_1}{\partial x_2}\right) + \left(\frac{\partial x'_2}{\partial x_1}\right)\left(\frac{\partial x'_2}{\partial x_2}\right) + \left(\frac{\partial x'_3}{\partial x_1}\right)\left(\frac{\partial x'_3}{\partial x_2}\right) = -\dot{\gamma}(t-t') \\
C_{22} &= \left(\frac{\partial x'_1}{\partial x_2}\right)\left(\frac{\partial x'_1}{\partial x_2}\right) + \left(\frac{\partial x'_2}{\partial x_2}\right)\left(\frac{\partial x'_2}{\partial x_2}\right) + \left(\frac{\partial x'_3}{\partial x_2}\right)\left(\frac{\partial x'_3}{\partial x_2}\right) = \dot{\gamma}^2(t-t')^2 + 1 \\
C_{33} &= 1 \\
C_{13} &= C_{31} = C_{23} = C_{32} = 0.
\end{aligned}\right\} \quad (6.1\text{-}66)$$

That is,

$$C_{ij} = \begin{bmatrix} 1 & -\dot{\gamma}(t-t') & 0 \\ -\dot{\gamma}(t-t') & 1 + \dot{\gamma}^2(t-t')^2 & 0 \\ 0 & 0 & 1 \end{bmatrix}, \quad (6.1\text{-}67)$$

which is identical to equation 6.1-57, with $\gamma = \dot{\gamma}(t-t')$.

6.2 Formulation of Constitutive Equations

A constitutive equation or rheological equation of state is a mathematical relationship that describes in a given material the stresses as a function of the deformation and/or rate-of-deformation history. In Chapters 2 and 5, we introduced special classes of constitutive equations, generally called rheological models. The generalized Newtonian fluid of Section 2.2 is a valid constitutive equation, but restricted to inelastic fluids. In contrast, all of the rheological models of Chapter 5 are restricted to small deformation flows and are therefore not valid for large deformation flows. These models do not obey the principle of material objectivity or frame invariance described below.

6.2-1 Material Objectivity and Formulation of Constitutive Equations

Following Lodge (1964, 1974), the formulation of admissible (or valid) constitutive equations should be based on the following three principles (largely established by Oldroyd, 1950, 1958, 1965). An alternative formulation has been provided by Coleman and Noll (1961).

(a) Material Objectivity
Any constitutive equation should be formulated in such a way that the relationship between the stress and deformation and rates of deformation is independent of any superposed rigid motion of the material relative to fixed coordinates: that is, independent of local position in space, local rigid rotation, and translation.

(b) Coordinate Indifference
The rheological equation of state should have a significance which is independent of the choice of the base vectors used to express the stress, deformation tensor and rate-of-deformation tensor.

In simpler terms, the results obtained by using a given constitutive equation should not depend on the frame of reference (material objectivity). The stress in a spring should not depend on the local position or rigid rotation, but uniquely on the imposed (internal) deformation. The simple Maxwell model defined in Section 5.2 does not satisfy material objectivity, since the time derivative of the shear stress is a function of the frame of reference used.

By using embedded coordinates, condition (a) is automatically satisfied. By writing the equation in tensor form, condition (b) is automatically satisfied.

(c) Determinism
This principle was initially introduced by Oldroyd (1950) as a third condition for invariance. The stress response in a material element should be invariant to a change of the rheological history of neighboring fluid elements. The concept was reformulated in a different framework by Coleman and Noll (1961). For example, a simple fluid is defined as an incompressible and isotropic material for which the stress at any time (in embedded coordinates) is given by a functional of the strain tensor. That is,

$$\hat{\sigma}^{ij}(t) = -\Phi_{-\infty}^{t}\{(\hat{g}^{ij}(t'))\} \qquad (6.2\text{-}1\text{a})$$

$$\text{and} \quad \hat{\sigma}_{ij}(t) = -\Phi_{-\infty}^{t}\{(\hat{g}_{ij}(t'))\}, \qquad (6.2\text{-}1\text{b})$$

where $\Phi_{-\infty}^{t}$ is a functional of the deformation history. We recall that a function maps a set of numbers to another set of numbers and a functional maps a set of numbers to a set of functions. In equations 6.2-1a and b, the values of the six components of the stress tensor at time t depend on the six functions $\hat{g}_{ij}$ (or $\hat{g}^{ij}$) over the interval $-\infty \leq t' \leq t$. If the functional is linear, it can be represented by an integral. In Section 6.1, we showed that $\hat{g}^{ij}(t')$ or $\hat{g}_{ij}(t')$ completely describes the deformation of a material element at time t'. Implicit in equation 6.2-1a or b is the assumption that the stress in a given element is not affected by the deformation history of neighboring elements. This notion of a simple fluid is obviously not a model of the human behavior!

6.2-2 Maxwell Convected Models

To illustrate these principles, we present here a few classical results. More important constitutive equations are discussed in Sections 6.3 and 6.4. Any of the rheological models introduced in Chapter 5 can be reformulated in embedded coordinates, using contravariant, covariant, or mixed components.

Writing equation 5.2-4 in embedded coordinates generates the following results:

Maxwell A:

$$\left(1 + \lambda_0 \frac{\partial}{\partial t}\right)\hat{\sigma}^{ij} = \eta_0 \frac{\partial \hat{g}^{ij}}{\partial t}, \tag{6.2-2}$$

and *Maxwell B*:

$$\left(1 + \lambda_0 \frac{\partial}{\partial t}\right)\hat{\sigma}_{ij} = -\eta_0 \frac{\partial \hat{g}_{ij}}{\partial t}. \tag{6.2-3}$$

Either equation can be easily integrated from $-\infty$ to t, using the integration factor e^{t/λ_0}. For example, equation 6.2-2 becomes

$$\hat{\sigma}^{ij}(t) = \frac{\eta_0}{\lambda_0}\hat{g}^{ij}(t) - \frac{\eta_0}{\lambda_0^2} e^{-\frac{t}{\lambda_0}} \int_{-\infty}^{t} e^{\frac{t'}{\lambda_0}} \hat{g}^{ij}(t') dt', \tag{6.2-4a}$$

which can be rewritten in a classical form:

Maxwell A':

$$\hat{\sigma}^{ij}(t) = -\frac{\eta_0}{\lambda_0^2}\int_{-\infty}^{t} e^{-\frac{(t-t')}{\lambda_0}} \{\hat{g}^{ij}(t') - \hat{g}^{ij}(t)\} dt'. \tag{6.2-4b}$$

Note that the stresses are assumed to be equal to zero at $t' = -\infty$. This is valid within an isotropic contribution. $\hat{\sigma}^{ij}$ are the deviatoric or extra stress components. Similarly, the integration of equation 6.2-3 yields

Maxwell B':

$$\hat{\sigma}_{ij}(t) = \frac{\eta_0}{\lambda_0^2}\int_{-\infty}^{t} e^{-\frac{(t-t')}{\lambda_0}} \{\hat{g}_{ij}(t') - \hat{g}_{ij}(t)\} dt'. \tag{6.2-5}$$

6.2 Formulation of Constitutive Equations

To obtain the *fixed components*, we use the transformation rules established in Section 6.1, taking the correspondence at $t'=t$. The differential models become

Maxwell A'':

$$\left(1 + \lambda_0 \frac{\delta}{\delta t}\right)\sigma^{ij} = -\eta_0 \dot{\gamma}^{ij} \tag{6.2-6}$$

and *Maxwell B''*:

$$\left(1 + \lambda_0 \frac{\delta}{\delta t}\right)\sigma_{ij} = -\eta_0 \dot{\gamma}_{ij} \tag{6.2-7}$$

where $(\delta \sigma^{ij}/\delta t)$ is the upper-convected derivative defined by equation 6.1-13, and $(\delta \sigma_{ij}/\delta t)$ the lower-convected derivative (equation 6.1-14); and $\dot{\gamma}^{ij}$ and $\dot{\gamma}_{ij}$ are respectively the contravariant and covariant components of the rate-of-deformation tensor (equations 6.1-11 and 6.1-12).

The fixed components of the integrated models are

Maxwell A''':

$$\sigma^{ij}(t) = -\frac{\eta_0}{\lambda_0^2} \int_{-\infty}^{t} e^{-\frac{(t-t')}{\lambda_0}} \{C^{-1\,ij}(t') - g^{ij}(t)\} dt'$$

$$= -\frac{\eta_0}{\lambda_0^2} \int_{-\infty}^{t} e^{-\frac{(t-t')}{\lambda_0}} \Gamma^{-1\,ij}(t') dt', \tag{6.2-8}$$

where the relative strain Finger tensor is

$$\underline{\underline{\Gamma}}^{-1} = \underline{\underline{C}}^{-1} - \underline{\underline{\delta}}. \tag{6.2-9}$$

$\underline{\underline{C}}^{-1}$ is defined by equation 6.1-8.

Maxwell B''':

$$\sigma_{ij} = \frac{\eta_0}{\lambda_0^2} \int_{-\infty}^{t} e^{-\frac{(t-t')}{\lambda_0}} (C_{ij}(t') - g_{ij}(t)) dt'$$

$$= \frac{\eta_0}{\lambda_0^2} \int_{-\infty}^{t} e^{-\frac{(t-t')}{\lambda_0}} \Gamma_{ij}(t') dt', \tag{6.2-10}$$

with the relative Cauchy-Green strain tensor defined by

$$\underline{\underline{\Gamma}} = \underline{\underline{C}} - \underline{\underline{\delta}}. \tag{6.2-11}$$

Obviously, not all these equations are different. Only two sets are different, the contravariant (models A) and the covariant (models B). The differences are illustrated in Example 6.2-1.

Example 6.2-1 Steady Shear Functions for the Convected Maxwell Models

Obtain the viscosity and primary and secondary normal stress functions for the convected Maxwell models in steady shear flow.

Solution

For steady shear flow, the Cartesian components of the velocity gradient tensor and its transpose are

$$V^j_{,i} = \dot{\gamma} \begin{pmatrix} 0 & 0 & 0 \\ 1 & 0 & 0 \\ 0 & 0 & 0 \end{pmatrix} \quad (6.2\text{-}12a)$$

and

$$V^i_{,j} = \dot{\gamma} \begin{pmatrix} 0 & 1 & 0 \\ 0 & 0 & 0 \\ 0 & 0 & 0 \end{pmatrix}. \quad (6.2\text{-}12b)$$

The upper-convected derivative of the stress tensor simplifies to

$$\frac{\delta}{\delta t} \sigma^{ij} = -V^i_{,k}\sigma^{kj} - V^j_{,k}\sigma^{ik} = -V_{i,k}\sigma_{kj} - V_{j,k}\sigma_{ki}$$

$$= -\dot{\gamma} \begin{bmatrix} 2\sigma_{21} & \sigma_{22} & \sigma_{23} \\ \sigma_{22} & 0 & 0 \\ \sigma_{23} & 0 & 0 \end{bmatrix}. \quad (6.2\text{-}13)$$

The lower-convected derivative is

$$\frac{\delta}{\delta t} \sigma_{ij} = V^k_{,i}\sigma_{kj} + V^k_{,j}\sigma_{ik} = V_{k,i}\sigma_{kj} + V_{k,ji}\sigma_{ik}$$

$$= \dot{\gamma} \begin{bmatrix} 0 & \sigma_{11} & 0 \\ \sigma_{11} & 2\sigma_{12} & \sigma_{13} \\ 0 & \sigma_{13} & 0 \end{bmatrix}. \quad (6.2\text{-}14)$$

To obtain these results, we have made use of the symmetry of the stress tensor. All of the superscripts are lowered since the covariant and contravariant components are identical in a Cartesian coordinate system.

The upper-convected Maxwell model (equation 6.2-6) reduces to

$$\begin{bmatrix} \sigma_{11} & \sigma_{12} & \sigma_{13} \\ \sigma_{21} & \sigma_{22} & \sigma_{23} \\ \sigma_{31} & \sigma_{32} & \sigma_{33} \end{bmatrix} - \lambda_0 \dot{\gamma} \begin{bmatrix} 2\sigma_{21} & \sigma_{22} & \sigma_{23} \\ \sigma_{22} & 0 & 0 \\ \sigma_{23} & 0 & 0 \end{bmatrix} = -\eta_0 \begin{bmatrix} 0 & \dot{\gamma} & 0 \\ \dot{\gamma} & 0 & 0 \\ 0 & 0 & 0 \end{bmatrix}, \quad (6.2\text{-}15)$$

which yields the following set of algebraic equations:

$$\sigma_{11} - 2\lambda_0 \dot{\gamma} \sigma_{21} = 0,$$
$$\sigma_{21} - \lambda_0 \dot{\gamma} \sigma_{22} = -\eta_0 \dot{\gamma},$$
$$\sigma_{22} = 0,$$
$$\sigma_{23} = \sigma_{32} = 0, \qquad (6.2\text{-}16)$$
$$\sigma_{31} - \lambda_0 \dot{\gamma} \sigma_{23} = 0,$$
$$\text{and} \quad \sigma_{33} = 0.$$

Simplifying, we obtain

$$\sigma_{21} = -\eta_0 \dot{\gamma},$$
$$\sigma_{11} = -2\lambda_0 \eta_0 \dot{\gamma}^2, \qquad (6.2\text{-}17\text{a})$$
$$\text{and} \quad \sigma_{22} = 0;$$

and

$$\eta = \eta_0,$$
$$\psi_1 = -\frac{(\sigma_{11} - \sigma_{22})}{\dot{\gamma}^2} = 2\lambda_0 \eta_0, \qquad (6.2\text{-}17\text{b})$$
$$\text{and} \quad \psi_2 = -\frac{(\sigma_{22} - \sigma_{33})}{\dot{\gamma}^2} = 0.$$

That is, the upper-convected Maxwell model predicts a constant viscosity, a constant (positive) primary normal stress coefficient, and a secondary normal stress coefficient equal to zero. This is unrealistic, and the model has to be generalized to predict, in particular, shear-thinning effects.

When using the lower-convected Maxwell model, the only difference for steady shear flow is the nonzero ψ_2. The simple shear results for the *lower-convected Maxwell* model are

$$\eta = \eta_0,$$
$$\psi_1 = 2\lambda_0 \eta_0, \qquad (6.2\text{-}18)$$
$$\text{and} \quad \psi_2 = -2\lambda_0 \eta_0.$$

We note that the equal magnitudes of ψ_1 and ψ_2 are not realistic. With respect to the bulk of available data on polymeric systems, the lower-convected Maxwell model is less acceptable than the upper-convected form, which is still far from being an acceptable equation, as the second example will show.

Example 6.2-2 Elongational Viscosity for the Upper-Convected Maxwell Model
Consider the steady uniaxial elongational flow as defined in Section 6.1. Determine the elongational viscosity predicted by the upper-convected Maxwell model.

Solution
From the flow field,

$$\dot{\underline{\underline{\gamma}}} = \dot{\varepsilon}\begin{bmatrix} 2 & 0 & 0 \\ 0 & -1 & 0 \\ 0 & 0 & -1 \end{bmatrix} \quad (6.2\text{-}19)$$

and

$$\nabla\underline{V} = \nabla\underline{V}^{+} = \dot{\varepsilon}\begin{bmatrix} 1 & 0 & 0 \\ 0 & -\frac{1}{2} & 0 \\ 0 & 0 & -\frac{1}{2} \end{bmatrix}. \quad (6.2\text{-}20)$$

As there is no shear component, we readily assume that all shear stress components are equal to zero. The upper-convected derivative of the stress tensor is simplified to

$$\frac{\delta}{\delta t}\sigma_{ij} = -2V_{i,k}\sigma_{kj} = -2\dot{\varepsilon}\begin{bmatrix} \sigma_{11} & 0 & 0 \\ 0 & -\frac{\sigma_{22}}{2} & 0 \\ 0 & 0 & -\frac{\sigma_{33}}{2} \end{bmatrix} \quad (6.2\text{-}21)$$

and the upper-convected Maxwell model, written in matrix form, is

$$\begin{bmatrix} \sigma_{11} & 0 & 0 \\ 0 & \sigma_{22} & 0 \\ 0 & 0 & \sigma_{33} \end{bmatrix} - \lambda_0\dot{\varepsilon}\begin{bmatrix} 2\sigma_{11} & 0 & 0 \\ 0 & -\sigma_{22} & 0 \\ 0 & 0 & -\sigma_{33} \end{bmatrix}$$

$$= -\eta_0\begin{bmatrix} 2\dot{\varepsilon} & 0 & 0 \\ 0 & -\dot{\varepsilon} & 0 \\ 0 & 0 & -\dot{\varepsilon} \end{bmatrix}. \quad (6.2\text{-}22)$$

This generates the following set of equations:

$$\sigma_{11} - 2\lambda_0\dot{\varepsilon}\sigma_{11} = -2\eta_0\dot{\varepsilon},$$
$$\sigma_{22} + \lambda_0\dot{\varepsilon}\sigma_{22} = \eta_0\dot{\varepsilon}, \quad (6.2\text{-}23)$$
$$\text{and} \quad \sigma_{33} + \lambda_0\dot{\varepsilon}\sigma_{33} = \eta_0\dot{\varepsilon},$$

which simplify to

$$\sigma_{22} = \sigma_{33},$$
$$\sigma_{11} = -\frac{2\eta_0\dot{\varepsilon}}{1 - 2\lambda_0\dot{\varepsilon}}, \quad (6.2\text{-}24)$$
$$\text{and} \quad \sigma_{22} = \frac{\eta_0\dot{\varepsilon}}{1 + \lambda_0\dot{\varepsilon}}.$$

6.2 Formulation of Constitutive Equations

The elongational viscosity is hence expressed by

$$\eta_E \equiv -\frac{\sigma_{11} - \sigma_{22}}{\dot{\varepsilon}} = \eta_0 \left[\frac{2}{1 - 2\lambda_0 \dot{\varepsilon}} + \frac{1}{1 + \lambda_0 \dot{\varepsilon}} \right]$$

$$= \frac{3\eta_0}{(1 - 2\lambda_0 \dot{\varepsilon})(1 + \lambda_0 \dot{\varepsilon})}. \quad (6.2\text{-}25)$$

The behavior is illustrated in Fig. 6.2-1, which shows the predictions to be strain-hardening and the elongational viscosity to become unbounded at $\dot{\varepsilon} = \lambda_0/2$. As $\dot{\varepsilon} \to \lambda_0/2$, a steady solution cannot be obtained. This behavior appears to be highly unrealistic, but some polymers, such as branched polyethylenes, have been shown to be strain-hardening (Meissner, 1971). It has been observed that the steady-state elongational viscosity for low density polyethylene at elongational rates as low as 10^{-2} s^{-1} could not be measured. Note that in the limit of λ_0 approaching zero, this result yields the classical *Trouton relation*

$$\eta_E = 3\eta_0. \quad (6.2\text{-}26)$$

The numerous rheological equations that we have presented in this section may appear to be superfluous. However, any one of these or their combinations may lead to more general and useful constitutive equations, as illustrated in the following sections.

6.2-3 Generalized Newtonian Models

The concept of a generalized Newtonian fluid has been introduced in Chapter 2, as a fluid that exhibits a shear-thinning viscosity, but no normal stress differences under simple shear

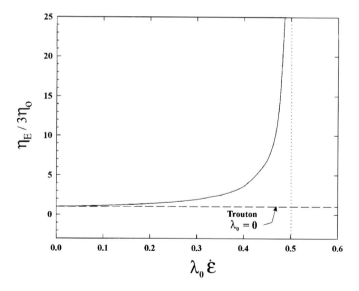

Figure 6.2-1 Steady elongational viscosity of an upper-convected Maxwell fluid

experiments. A more general definition of a purely viscous fluid is to state that the extra stress tensor, $\underline{\underline{\sigma}}$, depends only on the rate-of-deformation tensor $\dot{\gamma}$. Hence, from general principles of continuum mechanics, the total stress tensor of an isotropic material should be an isotropic function of the rate-of-deformation tensor. That is, for a second-order fluid, the total stress is given by

$$\pi^{ij} = \sigma^{ij} + P\delta^{ij} = A(I_{\dot{\gamma}}, II_{\dot{\gamma}}, III_{\dot{\gamma}})\delta^{ij} + B(I_{\dot{\gamma}}, II_{\dot{\gamma}}, III_{\dot{\gamma}})\dot{\gamma}^{ij}$$
$$+ C(I_{\dot{\gamma}}, II_{\dot{\gamma}}, III_{\dot{\gamma}})\dot{\gamma}^{ik}\dot{\gamma}_k^j \qquad (6.2\text{-}27)$$

where $I_{\dot{\gamma}}$, $II_{\dot{\gamma}}$, and $III_{\dot{\gamma}}$ are the three invariants of the rate-of-deformation tensor (equations 2.2-13 to 2.2-16). For incompressible materials, $I_{\dot{\gamma}} = 0$ and equation 6.2-27 becomes, in terms of the extra stress tensor,

$$\sigma^{ij} = B(II_{\dot{\gamma}}, III_{\dot{\gamma}})\dot{\gamma}^{ij} + C(II_{\dot{\gamma}}, III_{\dot{\gamma}})\dot{\gamma}^{ik}\dot{\gamma}_k^j \qquad (6.2\text{-}28)$$

Note that for incompressible materials the pressure or isotropic term cannot be specified in a given flow situation, and the pressure can be arbitrarily set to one third of the trace of the total stress tensor. That is, the trace of the extra (or deviatoric) stress tensor is equal to zero in absence of flow ($\dot{\gamma}^{ii} = 0$). Equation 6.2-28 includes as a special case the Newtonian fluid ($B = -\mu$, $C = 0$).

We can introduce derivatives of the rate-of-deformation tensor, but the resulting equation is then no longer consistent with the definition of a generalized Newtonian fluid. One of the well-known models in this class is the so-called second-order fluid of Rivlin and Ericksen (1955), which may be written as

$$\sigma^{ij} = -\eta_0\dot{\gamma}^{ij} + \beta_1\frac{\delta\dot{\gamma}^{ij}}{\delta t} - \beta_2\dot{\gamma}^{ik}\dot{\gamma}_k^j \qquad (6.2\text{-}29)$$

This second-order model describes primary and secondary normal stress differences, but it cannot predict elastic effects in transient flow experiments such as stress growth or relaxation or recoil, as discussed in Chapter 5. The model is then valid only for steady-state or slowly varying flows (creeping flows) or for low elasticity fluids for which the relaxation time is much smaller than the residence or process time (low Deborah number).

Example 6.2-3 Steady Shear Properties of a Rivlin-Ericksen Fluid
Obtain the material functions for steady simple shear of a Rivlin-Ericksen fluid (equation 6.2-29).

Solution
For steady shear flow, the rate-of-deformation tensor is

$$\dot{\gamma}^{ij} = \begin{bmatrix} 0 & \dot{\gamma} & 0 \\ \dot{\gamma} & 0 & 0 \\ 0 & 0 & 0 \end{bmatrix}. \qquad (6.2\text{-}30a)$$

Hence,

$$\dot{\gamma}^{ik}\dot{\gamma}^{j}_{k} = \begin{bmatrix} \dot{\gamma}^2 & 0 & 0 \\ 0 & \dot{\gamma}^2 & 0 \\ 0 & 0 & 0 \end{bmatrix} \qquad (6.2\text{-}30b)$$

and

$$\frac{\delta \dot{\gamma}^{ij}}{\delta t} = -V^i_{,k}\dot{\gamma}^{kj} - V^j_{,k}\dot{\gamma}^{ik}$$

$$= -\begin{bmatrix} 2\dot{\gamma}^2 & 0 & 0 \\ 0 & 0 & 0 \\ 0 & 0 & 0 \end{bmatrix}. \qquad (6.2\text{-}30c)$$

This result follows from equation 6.2-13, replacing the stress by the rate-of-deformation tensor and noting that only the 12- and 21-components do not vanish. Using results of equations 6.2-30, equation 6.2-29 can be written as

$$\begin{bmatrix} \sigma_{11} & \sigma_{12} & \sigma_{13} \\ \sigma_{21} & \sigma_{22} & \sigma_{23} \\ \sigma_{31} & \sigma_{32} & \sigma_{33} \end{bmatrix} = -\eta_0 \begin{bmatrix} 0 & \dot{\gamma} & 0 \\ \dot{\gamma} & 0 & 0 \\ 0 & 0 & 0 \end{bmatrix} - \beta_1 \begin{bmatrix} 2\dot{\gamma}^2 & 0 & 0 \\ 0 & 0 & 0 \\ 0 & 0 & 0 \end{bmatrix}$$

$$- \beta_2 \begin{bmatrix} \dot{\gamma}^2 & 0 & 0 \\ 0 & \dot{\gamma}^2 & 0 \\ 0 & 0 & 0 \end{bmatrix}. \qquad (6.2\text{-}31)$$

Therefore,

$$\sigma_{12} = \sigma_{21} = -\eta_0 \dot{\gamma},$$
$$\sigma_{11} - \sigma_{22} = -2\beta_1 \dot{\gamma}^2, \qquad (6.2\text{-}32)$$
$$\text{and} \quad \sigma_{22} - \sigma_{33} = -\beta_2 \dot{\gamma}^2,$$

and the material functions are given by

$$\eta = \eta_0,$$
$$\psi_1 = 2\beta_1, \qquad (6.2\text{-}33)$$
$$\text{and} \quad \psi_2 = \beta_2,$$

which are constants.

Example 6.2-4 Simulation of the Flow in a Journal Bearing
A slightly different form of equation 6.2-29 is known as the CEF equation (Criminale, Ericksen, Filbey, 1958). It can be written as

$$\sigma^{ij} = -\eta_0 \dot{\gamma}^{ij} + \frac{1}{2}\psi_1 \frac{D}{Dt}\dot{\gamma}^{ij} - \left(\frac{1}{2}\psi_1 + \psi_2\right)\dot{\gamma}^{ik}\dot{\gamma}^j_k. \qquad (6.2\text{-}34)$$

218   Nonlinear Viscoelasticity

Derdouri and Carreau (1989) used the CEF model to simulate the flow in a journal bearing, schematically shown in Fig. 6.2-2.

Experimental values for the exerted load, W, for two fluids under different conditions are reported in Fig. 6.2-3. The vitrea 320 is a Newtonian oil, whereas the 2.2% PIB L80 is a polyisobutylene solution in vitrea 32, for which the zero shear viscosity at 25 °C is equal to that of vitrea 320. The 2.2% PIB solution is a slightly shear-thinning viscoelastic fluid.

As shown in Fig. 6.2-3, the load increases with increasing eccentricity, ϵ ($=a/(R_2 - R_1)$), but levels off at larges values of ϵ. The lower load generated by the viscoelastic PIB solution compared to the Newtonian case is attributed to shear-thinning and heat dissipation effects, which are considerably more important for the viscoelastic fluid. Thermal effects and shear thinning are equally important in reducing the load capacity of the non-Newtonian oil with respect to the Newtonian oil as is shown in Fig. 6.2-3. The increase of normal stresses by a factor of 1.5 is not sufficient (since it is inhibited by the rise in temperature) to compensate for the loss of viscosity.

Figure 6.2-4 shows the Sommerfeld number for the Newtonian oil (Vitrea 320) and its equivalent non-Newtonian oil (2.2% PIB L80 in a less viscous oil Vitrea 32). The Sommerfeld number is defined by (Cameron, 1966)

$$So = \frac{W(aR_1)^2}{LR_1\Omega_1\eta_0}, \qquad (6.2\text{-}35)$$

where L is the length of the bearing. Up to a relative eccentricity, $\epsilon = 0.5$, both oils give a similar performance; beyond this value, the Newtonian oil has a higher load capacity than its non-Newtonian equivalent as mentioned before. The lower performance of the polymer solution is due to shear-thinning effects that cannot be compensated for by elastic effects. The temperature rise in the liquid film of the bearing results in a large decrease of the normal stresses developed by the polymer-added oil. The solid curve in the figure gives the results obtained for the 2.2% PIB L80 solution when the Reynolds equation, generalized

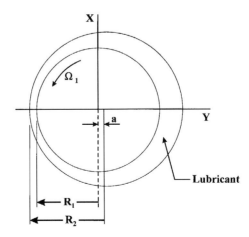

Figure 6.2-2   Schematic view of the journal bearing configuration

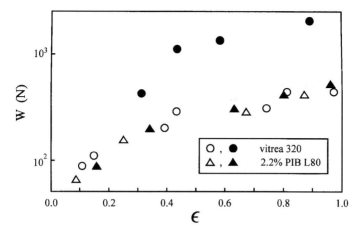

Figure 6.2-3 Applied load vs relative eccentricity (Adapted from Derdouri and Carreau, 1989)
100 rev/min: ○, △
600 rev/min: ▲
650 rev/min: ●

for a non-Newtonian fluid, and the thermal energy equation were solved numerically using a finite difference method by Derdouri (1985) for the CEF model.

The CEF model is strictly valid for viscometric flows, but has been used as a first approximation because of its relative simplicity, as opposed to other more elaborate models (see Davies and Walters, 1973) that used the Oldroyd constitutive equation. The theoretical results are in excellent agreement with the experimental results for low and moderate

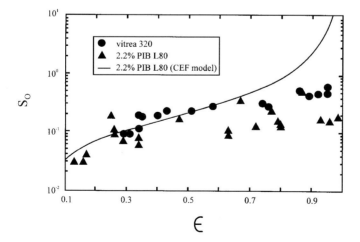

Figure 6.2-4 Sommerfeld number vs relative eccentricity for a journal bearing (From Derdouri and Carreau, 1989)

values of the relative eccentricity, which suggests that the above approximation is valid to some extent.

The solution for the isothermal flow of a second-order fluid in a journal bearing is presented by Bird, Armstrong, and Hassager (1977 and 1987). Their results, as well as those of Davies and Walters (1973) for the Oldroyd model, suggest viscoelastic lubricants will support greater loads, due to a direct contribution of the primary normal stress coefficient. However, the numerical calculations of the non-isothermal case (Derdouri and Carreau, 1989) have shown drastic temperature increases of the lubricant across the gap of the bearing, resulting in severe reductions of viscosity and normal stresses, explaining the little enhancement of the load on the bearing observed with a viscoelastic lubricant.

6.3 Differential Constitutive Equations

The choice of a rheological model depends on the following criteria:

- its capacity to adequately describe the rheological behavior of a given material under flow situations close to those encountered in the process to be studied;
- the possibility and ease of determining the various parameters of the models from simple rheological measurements. In the case of the 8-constant Oldroyd model (Section 6.3-2), only a few parameters can be determined from a simple shear experiment. The other parameters are usually obtained by fitting experimental data with the model predictions for a more complex flow situation. Extrapolation to other flow situations should be carried out with caution;
- the possibility or the ease of using the rheological model for numerical simulation. It is worth noting that the predictions cannot be more accurate than the chosen rheological model. The choice of the model is then necessarily a compromise between accuracy and simplicity.

The models discussed in this section contain a derivative of the stress tensor. Their main advantages (as for the integral models) over the generalized Newtonian models are that they can describe time effects as well as normal stresses in steady simple shear flow. Compared to the integral models, the differential models are usually simpler to use for solving complex flow problems by numerical techniques.

The simplest model of this group is the Maxwell model presented in Section 6.2. As mentioned above, the major drawback of the Maxwell model is its inability to predict shear-thinning properties. Most of the other forms proposed in the literature aim at alleviating this deficiency.

6.3-1 The De Witt Model

De Witt (1955) proposed a model that is similar to the convected Maxwell model in which the convected derivative is replaced by the Jaumann derivative. It can be expressed by

$$\sigma^{ij} + \lambda_0 \frac{\mathcal{D}\sigma^{ij}}{\mathcal{D}t} = -\eta_0 \dot{\gamma}^{ij} \qquad (6.3\text{-}1)$$

where $\mathcal{D}/\mathcal{D}t$ is the Jaumann derivative defined by equation 6.1-21 or 6.1-22. The material functions for steady simple shear are then

$$\eta = \frac{\eta_0}{1 + (\lambda_0 \dot{\gamma})^2},$$

$$\psi_1 = \frac{2\eta_0 \lambda_0}{1 + (\lambda_0 \dot{\gamma})^2}, \qquad (6.3\text{-}2)$$

$$\text{and} \quad \psi_2 = -\frac{\eta_0 \lambda_0}{1 + (\lambda_0 \dot{\gamma})^2}.$$

Shear-thinning effects are predicted, but the slope of -2 for the viscosity in the power-law region is unrealistic. (The shear stress is a decreasing function of the shear rate at large $\dot{\gamma}$.) Negative secondary normal stress differences are predicted, but the magnitude of ψ_2/ψ_1 is not reasonable. More realistic predictions can be obtained by using multimodes, that is, a series of terms as in the generalized Maxwell model

$$\sigma^{ij} = \sum_{p=1}^{M} \sigma^{ij}_{(p)}, \qquad (6.3\text{-}3a)$$

with

$$\sigma^{ij}_{(p)} + \lambda_p \frac{\mathcal{D}}{\mathcal{D}t} \sigma^{ij}_{(p)} = -\eta_p \dot{\gamma}^{ij}. \qquad (6.3\text{-}3b)$$

For small amplitude oscillatory flow, this model reduces to the generalized Maxwell model given by equations 5.2-28a and b, for which the real part of the complex viscosity η' is given by equation 5.2-35. This generalized De Witt equation yields the same expression for the shear viscosity, that is,

$$\eta = \sum_{p=1}^{M} \frac{\eta_p}{1 + (\lambda_p \dot{\gamma})^2}, \qquad (6.3\text{-}4)$$

which shows complete analogy between η and η'. This is quite different from the Cox-Merz analogy (η and η^*) discussed in Section 2.4.

The De Witt model is a special case of the more general Oldroyd model introduced in the next section.

6.3-2 The Oldroyd Models

Oldroyd proposed quite a few rheological models, but his most general one is known as the 8-constant model (Oldroyd, 1958), expressed in tensorial form by

$$\underline{\underline{\sigma}} + \lambda_1 \frac{D\underline{\underline{\sigma}}}{Dt} - \frac{1}{2}\mu_1\{\underline{\underline{\sigma}} \cdot \underline{\dot{\underline{\gamma}}} + \underline{\dot{\underline{\gamma}}} \cdot \underline{\underline{\sigma}}\} + \frac{1}{2}\mu_0\{\text{tr } \underline{\underline{\sigma}}\}\underline{\dot{\underline{\gamma}}} + \frac{1}{2}v_1 \text{ tr}\{\underline{\underline{\sigma}} \cdot \underline{\dot{\underline{\gamma}}}\}\underline{\underline{\delta}}$$
$$= -\eta_0\left[\underline{\dot{\underline{\gamma}}} + \lambda_2 \frac{D}{Dt}\underline{\dot{\underline{\gamma}}} - \mu_2\underline{\dot{\underline{\gamma}}}^2 + \frac{1}{2}v_2 \text{ tr}\{\underline{\dot{\underline{\gamma}}}^2\}\underline{\underline{\delta}}\right], \tag{6.3-5}$$

where tr stands for trace of the tensor and $\underline{\dot{\underline{\gamma}}}^2 = \underline{\dot{\underline{\gamma}}} \cdot \underline{\dot{\underline{\gamma}}}$. The eight model parameters are $\lambda_1, \lambda_2, \mu_0, \mu_1, \mu_2, v_1, v_2$, and η_0. Obviously, all the parameters cannot be determined from steady simple shear experiments, for which the model predictions are given by

$$\frac{\eta}{\eta_0} = \frac{1 + [\lambda_1\lambda_2 + \mu_0(\mu_2 - \frac{3}{2}v_2) - \mu_1(\mu_2 - v_2)]\dot{\gamma}^2}{1 + [\lambda_1^2 + \mu_0(\mu_1 - \frac{3}{2}v_1) - \mu_1(\mu_1 - v_1)]\dot{\gamma}^2}, \tag{6.3-6a}$$

$$\frac{\psi_1}{2\eta_0\lambda_1} = \frac{\eta(\dot{\gamma})}{\eta_0} - \frac{\lambda_2}{\lambda_1}, \tag{6.3-6b}$$

and $\quad \dfrac{\psi_2}{\eta_0\lambda_1} = -\left(1 - \dfrac{\mu_1}{\lambda_1}\right)\dfrac{\eta(\dot{\gamma})}{\eta_0} + \dfrac{\lambda_2}{\lambda_1}\left(1 - \dfrac{\mu_2}{\lambda_2}\right). \tag{6.3-6c}$

For small amplitude oscillatory shear flow, the Jaumann derivative reduces to the partial time derivative, and all nonlinear terms in equation 6.3-5 vanish. The real and imaginary parts of the complex viscosity are given by

$$\frac{\eta'}{\eta_0} = \frac{1 + \lambda_1\lambda_2\omega^2}{1 + (\lambda_1\omega)^2} \tag{6.3-7a}$$

and

$$\frac{\eta''}{\eta_0} = \frac{(\lambda_1 - \lambda_2)\omega}{1 + (\lambda_1\omega)^2}. \tag{6.3-7b}$$

This is the result obtained for the Jeffreys model in Section 5.2-3. Obviously, this result is acceptable only if $\lambda_2 \leq \lambda_1$. Note that the analogy between η' and η is in general not observed. The De Witt model is recovered by setting all the parameters, except η_0, λ_1, and λ_2, equal to zero. The upper-convected Maxwell model is recovered by setting $\lambda_1 = \mu_1$, and $v_1 = v_2 = \mu_0 = \lambda_2 = \mu_2 = 0$. Two classical three-parameter Oldroyd models are

Oldroyd A: with $\quad \mu_1 = -\lambda_1, \quad \mu_2 = -\lambda_2, \quad \mu_0 = v_1 = v_2 = 0$

and *Oldroyd B*: with $\quad \mu_1 = \lambda_1, \quad \mu_2 = \lambda_2, \quad \mu_0 = v_1 = v_2 = 0.$

Numerous numerical simulations of viscoelastic flows have been carried out, using the Oldroyd-B model (see Crochet, Davies, and Walters, 1984). The main reason for the Oldroyd-B model's popularity is its simplicity. It contains a retardation time, λ_2, which allows for predictions of creep and recovery behavior, which are more realistic than those of the Maxwell model. Also, the result for the steady uniaxial elongational viscosity, given by

$$\eta_E = 3\eta_0 \frac{1 - \lambda_2 \dot{\varepsilon} - 2\lambda_1 \lambda_2 \dot{\varepsilon}^2}{1 - \lambda_1 \dot{\varepsilon} - 2\lambda_1^2 \dot{\varepsilon}^2}, \qquad (6.3\text{-}8)$$

is slightly more flexible than the result obtained for the Maxwell model. The Maxwell model result is recovered by setting λ_2 equal to zero. For $\lambda_2 = \lambda_1$, Newtonian behavior is predicted, and equation 6.3-8 reduces to the Trouton relation. For $\lambda_2 < \lambda_1$, the behavior is similar to that shown in Fig. 6.2-1 for the Maxwell model, but the increase of the elongational viscosity with the elongational rate is not as sharp. Note that both the Oldroyd-A and B models predict a constant shear viscosity, which is a very severe limitation when we wish to simulate the flow of polymers, which are generally shear-thinning. A more realistic model in this respect is that proposed by White and Metzner (1963), discussed in Section 6.3-3.

6.3-3 The White-Metzner Model

White and Metzner (1963) have proposed a generalization of the Maxwell model by taking the viscosity and the relaxation time to be functions of the rate of deformation. Their model can be expressed by

$$\left(1 + \lambda(II_{\dot{\gamma}}) \frac{\delta}{\delta t}\right) \sigma^{ij} = -\eta(II_{\dot{\gamma}}) \dot{\gamma}^{ij} \qquad (6.3\text{-}9)$$

where the relaxation time and the viscosity term are unspecified functions of the second invariant of the rate-of-deformation tensor, which are determined from experimental data. For steady simple shear flow, the material functions are

$$\eta = \eta(II_{\dot{\gamma}}), \qquad (6.3\text{-}10\text{a})$$

$$\psi_1 = 2\eta(II_{\dot{\gamma}})\lambda(II_{\dot{\gamma}}), \qquad (6.3\text{-}10\text{b})$$

$$\text{and} \quad \psi_2 = 0. \qquad (6.3\text{-}10\text{c})$$

Except for the zero secondary normal stress coefficient, the model can adequately describe the steady shear behavior of any polymeric material. For transient flows, the White–Metzner equation as well as the upper-convected Maxwell model (Section 6.2-2) suffer from the following deficiencies. With a single relaxation time, stress growth and relaxation behavior cannot be adequately described: no overshoot is predicted in the start-up of steady shear flow, and on cessation of steady simple shear flow, the stress relaxation is independent of the initially imposed shear rate (see Problem 6.6-7).

Example 6.3-1 Steady-State Properties of a White–Metzner Fluid

Determine the steady-state simple shear functions and the uniaxial elongational viscosity of a White–Metzner fluid, for which the shear-thinning properties are described by the expression suggested by Ide and White (1977):

$$\lambda(II_{\dot{\gamma}}) = \frac{\lambda_0}{1 + a\lambda_0 \sqrt{\frac{1}{2} II_{\dot{\gamma}}}} \qquad (6.3\text{-}11a)$$

and $\quad \eta(II_{\dot{\gamma}}) = G_0 \lambda(II_{\dot{\gamma}}),\qquad$ (6.3-11b)

where a is a dimensionless parameter, and λ_0 and G_0 are a characteristic time and modulus respectively.

Solution

The steady shear functions are given by equations 6.3-10a, b, and c. Combining with equations 6.3-11a and b, we obtain

$$\eta = \frac{G_0 \lambda_0}{1 + a\lambda_0 |\dot{\gamma}|} = \frac{\eta_0}{1 + a\lambda_0 |\dot{\gamma}|}, \qquad (6.3\text{-}12a)$$

$$\psi_1 = \frac{2G_0 \lambda_0^2}{(1 + a\lambda_0 |\dot{\gamma}|)^2} = \frac{\psi_{10}}{(1 + a\lambda_0 |\dot{\gamma}|)^2}, \qquad (6.3\text{-}12b)$$

and $\quad \psi_2 = 0.\qquad$ (6.3-12c)

The steady uniaxial elongational properties can be readily calculated from the results of Example 6.2-2, replacing λ_0 and η_0 by $\lambda(II_{\dot{\gamma}})$ and $\eta(II_{\dot{\gamma}})$. The second invariant is $II_{\dot{\gamma}} = \text{tr}(\underline{\dot{\gamma}} \cdot \underline{\dot{\gamma}})$ where $\underline{\dot{\gamma}}$ is given by equation 6.2-19. It follows that $II_{\dot{\gamma}} = 6\dot{\varepsilon}^2$, and the steady elongational viscosity is expressed by

$$\eta_E = \frac{3G_0\lambda_0(1 + a\sqrt{3}\lambda_0\dot{\varepsilon})}{[1 + (a\sqrt{3} - 2)\lambda_0\dot{\varepsilon}][1 + (a\sqrt{3} + 1)\lambda_0\dot{\varepsilon}]}. \qquad (6.3\text{-}13)$$

Thus for $a > 2/\sqrt{3}$, the singularity is eliminated and the elongational viscosity is a decreasing function of the elongational rate.

The empiricism proposed in equations 6.3-11a and b leads to some useful results. The use of the model, in the form proposed by Ide and White (1977), should be restricted to small values of the rate of deformation, since the expression for the viscosity (equation 6.3-12a) predicts a power-law exponent of -1, which is unrealistic. Also, as pointed out previously, the White–Metzner model with a single relaxation time would not quantitatively describe transient experiments. Other interesting remarks on the White-Metzner model can be found in Larson (1988).

Example 6.3-2 Spinning of a Viscoelastic Fiber

Synthetic fibers are manufactured by extruding a polymer melt or solution through a small orifice known as a spinneret. The liquid extrudate is stretched as it travels to a take-up device. During the passage from the spinneret to the take-up device, the filament is cooled or the solvent evaporates to obtain a solidified polymeric filament. As a first approximation,

6.3 Differential Constitutive Equations

we will assume that the filament is isothermal and the fluid can be described by the White–Metzner equation. We will also neglect extrudate swelling at the exit of the spinneret and analyze the flow illustrated in Fig. 6.3-1.

The radius of the spinneret is R_0 and the polymer is extruded at the average velocity V_0. The take-up roller is at a distance L from the spinneret. The polymeric filament is drawn at the velocity V_L.

Obtain the expression for the velocity profile and for the force exerted on the filament, assuming negligible air drag, capillary, and gravitational effects. The viscosity term in the White–Metzner model is described by a power-law expression.

Solution

We assume steady-state conditions. The stress components and V_z are functions of z only. From an overall mass balance, we have

$$\frac{V_z}{V_0} = \left(\frac{R_0}{R(z)}\right)^2. \tag{6.3-14}$$

The equation of continuity for an incompressible fluid simplifies to

$$\frac{1}{r}\frac{\partial}{\partial r}(rV_r) + \frac{dV_z}{dz} = 0. \tag{6.3-15}$$

This equation can be integrated, using the boundary condition that $V_r = 0$ at $r = 0$, to obtain V_r:

$$V_r = -\frac{r}{2}\frac{dV_z}{dz}. \tag{6.3-16}$$

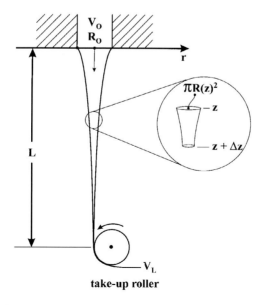

Figure 6.3-1 Fiber spinning

226   Nonlinear Viscoelasticity

The components of the velocity gradient tensor are then

$$\nabla \underline{V} = \begin{bmatrix} -\dfrac{1}{2}\dfrac{dV_z}{dz} & 0 & 0 \\ 0 & -\dfrac{1}{2}\dfrac{dV_z}{dz} & 0 \\ -\dfrac{1}{2}r\dfrac{d^2V_z}{dz^2} & 0 & \dfrac{dV_z}{dz} \end{bmatrix}. \qquad (6.3\text{-}17)$$

The order of magnitude of dV_z/dz is (V_L/L) and that of rd^2V_z/dz^2 is $(V_L R_0/L^2)$. Clearly, for $R_0/L \ll 1$, the term containing the second derivative in equation 6.3-17 can be neglected. The only nonzero components of the White–Metzner tensorial equation are the rr-, $\theta\theta$-, and zz-components. Since $\sigma_{rr} = \sigma_{\theta\theta}$, the two independent components are

$$\sigma_{rr} + \frac{\eta(\bar{\dot{\gamma}})}{G_0}\left[V_z \frac{d\sigma_{rr}}{dz} + \frac{dV_z}{dz}\sigma_{rr}\right] = \eta(\bar{\dot{\gamma}})\frac{dV_z}{dz} \qquad (6.3\text{-}18)$$

and

$$\sigma_{zz} + \frac{\eta(\bar{\dot{\gamma}})}{G_0}\left[V_z \frac{d\sigma_{zz}}{dz} - 2\frac{dV_z}{dz}\sigma_{zz}\right] = -2\eta(\bar{\dot{\gamma}})\frac{dV_z}{dz}, \qquad (6.3\text{-}19)$$

where the relaxation time for the White–Metzner model is $\lambda(II_{\dot{\gamma}}) = \eta(\bar{\dot{\gamma}})/G_0$, with $\bar{\dot{\gamma}} = \sqrt{II_{\dot{\gamma}}/2} = \sqrt{3}dV_z/dz$ (see Section 2.2-13), and $\eta(\bar{\dot{\gamma}}) = m|\bar{\dot{\gamma}}|^{n-1}$. Neglecting the capillary, drag, and gravitational forces, we write a momentum balance on a control volume $\pi R^2(z)\Delta z$ as follows:

$$(\rho V_z V_z + P + \sigma_{zz})\pi R^2|_z - (\rho V_z V_z + P + \sigma_{zz})\pi R^2|_{z+\Delta z} = 0. \qquad (6.3\text{-}20)$$

Note that the pressure of the surroundings acting on the filament has been arbitrarily set equal to zero, since it has no net contribution to the force balance. Dividing by Δz and taking the limit as $\Delta z \to 0$ (see Bird, Stewart, and Lighfoot, 1960), we get

$$2\rho V_z V'_z + \frac{d}{dz}(P + \sigma_{zz}) = -\frac{2}{R}R'(\rho V_z V_z + P + \sigma_{zz}), \qquad (6.3\text{-}21)$$

where $V'_z = dV_z/dz$ and $R' = dR/dz$. From equation 6.3-14,

$$V'_z = -\frac{2V_z}{R}R', \qquad (6.3\text{-}22)$$

and combining equations 6.3-21 and 6.3-22, we obtain

$$\rho V_z V'_z + \frac{d}{dz}(P + \sigma_{zz}) = \frac{V'_z}{V_z}(P + \sigma_{zz}). \qquad (6.3\text{-}23)$$

To eliminate the pressure term, unknown inside the polymeric filament, we make use of the r-component of the equation of motion,

$$P + \sigma_{rr} = \text{constant} = P_a = 0. \qquad (6.3\text{-}24)$$

6.3 Differential Constitutive Equations

For viscous polymer melts or solutions, the inertial term can be neglected ($\rho V_z V_z' \approx 0$) and equation 6.3-23 reduces to

$$\frac{d}{dz}(\sigma_{zz} - \sigma_{rr}) = \frac{V_z'}{V_z}(\sigma_{zz} - \sigma_{rr}). \tag{6.3-25}$$

Equations 6.3-18, 6.3-19, and 6.3-25 have to be solved subject to the following boundary conditions:

$$\text{B.C. 1:} \quad \text{at} \quad z = 0, \quad V_z = V_0, \tag{6.3-26a}$$

$$\text{B.C. 2:} \quad \text{at} \quad z = 0, \quad \sigma_{zz} - \sigma_{rr} = -\frac{F}{\pi R_0^2}, \tag{6.3-26b}$$

$$\text{and} \quad \text{B.C. 3:} \quad \text{at} \quad z = L, \quad V_z = V_L. \tag{6.3-26c}$$

Boundary condition (2) is not in fact a boundary condition, but is simply a condition imposed by the overall force balance over the entire filament. F, the drawing force exerted on the filament, is equal to the tension inside the filament multiplied by the cross section of the filament. This is correct since we have neglected gravitational, inertial, capillary, and drag forces.

The problem is solved using the following dimensionless variables:

$$\zeta = \frac{z}{L}, \quad \phi = \frac{V_z}{V_0}, \quad \text{and} \quad S_{ij} = \sigma_{ij}\frac{\pi R_0^2}{F}. \tag{6.3-27}$$

The three equations (6.3-18, 6.3-19, and 6.3-25) become

$$S_{rr} + De\phi'^{n-1}\left[\phi\frac{dS_{rr}}{d\zeta} + \phi' S_{rr}\right] = N\phi'^n, \tag{6.3-28}$$

$$S_{zz} + De\phi'^{n-1}\left[\phi\frac{dS_{zz}}{d\zeta} - 2\phi' S_{zz}\right] = -2N\phi'^n, \tag{6.3-29}$$

and

$$\frac{d}{d\zeta}(S_{zz} - S_{rr}) = \frac{1}{\phi}\phi'(S_{zz} - S_{rr}), \tag{6.3-30}$$

where $\phi' = d\phi/d\zeta$, the Deborah number is

$$De = \frac{\lambda_{\text{eff}} V_0}{L} = 3^{\frac{(n-1)}{2}}\left(\frac{m}{G_0}\right)\left(\frac{V_0}{L}\right)^n, \tag{6.3-31}$$

and the dimensionless number N is defined by

$$N = \frac{m\left(\sqrt{3}\frac{V_0}{L}\right)^{n-1}\left(\frac{V_0}{L}\right)}{\frac{F}{\pi R_0^2}}. \tag{6.3-32}$$

The Deborah number is a ratio of a fluid characteristic time ($\lambda_{eff} = (m/G_0)(\sqrt{3}V_0/L)^{n-1}$) to a characteristic process time (L/V_0). N is the ratio of the viscous stresses to the total tension in the filament. The boundary conditions are now expressed as

$$\text{B.C. 1:} \quad \text{at} \quad \zeta = 0, \quad \phi = 1, \tag{6.3-33a}$$

$$\text{B.C. 2:} \quad \text{at} \quad \zeta = 0, \quad S_{zz} - S_{rr} = -1, \tag{6.3-33b}$$

$$\text{and} \quad \text{B.C. 3:} \quad \text{at} \quad \zeta = 1, \quad \phi = \frac{V_L}{V_0} = DDR. \tag{6.3-33c}$$

The ratio V_L/V_0 is known as the draw-down ratio, DDR. Integrating equation 6.3-30 and using boundary condition (2), we obtain

$$S_{zz} - S_{rr} = C\phi = -\phi. \tag{6.3-34}$$

This result can be combined with equations 6.3-28 and 6.3-29 to yield

$$S_{zz} = \frac{-\phi}{3De\phi'^n} - \frac{2}{3}\phi + \frac{N}{De}. \tag{6.3-35}$$

Note that the ratio N/De is equal to $G_0/(F/\pi R_0^2)$, that is, the fluid modulus over the tensile stresses in the filament. Combining equations 6.3-29 and 6.3-35 yields

$$\phi + (De\phi - 3N)\phi'^n - 2De^2\phi\phi'^{2n} - nDe\phi^2\phi''\phi'^{n-2} = 0, \tag{6.3-36}$$

where $\phi'' = d^2\phi/d\zeta^2$. This equation has to be solved numerically with the boundary conditions (1) and (3). However, analytical solutions exist for the following two limiting cases:

(a) Purely viscous fluid ($De \to 0$)

$$\phi = \left[1 + (3N)^{-\frac{1}{n}}\left(\frac{n-1}{n}\right)\zeta\right]^{\frac{n}{n-1}} \tag{6.3-37}$$

The expression for N is then obtained from equation 6.3-37 using boundary condition 3:

$$N = \frac{1}{3}\left[\frac{DDR^{\frac{(n-1)}{n}} - 1}{\frac{(n-1)}{n}}\right]^{-n}. \tag{6.3-38}$$

The draw-down force can be calculated from this expression via equation 6.3-32, and the expression for the stress in the filament is obtained by combining equations 6.3-35 and 6.3-37:

$$S_{zz} = -\frac{2}{3}\phi. \tag{6.3-39}$$

This describes the purely viscous fluid solution in the absence of inertial, capillary, gravitational, and drag forces.

6.3 Differential Constitutive Equations

(b) High Deborah number ($N/De \to 0$)

The second limit is of more interest for the spinning of viscoelastic fluids. The term $3N\phi'^n$ in equation 6.3-36 can be neglected, and the solution is

$$\phi = 1 + \frac{\zeta}{De^{\frac{1}{n}}}. \qquad (6.3\text{-}40)$$

The velocity profile becomes a linear function of z. Combining equations 6.3-35 and 6.3-40, we obtain the following expression for the stress:

$$S_{zz} = -\left(1 + \frac{\zeta}{De^{\frac{1}{n}}}\right). \qquad (6.3\text{-}41)$$

The stress is also a linear function of the axial position. Note that using the third boundary condition, equation 6.3-40 gives

$$\phi|_{\zeta=1} = DDR = 1 + \frac{1}{De^{\frac{1}{n}}}, \qquad (6.3\text{-}42)$$

and since for any finite Deborah number, the limiting case considered here corresponds to infinite tension ($N \to 0$), equation 6.3-42 also represents a limiting value for the draw-down ratio. We should, therefore, require

$$DDR < 1 + De^{-\frac{1}{3}}. \qquad (6.3\text{-}43)$$

Fisher and Denn (1976) have solved equation 6.3-36 numerically for various values of the Deborah number. The initial condition chosen is $S_{zz}(0) = -1$, which is the result of equation 6.3-41 for the highly elastic case. Note that $S_{zz}(0) = -2/3$ for purely viscous fluids (equation 6.3-39). Figure 6.3-2 compares Fisher and Denn's (1976) calculations to velocity data of a polystyrene solution with a power-law index, n, equal to $1/3$. The figure shows that the velocity profile becomes more and more linear as the fluid elasticity (De) increases. The limiting highly elastic case is obtained for $De = 0.58$. The experimental data are well described for a value of De between 0.4 and 0.5.

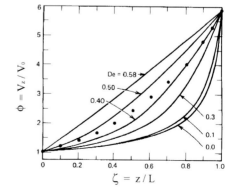

Figure 6.3-2 Predictions of the velocity profile in fiber spinning of a White-Metzner fluid compared to experimental data for a polystyrene melt at 170 °C; $n = 1/3$ and $DDR = 5.85$ (Adapted from Fisher and Denn, 1976)

230   Nonlinear Viscoelasticity

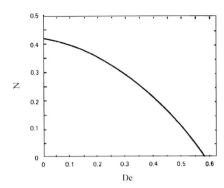

Figure 6.3-3 Numerical results of the force ratio as a function of the Deborah number for a White-Metzner fluid; $n = 1/3$, DDR $= 5.85$, and $S_{zz}(0) = -1$. N is defined in equation 6.3-32 (Adapted from Fisher and Denn, 1976)

The numerical results for the force ratio are reported in Fig. 6.3-3. The value of N decreases with increasing De, that is, the fluid elasticity contribution to the drawing force increases with De. The experimental conditions correspond to a value of De between 0.21 and 0.31 and $N = 0.08$. This value for N is considerably lower than the value of about 0.3 predicted from Fig. 6.3-3 for the experimental value of De. The velocity profile calculations (Fig. 6.3-2) are shown to be less sensitive to the model inadequacies. Some of the numerous assumptions made in this analysis may explain the differences between predictions and experimental results.

6.3-4 The Marrucci Model

The so-called Marrucci model was originally proposed by Acierno et al. (1976), based on the concept of network junctions discussed in Chapter 7. The model is given by the following set of equations:

$$\sigma^{ij} = \sum_P \sigma^{ij}_{(p)} \qquad (6.3\text{-}44\text{a})$$

$$\frac{1}{G_p}\sigma^{ij}_{(p)} + \lambda_p \frac{\delta}{\delta t}\left(\frac{\sigma^{ij}_{(p)}}{G_p}\right) = -\lambda_p \dot{\gamma}^{ij} \qquad (6.3\text{-}44\text{b})$$

$$G_p = G_{0p} x_p; \quad \lambda_p = \lambda_{0p} x_p^{1.4} \qquad (6.3\text{-}44\text{c})$$

and $\quad \dfrac{dx_p}{dt} = \dfrac{1}{\lambda_p}(1 - x_p) - a x_p \dfrac{1}{\lambda_p}\sqrt{\dfrac{E_p}{G_p}}. \qquad (6.3\text{-}44\text{d})$

G_{0p} and λ_{0p} are equilibrium values, describing the material behavior in the linear viscoelastic range. x_p describes the degree of structural non-equilibrium. The elastic modulus G_p is proportional to the junction concentration, and the time-dependent structural parameter x_p is a function of the junction creation due to Brownian motion, which in turn is proportional to the junction concentration difference between the equilibrium and the present states; and the rate of loss of junctions. The rate of loss of junctions (last term in equation

6.3-44d) is taken to be proportional to the number of junctions ($\sim x_p$) and a function of the excess free energy, E_p, given by

$$E_p = \frac{1}{2}\mathrm{tr}\,\underline{\underline{\sigma}}_{(p)} = \frac{1}{2}I_{\sigma(p)} \tag{6.3-45}$$

where $I_{\sigma(p)}$ is the first invariant of the contribution to the stress tensor of the pth mode polymeric chains and a is an empirical constant.

Conceptually, the Marrucci model is quite attractive and can describe fairly well most of the experimentally determined rheological functions of polymer melts or concentrated solutions. The model is, however, difficult to use for the calculation of transient flows, since equations 6.3-44 and 6.3-45 are implicit functions of the stress tensor.

For steady shear flow ($dx_p/dt = 0$) the model simplifies to

$$\frac{1-x_p}{x_p^{2.4}} = a\lambda_{0p}\dot\gamma. \tag{6.3-46}$$

The value of x_p can be determined as a function of $\dot\gamma$ for any value of the linear relaxation time, λ_{0p}. The parameter a is determined by fitting the steady viscosity and/or primary normal stress coefficient data using

$$\eta = \sum_p G_{0p}\lambda_{0p}x_p^{2.4} \tag{6.3-47a}$$

$$\text{and}\quad \psi_1 = 2\sum_p G_{0p}\lambda_{0p}^2 x_p^{3.8}. \tag{6.3-47b}$$

The secondary normal stress coefficient is equal to zero. The value of the parameter a is found to be around 0.4 for polymer melts. Figure 6.3-4 compares the model predictions for three values of a to steady shear viscosity and primary normal stress data of a low density polyethylene.

The stress growth function for polyethylene melts is well described by the Marrucci model, as illustrated in Fig. 6.3-5 for the shear stress growth function.

Other interesting comparisons of the model predictions and transient data can be found in Acierno et al. (1976). Deviations for stress growth are found to be appreciable at high

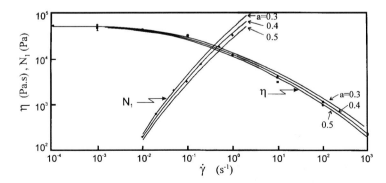

Figure 6.3-4 Marrucci model predictions compared to steady shear viscosity and primary normal stress data of a low density polyethylene melt (Adapted from Acierno et al., 1976)

232 Nonlinear Viscoelasticity

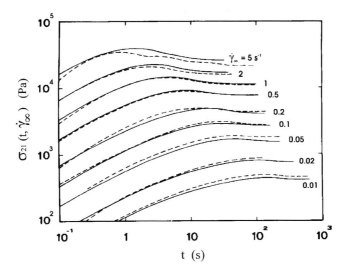

Figure 6.3-5 Marrucci model predictions compared to the shear stress growth data for a low density polyethylene melt at 150 °C (Adapted from Acierno et al., 1976)
——— Data from Meissner (1971)
– – – Model predictions for $a = 0.4$

shear rates. Also, Khan and Larson (1987) found that the Marrucci model showed pronounced oscillations in elongational stress growth and in shear stress relaxation after a step shearing.

Extensions or modifications of the Marrucci model have been proposed by De Cleyn and Mewis (1981) and by Mewis and Denn (1983). These authors used in their model the non-affine differential operator of Phan-Thien and Tanner (1977) instead of the upper-convected derivative. This allows for nonzero secondary normal stress differences, as discussed in Section 6.3-5. More recently, De Kee and Chan Man Fong (1994) introduced a single-mode constitutive equation, involving a structural parameter. In Chapter 2 we discussed the special cases of the steady and time-dependent behavior of inelastic materials. In Chapter 7 we will introduce the general elastic case through statistical mechanics. We note that the set of equations 6.3-44 can be considered to be a special case of De Kee and Chan Man Fong's model.

6.3-5 The Phan-Thien-Tanner Model

Independently to Acierno et al. (1976), Phan-Thien and Tanner (1977) derived a constitutive equation from a Lodge-Yamamoto type of network theory (see Chapter 7). The network junctions are no longer assumed to move as points of the continuum. That is, Phan-Thien and Tanner relaxed the hypothesis of affine deformation. The rates of creation and loss of junctions are assumed to depend on the instantaneous elastic energy of the network. This concept was also used by Acierno et al. (1976). The rheological model proposed by Phan-Thien and Tanner (1977) can be described by the following set of equations:

$$\sigma^{ij} = \sum_p \sigma^{ij}_{(p)} \qquad (6.3\text{-}48a)$$

$$\text{and} \quad \lambda_p \frac{\bar{\delta}}{\delta t}\sigma^{ij}_{(p)} + \sigma^{ij}_{(p)}\left(1 + \frac{a}{G_p}I_{\sigma_{(p)}}\right) = -\frac{G_p\lambda_p}{1-\xi}\dot{\gamma}^{ij} \qquad (6.3\text{-}48b)$$

where the non-affine differential operator is defined by

$$\frac{\bar{\delta}}{\delta t}\underline{\underline{\sigma}} = \frac{D}{Dt}\underline{\underline{\sigma}} - \underline{\underline{L}}\cdot\underline{\underline{\sigma}} - \underline{\underline{\sigma}}\cdot\underline{\underline{L}}^+, \qquad (6.3\text{-}49)$$

and $\underline{\underline{L}}$ is the effective velocity gradient tensor given by

$$\underline{\underline{L}} = \nabla\underline{V} - \frac{\xi}{2}\underline{\underline{\dot{\gamma}}}. \qquad (6.3\text{-}50)$$

$\bar{\delta}/\delta t$ is known as the Gordon-Schowalter (1972) derivative. Both a (ϵ in Phan-Thien and Tanner, 1977) and ξ are empirical parameters. The latter is referred to as the slip parameter. Following Acierno et al. (1976), the elastic energy is assumed to be proportional to the trace of the stress tensor (the first invariant, $I_{\sigma(p)}$). Hence, their rheological model is, as with the Marrucci model, also implicit in to the stress tensor. For $\xi=0$ (no slip or affine deformation), the derivative defined in equation 6.3-49 reduces to the upper-convected derivative (equation 6.1-13).

λ_p and G_p are obtained from complex viscosity data, and the two nonlinear parameters a and ξ are obtained by fitting simple shear and/or elongational viscosity data. Phan-Thien and Tanner (1977) obtained good agreement between their model predictions and viscometric data for a low density polyethylene melt. The predictions of the viscometric functions are not sensitive to the value of the parameter a. However, a steady elongational viscosity is predicted at all elongational rates. and its value depends on a. The slip parameter can be obtained from the normal stress differences, since the model yields

$$\frac{\psi_2}{\psi_1} = -\frac{\xi}{2}. \qquad (6.3\text{-}51)$$

Due to the lack of secondary normal stress data, Phan-Thien and Tanner (1977) assumed $\xi=0.2$. For $a=0$, the Phan-Thien Tanner (PTT) model is equivalent to the equation of Johnson and Segalman (1981). The main problem associated with the use of the non-affine differential operator is the oscillations predicted by the model for the stress growth functions at high shear rates. These oscillations appear, however, to be more reasonable than those predicted by corotational models (see Bird et al., 1974). The oscillations disappear if ξ is set to zero, that is, for affine deformation. A comparison of the PTT model with other differential models for polymer melts can be found in Tanner (1985) and Larson (1988). The PTT model is found to provide good descriptions of step strain, elongational growth, and step biaxial elongational experiments.

To stress the need of using a realistic rheological model in the computations of viscoelastic flows, we show in Fig. 6.3-6 the numerical results for the entry problem of

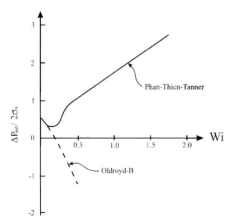

Figure 6.3-6 Excess pressure drop at the entrance of a tube as a function of the Weissenberg number, $Wi = \lambda \langle V \rangle / R$ (Adapted from Tanner, 1985) A single mode PTT model was used with $a = 0.015$ and $\xi = 0.1$

Keunings and Crochet (1984). They solved the 4:1 contraction problem for a modified Phan-Thien Tanner fluid, and the results for the excess pressure drop at the entrance are compared to the predictions for the Oldroyd-B model.

If we can assume that the exit pressure drop is negligible, $\Delta P_{ent}/2\sigma_R$ is the Bagley correction, e_0, introduced in Chapter 3 (equation 3.1-18). The exit pressure drop is believed to be about one quarter of the excess pressure drop at the entrance (see Choplin and Carreau, 1981), and in light of the results for the Bagley corrections published in the literature, the increase predicted by the PTT model at high Weissenberg numbers appears to be physically correct compared to the decrease and the negative values predicted when using the Oldroyd-B model.

6.4   Integral Constitutive Equations

In Section 6.2, we showed that the simple Maxwell model expressed in embedded coordinates leads to the different forms of convected Maxwell models. These can be integrated to obtain expressions of the stress tensor as functions of the Cauchy-Green and Finger deformation tensors. These equations have been at the origin of various integral constitutive equations proposed over the last few decades. Continuum mechanics principles and results of molecular theories have been guidelines for developing acceptable and useful forms, with the ultimate test being the assessment of the models with appropriate sets of experimental data.

In this section, we present a limited number of integral rheological equations that are useful in specific contexts. More exhaustive reviews of integral as well as differential constitutive equations can be found in Bird, Armstrong, and Hassager (1977 and 1987). Useful evaluations and comparisons of various models are presented in Larson (1988) and in Bird and Wiest (1995).

6.4-1 The Lodge Model

The Lodge (1964) model developed for rubber-like liquids can be considered as a direct extension of the upper-convected Maxwell model. It was later derived from network theory (Lodge, 1968, see also Chapter 7). It can be written as

$$\sigma^{ij}(t) = -\int_{-\infty}^{t} m(t-t')\Gamma^{-1\,ij}(t')dt', \qquad (6.4\text{-}1)$$

where $\underline{\underline{\Gamma}}^{-1} \; (= \underline{\underline{C}}^{-1} - \underline{\underline{\delta}})$ is the relative Finger tensor, defined by equations 6.1-8 and 6.2-9. $m(t - t')$ is an unspecified memory function. For

$$m(t-t') = \frac{\eta_0}{\lambda_0^2} e^{-\frac{(t-t')}{\lambda_0}}, \qquad (6.4\text{-}2)$$

we recover the upper-convected Maxwell model (integral form A''', equation 6.2-8). A multiple mode generalization leads to the following memory function:

$$m(t-t') = \sum_{p=1}^{M} \frac{\eta_p}{\lambda_p^2} e^{-\frac{(t-t')}{\lambda_p}}. \qquad (6.4\text{-}3)$$

Obviously, the Lodge model suffers from the same deficiencies as the upper-convected Maxwell model, that is, constant viscometric functions (the secondary normal stress coefficient is zero), transient functions independent of the shear rate, and an elongational viscosity that tends to infinity at a finite elongational rate. The Lodge model is nevertheless quite useful to qualitatively describe viscoelastic effects, and its predictions for the stress growth in elongational flow are surprisingly good for low density polyethylene melts, as shown in the following example.

Example 6.4-1 Elongational Stress Growth of a Lodge Rubber-Like Liquid
Obtain the expression for the stress growth at the start-up of uniaxial elongational flow of a fluid described by equation 6.4-1, with a memory function given by equation 6.4-3.

Solution
We assume a step change in the elongational rate, $\dot{\varepsilon}_\infty$ at $t = 0$, and the Finger tensor is obtained from equation 6.1-43b, taking the correspondence at t instead of at t_0. We write, for $0 \leq t' \leq t$,

$$\underline{\underline{C}}^{-1}(t') = \begin{bmatrix} \exp\{2\dot{\varepsilon}_\infty(t-t')\} & 0 & 0 \\ 0 & \exp\{\dot{\varepsilon}_\infty(t'-t)\} & 0 \\ 0 & 0 & \exp\{\dot{\varepsilon}_\infty(t'-t)\} \end{bmatrix}, \qquad (6.4\text{-}4a)$$

and for $t' \leq 0 \leq t$,

$$\underline{\underline{C}}^{-1} = \begin{bmatrix} \exp\{2\dot{\varepsilon}_\infty t\} & 0 & 0 \\ 0 & \exp\{-\dot{\varepsilon}_\infty t\} & 0 \\ 0 & 0 & \exp\{-\dot{\varepsilon}_\infty t\} \end{bmatrix}. \qquad (6.4\text{-}4b)$$

Combining with equations 6.4-1 and 6.4-3, we obtain

$$\sigma_{11}(t) = -\sum_{p=1}^{M} \frac{\eta_p}{\lambda_p^2} \int_{-\infty}^{0} \exp\left(-\frac{(t-t')}{\lambda_p}\right)[\exp(2\dot{\varepsilon}_\infty t) - 1]dt'$$

$$- \sum_{p=1}^{M} \frac{\eta_p}{\lambda_p^2} \int_{0}^{t} \exp\left(-\frac{(t-t')}{\lambda_p}\right)[\exp 2\dot{\varepsilon}_\infty(t-t') - 1]dt', \qquad (6.4\text{-}5a)$$

and

$$\sigma_{22}(t) = \sigma_{33}(t) = -\sum_{p=1}^{M} \frac{\eta_p}{\lambda_p^2} \int_{-\infty}^{0} \exp\left(-\frac{(t-t')}{\lambda_p}\right)[\exp(-\dot{\varepsilon}_\infty t) - 1]dt'$$

$$- \sum_{p=1}^{M} \frac{\eta_p}{\lambda_p^2} \int_{0}^{t} \exp\left(-\frac{(t-t')}{\lambda_p}\right)[\exp \dot{\varepsilon}_\infty(t'-t) - 1]dt'. \qquad (6.4\text{-}5b)$$

The elongational growth function is expressed by

$$\eta_E^+(t, \dot{\varepsilon}_\infty) = -\frac{(\sigma_{11} - \sigma_{22})}{\dot{\varepsilon}_\infty} = \frac{1}{\dot{\varepsilon}_\infty} \sum_{p=1}^{M} \frac{\eta_p}{\lambda_p} \left(e^{-\frac{t}{\lambda_p}(1 - 2\dot{\varepsilon}_\infty \lambda_p)} - e^{-\frac{t}{\lambda_p}(1 + \dot{\varepsilon}_\infty \lambda_p)}\right)$$

$$+ \sum_{p=1}^{M} \frac{\eta_p}{\lambda_p} \left\{ \frac{1}{1 - 2\dot{\varepsilon}_\infty \lambda_p}\left(1 - e^{-\frac{t}{\lambda_p}(1 - 2\dot{\varepsilon}_\infty \lambda_p)}\right) - \frac{1}{1 + \dot{\varepsilon}_\infty \lambda_p}\left(1 - e^{-\frac{t}{\lambda_p}(1 + \dot{\varepsilon}_\infty \lambda_p)}\right)\right\}.$$

$$(6.4\text{-}6a)$$

This result can be written as

$$\eta_E^+(t, \dot{\varepsilon}_\infty) = \eta_E + \frac{1}{\dot{\varepsilon}_\infty} \sum_{p=1}^{M} \frac{\eta_p}{\lambda_p} \{e^{-\frac{t}{\lambda_p}(1 - 2\dot{\varepsilon}_\infty \lambda_p)} - e^{-\frac{t}{\lambda_p}(1 + \dot{\varepsilon}_\infty \lambda_p)}\}$$

$$- \sum_{p=1}^{M} \frac{\eta_p}{\lambda_p} \left\{\frac{1}{1 - 2\dot{\varepsilon}_\infty \lambda_p} e^{-\frac{t}{\lambda_p}(1 - 2\dot{\varepsilon}_\infty \lambda_p)} - \frac{1}{1 + \dot{\varepsilon}_\infty \lambda_p} e^{-\frac{t}{\lambda_p}(1 + \dot{\varepsilon}_\infty \lambda_p)}\right\}, \qquad (6.4\text{-}6b)$$

where η_E is the steady-state elongational viscosity given by

$$\eta_E = 3 \sum_{p=1}^{M} \frac{\eta_p}{(1 - 2\dot{\varepsilon}_\infty \lambda_p)(1 + \dot{\varepsilon}_\infty \lambda_p)}, \qquad (6.4\text{-}7)$$

which, for one mode ($M=1$), is the result obtained for the upper-convected Maxwell model (see equation 6.2-25).

The Lodge model was found to describe the elongational properties of low density polyethylenes (LDPE) reasonably well. The model predictions using five modes are compared to the data of an LDPE in Fig. 6.4-1. The values of the parameters are listed in Table 6.4-1. The parameters were chosen to fit the linear viscoelastic properties of this LDPE.

The Lodge model predicts the monotonous increase of η_E^+ at low elongational rates very well. This is expected, since the parameters η_p and λ_p were obtained from the data in the

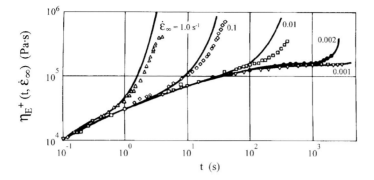

Figure 6.4-1 Predictions of the Lodge model compared to the elongational stress growth data obtained by Meissner (1971) for a LDPE (Adapted from Chang and Lodge, 1972)

linear regime. The model also describes well the strain hardening observed as the imposed elongational rate is increased. No steady-state values are predicted, and, unfortunately, no data at long enough times could be obtained to verify if the elongational viscosity of LDPE is bounded.

Table 6.4-1 Linear Viscoelastic Parameters for an LDPE Melt (From Chang and Lodge, 1972)

p	λ_p s	η_p Pa·s
1	100	1.903×10^4
2	10	2.306×10^4
3	1	6.241×10^3
4	0.1	3.299×10^3
5	0.01	3.72×10^2

Example 6.4-2 The Lodge-Meissner Relation
Consider the relaxation following a step shear strain as described in Example 5.2-5. Show that the Lodge model leads to

$$G(t) = G_N(t), \quad (6.4\text{-}8)$$

where $G(t) = -\sigma_{yx}(t)/\gamma_0$ is the shear relaxation modulus, and $G_N(t) = -(\sigma_{xx}(t) - \sigma_{yy}(t))/\gamma_0^2$ is the normal stress relaxation modulus. Equation 6.4-8 is known as the Lodge-Meissner relation (Lodge and Meissner, 1972).

Solution

Let us take the approach used in Example 5.2-5 and assume that a constant rate $\dot{\gamma}_0$ is applied at time $t=0$ for a very short time t_1. For $t > t_1$ (region III in Example 5.2-5), the Finger tensor is expressed by

For $t' < 0$:

$$C_{ij}^{-1}(t') = \begin{bmatrix} 1 + \dot{\gamma}_0^2 t_1^2 & \dot{\gamma}_0 t_1 & 0 \\ \dot{\gamma}_0 t_1 & 1 & 0 \\ 0 & 0 & 1 \end{bmatrix} \quad (6.4\text{-}9a)$$

For $0 \leq t' \leq t_1$:

$$C_{ij}^{-1}(t') = \begin{bmatrix} 1 + \dot{\gamma}_0^2 (t_1 - t')^2 & \dot{\gamma}_0 (t_1 - t') & 0 \\ \dot{\gamma}_0 (t_1 - t') & 1 & 0 \\ 0 & 0 & 1 \end{bmatrix} \quad (6.4\text{-}9b)$$

For $t_1 \leq t' \leq t$:

$$C_{ij}^{-1}(t') = \begin{bmatrix} 1 & 0 & 0 \\ 0 & 1 & 0 \\ 0 & 0 & 1 \end{bmatrix}. \quad (6.4\text{-}9c)$$

These results can be deduced from equation 6.1-64, noting that the shear rate is applied in the interval $0 \to t_1$ only. For $t > t_1$, the general Lodge model (equation 6.4-1) reduces to

$$\sigma_{11}(t) = -\dot{\gamma}_0^2 \int_{-\infty}^{0} \{m(t-t')\} t_1^2 dt' - \dot{\gamma}_0^2 \int_{0}^{t_1} \{m(t-t')\}(t_1 - t')^2 dt', \quad (6.4\text{-}10a)$$

$$\sigma_{21}(t) = -\dot{\gamma}_0 \int_{-\infty}^{0} \{m(t-t')\} t_1 dt' - \dot{\gamma}_0 \int_{0}^{t_1} \{m(t-t')\}(t_1 - t') dt', \quad (6.4\text{-}10b)$$

and $\sigma_{22} = \sigma_{33} = 0$. \quad (6.4-10c)

Hence, since $\gamma_0^2 = (\dot{\gamma}_0 t_1)^2$, we get

$$G_N(t) = \lim_{t_1 \to 0} -\frac{\sigma_{11} - \sigma_{22}}{\gamma_0^2}$$

$$= \lim_{t_1 \to 0} \frac{\int_{-\infty}^{0} \{m(t-t')\} t_1^2 dt' + \int_{0}^{t_1} \{m(t-t')\}(t_1 - t')^2 dt'}{t_1^2}$$

$$= \int_{-\infty}^{0} m(t-t') dt'. \quad (6.4\text{-}11)$$

Similarly,

$$G(t) = \lim_{t_1 \to 0} -\frac{\sigma_{21}}{\gamma_0} = \lim_{t_1 \to 0} \frac{\int_{-\infty}^{0} \{m(t-t')\} t_1 dt' + \int_{0}^{t_1} \{m(t-t')\}(t_1 - t') dt'}{t_1}$$

$$= \int_{-\infty}^{0} m(t-t') dt'. \quad (6.4\text{-}12)$$

The Lodge equation predicts that $G(t) = G_N(t)$. However, no strain dependence is predicted. The Lodge-Meissner relation appears to hold for polymer melts in general. Figure 6.4-2

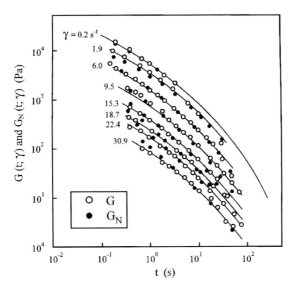

Figure 6.4-2 Shear relaxation modulus $G(t)$ (open symbols) and normal stress relaxation modulus G_N (filled symbols) for an LDPE melt (Data of Laun, 1978)
The time-temperature superposition principle has been used to generate this set of reliable data (Adapted from Larson, 1988)

compares the shear relaxation modulus and the normal stress relaxation modulus of a low density polyethylene. The relation is obeyed within experimental error. Note the very large reduction of the moduli as the imposed strain is increased. Reductions by a factor of 100 are observed as the strain is increased from 0.2 to 31. Clearly the Lodge model is strictly valid for the linear domain only, that is, for $\gamma_0 < 1$.

6.4-2 The Carreau Constitutive Equation

Many modifications of the Lodge model were developed at Wisconsin in the sixties, with the objective of obtaining a constitutive equation that would correctly predict shear-thinning and transient behavior of polymeric systems. These rheological models are called rate equations, and the most successful one is probably that derived by Carreau (1972) from the network theory (see Chapter 7). The general form can be written as

$$\sigma^{ij}(t) = -\int_{-\infty}^{t} M(t-t', II_{\dot{\gamma}}) \left\{ \left(1 + \frac{\epsilon}{2}\right) \Gamma^{-1\, ij}(t') + \frac{\epsilon}{2} \Gamma^{ij}(t') \right\} dt', \qquad (6.4\text{-}13)$$

where

$$M(t-t', II_{\dot{\gamma}}) = \sum_{p=1}^{M} \frac{\eta_p f_p(II_{\dot{\gamma}}(t'))}{\lambda_p^2} \exp\left[-\int_{t'}^{t} \frac{dt''}{\lambda_p g_p \{II_{\dot{\gamma}}(t'')\}}\right], \qquad (6.4\text{-}14)$$

240   Nonlinear Viscoelasticity

λ_p and η_p are material parameters describing the linear viscoelasticity, and f_p and g_p are functions that take into account the effects of the rate of deformation on the rates of creation and loss of entanglements. For different choices of f_p and g_p, various rheological models can be obtained: Carreau A and B (Carreau, 1972), MBC (Macdonald, 1975a, b), and De Kee (1977). The Bird-Carreau model (1968) is readily obtained by setting all g_p equal to 1, and the Lodge elastic model is obtained by taking $f_p = g_p = 1$ and $\epsilon = 0$.

The usual material functions are

(a) In steady simple shear flow

$$\eta(\dot{\gamma}) = \sum_{p=1}^{M} \eta_p f_p g_p^2, \tag{6.4-15}$$

$$\psi_1(\dot{\gamma}) = 2 \sum_{p=1}^{M} \eta_p \lambda_p f_p g_p^3, \tag{6.4-16}$$

and

$$\psi_2(\dot{\gamma}) = \frac{\epsilon}{2} \psi_1, \tag{6.4-17}$$

where f_p and g_p are functions of the shear rate $\dot{\gamma}$;

(b) For stress relaxation after steady shear flow

$$\eta^-(t, \dot{\gamma}_0) = \sum_{p=1}^{M} \eta_p f_p g_p^2 \exp\left(-\frac{t}{\lambda_p}\right) \tag{6.4-18}$$

and

$$\psi_1^-(t, \dot{\gamma}_0) = 2 \sum_{p=1}^{M} \eta_p f_p g_p^3 \lambda_p \exp\left(-\frac{t}{\lambda_p}\right), \tag{6.4-19}$$

where f_p and g_p are functions of the initially applied shear rate $\dot{\gamma}_0$;

(c) For stress growth at the start-up of steady shear flow

$$\eta^+(t, \dot{\gamma}_\infty) = \eta(\dot{\gamma}_\infty) + \sum_{p=1}^{M} \eta_p \left[\left(\frac{t}{\lambda_p}\right)(1 - f_p g_p) - f_p g_p^2 \right] \exp\left(-\frac{t}{\lambda_p g_p}\right) \tag{6.4-20}$$

and

$$\psi_1^+ = \psi_1(\dot{\gamma}_\infty) + \sum_{p=1}^{M} \eta_p \left[\left(\frac{t^2}{\lambda_p}\right)(1 - f_p g_p) - 2t f_p g_p^2 - 2\lambda_p f_p g_p^3 \right] \exp\left(-\frac{t}{\lambda_p g_p}\right), \tag{6.4-21}$$

where f_p and g_p are functions of the applied shear rate $\dot{\gamma}_\infty$;

(d) For steady uniaxial elongational flow (assuming $\epsilon = 0$)

$$\eta_E = 3 \sum_{p=1}^{M} \frac{(\eta_p f_p g_p^2)}{(1 + \dot{\epsilon} \lambda_p g_p)(1 - 2\dot{\epsilon} \lambda_p g_p)}, \tag{6.4-22}$$

where f_p and g_p are functions of the elongational rate $\dot{\epsilon}$.

In the limit, as the second invariant of the rate-of-strain tensor tends towards zero, $f_p = g_p = 1$. For *small amplitude oscillatory flow*, Carreau (1972) assumed implicitly that f_p and g_p remain equal to 1, although the rate of strain can reach large values at high frequencies. The real and imaginary parts of the complex viscosity are then

$$\eta' = \frac{1}{\omega}\int_0^\infty m(s)\sin(\omega s)ds = \sum_{p=1}^{M} \frac{\eta_p}{1+(\omega\lambda_p)^2} \qquad (6.4\text{-}23)$$

and

$$\eta'' = \frac{1}{\omega}\int_0^\infty m(s)\cos(\omega s)ds = \sum_{p=1}^{M} \eta_p\left[\frac{\omega\lambda_p}{1+(\omega\lambda_p)^2}\right]. \qquad (6.4\text{-}24)$$

The linear viscoelastic domain is believed to be defined by small deformation independently of the rate of deformation (see Chapter 5). However, it is well-known to experimentalists that tests at smaller strain values have to be conducted as the frequency is increased. As the functions f_p and g_p will, in general, not tend to 1 for small strain, large frequency situations, rate-dependent equations have been declared inadmissible by many authors. That is, rate-dependent equations will not reduce to the equation of linear viscoelasticity at small strains (Larson, 1988). However, rate-of-deformation effects predicted by rate-dependent models can be, in some cases, quite small (see Macdonald, 1975a, b), and negligible for specific choices of f_p and g_p. Although rate-dependent equations can be deficient for describing linear properties, they can be quite useful for engineering applications where deformations are large and linear viscoelastic behavior is of little significance.

Carreau (1972), following Bird and Carreau (1968), reduced the constants in a manner suggested by the molecular theory of Rouse (1953):

$$\eta_p = \eta_0 \frac{\lambda_p}{\sum_{p=1}^{M}\lambda_p} \qquad (6.4\text{-}25)$$

and

$$\lambda_p = \frac{2^\alpha \lambda}{(p+1)^\alpha} \qquad (6.4\text{-}26)$$

where η_0 is the zero shear viscosity, λ is the longest relaxation time, and α is a dimensionless parameter.

By choosing different functions, f_p and g_p, different rheological models can be obtained.

6.4-2.1 Carreau A

In this model, the functions f_p and g_p are assumed to be independent of p, that is,

$$\left.\begin{array}{l} f_p(II_{\dot\gamma}) = f(II_{\dot\gamma}) \\ g_p(II_{\dot\gamma}) = g(II_{\dot\gamma}) \end{array}\right\} \text{ for all } p. \qquad (6.4\text{-}27)$$

These functions are chosen in such a way that the viscosity and the primary normal stress coefficient given respectively by equations 6.4-15 and 6.4-16 become

$$\eta = \frac{\eta_0}{[1+(t_1\dot{\gamma})^2]^{\frac{1-n}{2}}} \qquad (6.4\text{-}28)$$

and

$$\psi_1 = \frac{\psi_{10}}{[1+(t_2\dot{\gamma})^2]^{1-n'}}, \qquad (6.4\text{-}29)$$

where

$$\psi_{10} = \frac{2^{\alpha+1}\lambda\eta_0[Z(2\alpha)-1]}{Z(\alpha)-1}. \qquad (6.4\text{-}30)$$

$Z(\alpha) = \sum_{p=1}^{\infty} 1/p^{\alpha}$ is the Riemann zeta function. Equation 6.4-28 is the three-parameter viscosity Carreau model presented in Chapter 2; n and n' are power-law parameters for the viscosity and the primary normal stress coefficient respectively. Carreau further reduced the number of parameters by taking $t_2 = t_1$.

Model A then contains seven parameters, η_0, α, and λ associated with the linear properties of the fluid, and n, n', t_1, and ϵ related to the nonlinear behavior. It can describe the functions η, ψ_1 (or N_1), η', and η'' (or G') very well, as shown in Fig. 6.4-3 for a polystyrene solution in aroclor. Infinite series ($M = \infty$) have been used, with their limiting

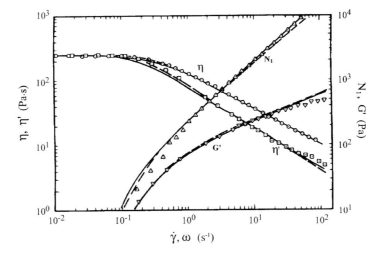

Figure 6.4-3 Viscosity, primary normal stress difference, dynamic viscosity, and storage modulus for a 4 mass % polystyrene solution ($M_w = 1.8 \times 10^6$ g/mol) in Aroclor 1248 (chlorinated oil) of viscosity equal to 0.30 Pa·s at 25 °C (Data of Ashare, 1968)
The model parameters in models A and B were chosen to obtain the same viscosity function (Adapted from Carreau, 1972)
---- Carreau A with $\eta_0 = 250$ Pa·s, $\alpha = 3.0$, $\lambda = 3.12$ s, $n = 0.456$, $n' = 0.428$, and $t_1 = 3.36$ s;
—— Carreau B with $\eta_0 = 250$ Pa·s, $\alpha = 2.7$, $\lambda = 4.08$ s, $S = 0.863$, $R = 0.161$, and $t_1 = 2.92$ s.

power-law expressions for high frequencies and taking $f_p = g_p = 1$, for the calculation of η' and G' (see Carreau, 1972).

Model A predicts too small overshoots for stress growth experiments, and the predicted stress relaxation at the cessation of steady shear flow is rate-independent. Nevertheless, Model A can be quite useful for engineering calculations of viscoelastic flows.

6.4-2.2 Carreau B

The Carreau model B is a compromise between the Bird-Carreau model, which predicts too large overshoots for stress growth, and model A, which predicts too small overshoots. The functions f_p and g_p are defined such that simple results are obtained for η and ψ_1:

$$f_{p-1}(II_{\dot{\gamma}}(t')) = \frac{1 + \frac{(\frac{1}{2}(2^\alpha t_1)^2 II_{\dot{\gamma}}(t'))^S}{p^{2\alpha}}}{\left[1 + \left(\frac{c^2}{2}\right)\lambda^2 II_{\dot{\gamma}}(t')\right]^{2R}} \tag{6.4-31}$$

$$g_{p-1}(II_{\dot{\gamma}}(t'')) = \frac{\left[1 + \left(\frac{c^2}{2}\right)\lambda^2 II_{\dot{\gamma}}(t'')\right]^R}{1 + \frac{(\frac{1}{2}(2^\alpha t_1)^2 II_{\dot{\gamma}}(t''))^S}{p^{2\alpha}}}$$

where R, S, t_1, and c are parameters related to nonlinear behavior. The expressions 6.4-15 and 6.4-16 become, after using 6.4-25 and 6.4-26 and rearranging the infinite series for low shear rate,

$$\frac{\eta}{\eta_0} = 1 - \frac{(2^\alpha t_1 \dot{\gamma})^{2S}}{Z(\alpha) - 1} \sum_{p=2}^{\infty} \frac{p^{-\alpha}}{p^{2\alpha} + (2^\alpha t_1 \dot{\gamma})^{2S}}, \tag{6.4-33}$$

and

$$\frac{\psi_1}{\psi_{10}} = (1 + c^2 \lambda^2 \dot{\gamma}^2)^R \left[1 - \frac{2(2^\alpha t_1 \dot{\gamma})^{2S}}{Z(2\alpha) - 1} \sum_{p=2}^{\infty} \frac{p^{-2\alpha}}{p^{2\alpha} + (2^\alpha t_1 \dot{\gamma})^{2S}} + \frac{(2^\alpha t_1 \dot{\gamma})^{4S}}{Z(2\alpha) - 1} \sum_{p=2}^{\infty} \frac{p^{-2\alpha}}{(p^{2\alpha} + (2^\alpha t_1 \dot{\gamma})^{2S})^2} \right], \tag{6.4-34}$$

where ψ_{10} is given by equation 6.4-30. At high shear rate, equations 6.4-33 and 6.4-34 simplify to (Macdonald et al., 1969):

$$\frac{\eta}{\eta_0} \rightarrow \frac{1}{Z(\alpha) - 1} \left[\frac{\pi |2^\alpha t_1 \dot{\gamma}|^{\frac{(1-\alpha)S}{\alpha}}}{2\alpha \sin\left\{\left[\frac{(1+\alpha)}{2\alpha}\right]\pi\right\}} - \frac{1}{(\alpha+1)(2^\alpha t_1 \dot{\gamma})^{2S}} - \frac{1 + \frac{\alpha}{6}}{2(1 + (2^\alpha t_1 \dot{\gamma})^{2S})} \right] \tag{6.4-35}$$

and

$$\frac{\psi_1}{\psi_{10}} \rightarrow \frac{(1 + c^2 \lambda^2 \dot{\gamma}^2)^R}{Z(2\alpha) - 1} \left[\frac{\pi |2^\alpha t_1 \dot{\gamma}|^{\frac{(1-2\alpha)S}{\alpha}}}{4\alpha^2 \sin\left(\frac{\pi}{2\alpha}\right)} - \frac{1}{(2\alpha+1)(2^\alpha t_1 \dot{\gamma})^{4S}} - \frac{1 + \frac{\alpha}{3}}{2(1 + (2^\alpha t_1 \dot{\gamma})^{2S})^2} \right]. \tag{6.4-36}$$

The expressions of η and η' are very similar; the parameter S is the ratio of the slopes of $\log \eta$ versus $\log \dot{\gamma}$ and of $\log \eta'$ versus $\log \omega$ in the power-law regions. For $S = 1$, the results are analogous, that is,

$$\eta(\dot{\gamma}) = \eta'(\omega) = \eta'\left[\left(\frac{t_1}{\lambda}\right)\dot{\gamma}\right], \quad (6.4\text{-}37)$$

where t_1/λ is a shift factor between η and η'. Carreau (1972) eliminated one parameter by taking $c = 1$ and obtained another 7-constant model.

Model B predictions for the steady shear viscosity and primary normal stress difference and the complex viscosity functions are compared in Fig. 6.4-3 to model A predictions for a 4% polystyrene solution. Slightly different values of the linear parameters were chosen for model B, but overall the fits are quite comparable. The advantages of model B over model A are illustrated in Figs. 6.4-4 and 6.4-5 for stress growth experiments on the same polystyrene solution. As discussed earlier, model A cannot predict any appreciable overshoot at high shear rates, as observed experimentally. The agreement between model B predictions and the data is excellent, especially if we remember that all the parameters were determined from data on steady shear flow, and small-amplitude, sinusoidal motion experiments. This is one of the few rheological models that adequately predict the overshoots and the times of maximum overshoot. The deviations observed between the predictions and the data are probably within experimental error. (See Section 3.3-4 for a discussion on problems associated with transient experiments.)

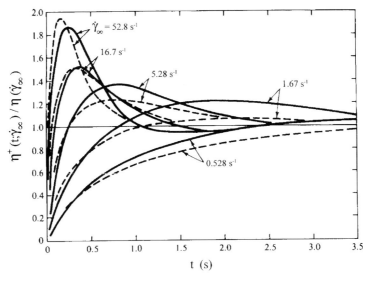

Figure 6.4-4   Shear stress growth function $\eta^+(t; \dot{\gamma}_\infty)/\eta(\dot{\gamma}_\infty)$ for 4.0 mass % polystyrene
——— Data from Macdonald (1968) (Adapted from Carreau, 1972)
- - - - Model B for parameter values given in Fig. 6.4-3

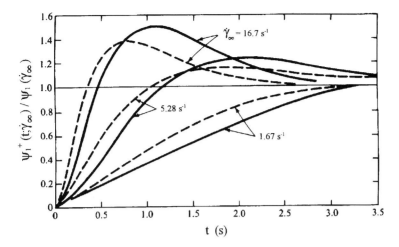

Figure 6.4-5 Normal stress growth function $\psi_1^+(t;\dot\gamma_\infty)/\psi_1(\dot\gamma_\infty)$ for 4.0 mass % polystyrene
——— Data from Macdonald (1968) (Adapted from Carreau, 1972)
- - - - Model B for parameter values given in Fig. 6.4-3

Note that the Bird-Carreau (1968) model, which is obtained by taking all g_p equal to 1, predicts stress overshoots and times for maximum stresses that are at least an order of magnitude too large at high shear rates (results not shown here).

The stress relaxation predictions of model B for the same polystyrene solution are compared to data in Fig. 6.4-6 for the viscosity function and in Fig. 6.4-7 for the normal stress function. Note that model B is capable of describing the more rapid relaxation of the stresses as the value of the initially applied shear rate ($\dot\gamma_0$) is increased. The η^- predictions at the lowest shear rates are not so good. This is due to the choice of the linear parameters (η_0, λ, and α). Nevertheless, the model slightly overpredicts the effect of $\dot\gamma_0$.

The predictions of model B for the normal stress relaxation are shown in Fig. 6.4-7. The model predicts correctly that the normal stresses relax more rapidly than the shear stress, but the effect of the initially imposed shear rate is too drastic. Obviously, better fits of the relaxation functions could be obtained by using a different set of parameters for this fluid. Another route was taken by Macdonald (1975a and b), who tried several empirical modifications of the functions f_p and g_p of the Carreau B model, in order to obtain more satisfactory predictions for high-amplitude data and relaxation functions. Better fits of relaxation data were obtained at the expense of poorer fits of the stress growth results.

246 Nonlinear Viscoelasticity

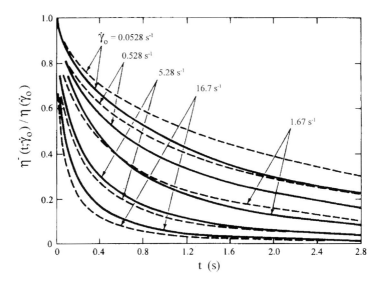

Figure 6.4-6 Shear stress relaxation function $\eta^-(t;\dot{\gamma}_0)/\eta(\dot{\gamma}_0)$ for 4.0 mass % polystyrene
—— Data from Macdonald (1968) (Adapted from Carreau, 1972)
- - - - Model B for parameter values given in Fig. 6.4-3

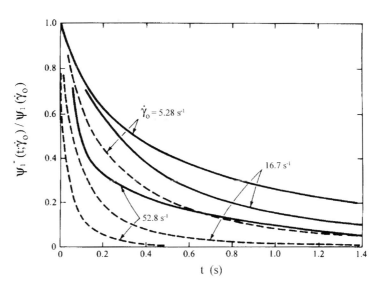

Figure 6.4-7 Normal stress relaxation function $\psi_1^-(t;\dot{\gamma}_0)/\psi_1(\dot{\gamma}_0)$ for 4.0 mass % polystyrene
—— Data from Macdonald (1968) (Adapted from Carreau, 1972)
- - - - Model B for parameter values given in Fig. 6.4-3

6.4-2.3 The De Kee Model

The De Kee model (De Kee, 1977 and De Kee and Carreau, 1979) is the result of efforts to simplify the functions f_p and g_p in the original Carreau model B. They are defined by simple exponential functions:

$$f_p = e^{-\sqrt{\frac{II_{\dot\gamma}}{2t_p}}(-2c+3)} \qquad (6.4\text{-}38)$$

and

$$g_p = e^{-\sqrt{\frac{II_{\dot\gamma}}{2t_p}}(c-1)}, \qquad (6.4\text{-}39)$$

where c is a dimensionless parameter and t_p represents time constants. The functions g_p and f_p are associated with rates of loss and formation of entanglements. Using only a few parameters, reasonable fits of the material functions discussed in this section can be obtained. This choice of f_p and g_p leads to the De Kee viscosity model presented in Section 2.2-7. This model predicts the so-called stress jump or discontinuity (Liang and Mackay, 1993; Gerhardt and Manke, 1994) associated with stress growth and stress relaxation experiments. This model has also been used successfully in biomechanics (Zhu et al., 1991).

6.4-3   The K-BKZ Constitutive Equation

A general class of integral constitutive equations can be appropriately referred to as the Kaye-Bernstein-Kearsley-Zapas (or K-BKZ) equation. It was independently proposed by Kaye (1962) and Bernstein et al. (1963) as an extension of the rubber-like elastic model. The equation can be written as

$$\sigma^{ij} = -\int_{-\infty}^{t}\{m_1(t-t', I_{c^{-1}}, II_{c^{-1}})\Gamma^{-1\,ij}(t') + m_2(t-t', I_{c^{-1}}, II_{c^{-1}})\Gamma^{ij}(t')\}dt'. \qquad (6.4\text{-}40)$$

Bernstein et al. (1963) derived the two memory functions from a strain energy which is allowed to be dependent on the first and second invariants of the strain tensor. m_1 and m_2 are defined as

$$m_1 = 2\frac{\partial u}{\partial I_{c^{-1}}} \qquad (6.4\text{-}41a)$$

and

$$m_2 = -2\frac{\partial u}{\partial II_{c^{-1}}}, \qquad (6.4\text{-}41b)$$

where $I_{c^{-1}}$ and $II_{c^{-1}}$ are the first and second invariants of the Finger tensor respectively, and u expresses the functional dependence of the free energy on the history of the strain tensor.

The K-BKZ constitutive equation is believed to be one of the most powerful and flexible rheological equations. However, the memory functions m_1 and m_2 are not easily determined from experimental data. For example, for steady shear flow, the Finger and the

Cauchy-Green tensors are given by equations 6.1-65 and 6.1-67 respectively, and the K-BKZ model yields

$$\eta = \int_{-\infty}^{t} \{m_1(t-t', I_{c^{-1}}, II_{c^{-1}}) - m_2(t-t', I_{c^{-1}}, II_{c^{-1}})\}(t-t')dt', \tag{6.4-42a}$$

$$\psi_1 = \int_{-\infty}^{t} \{m_1(t-t', I_{c^{-1}}, II_{c^{-1}}) - m_2(t-t', I_{c^{-1}}, II_{c^{-1}})\}(t-t')^2 dt', \tag{6.4-42b}$$

and $\quad \psi_2 = \int_{-\infty}^{t} \{m_2(t-t', I_{c^{-1}}, II_{c^{-1}})\}(t-t')^2 dt'. \tag{6.4-42c}$

The functions m_1 and m_2 cannot easily be extracted from viscosity data. In principle, we can use the stress relaxation following a step strain to determine $m_1 - m_2$. The result of example 6.4-2 can be readily applied to the K-BKZ equation. From equation 6.4-11, the shear and normal stress relaxation functions are given by

$$G(t, \gamma_0) = G_N(t, \gamma_0) = \int_{-\infty}^{0} \phi(t-t', \gamma_0) dt', \tag{6.4-43}$$

where

$$\phi(t-t', \gamma_0) = m_1(t-t', \gamma_0) - m_2(t-t', \gamma_0). \tag{6.4-44}$$

Note that the Lodge-Meissner relation holds for the K-BKZ equation for any imposed deformation, γ_0, but the model correctly predicts a strain dependence on the moduli as illustrated in Fig. 6.4-2. Equation 6.4-43 can be rewritten in terms of s ($=t-t'$) to obtain

$$G(t, \gamma_0) = G_N(t, \gamma_0) = \int_{t}^{\infty} \phi(s, \gamma_0) ds \tag{6.4-45}$$

and

$$\frac{\partial G}{\partial t} = \frac{\partial G_N}{\partial t} = -\phi(t, \gamma_0). \tag{6.4-46}$$

The function ϕ or $m_1 - m_2$ can be determined by taking the derivative of the relaxation function G or G_N.

The validity of the K-BKZ equation can be tested by performing double-step strain experiments, with a time interval between the two steps. The predictions of the model were found to be good when the two strain steps were in the same direction, but usually poor for steps in opposite directions (Osaki et al., 1981; Larson and Valesano, 1976). Other tests of the K-BKZ equation have been proposed by Attané et al. (1988) and by Chan Man Fong and De Kee (1992). Vrentas et al. (1990) also considered this type of flow in order to discriminate between constitutive equations. They proposed an extension of the K-BKZ model to include strain coupling.

The following specific relations between the transient viscosity $\eta^-(\dot{\gamma}_0, t)$ and primary normal stress coefficient can be obtained (Problem 6.6-12):

$$\dot{\gamma}_0^{-2} \int_0^\infty u\psi_1^-(u,t)du = \int_t^\infty \eta^-(\dot{\gamma}_0, s)ds \qquad (6.4\text{-}47a)$$

and $\quad -\dfrac{1}{2}\dfrac{\partial \psi_1^-}{\partial t}(\dot{\gamma}_0, t) = \eta^-(\dot{\gamma}_0, t) + \dfrac{\dot{\gamma}_0}{2}\dfrac{\partial \eta^-}{\partial \dot{\gamma}_0}(\dot{\gamma}_0, t). \qquad (6.4\text{-}47b)$

At the limit $t \to 0$, $\eta^-(\dot{\gamma}_0, 0) = \eta(\dot{\gamma}_0)$ and $\psi_1^-(\dot{\gamma}_0, 0) = \psi_1(\dot{\gamma}_0)$, and relations 6.4-47a and b become

$$\dot{\gamma}_0^{-2} \int_0^\infty u\psi_1(u)du = \int_0^\infty \eta^-(\dot{\gamma}_0, s)ds \qquad (6.4\text{-}48a)$$

and $\quad -\dfrac{1}{2}\dfrac{\partial \psi_1^-}{\partial t}(\dot{\gamma}_0, t)|_{t=0} = \eta(\dot{\gamma}_0) + \dfrac{\dot{\gamma}_0}{2}\dfrac{\partial \eta^-}{\partial \dot{\gamma}_0}|_{t=0}. \qquad (6.4\text{-}48b)$

Note that these relations hold for the K-BKZ model for any functions m_1 and m_2.

Carefully determined transient and steady shear data for polystyrene were used by Attané (1984) to assess the validity of relation 6.4-48a. Figure 6.4-8 compares the relation to one set of polystyrene data. The polystyrene 7 has a molecular weight average, M_w, of 1.6×10^6 g/mol, and $M_w/M_n = 1.3$. The solvent is dibutylphthalate. The relaxation time, λ_w, is defined by

$$\lambda_w = J_e^0 \eta_0 = \frac{\psi_{10}}{2\eta_0}, \qquad (6.4\text{-}49)$$

where the steady-state recoverable compliance, J_e^0, and the zero shear primary normal stress coefficient were determined using the Yamamoto relation (equations 6.4-70 and 6.4-72).

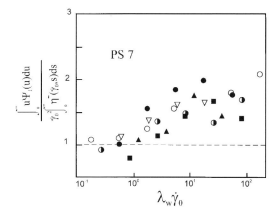

Figure 6.4-8 Experimental evaluation of relation 6.4-48a using a set of polystyrene solution data
The characteristics for the solutions and the symbols are defined in Table 6.4-2
– – – K-BKZ model (Adapted from Attané et al., 1988)

Table 6.4-2 Characteristics of PS7 Solutions Used to Generate Data of Fig. 6.4-8

Solution code	Symbol	Concentration mass %	η_0 Pa·s	λ_w s
PS705	▽	5	4.8	0.025
PS707	●	7	20	0.082
PS710	○	10	110	0.25
PS715	■	15	1450	1.5
PS720	▲	20	13500	7.0
PS725	◐	25	42600	15

The data of Fig. 6.4-8 show quite a bit of scatter, but clearly the ratio departs from the value of 1 predicted by the K-BKZ model. Similar trends were observed for other polystyrene solutions, but as far as we know, relation 6.4-48a has not been tested for polymer melts. In general, the K-BKZ equation tends to overestimate normal stress effects in simple shear flow, at least for polymer solutions. This appears to be inherent in the choice of strain-dependent functional in the model, but this lack of flexibility for predicting normal stress differences is more evident when we examine the similar and particular case proposed by Wagner (1976).

In the Wagner model, the time and strain effects are assumed to be separable. In its simplest form, ignoring m_2, the stress tensor can be expressed by

$$\underline{\underline{\sigma}} = -\int_{-\infty}^{t} m(t-t')h(I_{c^{-1}}, II_{c^{-1}})\underline{\underline{\Gamma}}^{-1}(t')dt', \qquad (6.4\text{-}50)$$

where $m(t - t')$ is a memory function as used in linear viscoelasticity, and $h(I_{c^{-1}}, II_{c^{-1}})$ is a damping function which is to be determined from shear-thinning viscosity data or from step strain experiments.

Wagner (1976, 1977) proposed a simple damping function which is defined for simple shear by

$$h(I_{c^{-1}}, II_{c^{-1}}) = h(\gamma) = \exp[-a\sqrt{\gamma^2}], \qquad (6.4\text{-}51)$$

where a is the unique nonlinear parameter. This simple form was shown by Wagner and others to yield reasonable predictions for the transient and steady shear data of polymer melts. Equation 6.4-51 does not allow enough flexibility, as shown next.

Example 6.4-3 Stress Growth and Steady Shear Behavior of a Wagner Fluid

Obtain the shear and normal stress growth functions for a Wagner fluid described by equations 6.4-50 and 6.4-51 in terms of integrals of the relaxation function. Show that the primary normal stress difference can be expressed in terms of the derivative of the viscosity with respect to the applied shear rate.

6.4 Integral Constitutive Equations

Solution

For stress growth at the inception of steady shear flow (at $t=0$), the relative Finger tensor is expressed by

For $t' < 0 \leq t$:

$$\underline{\underline{\Gamma}}^{-1}(0) = \begin{bmatrix} \dot{\gamma}_\infty^2 t^2 & \dot{\gamma}_\infty t & 0 \\ \dot{\gamma}_\infty t & 0 & 0 \\ 0 & 0 & 0 \end{bmatrix}; \qquad (6.4\text{-}52a)$$

For $0 \leq t' \leq t$:

$$\underline{\underline{\Gamma}}^{-1}(t') = \begin{bmatrix} \dot{\gamma}_\infty^2 (t-t')^2 & \dot{\gamma}_\infty (t-t') & 0 \\ \dot{\gamma}_\infty (t-t') & 0 & 0 \\ 0 & 0 & 0 \end{bmatrix}. \qquad (6.4\text{-}52b)$$

Equation 6.4-50 can now be written using equation 6.4-51 as

$$\underline{\underline{\sigma}}(t) = -\int_{-\infty}^{0} m(t-t')h(\dot{\gamma}_\infty t)\underline{\underline{\Gamma}}^{-1}(0)dt'$$

$$-\int_{0}^{t} m(t-t')h(\dot{\gamma}_\infty [t-t'])\underline{\underline{\Gamma}}^{-1}(t')dt'. \qquad (6.4\text{-}53a)$$

This result becomes, in terms of $s = t - t'$,

$$\underline{\underline{\sigma}}(t) = -h(\dot{\gamma}_\infty t)\underline{\underline{\Gamma}}^{-1}(0)\int_{t}^{\infty} m(s)ds$$

$$-\int_{0}^{t} m(s)h(\dot{\gamma}_\infty s)\underline{\underline{\Gamma}}^{-1}(t-s)ds. \qquad (6.4\text{-}53b)$$

We note that the relaxation function defined in linear viscoelasticity (see Chapter 5) is related to the memory function by

$$G(t) = \int_{t}^{\infty} m(s)ds. \qquad (6.4\text{-}54)$$

The second integral of equation 6.4-53b can be integrated by parts, and equation 6.4-53b can then be written, using 6.4-54, as

$$\underline{\underline{\sigma}}(t) = -h(\dot{\gamma}_\infty t)\underline{\underline{\Gamma}}^{-1}(0)G(t)$$

$$+ G(t)h(\dot{\gamma}_\infty t)\underline{\underline{\Gamma}}^{-1}(0) - \int_{0}^{t} G(s)d[h(\dot{\gamma}_\infty s)\underline{\underline{\Gamma}}^{-1}(t-s)]$$

$$= -\int_{0}^{t} G(s)\left\{\frac{dh}{ds}(\dot{\gamma}_\infty s)\underline{\underline{\Gamma}}^{-1}(t-s) + h(\dot{\gamma}_\infty s)\frac{d\underline{\underline{\Gamma}}^{-1}}{ds}(t-s)\right\}ds. \qquad (6.4\text{-}55)$$

From equation 6.4-51, $h(\dot{\gamma}_\infty s) = \exp(-a\dot{\gamma}_\infty s)$, and the growth functions are

$$\eta^+(t, \dot{\gamma}_\infty) = -\frac{\sigma_{21}}{\dot{\gamma}_\infty} = \int_0^t G(s) \exp(-a\dot{\gamma}_\infty s)[1 - a\dot{\gamma}_\infty s]ds \qquad (6.4\text{-}56a)$$

and $\quad \psi_1^+(t, \dot{\gamma}_\infty) = -\frac{\sigma_{11} - \sigma_{22}}{\dot{\gamma}_\infty^2} = \int_0^t G(s) \exp(-a\dot{\gamma}_\infty s)(2 - a\dot{\gamma}_\infty s)s\, ds. \qquad (6.4\text{-}56b)$

The two growth functions are interrelated, as we observe by taking the derivative of η^+ with respect to $\dot{\gamma}_\infty$:

$$\frac{\partial \eta^+}{\partial \dot{\gamma}_\infty} = \int_0^t G(s) \exp(-a\dot{\gamma}_\infty s)[a^2 \dot{\gamma}_\infty s^2 - 2as]ds. \qquad (6.4\text{-}57)$$

Hence, from equation 6.4-56b,

$$\psi_1^+(t, \dot{\gamma}_\infty) = -\frac{1}{a}\frac{\partial \eta^+}{\partial \dot{\gamma}_\infty}(t, \dot{\gamma}_\infty), \qquad (6.4\text{-}58)$$

which is also valid at the limit of steady shear flow $\psi_1^+(t, \dot{\gamma}_\infty) \to \psi_1(\dot{\gamma})$ and $\eta^+(t, \dot{\gamma}_\infty) \to \eta(\dot{\gamma})$:

$$\psi_1(\dot{\gamma}) = -\frac{1}{a}\frac{d\eta(\dot{\gamma})}{d\dot{\gamma}}. \qquad (6.4\text{-}59)$$

Equation 6.4-59 provides a relation which can be used, though with caution, to estimate normal stress differences in shear flow. Wagner (1977) has determined the damping parameter a (n in the Wagner notation) by comparing the predictions to the normal stress and viscosity data. The value of a was found to vary from 0.10 to 0.20 for a series of polymer solutions and melts. Using relation 6.4-59 and analytical expressions for the viscosity, Wagner (1977) has shown reasonable or fair predictions for the primary normal stress differences. However, it is clear from equation 6.4-59 that the slope for the primary normal stress coefficient is uniquely related to that of the viscosity in the power-law region. That is, for a power-law fluid ($\eta = m|\dot{\gamma}|^{n-1}$), $\psi_1 \propto |\dot{\gamma}|^{n-2}$. De Kee and Stastna (1986) and Ait-Kadi et al. (1989) provide further discussions regarding the evaluation of elasticity from viscometric data.

Examining a variety of polymer solutions, Grmela and Carreau (1987) have observed that the primary normal stress coefficient is proportional to the square of the shear viscosity, as predicted by conformation tensor rheological models (see Chapter 7). This is illustrated for five polymer solutions in Fig. 6.4-9 where the solid lines are of slope 2. The plots suggest that $\psi_1 \propto |\dot{\gamma}|^{2n-2}$, which is considerably different from the prediction of the Wagner model. The solvent viscosity η_s is in most cases negligible with respect to the solution viscosity. Its inclusion, suggested by the conformation tensor theory, has no significant effect on the slope.

Using a discrete relaxation spectrum, we can show that the steady shear viscosity and primary normal stress functions can be expressed as

$$\eta(\dot{\gamma}) = \sum_{p=1}^{M} \frac{\eta_p}{(1 + a\lambda_p \dot{\gamma})^2} \qquad (6.4\text{-}60)$$

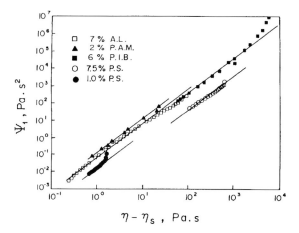

Figure 6.4-9 Primary normal stress coefficient as a function of the relative viscosity for five polymer solutions. The solutions are defined in Fig. 2.1-3. The solid lines are fitted power-law relations with exponent equal to 2

and

$$\psi_1(\dot{\gamma}) = 2 \sum_{p=1}^{M} \frac{\eta_p \lambda_p}{(1 + a\lambda_p \dot{\gamma})^3}. \tag{6.4-61}$$

Quantitative agreement with the viscosity function can be obtained with a single nonlinear parameter a, provided a sufficient number of linear parameters (η_p and λ_p) are used. However, quantitative agreement with normal stress data is usually obtained using a different value for the parameter a.

The Wagner model is a very useful constitutive equation for engineering applications. It is useful for predicting extensional properties and rheological behavior of polymer melts in complex flow situations. The lack of flexibility shown here can be partly resolved by using more than one nonlinear parameter. The damping function originally proposed by Osaki (1976),

$$h(\gamma) = f_1 \exp(-a_1 \gamma) + f_2 \exp(-a_2 \gamma), \tag{6.4-62}$$

has been shown to be quite flexible for describing shear and normal stress relaxation functions following a step strain (Laun, 1978), and nonlinear creep and recovery data (Wagner, 1978a).

Another difficulty encountered with the Wagner model as well as with other integral equations, which assume affine deformation, is the reversibility predicted under a double-step strain experiment. Using network theory concepts, Wagner (1978b) suggested that the damping function associated with a loss of entanglements should be irreversible. He modified his original equation as follows:

$$\underline{\underline{\sigma}}(t) = -\int_{-\infty}^{t} m(t-t')H\{I_{c^{-1}}, II_{c^{-1}}\}\underline{\underline{C}}^{-1}(t')dt', \qquad (6.4\text{-}63)$$

where the irreversible function H is defined by

$$H\{I_{c^{-1}}, II_{c^{-1}}\} = \min_{t''=t'}^{t''=t} h[I_{c^{-1}}(t'',t'), II_{c^{-1}}(t'',t')]. \qquad (6.4\text{-}64)$$

This irreversible constitutive equation has been shown to correctly predict the recoverable strain following steady elongational flow, which attains a plateau with increasing deformation (Wagner and Meissner, 1980). Another form for the memory function has recently been proposed by Zdilar and Tanner (1994) to account for irreversibility in recoil of rigid PVC.

6.4-4 The LeRoy-Pierrard Equation

The constitutive equation proposed by LeRoy and Pierrard (1973) is an extension of the quasi-linear corotational equation proposed by Goddard and Miller (1966). The equation can be written as

$$\underline{\underline{\sigma}} = -\int_{-\infty}^{t} \{G_1(t-t', II_{\dot{\gamma}}(t'), III_{\dot{\gamma}}(t'))\underline{\underline{\dot{\Gamma}}}(t') \\ + G_2(t-t', II_{\dot{\gamma}}(t'), III_{\dot{\gamma}}(t'))\underline{\underline{\dot{\Gamma}}}(t') \cdot \underline{\underline{\dot{\Gamma}}}(t')\}dt', \qquad (6.4\text{-}65)$$

where G_1 and G_2 are kernel functions of the second and third invariant of the rate-of-strain tensor; and $\underline{\underline{\dot{\Gamma}}}$ is the rate-of-strain tensor written in a corotational frame (Bird, Armstrong, and Hassager, 1977):

$$\underline{\underline{\dot{\Gamma}}}(t') = \underline{\underline{\omega}}(t') \cdot \underline{\underline{\dot{\gamma}}}(t') \cdot \underline{\underline{\omega}}^+(t'), \qquad (6.4\text{-}66)$$

where $\underline{\underline{\omega}}$ is the vorticity tensor defined by equation 6.1-23.

The Goddard-Miller equation is obtained by taking $G_1 = G_1(t-t')$ and $G_2 = 0$. The Goddard-Miller or the LeRoy-Pierrard equations can be seen as another generalization of the Maxwell model, writing the integrated form of equation 5.2-6 in a corotational frame of reference instead of an embedded frame as was done in Section 6.2-2. The results are quite different. For example, the viscometric functions for the *Goddard-Miller* equation are

$$\eta = \int_0^\infty G(s)\cos\dot{\gamma}s\,ds, \qquad (6.4\text{-}67\text{a})$$

$$\psi_1 = \frac{2}{\dot{\gamma}}\int_0^\infty G(s)\sin\dot{\gamma}s\,ds, \qquad (6.4\text{-}67\text{b})$$

$$\text{and}\quad \psi_2 = -\frac{1}{2}\psi_1. \qquad (6.4\text{-}67\text{c})$$

These results are analogous to the expressions for the real and imaginary parts of the complex viscosity of a generalized Maxwell fluid (equations 5.3-3 and 5.3-4), provided we set

$$\eta'(\omega = \dot{\gamma}) = \eta(\dot{\gamma}) \tag{6.4-68a}$$

$$\text{and} \quad \eta''(\omega = \dot{\gamma}) = \frac{\dot{\gamma}}{2}\psi_1(\dot{\gamma}). \tag{6.4-68b}$$

Hence, without introducing a single nonlinear parameter, shear effects are predicted. The secondary normal stress coefficient is nonzero and negative as experimentally observed. Qualitatively, the results are of considerable interest, but the analogy between the shear functions and the linear functions η' and η'' is unrealistic. The magnitude of ψ_2 is too large. The Goddard-Miller model predicts that the uniaxial elongational viscosity is equal to three times the zero shear viscosity (Trouton relation). This is to be expected, since elongational flow is irrotational. The generalization proposed by LeRoy and Pierrard (1973) aimed at correcting the main deficiencies of the Goddard-Miller equation, that is, the LeRoy-Pierrard model can describe shear viscosity and normal stress data which are not linked to complex viscosity data as suggested by equations 6.4-68.

Corotational models predict oscillations for stress growth experiments at large values of the imposed shear rate. This feature is largely responsible for the recent loss of interest in corotational models, which were highly popular in the seventies. The following test also shows that the LeRoy-Pierrard model fails to predict the correct relaxation behavior of polymer solutions.

We can show that the general LeRoy-Pierrard equation gives the following relations between material functions defined for stress relaxation experiments (Attané et al., 1988):

$$\psi_1^-(\dot{\gamma}_0, t) = 2\int_t^\infty \eta^-(\dot{\gamma}_0, \tau)d\tau \tag{6.4-69a}$$

or

$$-\frac{1}{2}\frac{\partial \psi_1^-}{\partial t}(\dot{\gamma}_0, t) = \eta^-(\dot{\gamma}_0, t). \tag{6.4-69b}$$

In the limit $t \to 0$, $\eta^-(\dot{\gamma}_0, t) = \eta(\dot{\gamma}_0)$ and $\psi_1^-(\dot{\gamma}_0, t) = \psi_1(\dot{\gamma}_0)$, and relations 6.4-69a and b become

$$\psi_1(\dot{\gamma}_0) = 2\int_0^\infty \eta^-(\dot{\gamma}_0, \tau)d\tau \tag{6.4-70a}$$

$$\text{and} \quad -\frac{1}{2}\frac{\partial \psi_1^-}{\partial t}(\dot{\gamma}_0, t)|_{t=0} = \eta(\dot{\gamma}_0). \tag{6.4-70b}$$

Equation 6.4-70 is known as the Yamamoto relation (1971). It holds for the Lodge and the Bird-Carreau equations, but it does not hold for the Carreau equation presented in Section 6.4-2, for which it can be shown that

$$\psi_1^-(\dot{\gamma}_0, t) \leq 2\int_t^\infty \eta^-(\dot{\gamma}_0, \tau)d\tau \tag{6.4-71a}$$

256   Nonlinear Viscoelasticity

or

$$-\frac{1}{2}\frac{\partial \psi_1^-}{\partial t}(\dot{\gamma}_0, t) \leq \eta^-(\dot{\gamma}_0, t), \qquad (6.4\text{-}71\text{b})$$

where the equality holds at the limit of $\dot{\gamma}_0 \to 0$. Note that the Carreau, K-BKZ, and LeRoy-Pierrard constitutive equations predict the same behavior at zero shear rate. For instance, equations 6.4-48a, 6.4-70a, and 6.4-71a reduce to

$$\psi_{10} = 2\int_0^\infty \eta_0^-(t)dt. \qquad (6.4\text{-}72)$$

These relations can be used as rheological tests to discriminate between various classes of constitutive equations. It should be emphasized that stress relaxation following steady shear rate experiments are not necessarily very accurate, and errors larger than 15% are not uncommon. Figure 6.4-10 shows clearly that the Yamamoto relation (equation 6.4-70a) is verified for the PS7 solutions only at low shear rates. The fluid characteristics are presented in Table 6.4-2. At large dimensionless shear rates ($\lambda_w \dot{\gamma}_0$), the ratio of the right side to the left side of equation 6.4-70a, is considerably larger than the theoretical value of 1. Similar results were obtained for solutions of the lower molecular weight polymers. Hence, the LeRoy-Pierrard and Bird-Carreau models are not flexible enough to predict the shear rate dependence at high shear rates. This is similar to the failure observed for the K-BKZ model in Fig. 6.4-8 using a different relation. In principle, the Carreau equation can describe the high shear rate dependence observed in Fig. 6.4-10.

In Fig. 6.4-11, the relaxation data are plotted in the form suggested by relation 6.4-69b. The same limits at $t=0$ for both the shear and normal stress functions are correctly

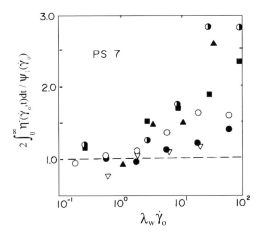

Figure 6.4-10 Assessment of the Yamamoto relation (equation 6.4-70a) for a series of polystyrene solutions (Adapted from Attané et al., 1988)
The solution characteristics and symbols are defined in Table 6.4-2
- - - - equation 6.4-70a

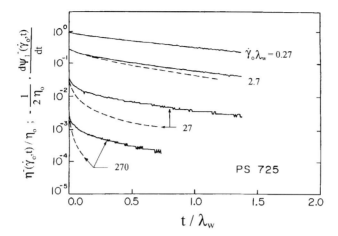

Figure 6.4-11 Assessment of relation 6.4-69b for a series of polystyrene solutions
The solution characteristics are presented in Table 6.4-2 (Adapted from Attané et al., 1988)
——— shear relaxation function
– – – derivative of the normal stress relaxation function

predicted by the LeRoy-Pierrard or Bird-Carreau models. However, large departures between the relaxation functions are observed at finite times for large shear rates. Clearly these two models cannot describe the nonlinear data. As equation 6.4-70a is an integral form of equation 6.4-69b, the observations of Fig. 6.4-11 confirm the pattern shown by Fig. 6.4-10, although different transient experimental data were used to evaluate relation 6.4-69b. On the other hand, the K-BKZ and Carreau equations can describe the differences observed at finite time (see equations 6.4-47b and 6.4-71b), but incorrectly predict different limits at $t=0$ for large shear rates (see equations 6.4-48b and 6.4-71b respectively).

Other tests for stress growth experiments have been used by Attané et al. (1988) to show that the Carreau and K-BKZ equations are capable of describing the behavior of polystyrene solutions, whereas the LeRoy-Pierrard and the Carreau-Bird equations fail drastically.

It is worth mentioning that stress relaxation following steady shear flow, and stress growth at the inception of steady shear flow are very severe tests for constitutive equations. In engineering applications, the flow field is frequently quite complex, but the velocity changes with time are rarely as severe as those encountered in the transient rheological tests. In that context, the LeRoy-Pierrard equation may be of considerable value.

6.5 Concluding Remarks

In this chapter, we have presented a variety of constitutive equations, starting from the simplest Maxwell convected model to the more complex integral Carreau, K-BKZ, and

LeRoy-Pierrard equations. These constitute only a small selection of constitutive equations which have been proposed in the literature over the last three decades. In consulting the recent volumes of the journals in the field (Journal of Rheology, Journal of Non-Newtonian Fluid Mechanics, Rheologica Acta, etc.), the reader will find a considerable amount of new ideas on constitutive equations and new rheological tests which can be used to discriminate between various models. These results are useful, and the efforts should eventually lead to better and more powerful constitutive equations.

It is doubtful, however, that a simple and universal constitutive equation that could describe the entire spectrum of rheological tests will eventually be developed. The behavior of polymers, especially polymeric multiphase systems, is quite complex, and a rheological equation should be considered as a tool for characterizing the material under very specific flow conditions, or for predicting the flow behavior in processing from the knowledge of the rheological properties under well-defined flow kinematics.

The most striking recent progress has been in computational rheology. With the easy access to fast and powerful computing facilities, numerical calculations that were unthinkable a few years ago are now possible. Complex integral constitutive equations, such as those presented in Section 6.4, are now used in finite-element code to compute flow and stress fields encountered in polymer processing. These efforts will help clarify trends and future orientations in the development of constitutive equations. In parallel, molecular theories and direct computer simulation using non-equilibrium molecular dynamics (see for example Dlugogorski et al., 1993) are fast growing areas. The results initially obtained for equilibrium (non-flow) conditions are now extended to non-equilibrium and large strain flows. These theories provide not only useful guidelines, but in some cases, as shown in Chapter 7, yield usable and powerful constitutive equations.

6.6 Problems*

6.6-1$_a$ Planar Elongational Flow

Show that the three invariants of the rate-of-deformation tensor for planar elongational flow are given by equation 6.1-50. Compare to the invariants for uniaxial elongational and biaxial deformation, defined in Section 2.1-4.

6.6-2$_a$ Elongational Viscosity of a Lower-Convected Maxwell Fluid

Following the procedure developed in Example 6.2-2, determine the expression for the elongational viscosity of a lower-convected Maxwell fluid. Discuss the differences between this result and that of Example 6.2-2.

*Problems are labelled by a subscript a or b indicating the level of difficulty.

6.6-3$_b$ Biaxial Elongation

Biaxial elongation can be defined by the flow field

$$V_1 = \dot{\varepsilon}_B x_1 \tag{6.6-1}$$

and

$$V_2 = \dot{\varepsilon}_B x_2, \tag{6.6-2}$$

where $\dot{\varepsilon}_B$ is the elongational rate. Consider a constant $\dot{\varepsilon}_B$ and an incompressible fluid.

(a) Obtain
   (i) the velocity component V_3,
   (ii) the three invariants of the rate-of-deformation tensor, and
   (iii) the expressions for the Finger tensor.
(b) Compare this flow situation to uniaxial and planar elongational flows discussed in Section 6.1.
(c) Obtain the expression of the elongational viscosity for an upper-convected Maxwell fluid.

6.6-4$_a$ Admissible Constitutive Equations

Justify which of the following are admissible constitutive equations.

(a) $$\sigma_{ij} = G_1 \left(\frac{\partial x^i}{\partial x'^r}\right)\left(\frac{\partial x^j}{\partial x'^s}\right) g^{rs}(t') \tag{6.6-3}$$

(b) $$\sigma^{ij} + \lambda \frac{\delta}{\delta t} \sigma^{ij} = -\eta \dot{\gamma}^{ij} - \beta(\sigma_k^i \dot{\gamma}^{kj} + \dot{\gamma}^{ik} \sigma_k^j) \tag{6.6-4}$$

(c) $$\sigma^{ij} + \lambda_1 \frac{\partial}{\partial t} \sigma^{ij} = -\eta \left(\dot{\gamma}^{ij} + \lambda_2 \frac{\partial}{\partial t} \dot{\gamma}^{ij}\right) \tag{6.6-5}$$

(d) $$\sigma^{ij} = -\int_{-\infty}^{t} m_1(t, t') C^{-1_{ij}}(t') dt'$$
$$- \int_{-\infty}^{t} m_2(t, t') C^{-1_{ik}}(t') C_k^{-1_j}(t') dt' \tag{6.6-6}$$

6.6-5_b Second-Order Fluid

A second-order fluid can be modeled for homogeneous flow by the following equation, expressed in embedded coordinates as

$$\hat{\sigma}^{ij} = -P\hat{g}^{ij} - \beta_1 \frac{d}{dt}\hat{g}^{ij} + \beta_2 \hat{g}^{ir}\hat{g}^{js}\frac{d^2\hat{g}_{rs}}{dt^2}$$
$$- \frac{1}{2}\beta_3 \frac{d^2\hat{g}^{ij}}{dt^2}. \tag{6.6-7}$$

(a) Obtain the corresponding equation in fixed coordinates.
(b) Obtain the material functions for steady shear flow and show that these results are equivalent to those of Example 6.2-3.

6.6-6_b Elongational Viscosity of an Oldroyd-B Fluid

(a) Verify that the uniaxial elongational viscosity for an Oldroyd-B fluid is given by equation 6.3-8.
(b) Calculate values of the elongational viscosity and illustrate the variation of η_E/η_0 as a function of $\lambda_1\dot{\varepsilon}$ for $\lambda_2/\lambda_1 = 0$, 1, 0.3, and 0.6.

6.6-7_b Transient Behavior of a White-Metzner Fluid

Obtain the transient growth functions, $\eta^+(t,\dot{\gamma}_\infty)$ and $\psi^+(t,\dot{\gamma}_\infty)$, and relaxation functions, $\eta^-(t,\dot{\gamma}_\infty)$ and $\psi^-(t,\dot{\gamma}_0)$, for a White-Metzner fluid described by equation 6.3-9. Illustrate the behavior and discuss the influence of $\dot{\gamma}_\infty$ or $\dot{\gamma}_0$ on these functions.

6.6-8_b Flow of a White-Metzner Fluid in a Tube Under an Oscillatory Pressure Gradient

Consider the flow of a viscous polymeric fluid in a circular tube of length L and radius R. The pressure gradient varies sinusoidally as described by the following equation:

$$\frac{\Delta P}{L} = \frac{\Delta P_0}{L}(1 + \varepsilon \text{Re}\{e^{i\omega t}\}), \tag{6.6-8}$$

where ΔP_0 is the mean pressure drop, ε is the amplitude of the pressure gradient variation, Re is the real part, and ω is the frequency of the oscillations.

Neglect inertial effects and consider that the fluid is described by a White-Metzner equation (6.3-9) with a viscosity term expressed by a power-law, and $\lambda(II_{\dot\gamma})$ given by

$$\lambda(II_{\dot\gamma}) = \lambda(\dot\gamma) = \frac{\eta(\dot\gamma)}{G_0} = \frac{m|\dot\gamma|^{n-1}}{G_0}. \tag{6.6-9}$$

(a) Solve the equation of motion to obtain an expression for the shear rate as a function of r and t.
(b) Use the previous result to obtain an integral expression for the flow rate.
(c) Defining a flow enhancement function by

$$I = \frac{\langle Q \rangle - Q_0}{Q_0}, \tag{6.6-10}$$

where $\langle Q \rangle$ is the mean flow rate under oscillation and Q_0 is the flow rate in the absence of oscillations, show how I can be calculated. (The case of small amplitude is discussed in Bird, Armstrong, and Hassager, 1987).

6.6-9$_a$ Viscometric Functions for a Marrucci Fluid

(a) Verify that the viscosity and the primary normal stress function for a Marrucci fluid are given by equations 6.3-47a and b respectively. Show also that $\psi_2 = 0$.
(b) Using a single relaxation time and a single modulus, illustrate the behavior of η/η_0 and $\psi_1/2\lambda_0$ as a function of $(\lambda_0 \dot\gamma)$ for $a = 0.4$.

6.6-10$_b$ Material Functions for a Carreau Fluid

(a) Verify the expressions of the material functions for the general Carreau equation given by results 6.4-15 to 6.4-22.
(b) Obtain the stress growth function and the relaxation function (after steady shear) for a Carreau-A fluid. Show that $\eta^-(t, \dot\gamma_0)/\eta(\dot\gamma_0)$ is a unique function of time.

6.6-11$_b$ Material Functions for a Maxwell Model Involving Slip

Calculate the viscosity $\eta(\dot\gamma)$ and normal stress coefficients $\psi_1(\dot\gamma)$ and $\psi_2(\dot\gamma)$ for a generalized Maxwell fluid, given by the expression

$$\underline{\underline{\sigma}} + \lambda_m(\dot\gamma) \frac{\bar{\delta}}{\delta t} \underline{\underline{\sigma}} = -\eta_m(\dot\gamma) \underline{\underline{\dot\gamma}}, \tag{6.6-11}$$

where $\bar{\delta}/\delta t$ is the Gordon-Schowalter derivative and λ_m and η_m are functions of $\dot\gamma$.

6.6-12$_b$ Relations Between Material Functions

(a) Verify relations 6.4-47 and 6.4-48 for the K-BKZ model.
(b) Show that the Yamamoto relation (equation 6.4-70a) is verified for the Bird-Carreau equation (obtained by setting $g_p = 1$ in the Carreau model given by equations 6.4-13 and 6.4-14).
(c) Show that the Yamamoto relation also holds for the LeRoy-Pierrard model (equation 6.4-65) if G_2 is set equal to zero.

6.6-13$_b$ Flow Above an Oscillating Plate

The upper half of the (x_1, x_2) plane is filled with an incompressible fluid, and the rigid boundary at $x_2 = 0$ is oscillating with a velocity $V_0 \cos \omega t$, where V_0 is a constant and ω is the frequency. Determine the velocity $V_1(x_2, t)$ if

(a) The fluid obeys the Rivlin-Ericksen model (equation 6.2-29), and
(b) The fluid obeys the Lodge elastic liquid model (equation 6.4-1).
(c) Compare the results and show under which conditions the result in (a) can be obtained from the result in (b).

Assume that the velocity distribution is given by

$$V_1 = \text{Re}[V_0 \exp(\alpha x_2 + i\omega t)] \qquad (6.6\text{-}12)$$

and

$$V_2 = V_3 = 0, \qquad (6.6\text{-}13)$$

where α is a constant to be determined.

7 Constitutive Equations from Molecular Theories

7.1 Bed- and Spring-Type Models 264
    7.1-1 Hookean Elastic Dumbbell 265
        7.1-1.1 Relation Between the Connector Force and the Stress Tensor 265
        7.1-1.2 Distribution Function 267
        7.1-1.3 Distribution Function $\psi(\underline{R}, t)$ 268
        7.1-1.4 Force Balance on Dumbbells 269
    7.1-2 Finitely Extensible Nonlinear Elastic (FENE) Dumbbell 272
    7.1-3 Rouse and Zimm Models 276

7.2 Network Theories 284
    7.2-1 General Network Concept 284
    7.2-2 Rubber-Like Solids 286
    7.2-3 Elastic Liquids 288
    7.2-4 Recent Developments 290

7.3 Reptation Theories 294
    7.3-1 The Tube Model 294
    7.3-2 The Doi-Edwards Model 296
    7.3-3 The Curtiss-Bird Kinetic Theory 300

7.4 Conformation Tensor Rheological Models 304
    7.4-1 Basic Description of the Conformation Model 304
    7.4-2 FENE-Charged Macromolecules 307
    7.4-3 Rod-Like and Worm-Like Macromolecules 312
    7.4-4 Generalization of the Conformation Tensor Model 320

7.5 Problems 327
    7.5-1$_b$ Hookean Dumbbell Model 327
    7.5-2$_b$ Tanner Equation 327
    7.5-3$_a$ Complex Viscosity of Rouse Fluid 327
    7.5-4$_b$ Network Model 327
    7.5-5$_b$ Conformation Model 328
    7.5-6$_b$ FENE Conformation Model 328
    7.5-7$_b$ Rod-Like Macromolecules 328

In the previous chapters, we have shown that an adequate description of the rheological behavior (stress-strain rate relationship) is required in order to solve flow problems that arise in polymer manufacturing and processing. As is also evident from Chapter 6, the choice of a constitutive equation or rheological model may not be obvious. A computer simulation cannot describe a given flow process better than the rheological model on which the algorithm is based. The use of a complex and fully predictive model will inevitably lead to unacceptably long computing times. Therefore, at least for engineering applications, the choice of a rheological equation will necessarily be a compromise between completeness and computation facility. Rheological equations also find applications in the characterization of polymer systems. A rheological model may be needed to organize, interpret, and understand results from various experiments conducted on a given class of polymeric liquids. A carefully chosen rheological model should make it possible to extract the pertinent material characteristics and predict the rheological behavior under other flow conditions. If molecular considerations are built into the model, model parameters evaluated from rheological measurements should relate directly to the molecular structure of the polymer.

The development of constitutive equations discussed in Chapter 6 has been guided by principles of continuum mechanics and thermodynamics, and inspired by experimental investigations. In this chapter, we introduce molecular approaches which involve simple descriptions of the macromolecules. Results of non-equilibrium statistical mechanics are used to obtain relationships between measurable quantities such as viscosity and normal stress coefficients, and parameters related to the description of the macromolecules. We consider some of the most important theories based on molecular considerations: (i) dumbbell-type models developed for dilute polymer solutions, and (ii) network and reptation theories proposed for polymer melts and concentrated solutions. The advantages and limitations of each group of theories are discussed. In the last section, we present a different approach proposed by Grmela (1985) and Grmela and Carreau (1987), which combines molecular arguments with principles of continuum mechanics and thermodynamics. This so-called mesoscopic approach based on molecular state variables leads to models which are highly flexible and can describe most of the rheological properties encountered with polymeric liquids.

This chapter does not review many of the new and fascinating molecular theories that have been proposed recently. This is outside the scope of this textbook, and the interested reader is referred to the highly specialized books by Bird et al. (1987) and Doi and Edwards (1986).

7.1 Bead- and Spring-Type Models

In this section we present a detailed analysis of the simple dumbbell followed by a summary of the extended results of Rouse (1953) and Zimm (1956) to chain-like molecules.

7.1-1 Hookean Elastic Dumbbell

We start by considering the dumbbell model, where a polymer molecule is represented by two beads connected by a spring as shown in Fig. 7.1-1. We consider the dumbbells to be suspended in a Newtonian fluid of viscosity η_s. We further consider the system to be dilute. That is, the dumbbells do not interact and we can therefore examine each dumbbell on its own. We neglect inertial forces due to the masses m_1 and m_2 of beads (1) and (2) respectively. We assume that the force $\underline{F}^c$, acting on the spring (connector) is proportional to the end-to-end vector, $\underline{R}$. That is,

$$\underline{F}^c = H\underline{R}, \tag{7.1-1}$$

where H for infinitely extensible dumbbells is a constant. The superscript c stands for connector.

7.1-1.1 Relation Between the Connector Force and the Stress Tensor

We look for a relation between the connector force and the total stress tensor $\underline{\underline{\pi}} = \underline{\underline{\sigma}} + P\underline{\underline{\delta}}$. In the absence of hydrodynamic interactions, we assume that the total stress in the solution is the sum of the contribution of the dumbbells (polymer) and the contribution of the solvent:

$$\pi_{ij}(\underline{r}, t) = \pi_{ij}^p(\underline{r}, t) + \pi_{ij}^s(\underline{r}, t). \tag{7.1-2}$$

Recall that π_{ij} represents the j-component of a force per unit surface acting on a surface normal to the i-direction. It can be also considered as the flux in the i-direction of the momentum in the j-direction (see Section 2.1).

Let us now consider an elementary cubic volume of fluid of side equal to $(1/n)^{1/3}$, where n is the dumbbell density (number of dumbbells per unit volume). The volume is moving at the local fluid velocity $\underline{V}$. The situation is illustrated in Fig. 7.1-2.

The probability that any dumbbell will cross a plane of area $(1/n)^{2/3}$ is

$$\frac{\underline{n} \cdot \underline{R}}{\left(\dfrac{1}{n}\right)^{\frac{1}{3}}} = \frac{\text{projection of } \underline{R} \text{ on } \underline{n}}{\text{length of one side}}, \tag{7.1-3}$$

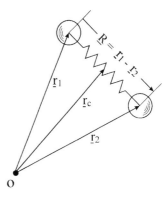

Figure 7.1-1 Representation of a dumbbell: beads (1) and (2) are connected by a spring

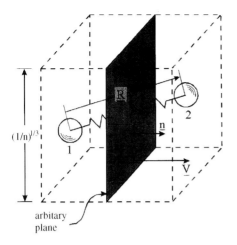

Figure 7.1-2 Dumbbell intersecting an arbitrary plane in cubic fluid volume equal to $1/n$

where $\underline{n}$ is the unit vector normal to the plane. The force per unit surface on the right side of the plane is given by

$$\underline{F} = -\frac{n^{\frac{1}{3}}(\underline{n} \cdot \underline{R})\underline{F}^c}{\left(\dfrac{1}{n}\right)^{\frac{2}{3}}}, \tag{7.1-4}$$

or in component form,

$$F_i = -n n_j R_j F_i^c. \tag{7.1-5}$$

Any force is related to the stress tensor by

$$F_i = n_j \pi_{ji}. \tag{7.1-6}$$

Hence, from equations 7.1-5 and 7.1-6, we obtain

$$\pi_{ji}^c = -n R_j F_i^c, \tag{7.1-7}$$

which is the contribution of the connector force to the total stress tensor.

The contribution of the beads to the total stress tensor is obtained by noting that the number of beads (1) that cross the arbitrary plane per unit surface and per unit time is

$$n(\underline{\dot{r}}_1 - \underline{V}) \cdot \underline{n}.$$

Hence, the momentum flux due to beads (1) is expressed by

$$Q_i^{(1)} = n m_1 (\dot{r}_{1j} - V_j) n_j (\dot{r}_{1i} - V_i). \tag{7.1-8}$$

Similarly for beads (2):

$$Q_i^{(2)} = n m_2 (\dot{r}_{2j} - V_j) n_j (\dot{r}_{2i} - V_i). \tag{7.1-9}$$

The momentum transferred is the sum $\underline{Q} = \underline{Q}^{(1)} + \underline{Q}^{(2)}$ and

$$Q_i = n_j \pi_{ij}^b. \tag{7.1-10}$$

From equations 7.1-8 and 7.1-9, we obtain the contributions of the beads to the total stress tensor (for $m_1 = m_2 = m$):

$$\pi_{ij}^b = nm[(\dot{r}_{1i} - V_i)(\dot{r}_{1j} - V_j) + (\dot{r}_{2i} - V_i)(\dot{r}_{2j} - V_j)]. \tag{7.1-11}$$

7.1-1.2 Distribution Function

The results obtained so far are functions of the end-to-end vector R and of the position vector r_1 and r_2. We introduce a distribution function for the dumbbells such that

$$\iiiint f(r_1, \dot{r}_1, r_2, \dot{r}_2, t) dr_1 dr_2 d\dot{r}_1 d\dot{r}_2 = n. \tag{7.1-12}$$

$f(r_1, \dot{r}_1, r_2, \dot{r}_2, t) dr_1 dr_2 d\dot{r}_1 d\dot{r}_2$ represents the number of dumbbells per unit volume at time t for which bead (1) is in the range dr_1 about r_1 and at velocity in the range $d\dot{r}_1$ about $\dot{r}_1$. Similarly, bead (2) is in the interval dr_2 about r_2 and at velocity in the range $d\dot{r}_2$ about $\dot{r}_2$. Note that in equation 7.1-12 we are not referring to four integrations. The vector quantity dr_1 represents three integrations. The same is true for the other vector quantities. We make the following assumptions:

(i) We assume that the spatial and velocity distributions can be separated as follows:

$$f(r_1, r_2, \dot{r}_1, \dot{r}_2, t) = \phi(r_1, r_2, t)\Theta(\dot{r}_1, \dot{r}_2), \tag{7.1-13}$$

with

$$\iint \Theta(\dot{r}_1, \dot{r}_2) d\dot{r}_1 d\dot{r}_2 = 1. \tag{7.1-14}$$

ϕ is the distribution in the configuration space and f is the distribution in the phase space.

(ii) The distribution function for the velocity is assumed to be Maxwellian. That is,

$$\Theta(\dot{r}_1, \dot{r}_2) = \frac{\exp\left[-\frac{m|\dot{r}_1 - V|^2}{2 \, k_B T} - \frac{m|\dot{r}_2 - V|^2}{2 \, k_B T}\right]}{\iint \exp\left[-\frac{m|\dot{r}_1 - V|^2}{2 \, k_B T} - \frac{m|\dot{r}_2 - V|^2}{2 \, k_B T}\right] d\dot{r}_1 d\dot{r}_2}, \tag{7.1-15}$$

where T is the absolute temperature and k_B is the Boltzmann constant. Obviously, equation 7.1-15 satisfies equation 7.1-14.

(iii) We assume that the dumbbells are distributed homogeneously and thus $\phi(r_1, r_2, t)$ depends only on the vector $R = r_2 - r_1$ connecting the beads in a dumbbell. We write it in the form

$$\phi(r_1, r_2, t) = n\psi(R, t). \tag{7.1-16}$$

In order to satisfy 7.1-12 we require that

$$\int \psi(R, t) dR = 1. \tag{7.1-17}$$

With the help of the distribution function $f(r_1, r_2, \dot{r}_1, \dot{r}_2, t)$, we can now write the contribution of all dumbbells to the total stress tensor as

$$\underline{\underline{\pi}}^p = \iiiint \underline{\underline{\pi}}^{p_1} f(r_1, r_2, \dot{r}_1, \dot{r}_2, t) dr_1 dr_2 d\dot{r}_1 d\dot{r}_2, \tag{7.1-18}$$

where $\underline{\underline{\pi}}^{p_1} = (\underline{\underline{\pi}}^{c_1} + \underline{\underline{\pi}}^{b_1})$ is the contribution of a typical dumbbell. Using results 7.1-7 and 7.1-11 as well as the simplifications implied by equations 7.1-13 to 7.1-17, we obtain (see Problem 7.5-1)

$$\pi_{ij}^p = -n\langle R_i F_j^c \rangle + 2nk_B T \delta_{ij}, \quad (7.1\text{-}19)$$

where

$$\langle R_i F_j^c \rangle = \int \psi(\underline{R}, t) R_i F_j^c d\underline{R}. \quad (7.1\text{-}20)$$

The polymer contribution may be assumed to consist of two parts, an isotropic or pressure term and a deviatoric part. That is,

$$\pi_{ij}^p = P^p \delta_{ij} + \sigma_{ij}^p, \quad (7.1\text{-}21)$$

where P^p is the equilibrium pressure of an ideal gas composed of n particles, that is,

$$P^p = nk_B T. \quad (7.1\text{-}22)$$

The total equilibrium pressure is thus

$$P = P^s + nk_B T, \quad (7.1\text{-}23)$$

where P^s is the solvent contribution to the pressure. Combining equations 7.1-19, 7.1-21, and 7.1-22, we obtain the following expression for the polymer contribution to the extra stress:

$$\sigma_{ij}^p = -n\langle R_i F_j^c \rangle + nk_B T \delta_{ij}. \quad (7.1\text{-}24)$$

Adding the solvent contribution to equation 7.1-24, we can write

$$\sigma_{ij} = -\eta_s \dot{\gamma}_{ij} - n\langle R_i F_j^c \rangle + nk_B T \delta_{ij}, \quad (7.1\text{-}25)$$

where $\dot{\gamma}_{ij}$ are the components of the rate-of-deformation tensor. This is the *Kramers expression* for the stress tensor, which in principle can be evaluated if the distribution function $\psi(\underline{R}, t)$ is known.

7.1-1.3 Distribution Function $\psi(\underline{R}, t)$

The principle of conservation of dumbbells, or the equation of continuity for the distribution function $\psi(\underline{R}, t)$, is expressed as

$$\frac{\partial \psi}{\partial t}(\underline{R}, t) = -\nabla \cdot \psi(\underline{R}, t)\underline{V}, \quad (7.1\text{-}26a)$$

or

$$\frac{\partial \psi}{\partial t}(\underline{R}, t) = -\frac{\partial}{\partial R_i}(\dot{R}_i \psi(\underline{R}, t)). \quad (7.1\text{-}26b)$$

7.1-1.4 Force Balance on Dumbbells

Newton's second law of motion for bead (1) is

$$m\ddot{\underline{r}}_1 = \underline{F}_1^d + \underline{F}_1^c + \underline{F}_1^b, \qquad (7.1\text{-}27a)$$

where $\underline{F}_1^d$, $\underline{F}_1^c$, and $\underline{F}_1^b$ are the drag, connector, and Brownian forces respectively. Also, we have for bead (2):

$$m\ddot{\underline{r}}_2 = \underline{F}_2^d + \underline{F}_2^c + \underline{F}_2^b. \qquad (7.1\text{-}27b)$$

We assume that the drag, force is expressed by Stokes' law:

$$\underline{F}_1^d = -\zeta(\dot{\underline{r}}_1 - \underline{V}_1) \qquad (7.1\text{-}28a)$$

$$\text{and} \quad \underline{F}_2^d = -\zeta(\dot{\underline{r}}_2 - \underline{V}_2), \qquad (7.1\text{-}28b)$$

where the fluid velocity components at beads (1) and (2) are given respectively by

$$V_{1i}(\underline{r}_1, t) = V_i(\underline{r}_c, t) + \kappa_{ij}(\underline{r}_c, t)(r_{1j} - r_{cj}). \qquad (7.1\text{-}29a)$$

and

$$V_{2i}(\underline{r}_2, t) = V_i(\underline{r}_c, t) + \kappa_{ij}(\underline{r}_c, t)(r_{2j} - r_{cj}). \qquad (7.1\text{-}29b)$$

$\underline{\underline{\kappa}}$ is the transpose of the velocity gradient tensor, that is,

$$\kappa_{ij}(\underline{r}_c, t) = \frac{\partial}{\partial r_j} V_i(\underline{r}_c, t), \qquad (7.1\text{-}30)$$

and r_c is the position of the center of mass of the dumbbell (Fig. 7.1-1).

The connector force is expressed by equation 7.1-1, and the Brownian forces (forces resulting from collisions of the beads with the solvent molecules) are given by

$$\underline{F}_1^b = -k_B T \frac{\partial}{\partial \underline{r}_1} \ln \psi(\underline{R}, t) \qquad (7.1\text{-}31a)$$

and

$$\underline{F}_2^b = -k_B T \frac{\partial}{\partial \underline{r}_2} \ln \psi(\underline{R}, t). \qquad (7.1\text{-}31b)$$

We neglect the acceleration (inertial) terms and write the force balance as

$$\underline{F}^c = \underline{F}_2^c - \underline{F}_1^c = H\underline{R} = \zeta(\dot{\underline{r}}_2 - \underline{V}_2) - \zeta(\dot{\underline{r}}_1 - \underline{V}_1)$$
$$+ k_B T \frac{\partial}{\partial \underline{r}_2} \ln \psi - k_B T \frac{\partial}{\partial \underline{r}_1} \ln \psi. \qquad (7.1\text{-}32)$$

To simplify this result we make the following change of coordinates (or variables):

$$(\underline{r}_1, \underline{r}_2) \rightarrow (\underline{r}_c, \underline{R}), \qquad (7.1\text{-}33a)$$

where the vector position of the center of mass and the end-to-end vectors are given respectively by (see Fig. 7.1-1)

$$\underline{r}_c = \frac{1}{2}(\underline{r}_1 + \underline{r}_2) \qquad (7.1\text{-}33b)$$

and
$$R = r_2 - r_1. \tag{7.1-33c}$$

It follows from the chain rule that:

$$\frac{\partial}{\partial r_1} = \frac{\partial}{\partial r_c}\frac{\partial r_c}{\partial r_1} + \frac{\partial}{\partial R}\frac{\partial R}{\partial r_1} = \frac{1}{2}\frac{\partial}{\partial r_c} - \frac{\partial}{\partial R} \tag{7.1-34a}$$

and

$$\frac{\partial}{\partial r_2} = \frac{\partial}{\partial r_c}\frac{\partial r_c}{\partial r_2} + \frac{\partial}{\partial R}\frac{\partial R}{\partial r_2} = \frac{1}{2}\frac{\partial}{\partial r_c} + \frac{\partial}{\partial R}. \tag{7.1-34b}$$

We finally assume affine deformation, that is, the center of mass of the dumbbell moves at the solvent velocity, $\dot{r}_{ci} = V_i$. Equation 7.1-32 then simplifies to (see Problem 7.5-1)

$$\dot{R}_i = \kappa_{ij}R_j - \frac{2}{\zeta}\left(k_B T \frac{\partial}{\partial R_i}\ln\psi(R,t) + HR_i\right) \tag{7.1-35a}$$

$$= \kappa_{ij}R_j - \frac{2}{\zeta}(F_i^b + F_i^c). \tag{7.1-35b}$$

Substituting equation 7.1-35b in the equation of continuity for $\psi(R,t)$, equation 7.1-26b, yields

$$\frac{\partial\psi(R,t)}{\partial t} = -\frac{\partial}{\partial R_i}\left[\left(\kappa_{ij}R_j - \frac{2}{\zeta}(F_i^b + F_i^c)\right)\psi(R,t)\right]. \tag{7.1-36}$$

This is the evolution or diffusion equation for $\psi(R,t)$. Note that the first term on the right side represents the contribution of the hydrodynamic forces; and the second and third terms describe the effects of the Brownian motion and of the connector (elastic force) respectively. This equation has to be solved simultaneously with equation 7.1-25 for the stress. There exist only a few analytical solutions for these equations, mainly for Hookean dumbbells ($H = H_0$ = constant), presented in this section.

The diffusion equation 7.1-36 leads to the so-called Giesekus equation (Giesekus, 1966). We multiply each term by $R_i R_j$ and integrate over R:

$$\underbrace{\int \frac{\partial}{\partial t}\psi(R,t)R_i R_j dR}_{(a)} = \underbrace{-\int R_i R_j \frac{\partial}{\partial R_k}(\kappa_{kl}R_l\psi(R,t))dR}_{(b)}$$

$$+ \underbrace{\frac{2}{\zeta}k_B T \int R_i R_j \frac{\partial}{\partial R_k}\left(\frac{\partial}{\partial R_k}(\ln\psi(R,t)\psi(R,t))\right)dR}_{(c)} \tag{7.1-37}$$

$$+ \frac{2}{\zeta}\int R_i R_j \frac{\partial}{\partial R_k}(F_k^c\psi(R,t))dR.$$

The integral (a) is

$$\frac{D}{Dt}\int \psi(R,t)R_i R_j dR = \frac{D}{Dt}\langle R_i R_j\rangle. \tag{7.1-38a}$$

7.1 Bead- and Spring-Type Models

This result follows from the Leibnitz formula for differentiating an integral (see Bird, Stewart, and Lightfoot, 1960, Appendix A-5). The substantial derivative is used since we are following the distribution of dumbbells at the average velocity of the fluid. Integrating (b), (c), and (d) by parts yields (see Problem 7.5-1)

$$(b) = \int (\delta_{ki} R_j R_l \kappa_{kl} + \delta_{jk} R_i R_l \kappa_{kl}) \psi(\underline{R}, t) d\underline{R}$$

$$= \langle R_j R_l \rangle \kappa_{il} + \langle R_i R_l \rangle \kappa_{jl} \tag{7.1-38b}$$

$$(c) = -\frac{2}{\zeta} k_B T \int \left(\delta_{ki} R_j \frac{\partial}{\partial R_k} \psi(\underline{R}, t) - \delta_{kj} R_i \frac{\partial}{\partial R_k} \psi(\underline{R}, t) \right) d\underline{R}$$

$$= \frac{4}{\zeta} k_B T \delta_{ij} \tag{7.1-38c}$$

and $(d) = -\frac{2}{\zeta} \int (\delta_{ik} R_j F_k^c + \delta_{jk} R_i F_k^c) \psi(\underline{R}, t) d\underline{R}$

$$= -\frac{2}{\zeta} (\langle R_j F_i^c \rangle + \langle R_i F_j^c \rangle)$$

$$= -\frac{4}{\zeta} \langle R_i F_j^c \rangle \tag{7.1-38d}$$

since vectors $\underline{R}$ and $\underline{F}^c$ are in the same direction. Equation 7.1-37 can now simply be written as

$$\frac{D}{Dt} \langle R_i R_j \rangle = \langle R_j R_k \rangle \kappa_{ik} + \langle R_i R_k \rangle \kappa_{jk}$$
$$- \frac{4}{\zeta} \langle R_i F_j^c \rangle + \frac{4}{\zeta} k_B T \delta_{ij}. \tag{7.1-39}$$

The first two terms on the right side, along with the left side of equation 7.1-39, correspond to the upper-convected derivative for $\langle R_i R_j \rangle$, written here in Cartesian coordinates. Hence, equation 7.1-39 can be simplified as

$$\frac{\delta}{\delta t} \langle R_i R_j \rangle = -\frac{4}{\zeta} \langle R_i F_j^c \rangle + \frac{4}{\zeta} k_B T \delta_{ij}. \tag{7.1-40}$$

Comparing this result with equation 7.1-24, we deduce

$$\sigma_{ij}^p = \frac{n\zeta}{4} \frac{\delta}{\delta t} \langle R_i R_j \rangle, \tag{7.1-41}$$

where $\delta/\delta t$ is the upper-convected derivative. This result is known as the *Giesekus (1966) equation*.

For a Hookean dumbbell, $F_i^c = H_0 R_i$, where H_0 is a constant. Equation 7.1-24 reduces to

$$\langle R_i R_j \rangle = -\frac{\sigma_{ij}^p}{n H_0} + \frac{k_B T}{H_0} \delta_{ij}. \tag{7.1-42}$$

Combining this result with the Giesekus equation 7.1-41, we obtain, including the solvent contribution, the following constitutive equation:

$$\sigma_{ij} + \frac{\zeta}{4H_0}\frac{\delta}{\delta t}\sigma_{ij} = -\left(nk_BT\frac{\zeta}{4H_0} + \eta_s\right)\dot{\gamma}_{ij}. \quad (7.1\text{-}43)$$

This is identical to the upper-convected Maxwell model with

$$\lambda_0 = \frac{\zeta}{4H_0} \quad (7.1\text{-}44)$$

and

$$\eta_0 = \frac{nk_BT\zeta}{4H_0} + \eta_s. \quad (7.1\text{-}45)$$

We recall that the upper-convected Maxwell model predicts for steady simple shear flow a constant viscosity, $\eta = \eta_0$, a constant primary normal stress coefficient, $\psi_1 = 2\eta_0\lambda_0$, and a zero secondary normal stress coefficient. We note with interest that this simple molecular approach leads to the upper-convected form of the Maxwell model, and not the lower-convected form which was rejected in Chapter 6 on phenomenological considerations.

7.1-2 Finitely Extensible Nonlinear Elastic (FENE) Dumbbell

The final results of Section 7.1-1 are restricted to infinitely extensible dumbbells. This is not realistic for large deformation flow situations. A more reasonable form for the connector force has been proposed by Warner (1972). It can be expressed as

$$\underline{F}^c = \frac{H_0\underline{R}}{1 - \frac{R^2}{R_\infty^2}}, \quad (7.1\text{-}46)$$

where R_∞ is the maximum dumbbell (polymeric chain) extension. We note that for small deformations or large values of R_∞ ($R_\infty \to \infty$) the connector force reduces to $\underline{F}^c = H_0\underline{R}$, which is the original Hookean (infinitely extensible dumbbell) expression for the spring force. Equation 7.1-46 leads to the so-called FENE dumbbell model. All the other assumptions of Section 7.1-1 are retained and the rheological properties or material functions are calculated using equations 7.1-40 and 7.1-41, together with the diffusion equation 7.1-36. Unfortunately, there is no exact analytical solution, and we have to rely on approximate solutions. For engineering purposes, we will restrict this section to the approximate approach developed by Bird et al. (1980), where no sophisticated numerical techniques are required. The approximation consists of pre-averaging the term $(1 - (R/R_\infty)^2)$ in equation 7.1-46 with respect to the distribution function $\psi(\underline{R}, t)$. Equation 7.1-46 is thus replaced by

$$\underline{F}^c = \frac{H_0\underline{R}}{\left[1 - \left\langle\frac{R^2}{R_\infty^2}\right\rangle\right]}, \quad (7.1\text{-}47a)$$

where $\langle R^2/R_\infty^2 \rangle$ is the average defined earlier as

$$\left\langle \frac{R^2}{R_\infty^2} \right\rangle = \int \psi(\underline{R}, t) \left(\frac{R^2}{R_\infty^2}\right) d\underline{R}. \tag{7.1-47b}$$

This is based on an idea of Peterlin (1966), and the model described in the example below is referred to as the FENE-P dumbbell model. This leads to a model different from equation 7.1-43. In particular, the average $\langle R^2/R_\infty^2 \rangle$ can be evaluated without knowing ψ, as in the case of the Hookean connector.

To obtain a rheological equation, we substitute $\underline{F}^c$ from equation 7.1-47a into equation 7.1-24. We take the trace to obtain the following expression for $\langle R^2/R_\infty^2 \rangle$ in terms of the trace of the stress tensor (tr $\underline{\underline{\sigma}}^p$):

$$\left\langle \frac{R^2}{R_\infty^2} \right\rangle = 1 - \frac{1}{Z(\text{tr } \underline{\underline{\sigma}}^p)}, \tag{7.1-48a}$$

where $Z(\text{tr } \underline{\underline{\sigma}}^p)$ is a function of the trace of $\underline{\underline{\sigma}}^p$ defined by

$$Z(\text{tr } \underline{\underline{\sigma}}^p) = 1 + \frac{3}{b}\left[1 - \frac{\text{tr } \underline{\underline{\sigma}}^p}{3nk_BT}\right], \tag{7.1-48b}$$

and

$$b = \frac{2H_0 R_\infty^2}{k_B T}. \tag{7.1-48c}$$

It follows that

$$\underline{F}^c = H_0 Z \underline{R}, \tag{7.1-49}$$

and the Kramers expression (equation 7.1-24) for the polymer contribution to the stress is

$$\sigma_{ij}^p = -nH_0 Z \langle R_i R_j \rangle + nk_B T \delta_{ij}. \tag{7.1-50}$$

The term $\langle R_i R_j \rangle$ can be eliminated by taking the upper-convective derivative, $\delta/\delta t$, of equation 7.1-50 and combining the result with the Giesekus equation 7.1-41 (see Bird et al., 1987). We obtain the following rheological equation:

$$Z\sigma_{ij}^p + \lambda_H \frac{\delta \sigma_{ij}^p}{\delta t} - \lambda_H (\sigma_{ij}^p - nk_B T \delta_{ij}) \frac{D}{Dt} \ln Z = -nk_B T \lambda_H \dot{\gamma}_{ij}, \tag{7.1-51}$$

where $\delta/\delta t$ is the upper-convected derivative (written here in Cartesian coordinates) and $\lambda_H (= \zeta/4H_0)$ is a characteristic time. Note that since Z is a scalar function, its upper-convected derivative is in fact the substantial derivative. Result 7.1-51 is known as the Tanner equation, since it was previously derived by Tanner (1975) via another route. In Section 7.4, we will show that the same result can be obtained using the conformation tensor approach.

As Z depends on tr $\underline{\underline{\sigma}}^p$, equation 7.1-51 is nonlinear. For $b \to \infty$ ($R_\infty \to \infty$), Z becomes unity and the Hookean elastic dumbbell model is recovered.

Despite the oversimplified idealization of polymer molecules, the idea of finite extensibility results into qualitative predictions of nonlinear effects observed with polymer solutions and melts. The predictions for steady shear flow are discussed in the following example.

Example 7.1-1 Steady Shear Material Functions for the FENE-P Model
Obtain the material functions for the FENE-P model under steady shear flow.

Solution
For steady shear flow, $V_1 = \dot{\gamma} x_2$, and $V_2 = V_3 = 0$. Following the approach outlined in Example 6.2-1 for the upper-convected Maxwell model, equation 7.1-51 reduces to

$$Z\begin{bmatrix} \sigma^p_{11} & \sigma^p_{12} & \sigma^p_{13} \\ \sigma^p_{21} & \sigma^p_{22} & \sigma^p_{23} \\ \sigma^p_{31} & \sigma^p_{32} & \sigma^p_{33} \end{bmatrix} - \lambda_H \dot{\gamma} \begin{bmatrix} 2\sigma^p_{21} & \sigma^p_{22} & \sigma^p_{23} \\ \sigma^p_{22} & 0 & 0 \\ \sigma^p_{23} & 0 & 0 \end{bmatrix} = -nk_B T \lambda_H \begin{bmatrix} 0 & \dot{\gamma} & 0 \\ \dot{\gamma} & 0 & 0 \\ 0 & 0 & 0 \end{bmatrix}. \quad (7.1\text{-}52)$$

We note that for $Z = 1$, $\lambda_H = \lambda_0$, and $nk_B T \lambda_H = \eta_0$, we recover result 6.2-15. As Z is a scalar function of σ^p_{ij}, which in turn is a function of x_2, $D\ln Z/Dt = 0$, since V_1 is the only nonzero component of velocity. Equation 7.1-52 leads to the following set of algebraic equations:

$$\begin{aligned} Z\sigma^p_{11} - 2\lambda_H \dot{\gamma} \sigma^p_{21} &= 0, \\ Z\sigma^p_{21} - \lambda_H \dot{\gamma} \sigma^p_{22} &= -nk_B T \lambda_H \dot{\gamma}, \\ Z\sigma^p_{22} &= 0, \\ Z\sigma^p_{23} = Z\sigma^p_{32} &= 0, \\ Z\sigma^p_{31} - \lambda_H \dot{\gamma} \sigma^p_{23} &= 0, \\ \text{and} \quad Z\sigma^p_{33} &= 0. \end{aligned} \quad (7.1\text{-}53)$$

Simplifying, we obtain

$$Z\sigma^p_{21} = -nk_B T \lambda_H \dot{\gamma} \quad (7.1\text{-}54a)$$

$$\text{and} \quad Z\sigma^p_{11} = 2\lambda_H \sigma^p_{21} \dot{\gamma}, \quad (7.1\text{-}54b)$$

and equation 7.1-48b reduces to

$$Z = 1 + \frac{3}{b}\left[1 - \frac{\sigma^p_{11}}{3nk_B T}\right]. \quad (7.1\text{-}55)$$

As $\dot{\gamma} \to 0$, $\sigma^p_{11} \to 0$, and $Z \to 1 + 3/b$. Hence,

$$\sigma^p_{21} = -\frac{nk_B T \lambda_H \dot{\gamma}}{1 + \frac{3}{b}} = -(\eta_0 - \eta_s)\dot{\gamma}, \quad (7.1\text{-}56)$$

where η_0 and η_s are the zero shear and the solvent viscosities respectively. As in the case of the Hookean dumbbell model, the solvent is assumed to contribute independently to the stress tensor. Hence,

$$\eta_0 - \eta_s = \frac{nk_B T \lambda_H b}{b + 3}. \quad (7.1\text{-}57)$$

7.1 Bead- and Spring-Type Models

To obtain the solution for finite shear rates, we introduce the following dimensionless variables:

$$\xi = \frac{\sigma^p_{11}}{3nk_BT} \tag{7.1-58a}$$

and

$$v = \frac{\sigma^p_{21}}{3nk_BT}. \tag{7.1-58b}$$

Equations 7.1-54 reduce to

$$\left[1 + \frac{3}{b}(1-\xi)\right]v = -\frac{1}{3}\lambda_H\dot\gamma \tag{7.1-59a}$$

and

$$\left[1 + \frac{3}{b}(1-\xi)\right]\xi = 2\lambda_H\dot\gamma v, \tag{7.1-59b}$$

from which we get

$$-\xi = 6v^2 \tag{7.1-60a}$$

and

$$v^3 + \frac{1}{6}\left(1 + \frac{b}{3}\right)v + \frac{b}{54}(\lambda_H\dot\gamma) = 0. \tag{7.1-60b}$$

The real solution of equation 7.1-60b is

$$v = (-C_1 + C_2)^{\frac{1}{3}} + (-C_1 - C_2)^{\frac{1}{3}}, \tag{7.1-61a}$$

where

$$C_1 = \frac{b}{108}(\lambda_H\dot\gamma) \tag{7.1-61b}$$

and

$$C_2 = \left(C_1^2 + \left(\frac{3+b}{54}\right)^3\right)^{\frac{1}{2}}. \tag{7.1-61c}$$

We note from equation 7.1-60a that the primary normal stress coefficient for any shear rate is uniquely related to the viscosity function by

$$\psi_1 = -\frac{(\sigma^p_{11} - \sigma^p_{22})}{\dot\gamma^2} = \frac{2(\eta - \eta_s)^2}{nk_BT}. \tag{7.1-62}$$

For $\dot\gamma \to \infty$, the second term in equation 7.1-60b becomes negligible with respect to the other two, and the solution for v is

$$v = \frac{\sigma^p_{21}}{3nk_BT} = \left|\frac{b}{54}\lambda_H\dot\gamma\right|^{\frac{1}{3}}. \tag{7.1-63}$$

Hence, at high shear rates, the FENE-P model predicts that

$$\eta - \eta_s \propto |\lambda_H\dot\gamma|^{-\frac{2}{3}}, \tag{7.1-64}$$

276   Constitutive Equations from Molecular Theories

and from result 7.1-62,

$$\psi_1 \propto |\lambda_H \dot{\gamma}|^{-\frac{4}{3}}. \tag{7.1-65}$$

It is clear from equation 7.1-64 that the model predicts shear-thinning behavior with a constant slope equal to $-2/3$ on a log-log plot. It follows that ψ_1 also decreases following a power-law model with a constant slope of $-4/3$. These predictions are too restrictive and are considered to be major drawbacks of the model. Grmela and Carreau (1987) have shown that these specific drawbacks are a consequence of the finite extensibility concept.

Considerable improvement in the model predictions could be achieved by using a set of relaxation times to account for molecular weight distribution or various relaxation modes of the polymeric chains. However, the simplicity would definitely be lost. Further discussion on the FENE-P model predictions can be found in Section 7.4, where conformation tensor model predictions are presented and compared to FENE-P predictions.

7.1-3   Rouse and Zimm Models

In this section, we present more realistic models by considering the macromolecules to be made up of N beads connected by $N-1$ springs, as illustrated in Fig. 7.1-3.

Figure 7.1-4 identifies the position vectors for bead (v) relative to the origin ($\underline{r}_v$) and to the center of mass ($\underline{Q}_v$). The vector $\underline{r}_c$, from the origin (0) to the center of mass (c), is given by

$$\underline{r}_c = \frac{1}{N} \sum_{v=1}^{N} \underline{r}_v. \tag{7.1-66}$$

The vector $\underline{Q}_v$ is given by

$$\underline{Q}_v = \underline{r}_v - \underline{r}_c, \tag{7.1-67}$$

and the connector vector $\underline{R}_v$ connecting bead ($v+1$) to bead (v) is given by

$$\underline{R}_v = \underline{r}_{v+1} - \underline{r}_v. \tag{7.1-68}$$

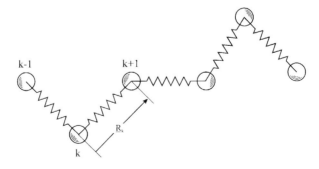

Figure 7.1-3 N beads connected by $N-1$ springs

7.1 Bead- and Spring-Type Models

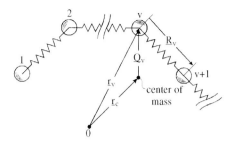

Figure 7.1-4 Bead-spring model, identifying the position vectors

To deduce a constitutive equation, we have to consider the motion of each bead. The equations governing the motion are written in terms of the vector r_ν. The connector forces are given in terms of the vectors $\underline{R}_\nu$. For uniformity, we wish to work in terms of vectors $\underline{R}_\nu$ only. Next, we deduce the linear relations between the vectors r_ν and $\underline{R}_\nu$. These relations are given by a matrix $B_{k\nu}$ as follows:

$$\underline{R}_k = B_{k\nu} \underline{r}_\nu \qquad k = 1, \ldots, N-1, \tag{7.1-69}$$

where the matrix $B_{k\nu}$ has $N-1$ rows and N columns. We first illustrate this via an example involving three beads and two connectors.

Example 7.1-2 Relation Between Vectors $\underline{r}_\nu$ and $\underline{R}_\nu$
From Fig. 7.1-5 we deduce

$$\underline{R}_1 = \underline{r}_2 - \underline{r}_1 \tag{7.1-70a}$$

$$\text{and} \quad \underline{R}_2 = \underline{r}_3 - \underline{r}_2. \tag{7.1-70b}$$

We can write this in matrix form as

$$\begin{bmatrix} \underline{R}_1 \\ \underline{R}_2 \end{bmatrix} = \begin{bmatrix} -1 & 1 & 0 \\ 0 & -1 & 1 \end{bmatrix} \begin{bmatrix} \underline{r}_1 \\ \underline{r}_2 \\ \underline{r}_3 \end{bmatrix}, \tag{7.1-71}$$

or in abbreviated form,

$$B_{k\nu} = \delta_{k+1,\nu} - \delta_{k\nu}. \tag{7.1-72}$$

Note that the diagonal elements are -1 and the off-diagonal elements are $+1$.

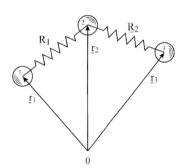

Figure 7.1-5 Three beads connected by two springs

278   Constitutive Equations from Molecular Theories

We can now generalize this result and write

$$B_{kv}\underline{r}_v = (\delta_{k+1,v} - \delta_{kv})\underline{r}_v \tag{7.1-73}$$
$$= \delta_{k+1,v}\underline{r}_v - \delta_{kv}\underline{r}_v = \underline{r}_{k+1} - \underline{r}_k = \underline{R}_k.$$

That is,

$$\underline{\underline{B}} = \begin{bmatrix} -1 & 1 & 0 & . & . & . & 0 \\ 0 & -1 & 1 & 0 & . & . & 0 \\ . & . & . & . & . & . & . \\ . & . & . & . & . & . & . \\ . & . & . & . & . & . & . \\ 0 & . & . & . & 0 & -1 & 1 \end{bmatrix}. \tag{7.1-74}$$

Furthermore, it is also useful to introduce relations between vectors $\underline{r}_c$, $\underline{r}_v$, and connector vectors $\underline{R}_k$. That is, we determine the matrix B^*_{vk} relating vectors $\underline{Q}_v$ and $\underline{R}_k$ as follows:

$$\underline{Q}_v = B^*_{vk}\underline{R}_k. \tag{7.1-75}$$

The matrix B^* is an $N \times (N-1)$ matrix, since $v = 1, \ldots, N$ and $k = 1, \ldots, N-1$. The $*$ is introduced to draw attention to the fact that $B^*_{vk} \neq B^+_{kv}$ (transpose of B_{kv}). In order to determine the components of matrix B^*_{vk}, we approach the problem by considering a fixed value of v in equation 7.1-75. We express both vectors $\underline{Q}_v$ and $\underline{R}_k$ in terms of vectors $\underline{r}_k$, and then we compare the coefficients of $\underline{r}_k$ for various values of k. The right side of equation 7.1-75 is

$$B^*_{vk}\underline{R}_k = B^*_{v1}\underline{R}_1 + B^*_{v2}\underline{R}_2 + \ldots + B^*_{vv}\underline{R}_v + \ldots + B^*_{v,N-1}\underline{R}_{N-1}$$
$$= B^*_{v1}(\underline{r}_2 - \underline{r}_1) + B^*_{v2}(\underline{r}_3 - \underline{r}_2) + \ldots + B^*_{vv}(\underline{r}_{v+1} - \underline{r}_v)$$
$$+ \ldots + B^*_{v,N-1}(\underline{r}_N - \underline{r}_{N-1}) \tag{7.1-76}$$
$$= -B^*_{v1}\underline{r}_1 + \underline{r}_2(B^*_{v1} - B^*_{v2}) + \ldots + \underline{r}_v(B^*_{v,v-1} - B^*_{vv})$$
$$+ \underline{r}_{v+1}(B^*_{vv} - B^*_{v,v+1}) + \ldots.$$

The left side of equation 7.1-75, $\underline{Q}_v$, is

$$\underline{Q}_v = \underline{r}_v - \frac{1}{N}\sum_{\mu=1}^{N} \underline{r}_\mu \tag{7.1-77}$$
$$= \underline{r}_v\left(1 - \frac{1}{N}\right) - \frac{1}{N}[\underline{r}_1 + \ldots + \underline{r}_{v-1} + \underline{r}_{v+1} + \ldots + \underline{r}_N].$$

We now compare the coefficients of $\underline{r}_k$ via equations 7.1-76 and 7.1-77.

For $\underline{r}_1$ we have

$$B^*_{v1} = \frac{1}{N}. \tag{7.1-78a}$$

For $\underline{r}_2$ we have

$$B^*_{v1} - B^*_{v2} = -\frac{1}{N},$$

or
$$B^*_{v2} = B^*_{v1} + \frac{1}{N} = \frac{2}{N}. \tag{7.1-78b}$$

Similarly, we determine
$$B^*_{v\mu} = \frac{\mu}{N}, \quad \text{for} \quad v > \mu$$

or
$$B^*_{vk} = \frac{k}{N}, \quad \text{for} \quad v > k. \tag{7.1-78c}$$

Note that in equation 7.1-77, the coefficient of $\underline{r}_v$ is $(1 - 1/N)$, whereas for all other values of v it is $(-1/N)$. Therefore we consider the values of $k < v$ and $k > v$ separately. When $k = v$,
$$B^*_{vv-1} - B^*_{vv} = 1 - \frac{1}{N},$$

where
$$B^*_{vv-1} = \frac{v-1}{N}. \tag{7.1-78d}$$

It follows that
$$B^*_{vv} = \frac{v-1}{N} - 1 + \frac{1}{N} = -\left(1 - \frac{v}{N}\right). \tag{7.1-78e}$$

For $\underline{r}_{v+1}$, we have $B^*_{vv} - B^*_{vv+1} = -1/N$. This leads to
$$B^*_{vv+1} = B^*_{vv} + \frac{1}{N} = -1 + \frac{v}{N} + \frac{1}{N} = -\left(1 - \frac{v+1}{N}\right).$$

Replacing $v+1$ by k we obtain
$$B^*_{vk} = -\left(1 - \frac{k}{N}\right), \quad k > v. \tag{7.1-78f}$$

Therefore,
$$B^*_{vk} = \begin{cases} \dfrac{k}{N}, & v > k \\ -\left(1 - \dfrac{k}{N}\right) & k \geq v \end{cases}. \tag{7.1-78g}$$

The matrix $\underline{\underline{B}}^*$ is now
$$\underline{\underline{B}}^* = \begin{bmatrix} -\dfrac{N-1}{N} & -\dfrac{N-2}{N} & \cdot & \cdot & \cdot & -\dfrac{1}{N} \\ \dfrac{1}{N} & -\dfrac{N-2}{N} & \cdot & \cdot & \cdot & -\dfrac{1}{N} \\ \dfrac{1}{N} & \dfrac{2}{N} & -\dfrac{N-3}{N} & \cdot & \cdot & -\dfrac{1}{N} \end{bmatrix}. \tag{7.1-79}$$

Next, we multiply equation 7.1-75 by B_{vk}^{*-1} to obtain:

$$\underline{R}_k = B_{vk}^{*-1}\underline{Q}_v = B_{vk}^{*-1}(\underline{r}_v - \underline{r}_c). \tag{7.1-80}$$

Comparing equations 7.1-69 and 7.1-80 we observe B_{vk}^{*-1} to be related to the matrix B_{vk}. Note that since B_{vk}^* is not a square matrix, B_{vk}^{*-1} cannot be obtained by the inverse of a square matrix. We therefore proceed as follows. We define the so-called Rouse matrix A_{ij} as

$$A_{ij} = B_{iv}B_{jv} = B_{i1}B_{j1} + B_{i2}B_{j2} + \ldots + B_{ii}B_{ji} + B_{ii+1}B_{ji+1} + \ldots. \tag{7.1-81}$$

We note that the only nonzero terms are $B_{ii}B_{ji} + B_{ii+1}B_{ji+1}$ (see equation 7.1-72). These are given by

$$(\delta_{i+1i} - \delta_{ii})(\delta_{j+1i} - \delta_{ji}) + (\delta_{i+1i+1} - \delta_{ii+1})(\delta_{j+1i+1} - \delta_{ji+1}),$$

and from the definition of δ_{ij} we obtain:

$$\text{For} \quad i = j, \quad A_{ii} = 2.$$
$$\text{For} \quad i = j - 1, \quad A_{ij} = -1.$$
$$\text{For} \quad i = j + 1, \quad A_{ij} = -1.$$

The Rouse matrix is a square tridiagonal matrix defined by

$$A_{ij} = \begin{cases} 2, & i = j \\ -1, & i = j \pm 1 \\ 0, & \text{otherwise} \end{cases}$$

$$\underline{\underline{A}} = \begin{bmatrix} 2 & -1 & 0 & \cdot & \cdot & \cdot & 0 \\ -1 & 2 & -1 & \cdot & \cdot & \cdot & 0 \\ 0 & -1 & 2 & -1 & \cdot & \cdot & 0 \\ \cdot & \cdot & \cdot & \cdot & \cdot & \cdot & \cdot \\ \cdot & \cdot & \cdot & \cdot & \cdot & \cdot & \cdot \\ \cdot & \cdot & \cdot & \cdot & \cdot & \cdot & \cdot \\ \cdot & \cdot & \cdot & \cdot & \cdot & \cdot & 2 \end{bmatrix}. \tag{7.1-82}$$

We note here that it is more convenient to work with square matrices. In that case, the inverse of the matrix is uniquely defined, the eigenvalues are real, and a determinant exists.

In addition to the Rouse matrix, we can define the Kramers matrix, which is another useful square matrix, given by

$$K_{ij} = B_{vi}^* B_{vj}^*. \tag{7.1-83}$$

This matrix relates the vectors $\underline{R}_k$ and $\underline{Q}_v$, whereas the Rouse matrix relates the vectors $\underline{R}_k$ and $\underline{r}_v$. Equation 7.1-83 can be written as

$$K_{ij} = B_{1i}^* B_{1j}^* + B_{2i}^* B_{2j}^* + B_{3i}^* B_{3j}^* + \ldots + B_{ii}^* B_{ij}^*$$
$$+ B_{i+1i}^* B_{i+1j}^* + \ldots + B_{ji}^* B_{jj}^{**} + (N - j) \quad \text{terms}. \tag{7.1-84}$$

7.1 Bead- and Spring-Type Models

This equation consists of three different sets of terms, which we identify through the following example. The matrix element K_{25} is given by

$$K_{25} = \underbrace{B^*_{12}B^*_{15} + B^*_{22}B^*_{25}}_{(a)} + \underbrace{B^*_{32}B^*_{35} + B^*_{42}B^*_{45} + B^*_{52}B^*_{55}}_{(b)} + \underbrace{B^*_{62}B^*_{65}}_{(c)} + \ldots$$

We recall that

$$B^*_{ij} = \begin{cases} \dfrac{j}{N} & i > j \\ -\left(1 - \dfrac{j}{N}\right) & j \geq i \end{cases}.$$

The first (i) terms ($i=2$ in this example) are represented by set (a) and are given by

$$\left(1 - \frac{2}{N}\right)\left(1 - \frac{5}{N}\right) + \left(1 - \frac{2}{N}\right)\left(1 - \frac{5}{N}\right).$$

The next $(j - i)$ terms ($j - i = 3$ in this example) are represented by set (b) and are given by

$$-\left(\frac{2}{N}\right)\left(1 - \frac{5}{N}\right) - \left(\frac{2}{N}\right)\left(1 - \frac{5}{N}\right) - \left(\frac{2}{N}\right)\left(1 - \frac{5}{N}\right).$$

The remaining $(N - j)$ terms ($N - 5$ in this example) are represented by set (c) and are given by

$$\left(\frac{2}{N}\right)\left(\frac{5}{N}\right) + \left(\frac{2}{N}\right)\left(\frac{5}{N}\right) + \ldots$$

Generalizing the above example, we observe that in equation 7.1-84, the first i terms contribute $i(1 - i/N)(1 - j/N)$. The next $(j - i)$ terms contribute $(j - i)(i/N)[-(1 - j/N)]$ and the last $(N - j)$ terms contribute $(N - j)(i/N)(j/N)$.

The Kramers matrix K_{ij} is then given by

$$K_{ij} = \begin{cases} \dfrac{i(N - j)}{N}, & j \geq i \\ \dfrac{j(N - i)}{N}, & j \leq i \end{cases} \tag{7.1-85}$$

K_{ij} is a symmetric matrix. These matrices are useful in formulating constitutive equations for various bead-spring systems.

The equation of motion for bead (v) corresponding to equation 7.1-32 for a single dumbbell, omitting the acceleration term, is

$$0 = -\zeta\{\dot{r}_v - V - [\kappa \cdot r_v]\} - k_B T \frac{\partial}{\partial r_v} \ln \psi + F_v, \tag{7.1-86}$$

where F_v is the net force acting on bead (v), the other symbols have the same meaning as in equation 7.1-32, and the distribution function is $\psi(R_1, R_2, \ldots, R_N, t)$.

282   Constitutive Equations from Molecular Theories

Equation 7.1-86 can be transformed into an equation of motion for the center of mass and equations of motion for the connectors. Via the relations between $\underline{r}_v$, $\underline{r}_c$, and $\underline{R}_k$ we obtain

$$\dot{\underline{r}}_c = \underline{V} + \underline{\kappa} \cdot \underline{r}_c \qquad (7.1\text{-}87)$$

and

$$\dot{\underline{R}}_j = \underline{\kappa} \cdot \underline{R}_j - \frac{1}{\zeta} A_{jk} \left(k_B T \frac{\partial}{\partial \underline{R}_k} \ln \psi + \underline{F}_k^c \right), \qquad (7.1\text{-}88)$$

where $\underline{F}_k^c$ is the tension in the kth connector and the A_{jk} are the elements of the Rouse matrix.

Proceeding as in the case of a single dumbbell, we find the following form for the stress tensor:

$$\underline{\underline{\sigma}} = -\eta_s \dot{\underline{\gamma}} + \sum_k \underline{\underline{\sigma}}_k \qquad (7.1\text{-}89\text{a})$$

and

$$\underline{\underline{\sigma}}_k = -\frac{n\zeta}{2} \frac{\delta}{\delta t} \langle K_{ik} \underline{R}_i \underline{R}_k \rangle \qquad (7.1\text{-}89\text{b})$$

which is the Giesekus form.

We can obtain the corresponding Kramers expression and then the following constitutive equation (see Bird et al., 1987, for details):

$$\underline{\underline{\sigma}}_k + \lambda_k \frac{\delta}{\delta t} \underline{\underline{\sigma}}_k = -nk_B T \lambda_k \dot{\underline{\gamma}}, \qquad (7.1\text{-}90\text{a})$$

where the stress tensor is given by equation 7.1-89b, and

$$\lambda_k = \frac{\zeta}{2Ha_k} = \frac{\dfrac{\zeta}{2H_0}}{4\sin^2\left(\dfrac{k\pi}{2N}\right)}. \qquad (7.1\text{-}90\text{b})$$

The a_k are the eigenvalues of the Rouse matrix and H_0 is the constant modulus of the connector. Equations 7.1-89a and 7.1-90 represent the constitutive equation for the Rouse (1953) model. Note that for a large number of beads, N, the long relaxation times (small k) are approximated by (Rouse, 1953)

$$\lambda_k \approx \frac{\dfrac{\zeta}{2H_0}}{4\left(\dfrac{k\pi}{2N}\right)^2} = \frac{6(\eta_0 - \eta_s)M}{c\pi^2 RTk^2}, \qquad (7.1\text{-}91)$$

where c is the polymer concentration, M is the polymer molecular weight, and R is the universal gas constant. This result is equivalent to the empirical definition given by equation 5.2-33 with $\alpha = 2$. Integral forms of the Rouse model can be obtained following the steps outlined in Section 6.2-2 for the Maxwell model. In terms of the relative strain tensor, equations 7.1-89a and 7.1-90 can be written in Cartesian coordinates as

$$\sigma_{ij} = -\eta_s \dot{\gamma}_{ij} - \int_{-\infty}^{t} nk_B T \sum_k \frac{e^{-\frac{(t-t')}{\lambda_k}}}{\lambda_k} \Gamma_{ij}^{-1}(t') dt', \qquad (7.1\text{-}92)$$

where Γ_{ij}^{-1} are the components of the relative Finger tensor ($= C_{ij}^{-1} - \delta_{ij}$) defined in Section 6.2-2.

The main advantage of the Rouse model over the simple Hookean dumbbell or Maxwell model is the discrete relaxation spectrum which allows for a better description of the linear viscoelastic properties. For example, the slope of the reduced real part of the complex viscosity, $\eta' - \eta_s$, as a function of the frequency at high frequencies (on log-log scales) is equal to $-1/2$. This is a substantial improvement over the slope of -2 predicted by the Maxwell model. However, this model is still not flexible enough for most polymeric systems. Furthermore, the Rouse model is not capable of predicting nonlinear effects such as shear thinning and it is restricted to dilute solutions (i.e., no polymer-polymer interactions). Finally, the effect of hydrodynamic interactions is neglected in the Rouse model.

In all the theories presented so far, we have implicitly assumed that the dumbbells or chains move in the solvent without disturbing the velocity field. This is the so-called free-draining assumption. The reader who wishes to learn how hydrodynamic interaction is introduced, should refer to the more specialized books on polymer kinetic theories, such as those of Bird et al. (1987). Here, we simply report the result of Zimm (1956) who has extended the result of Rouse (1953) to include equilibrium-averaged hydrodynamic interactions.

For the Zimm model, the λ_k in equation 7.1-90b are replaced by $\tilde{\lambda}_k$ defined by

$$\tilde{\lambda}_k = \frac{\zeta}{2H_0 \tilde{a}_k}, \qquad (7.1\text{-}93)$$

where $\tilde{a}_k$ are the eigenvalues of the modified Rouse matrix $\tilde{A}_{ik}$. The matrix $\tilde{A}_{ik}$ is given by

$$\tilde{A}_{ik} = B_{il} H_{lm} B_{km}, \qquad (7.1\text{-}94\text{a})$$

with

$$H_{lm} = \delta_{lm} + (1 - \delta_{lm}) h^* \sqrt{\frac{2}{|l-m|}}, \qquad (7.1\text{-}94\text{b})$$

and

$$h^* = \frac{\zeta}{\eta_s} \sqrt{\frac{H_0}{36\pi^3 k_B T}}. \qquad (7.1\text{-}94\text{c})$$

The quantity H_{lm} is the Zimm hydrodynamic interaction matrix, and h^* is referred to as the hydrodynamic interaction parameter. If hydrodynamic interaction is neglected, H_{lm} reduces to the unit matrix and $\tilde{A}_{ik}$ simplifies to the Rouse matrix.

The inclusion of the equilibrium-averaged hydrodynamic interaction leads to more realistic results for linear viscoelasticity. Mainly it predicts that the intrinsic viscosity is given by

$$[\eta] = KM^\alpha, \quad (7.1\text{-}95)$$

with a value of $\alpha < 1$, which is in agreement with experimental data. Equation 7.1-95 is the Mark-Houwink equation discussed in Section 3.1-4. Also, the slope of $\eta' - \eta_s$ versus the frequency ω at high ω on log-log scales is given by $-(1+\sigma)/(2+\sigma)$, where

$$\sigma = -1.40 h^{*0.18}. \quad (7.1\text{-}96)$$

Hence, the Zimm model is much more flexible than the Rouse model in describing linear viscoelasticity. It suffers from the same drawbacks as far as nonlinear effects are concerned.

Results for other molecular models such as rod-like and bead-rod molecules can be found in Bird et al. (1987) and in Doi and Edwards (1986).

7.2   Network Theories

Network theories have been originally developed by Green and Tobolsky (1946) and James (1947) for solid rubber-like materials. The concepts were later extended to liquid polymeric materials by Lodge (1956, 1960, 1968) and by Yamamoto (1956, 1957, and 1958). These theories lead to very useful constitutive equations when stress, strain, or rate-dependent kinetics of entanglements are included (Kaye, 1966; Carreau, 1972; De Kee and Carreau, 1979). Unfortunately, these theories are currently less popular than reptation theories (Section 7.3). In this section, we present the general concept of network theories, the extension of Lodge to elastic liquids, and generalizations to include stress- or rate-dependent effects.

7.2-1   General Network Concept

The polymer molecules are considered to form a network of segments of various lengths linked together by junctions or entanglements. Some of the macromolecules are unattached and are considered as free or loose chains, as illustrated in Fig. 7.2-1a.

The entanglements or junctions can be of different nature: mechanical, physical, or chemical bonds. For polymeric liquids, the entanglements have a temporary life that is assumed to be much shorter than a characteristic time associated with the macroscopic flow. The contribution to the extra stress tensor is then assumed to arise from the deformation of the segments formed by adjacent entanglements. Such a segment consisting of n links with end-to-end vector $\underline{R}$ is shown in Fig. 7.2-1b. The Helmholtz free energy per unit

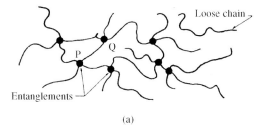
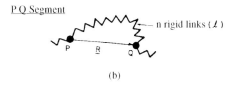

Figure 7.2-1 Polymer representation in network theories
(a) network of entangled chains
(b) detail of a n-segment

volume of a network of Gaussian segments consisting of n freely-rotating jointed rigid links is given by (see Lodge, 1968)

$$A(t) = \text{constant} + k_B T \sum_n b_n \int \psi_n(\underline{R}, t) R^2 d^3 x, \qquad (7.2\text{-}1)$$

where, for Gaussian segments ($nl \gg r$),

$$b_n = \frac{3}{2nl^2}. \qquad (7.2\text{-}2)$$

The constant in equation 7.2-1 represents the contributions of the solvent and of the loose chains; ψ_n is the probability of the n-segments to have end-to-end distance in the range $(\underline{R}, \underline{R} + d\underline{R})$. Lodge (1968) introduced another index to account for various complexities (or different nature) of the entanglements. This, however, does not lead to more flexible results and will not be further discussed here. Using the assumption of affine deformation (no-slip hypothesis), equation 7.2-1 yields the following expression for the contribution of the segments to the stress tensor:

$$\sigma_{ij} = -\sum_n H_n \langle R_i R_j \rangle_n, \qquad (7.2\text{-}3)$$

where H_n are moduli defined by

$$H_n = \frac{3k_B T}{nl^2}, \qquad (7.2\text{-}4)$$

and $\langle \ \rangle$ refers to an average with respect to the probability function ψ_n (see Section 7.1). In comparison with the kinetic theories for dilute solutions, the segments are taken as Hookean springs, and there is no drag force acting on the segments. The loose chains, as well as the solvent, do not contribute to the extra stress tensor. Details of the results are given below.

As discussed in Chapter 6, it is easier to formulate constitutive equations using embedded coordinate systems. Equation 7.2-1 for the Helmholtz free energy can be written as

$$A(t) = \text{constant} + k_B T \sum_n b_n \int \hat{\varphi}_n(\hat{x}, t) \hat{x}^i \hat{x}^j \hat{g}_{ij}(t) d^3\hat{x}, \quad (7.2\text{-}5)$$

where (see Appendix A and Section 6.1)

$$\hat{x}^i \hat{x}^j \hat{g}_{ij}(t) = R^2. \quad (7.2\text{-}6)$$

Note that the distribution function $\hat{\varphi}_n$ is not the same as ψ_n written in fixed coordinates. The covariant components of the metric tensor in embedded coordinates, $\hat{g}_{ij}(t)$, are unique functions of time for homogeneous deformations. Without loss of generality, the flow may be assumed to be locally homogeneous.

7.2-2 Rubber-Like Solids

For rubber-like solid materials, we postulate that the entanglements arise from crosslinking or other strong chemical bonds. Hence, in embedded coordinates, the distribution function is constant (no loss or creation of entanglements):

$$\frac{\partial}{\partial t} \hat{\varphi}_n(\hat{x}, t) = 0. \quad (7.2\text{-}7)$$

Using the results of Wall (1942), obtained for Gaussian segments, we write for isotropic conditions at t_0,

$$\psi_n(x, t_0) = C_n \left(\frac{b_n}{\pi}\right)^{\frac{3}{2}} \exp(-b_n R^2), \quad (7.2\text{-}8\text{a})$$

with

$$C_n = \int \psi_n(x, t_0) d^3 x. \quad (7.2\text{-}8\text{b})$$

At time t_0, we consider both fixed and embedded coordinates to coincide. That is,

$$\hat{\varphi}_n(\hat{x}, t_0) = \psi_n(x, t_0), \quad (7.2\text{-}9)$$

and since $\hat{\varphi}_n(\hat{x}, t) = \hat{\varphi}_n(\hat{x}, t_0)$, equations 7.2-8a and 7.2-9 lead to

$$\hat{\varphi}_n(\hat{x}, t_0) = C_n \left(\frac{b_n}{\pi}\right)^{\frac{3}{2}} \exp\{-b_n \hat{x}^i \hat{x}^j \hat{g}_{ij}(t_0)\}, \quad (7.2\text{-}10)$$

and the Helmholtz free energy is now expressed by

$$A(t) = \text{constant} + k_B T \sum_n b_n C_n \left(\frac{b_n}{\pi}\right)^{\frac{3}{2}} \int \exp\{-b_n \hat{x}^i \hat{x}^j \hat{g}_{ij}(t_0)\} \hat{g}_{kl}(t) \hat{x}^k \hat{x}^l d^3 x. \quad (7.2\text{-}11)$$

(a)

We note that the integral in equation 7.2-11, (a), can be evaluated using spherical coordinates as follows:

$$(a) = \int \exp\{-b_n R^2\} R^2 d^3x = 4\pi \int_0^\infty \exp\{-b_n R^2\} R^4 dR$$

$$= \frac{3}{2} \frac{\pi^{\frac{3}{2}}}{(b_n)^{\frac{5}{2}}} \quad (7.2\text{-}12)$$

We multiply (a) by $\hat{g}^{kl}(t_0)$, and noting that

$$\hat{g}^{kl}(t_0)\hat{g}_{kl}(t_0) = \delta_k^k = 3, \quad (7.2\text{-}13)$$

we obtain

$$\int \exp\{-b_n \hat{x}^i \hat{x}^j \hat{g}_{ij}(t_0)\} \hat{x}^k \hat{x}^l d^3\hat{x} = \frac{\pi^{\frac{3}{2}} \hat{g}^{kl}(t_0)}{2(b_n)^{\frac{5}{2}}}. \quad (7.2\text{-}14)$$

Result 7.2-11 can now be written in a simpler form:

$$A(t) = \text{constant} + \frac{1}{2} k_B T N_0^* \hat{g}_{kl}(t) \hat{g}^{kl}(t_0), \quad (7.2\text{-}15a)$$

where

$$N_0^* = \sum_n C_n \quad (7.2\text{-}15b)$$

is the total concentration of segments.

To obtain a constitutive equation, we note that for a small and reversible isothermal deformation, the work done is related to the Helmholtz free energy by the following relation (see Lodge, 1964):

$$dW = \frac{1}{2} \hat{\sigma}^{kl} d\hat{g}_{kl} = dA. \quad (7.2\text{-}16)$$

Considering equations 7.2-15a and 7.2-16, we get

$$\hat{\sigma}^{kl} d\hat{g}_{kl}(t) = k_B T N_0^* \hat{g}^{kl}(t_0) d\hat{g}_{kl}(t). \quad (7.2\text{-}17)$$

This result can be written as

$$d\hat{g}_{kl}(t)(\hat{\sigma}^{kl}(t) - k_B T N_0^* \hat{g}^{kl}(t_0)) = 0. \quad (7.2\text{-}18)$$

For incompressible materials (see Lodge, 1964),

$$\hat{g}^{kl} d\hat{g}_{kl}(t) = 0, \quad (7.2\text{-}19)$$

and we obtain the following constitutive equation:

$$\hat{\sigma}^{ij}(t) = -P_0 \hat{g}^{ij}(t) + k_B T N_0^* \hat{g}^{ij}(t_0), \quad (7.2\text{-}20)$$

where P_0 is an arbitrary isotropic term.

In fixed coordinates, result 7.2-20 can be written as (see Section 6.1)

$$\pi^{ij}(t) = \sigma^{ij} + P_0 g^{ij} = k_B T N_0^* C^{-1\,ij}(t_0), \quad (7.2\text{-}21a)$$

which in Cartesian coordinates becomes

$$\pi_{ij}(t) = \sigma_{ij} + P_0\delta_{ij} = k_B T N_0^* C_{ij}^{-1}(t_0), \qquad (7.2\text{-}21\text{b})$$

where $\underline{\underline{C}}^{-1}$ is the Finger tensor.

This theory predicts the shear modulus of rubber within a factor of 2.

7.2-3 Elastic Liquids

For concentrated polymer solutions or melts, the entanglements or junctions are assumed to be temporary and of a relatively short life compared to a characteristic time for the flow. Kinetic models for the creation and loss of junctions have to be proposed. Characteristic times associated with the creation and loss of junctions are assumed to be much larger than the characteristic times associated with Brownian motion. Hence, equilibrium results of statistical mechanics can be applied to obtain the stress from the free energy expression. However, the times related to the creation and loss of segments are assumed to be significantly shorter than the characteristic times of the flow, so that the network can deform with the fluid (affine deformation). We consider that all the $(\hat{x}, n)$-segments have the same probability λ_n^{-1} of leaving the network. The n-segments are created at a rate L_n per unit volume, and we assume that, at the time of creation, the distribution of the n-segments is Gaussian. Hence, a balance of $(\hat{x}, n)$-segments leads to

$$\frac{\partial}{\partial t'}(\hat{\varphi}_n(\hat{x}, t')d^3\hat{x}) = L_n\left(\frac{b_n}{\pi}\right)^{\frac{3}{2}} \exp\{-b_n\hat{x}^k\hat{x}^l\hat{g}_{kl}(t')\}d^3\hat{x}$$
$$- \frac{\hat{\varphi}_n(\hat{x}, t')d^3\hat{x}}{\lambda_n}, \qquad (7.2\text{-}22)$$

where $\hat{\varphi}_n(\hat{x}, t')d^3\hat{x}$ is the number of $(\hat{x}, n)$-segments in volume $d^3\hat{x}$. The first term on the right side represents the rate of creation of $(\hat{x}, n)$-segments (input) and the second term expresses the rate of loss (output). This is a first-order differential equation that can be integrated between $-\infty$ and t to obtain

$$\hat{\varphi}_n(\hat{x}, t) = \frac{1}{e^{\frac{t}{\lambda_n}}} \int_{-\infty}^{t} e^{\frac{t'}{\lambda_n}} L_n \left(\frac{b_n}{\pi}\right)^{\frac{3}{2}} e^{\{-b_n\hat{x}^k\hat{x}^l\hat{g}_{kl}\}} dt' + \frac{C_1}{e^{\frac{t}{\lambda_n}}}. \qquad (7.2\text{-}23)$$

Considering that the distribution function $\hat{\varphi}_n$ must remain finite at all times, including $t = -\infty$, the constant C_1 in equation 7.2-23 has to vanish. Hence, we have

$$\hat{\varphi}_n(\hat{x}, t) = L_n\left(\frac{b_n}{\pi}\right)^{\frac{3}{2}} \int_{-\infty}^{t} \exp\left\{-b_n\hat{x}^k\hat{x}^l\hat{g}_{kl}(t') + \frac{t'-t}{\lambda_n}\right\} dt'. \qquad (7.2\text{-}24)$$

Using results 7.2-14 and 7.2-24, the Helmholtz free energy (equation 7.2-5) is expressed by

$$A(t) = \text{constant} + \frac{1}{2}k_B T \hat{g}_{ij}(t) \int_{-\infty}^{t} N^*(t-t')\hat{g}^{ij}(t')dt', \qquad (7.2\text{-}25\text{a})$$

where

$$N^*(t - t') = \sum_n L_n \exp\left(-\frac{t - t'}{\lambda_n}\right). \tag{7.2-25b}$$

We note that $N^*(t - t')$ plays the role of a memory function. It physically represents the net creation rate of segments during the time interval $t - t'$.

Following the procedure discussed in Section 7.2-2, that is, assuming incompressibility and small reversible deformation, we obtain the following constitutive equation:

$$\hat{\sigma}^{ij}(t) = -P_0 \hat{g}^{ij}(t) - k_B T \int_{-\infty}^{t} N^*(t - t') \hat{g}^{ij}(t') dt', \tag{7.2-26}$$

which can be written in fixed coordinates as

$$\sigma^{ij}(t) = -P_0 g^{ij}(t) - k_B T \int_{-\infty}^{t} N^*(t - t') C^{-1\,ij}(t') dt'. \tag{7.2-27}$$

This result is equivalent to the result obtained for the elastic dumbbell model, although the assumptions are quite different. The dumbbell approach is restricted to dilute solutions, whereas the network theory is designed for concentrated polymer solutions or melts. Equation 7.2-27 is equivalent to the Lodge elastic model presented in Chapter 6, and it leads to the same results. That is, for steady shear flow, we obtain

$$\eta = k_B T \int_{-\infty}^{t} N^*(t - t') \times (t - t') dt' = \eta_0 \quad \text{(constant)}, \tag{7.2-28a}$$

$$\psi_1 = k_B T \int_{-\infty}^{t} N^*(t - t') \times (t - t')^2 dt' = \psi_{10} \quad \text{(constant)}, \tag{7.2-28b}$$

and $\psi_2 = 0$. $\tag{7.2-28c}$

The interest of this approach in embedded coordinates is that it can be easily extended to include effects of the rate of deformation or stress on the creation and loss of segments. With the objective of obtaining constitutive equations that are explicit with respect to the stress tensor, Carreau (1972) and De Kee and Carreau (1979) considered the rates of creation and loss of segments in equation 7.2-22 to be functions of the second invariant of the rate-of-deformation tensor. This leads to the following result for $N^*(t - t')$:

$$N^*(t - t') = \sum_n L_n(II_{\dot{\gamma}}(t')) \exp\left\{-\int_{-t'}^{t} \frac{dt''}{\lambda_n(II_{\dot{\gamma}}(t''))}\right\}, \tag{7.2-29}$$

and the constitutive equation in fixed coordinates can be expressed as

$$\sigma^{ij}(t) = P_0 g^{ij}(t) - \int_{-\infty}^{t} m(t - t', II_{\dot{\gamma}}(t - t')) C^{-1\,ij}(t') dt', \tag{7.2-30a}$$

with

$$m(t - t', II_{\dot{\gamma}}(t, t')) = k_B T \sum_n L_n(II_{\dot{\gamma}}(t')) \exp\left\{-\int_{t'}^{t} \frac{dt''}{\lambda_n(II_{\dot{\gamma}}(t''))}\right\}. \tag{7.2-30b}$$

This constitutive equation, as well as specific rheological models are discussed in detail in Section 6.4. These are shown to be quite powerful for describing rheological properties of polymeric systems. It is our belief that even more realistic results could be obtained if invariants of the stress tensor were used instead of the second invariant of the rate-of-deformation tensor. This idea was suggested by Kaye (1966), but it leads to difficult-to-use implicit equations for the stress tensor. The equation developed by Acierno et al. (1976) is also implicit in terms of the stress tensor (see Section 6.3-4). However, this so-called Marrucci model cannot be derived via this network framework. Some of the most successful models have been proposed by Wagner (1976, 1977, 1978a, b), who considered the rate of loss and creation of segments to be functions of the invariants of the strain tensor (see Section 6.4-3).

Lodge (1968) has examined the effects of using a non-Gaussian distribution for the segments. This leads to a nonzero secondary normal stress coefficient, but the resulting shear-dependence of the viscosity and of the normal stress coefficients is unrealistic. In the work of Yamamoto (1956, 1957, 1958), the rates of creation and loss of segments are allowed to depend on the segment length and orientation. His theory does not yield an explicit constitutive equation, and it is difficult to assess the merits of the choices made by Yamamoto.

7.2-4 Recent Developments

The network theory has been extended to include non-affine deformation as well as the effect of electrostatic charges in the case of polyelectrolytes. Chan Man Fong and De Kee (1995a) have obtained constitutive equations for charged networks via the embedded coordinate system. We can also work in terms of a fixed coordinate system, as shown next.

In a fixed coordinate system, we can denote the distribution of segments consisting of n links and with end-to-end vector $\underline{R}$ by $\psi_n(\underline{R}, t)$. The number density, N_n, of the segments containing n links is given by

$$N_n = \int_{\underline{R}} \psi_n(\underline{R}, t) d\underline{R}, \qquad (7.2\text{-}31)$$

where the integration is to be taken over the whole $\underline{R}$-space (configuration space). The $\underline{R}$-space must enclose all possible directions and magnitudes of $\underline{R}$, and therefore, the infinite sphere is appropriate.

A balance of segments leads to

$$\frac{\partial \psi_n}{\partial t} + \frac{\partial}{\partial \underline{R}} \cdot (\underline{\dot{R}} \psi_n) = c_n - l_n, \qquad (7.2\text{-}32)$$

where c_n and l_n are the rates of creation and loss of segments. This is equivalent to the balance written in embedded coordinates of equation 7.2-22. The second term on the left side of equation 7.2-32 represents the convective contribution and is absent in an embedded coordinate system.

The rates of creation and loss can be attributed to Brownian motion and to the imposed flow. We can assume that the contribution to c_n due to Brownian motion is proportional to the equilibrium distribution function ψ_{n0}, and we can express c_n and l_n as

$$c_n = k_c \psi_{n0} + k_1 f_2(II_{\dot{\gamma}}) \psi_n \qquad (7.2\text{-}33a)$$

$$\text{and} \quad l_n = k_c \psi_n + k_l f_1(II_{\dot{\gamma}}) \psi_n, \qquad (7.2\text{-}33b)$$

where k_c, k_1, and k_l are constants, and f_1 and f_2 are functions of the second invariant of the rate-of-deformation tensor, $II_{\dot{\gamma}} = \underline{\underline{\dot{\gamma}}} : \underline{\underline{\dot{\gamma}}}$. At equilibrium ($II_{\dot{\gamma}} = 0$), we require that $c_n = l_n$ and that

$$f_1(0) = f_2(0) = 0. \qquad (7.2\text{-}34)$$

Combining equations 7.2-32 and 33, we obtain

$$\frac{\partial \psi_n}{\partial t} + \frac{\partial}{\partial \underline{R}} \cdot (\underline{\dot{R}} \psi_n) = k_c(\psi_{n0} - \psi_n) + [k_1 f_2(II_{\dot{\gamma}}) - k_l f_1(II_{\dot{\gamma}})]\psi_n. \qquad (7.2\text{-}35)$$

Integration over the $\underline{R}$-space yields

$$\frac{\partial}{\partial t} \int_R \psi_n d\underline{R} + \int_R \frac{\partial}{\partial \underline{R}} \cdot (\underline{\dot{R}} \psi_n) d\underline{R} = k_c \int_R (\psi_{n0} - \psi_n) d\underline{R} + [k_1 f_2 - k_l f_1] \int_R \psi_n d\underline{R}. \qquad (7.2\text{-}36)$$

The divergence theorem yields

$$\int_R \frac{\partial}{\partial \underline{R}} \cdot (\underline{\dot{R}} \psi_n) d\underline{R} = \int_S (\underline{\dot{R}} \psi_n) \cdot \underline{n} dS, \qquad (7.2\text{-}37)$$

where S is the surface enclosing the $\underline{R}$-space, and the vector $\underline{n}$ is the unit normal outward to S. The surface S is the infinite surface, and the probability of a segment having an infinite length is zero. This means that ψ_n is zero on S and the surface integral is zero. Combining equations 7.2-31, 7.2-36, and 7.2-37 yields

$$\frac{dN_n}{dt} = k_c(N_{n0} - N_n) + (k_1 f_2 - k_l f_1)N_n. \qquad (7.2\text{-}38)$$

Solving for N_n, assuming that the viscosity η is proportional to N_n (and η_0 to N_{n0}), we recover equation 2.3-1 with $b = k_l/k_c$ and $c = k_1/k_c$. In the network model, only the segments contribute to the extra stress tensor. Following an analysis similar to that given for a dumbbell, we deduce that the extra stress tensor $\underline{\underline{\sigma}}$ is given by (see equation 7.1-25)

$$\underline{\underline{\sigma}} = -[N_n \langle \underline{RF} \rangle - N_{n0} \langle \underline{RF} \rangle_0], \qquad (7.2\text{-}39)$$

where $\underline{F}$ is the force on the segment and where $\langle\,\rangle$ and $\langle\,\rangle_0$ denote the average over the $\underline{R}$-space at time t and at equilibrium respectively. We assume the segments to be Hookean, and a charge q is placed at each end of the end-to-end vectors. A similar idea has been proposed by Ait-Kadi et al. (1988) in the framework of conformation tensors (see Section 7.4-2). $\underline{F}$ can then be written as

$$\underline{F} = H_0 \underline{R} - \frac{2q^2 \underline{R}}{(\epsilon R^3)}, \qquad (7.2\text{-}40)$$

where H_0 is the constant elastic modulus and ϵ is the dielectric constant of the medium. Combining equations 7.2-39 and 7.2-40 we obtain

$$\underline{\underline{\sigma}} = -N_n\left[H_0\langle \underline{R}\underline{R}\rangle - \frac{2q^2}{\epsilon}\left\langle\frac{\underline{R}\underline{R}}{R^3}\right\rangle\right] + N_{n0}\left[H_0\langle \underline{R}\underline{R}\rangle - \frac{2q^2}{\epsilon}\left\langle\frac{\underline{R}\underline{R}}{R^3}\right\rangle_0\right]. \quad (7.2\text{-}41)$$

To evaluate $\langle \underline{R}\underline{R}/R^3\rangle$, we need to solve for ψ_n via equation 7.2-35. As mentioned earlier, a general closed-form solution is not possible. Chan Man Fong and De Kee (1995b) obtained a perturbed solution about the equilibrium state. While the lengthy algebraic details are not presented here, we derive a constitutive equation for the case where $q=0$ by the method used to obtain the Giesekus equation 7.1-41. We multiply equation 7.2-35 by $\underline{R}\underline{R}$, integrate over the whole $\underline{R}$-space, use equation 7.2-38 and obtain

$$\frac{D}{Dt}\langle \underline{R}\underline{R}\rangle - \langle \underline{R}\underline{R}\rangle \cdot \underline{\underline{\kappa}}^+ - \underline{\underline{\kappa}}\cdot\langle \underline{R}\underline{R}\rangle \qquad (7.2\text{-}42)$$
$$= k_c(\langle \underline{R}\underline{R}\rangle_0 - \langle \underline{R}\underline{R}\rangle) + [k_1 f_2 - k_l f_1]\langle \underline{R}\underline{R}\rangle.$$

The left side, as stated earlier, is the upper-convected derivative, and equation 7.2-42 can be written as

$$\frac{\delta}{\delta t}\langle \underline{R}\underline{R}\rangle = k_c(\langle \underline{R}\underline{R}\rangle_0 - \langle \underline{R}\underline{R}\rangle) + [k_1 f_2 - k_1 f_1]\langle \underline{R}\underline{R}\rangle. \quad (7.2\text{-}43)$$

At equilibrium ψ_{n0} is proportional to $\exp[-\frac{1}{2}(H_0 R^2)/k_B T]$ (see equation 7.2-8a) and $\langle \underline{R}\underline{R}\rangle_0$ is given by

$$\langle \underline{R}\underline{R}\rangle_0 = \frac{\int_R \exp\left[-\frac{(H_0 R^2)}{2k_B T}\right]\underline{R}\underline{R}d\underline{R}}{\int_R \exp\left[-\frac{(H_0 R^2)}{2k_B T}\right]d\underline{R}}. \quad (7.2\text{-}44)$$

To evaluate the integral, it is usual to use the spherical polar coordinate system (see also equation 7.2-14). Following Bird et al. (1987) we write

$$\underline{R} = R\underline{n} \quad (7.2\text{-}45a)$$

and $\quad d\underline{u} = \sin\theta d\theta d\phi, \quad 0 \le \theta \le \pi, \quad 0 \le \phi \le 2\pi. \quad (7.2\text{-}45b)$

Note that the volume element $d\underline{R}$ is given by $R^2 \sin\theta dRd\theta d\phi$. Equation 7.2-44 can now be written as

$$\langle \underline{R}\underline{R}\rangle_0 = \frac{\int_0^\infty \exp\left[-\frac{(H_0 R^2)}{2k_B T}\right]dR \int \underline{n}\underline{n}d\underline{u}}{\int_0^\infty \exp\left[-\frac{(H_0 R^2)}{2k_B T}\right]dR \int d\underline{u}}. \quad (7.2\text{-}46)$$

It also follows that

$$\int \underline{n}\,\underline{n}\,du = \frac{4\pi}{3}\underline{\underline{\delta}} \qquad (7.2\text{-}47a)$$

$$\text{and} \quad \int du = 4\pi. \qquad (7.2\text{-}47b)$$

The other integrals can be found in a table of integrals or can be evaluated by parts to obtain

$$\langle \underline{R}\,\underline{R}\rangle_0 = \frac{k_B T}{2H_0}\underline{\underline{\delta}}. \qquad (7.2\text{-}48)$$

Substituting equation 7.2-48 into 7.2-41, with $q=0$, yields

$$\underline{\underline{\sigma}} = -N_n H_0 \langle \underline{R}\,\underline{R}\rangle + \frac{1}{2}N_{n0}k_B T\underline{\underline{\delta}}. \qquad (7.2\text{-}49)$$

Taking the upper-convected derivative and noting that (see Section 6.1-1)

$$\frac{\delta}{\delta t}\underline{\underline{\delta}} = -\underline{\dot{\gamma}},$$

we have

$$\frac{\delta}{\delta t}\underline{\underline{\sigma}} = -N_n H_0 \frac{\delta}{\delta t}\langle \underline{R}\,\underline{R}\rangle - \frac{1}{2}N_{n0}k_B T\underline{\dot{\gamma}}. \qquad (7.2\text{-}50)$$

Combining equations 7.2-42, 7.2-48, 7.2-49 and 7.2-50 yields

$$\underline{\underline{\sigma}} + (k_c + k_l f_1 - k_1 f_2)^{-1}\frac{\delta}{\delta t}\underline{\underline{\sigma}} = -\frac{1}{2}N_{n0}k_B T(k_c + k_l f_1 - k_1 f_2)^{-1}\underline{\dot{\gamma}}. \qquad (7.2\text{-}51)$$

We have assumed that f_1 and f_2 are functions of $II_{\dot{\gamma}}$, and this implies that the viscosity and the primary normal stress coefficient are functions of $II_{\dot{\gamma}}$. Equation 7.2-51 is the White-Metzner model (see equation 6.3-9). In the earlier model, we have assumed that the rates of creation and loss of segments are independent of $II_{\dot{\gamma}}$, and this corresponds to setting

$$k_l = k_1 = 0 \qquad (7.2\text{-}52a)$$

$$\text{and} \quad k_c = \frac{1}{\lambda_n}. \qquad (7.2\text{-}52b)$$

In this special case, equation 7.2-52 becomes

$$\underline{\underline{\sigma}} + \lambda_n \frac{\delta}{\delta t}\underline{\underline{\sigma}} = -\frac{1}{2}N_{n0}k_B T\lambda_n\underline{\dot{\gamma}}. \qquad (7.2\text{-}53)$$

Equation 7.2-53 is the differential form of 7.2-27 and is similar to 7.1-43. Chan Man Fong and De Kee (1995b) considered the deformation to be non-affine, and the upper-convected derivative is then replaced by the Gordon-Schowalter derivative.

7.3 Reptation Theories

This molecular approach is currently quite popular in modeling the rheological behavior of polymer melts and concentrated solutions. The initial concept was introduced by de Gennes (1971), which led to constitutive equations developed by Doi and Edwards (1978, 1979).

In this section we present a brief introduction to reptation theories. We review the tube model, as well as some of the important results obtained by Doi and Edwards (1978, 1979), and we close by summarizing the theory of Curtiss and Bird (1981).

7.3-1 The Tube Model

The tube model is used to represent a polymeric chain, surrounded by or entangled with other chains as illustrated in Fig. 7.3-1. Part (a) of the figure shows a typical network with two entanglement points A and B. These entanglements could be of a permanent nature for crosslinked polymers or of a temporary nature as in the case of a polymer melt or a concentrated solution. In part (b) of the figure, the chain AB is seen as enclosed in a tube, representing the constraints of the surrounding polymeric chains (and solvent molecules in the case of solutions). Due to this topological constraint, long range displacements are only possible by motion along the tube axis. This was called reptation by de Gennes (1971) after the Latin word *reptare*, from which the word reptile is derived. This motion is described in Fig. 7.3-2.

De Gennes (1971) discussed the Brownian motion of an unattached chain confined in a tube-like region. Since the polymeric chain is rather long compared to the tube, it forms a series of loops or defects which can flow up and down the tube. As a result of the reptation, the tube itself will change configuration with time. Figure 7.3-2 shows an example in which the chain moves to the right creating a random extension of the tube ($B' - B$). The left part $A' - A$ disappears.

The time required for a chain to vacate the original tube (reptation time) is proportional to the square of the contour length divided by the reptation velocity. The contour length is proportional to the polymer molecular weight, and because the drag force is also assumed to be proportional to the molecular weight, it follows that the reptation rate is inversely proportional to the molecular weight. Hence, the reptation time is

$$\lambda_{rep} \propto \frac{M^2}{M^{-1}} = M^3. \tag{7.3-1}$$

De Gennes (1971) derived the expression for the probability of a Rouse chain to vacate the original tube. The fraction of the original tube that remains occupied is given by

$$P(t) = \sum_{k \text{ odd}} \frac{8}{\pi^2 k^2} \exp\left(-\frac{k^2 t}{\lambda}\right), \tag{7.3-2}$$

7.3 Reptation Theories 295

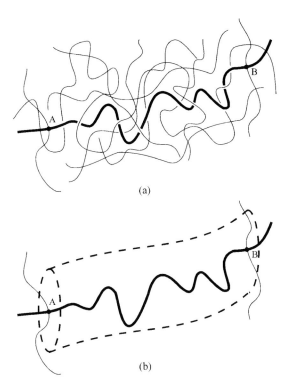

Figure 7.3-1 The tube model (Adapted from Doi and Edwards, 1986)
(a) a polymeric chain in a network
(b) the same polymeric chain in an imaginary tube

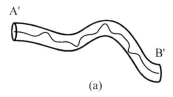

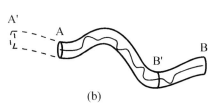

Figure 7.3-2 Chain reptation inside a tube
(Adapted from Doi and Edwards, 1986)
(a) initial position
(b) position and tube extension at a later time

where λ is a characteristic reptation time. If we consider a Rouse chain without constraint, $\lambda(\lambda_R)$ is the Rouse constant, the longest relaxation time in the Rouse model. It is obtained for $k=1$ from equation 7.1-91:

$$\lambda_R = \frac{\frac{\zeta}{2H_0}}{4\left(\frac{\pi}{2N}\right)^2} = \frac{6(\eta_0 - \eta_s)M}{c\pi^2 RT}. \tag{7.3-3}$$

Since N is proportional to M (the molecular weight), $\lambda_R \sim M^2$.

7.3-2 Doi-Edwards Model

Doi and Edwards (1978, 1979) considered two mechanisms for changes of configuration of polymeric chains. At very short times, only redistribution of the conformation of segments between entanglements (topological constraints) is assumed. This rapid relaxation process is characterized by λ_e, called the equilibrium relaxation time. The other process is related to the reptation of the primitive chain, as illustrated in Fig. 7.3-2. The reptation process is much slower than the relaxation process of chain segments, that is, $\lambda_d \gg \lambda_e$. Also considered by Doi and Edwards is the relaxation of the contour length or retraction of the tube. This relaxation process does not make a significant contribution to the linear viscoelastic properties (Doi and Edwards, 1986).

For small t ($t \leq \lambda_e$), the dynamics of the chain segments is described by the Rouse model. The relaxation function is then given by

$$\begin{aligned} G(t) &= \frac{c}{N} k_B T \sum_{k=1}^{\infty} \exp\left(-2\frac{tk^2}{\lambda_R}\right) \\ &\approx \frac{c}{N} k_B T \int_0^{\infty} \exp\left(-2\frac{tk^2}{\lambda_R}\right) dk \\ &= \frac{c}{2\sqrt{2}N} k_B T \left(\frac{\lambda_R}{t}\right)^{\frac{1}{2}} \approx G_N^0 \left(\frac{\lambda_e}{t}\right)^{\frac{1}{2}}, \end{aligned} \tag{7.3-4}$$

where G_N^0 is the plateau modulus taken by Doi and Edwards (1986) as the value of $G(t)$ at $t = \lambda_e$.

The plateau modulus refers to the plateau observed in the plot of G' versus ω at intermediate frequencies for high molecular weight with narrow molecular weight distribution polymers (Onogi et al., 1970; see also Ferry, 1980; and Larson, 1988). The molecular weight at which a plateau is first observed corresponds roughly to M_c, the critical molecular weight, above which the zero shear viscosity starts to increase as $M^{3.4}$. The existence of a plateau implies that little relaxation occurs within the given frequency range. High molecular weight polymers, in solution or in molten state, tend to behave as a perfect rubber. As a first approximation (see Larson, 1988),

$$G_N^0 = \nu k_B T, \tag{7.3-5}$$

where ν is the number of entanglements per unit volume. G_N^0 is usually independent of molecular weight and can be used to calculate ν. Knowing the polymer density, we can estimate M_e, the molecular weight of chain segments between entanglements. M_e is found to be approximately equal to $M_c/2$ (see Ferry, 1980).

For $t \geq \lambda_e$, the Rouse dynamics is stopped by topological constraints and the reptation dynamics takes over, characterized by a relaxation time λ_d expressed as

$$\lambda_d = \frac{\zeta N^3 l^4}{\pi^2 k_B T a}, \qquad (7.3\text{-}6)$$

where l is a characteristic length of the polymeric chain, related to the end-to-end distance, and a is a parameter associated with the statistical nature of the network (Doi and Edwards, 1986). The contour length of the primitive chain, L, is given by

$$L = \frac{Nl^2}{a}. \qquad (7.3\text{-}7)$$

The relaxation modulus can be expressed as the product of the plateau modulus, G_N^0, and the probability function $P(t)$ of equation 7.3-2 with λ replaced by λ_d. The result can be written as

$$G(t) = \frac{8 G_N^0}{\pi^2} \sum_{k \text{ odd}} \frac{1}{k^2} e^{-\frac{k^2 t}{\lambda_d}}. \qquad (7.3\text{-}8)$$

The general behavior predicted by the Doi-Edwards theory is illustrated in Fig. 7.3-3. Part (a) of the figure shows the results of equations 7.3-4 and 7.3-8 for two molecular weights. The theory correctly predicts the shift of the terminal zone to longer times as the molecular weight is increased. Part (b) of the figure illustrates the G' behavior. ($G'(\omega)$ and $G''(\omega)$ can be calculated from $G(t)$ via equations 5.3-3 and 5.3-4.) Note the shift of the terminal zone to lower frequencies as M increases. The value of the plateau modulus is independent of M but the range of frequencies for which the plateau modulus is observed becomes wider as M increases.

The Doi-Edwards theory describes the linear viscoelastic behavior of polymer melts of narrow molecular weight distribution very well. A striking example is shown in Fig. 7.3-4, which reports on the storage modulus data of a polystyrene. All the features of the theory are respected, except for the shift of the terminal zone, which follows a molecular weight dependence to the power 3.4, compared to 3.0 predicted by the theory. The molecular weight at which the plateau first appears is roughly equal to $M_c (\approx 33\,000$ kg/kmol) according to Ferry, 1980). For the highest molecular weight (580 000 kg/kmol), the plateau covers about six decades of frequency.

The zero shear viscosity η_0 and the steady-state recoverable compliance can be calculated as

$$\eta_0 = \int_0^\infty G(t) dt = \frac{\pi^2}{12} G_N^0 \lambda_d \qquad (7.3\text{-}9)$$

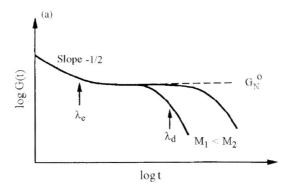

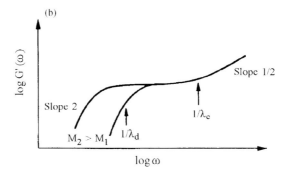

Figure 7.3-3 Behavior predicted by the Doi-Edwards theory
(Adapted from Doi and Edwards, 1986)
(a) $G(t)$
(b) $G'(\omega)$

and

$$J_e^0 = \frac{\int_0^\infty G(t)t\,dt}{\left[\int_0^\infty G(t)\,dt\right]^2} = \frac{6}{5G_N^0}. \tag{7.3-10}$$

Since G_N^0 is independent of molecular weight and $\lambda_d \propto M^3$, it follows that the theory predicts η_0 to be proportional to $M^{3.0}$, compared to $M^{3.4}$, which is generally observed for linear high molecular weight polymers of narrow molecular weight distribution. The theory predicts a constant value for J_e^0 in accord with data for high molecular weight polymers, but the product $J_e^0 G_N^0$ is equal to 6/5, which is below the value of about 2 observed for many polymers (Graessley, 1980). High frequency storage and loss moduli predicted by the Doi-Edwards theory are not in general in accord with experimental data. The failure is attributed to the very narrow relaxation spectrum: a single value, λ_e, at very short times. For monodisperse polymers, the time required for the tube to lose its identity is much longer than the reptation time, λ_d. However, for polydisperse polymers, the tube renewal

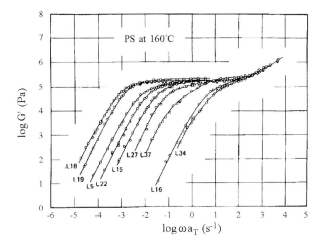

Figure 7.3-4 Storage modulus G' vs frequency ω, for polystyrenes of narrow molecular weight distribution
The molecular weight decreases from left to right (580 000 to 47 000 kg/kmol). The data have been reduced to 160 °C (Adapted from Onogi et al., 1970)

time may be much shorter due to the presence of low molecular weight chains. Thus, the tube renewal process may provide an additional relaxation mechanism, as discussed by Doi et al. (1987).

Long chain branching introduces considerable complications. Reptation is no longer possible and relaxation occurs primarily through the mechanism of tube retraction. De Gennes (1979) has shown that the longest relaxation time varies exponentially with the molecular weight, instead of the power 3 predicted for linear polymers.

The results discussed so far are restricted to linear viscoelasticity. Doi and Edwards (1978, 1979) derived an approximate nonlinear constitutive equation for the reptation dynamics. They assumed that the contour length of the primitive chain remains at the equilibrium value under large strain flow. This is valid provided that the magnitude of the velocity gradient is less than the inverse of the Rouse time, that is,

$$II_{\dot{\gamma}}^{\frac{1}{2}}\lambda_R < 1. \tag{7.3-11}$$

They used what they called the independent alignment (IA) approximation (Doi and Edwards, 1978) for the calculation of the contour length of the primitive chain, to arrive at a less complex constitutive equation. This assumption appears to be equivalent to the mild curvature assumption used by Curtiss and Bird (1981) in their phase space kinetic approach (see Section 7.3-3). The simplified Doi-Edwards constitutive equation can be written as

$$\sigma_{ij}(t) = -\frac{8G_e}{\lambda_d}\int_{-\infty}^{t}\sum_{k\,\mathrm{odd}}e^{\frac{-k^2(t-t')}{\lambda_d}}Q_{ij}^{IA}(\underline{\gamma})dt', \tag{7.3-12}$$

where Q_{ij}^{IA} is an average strain measured over the contour length of the primitive chain, given by (Doi and Edwards, 1986)

$$Q_{ij}^{IA} = \frac{1}{4\pi} \int \left(\frac{(\underline{\underline{\gamma}} \cdot \underline{u})_i (\underline{\underline{\gamma}} \cdot \underline{u})_j}{|\underline{\underline{\gamma}} \cdot \underline{u}|^2} - \frac{1}{3} \delta_{ij} \right) d\underline{u}, \qquad (7.3\text{-}13)$$

where $\underline{\underline{\gamma}}$ is the linear strain tensor (defined in Example 6.1-1) and $\underline{u}$ is the unit vector tangent to the primitive chain backbone. The equilibrium modulus in equation 7.3-12 is related to the plateau modulus by

$$G_e = \frac{15}{4} G_N^0 = 3 k_B T \frac{cl^2}{a^2}. \qquad (7.3\text{-}14)$$

The simplified Doi-Edwards constitutive equation qualitatively predicts the nonlinear phenomena observed for polymeric liquids. It predicts a negative secondary normal stress coefficient, given by

$$\psi_2 = -\frac{2}{7} \psi_1. \qquad (7.3\text{-}15)$$

This result appears to be in agreement with the scarce secondary normal stress data available in the literature. The ratio ψ_2/ψ_1 is of the order of 0.10, quoted in Chapter 2. The major drawback of the simplified theory is that it predicts a non-Newtonian viscosity that depends too strongly on shear rate ($\eta \propto \dot{\gamma}^{-3/2}$ at high shear rates). This failure has been attributed to the assumptions and simplifications made in the original work of Doi and Edwards. Improvements and more rigorous constitutive equations have been proposed (see for example Marrucci and Grizzuti, 1986; Marrucci, 1986). These modifications introduce considerable mathematical complexity, which is beyond the scope of this textbook.

7.3-3 The Curtiss-Bird Kinetic Theory

In an attempt to derive the Doi-Edwards results, without using the assumptions and simplifications made by Doi and Edwards (1978, 1979), Curtiss and Bird (1981) developed a theory in the framework of their general phase-space theory. We summarize here their main ideas and some of the important results of their theory.

Curtiss and Bird did not require the molecules to be confined within tubes. A linear polymeric chain, surrounded by other polymeric chains, is idealized as a Kramers freely jointed bead-rod chain as illustrated in Fig. 7.3-5. Such a chain, depicted in Fig. 7.3-6, consists of N identical beads of mass m joined by $N-1$ massless rigid rods of length l. Such a chain can be described by the Kramers and Rouse matrices, presented in Section 7.1-3.

Due to the constraint of surrounding chains, the mobility of the primitive chain is larger in the direction of the chain backbone. This is expressed by a drag tensor given by

$$\underline{\underline{\zeta}}_k = \zeta [\underline{\underline{\delta}} - (1-\epsilon) \underline{u}_k \underline{u}_k], \qquad (7.3\text{-}16)$$

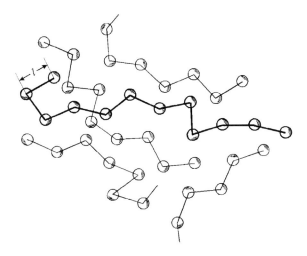

Figure 7.3-5 A linear polymeric chain surrounded by others

where ζ is the scalar drag coefficient used in the models discussed so far, and $\underline{u}_k$ the unit vector tangent to the chain backbone at bead (k) (see Fig. 7.3-6). ϵ is called the link-tension coefficient. For $\epsilon = 1$ the drag tensor is isotropic and the results reduce to those obtained for dilute solutions with no hydrodynamic interaction (Bird et al., 1987). Curtiss and Bird (1981) introduced another parameter to account for the increase in constraint by surrounding chains. The force acting on the kth link is assumed to be proportional to N^β (or M^β), where β is an empirical constant found to be about 0.4 to 0.5 by comparing the results to experimental data. For homogeneous flow, the net force acting on the kth link is

$$(\underline{F}_{k+1} - \underline{F}_k) \cdot \underline{u}_k = \epsilon N^\beta l \zeta (\underline{\underline{\kappa}} : \underline{u}_k \underline{u}_k). \tag{7.3-17}$$

Hence, for $\epsilon = 0$, the tension is zero in all links. Finally, Curtiss and Bird introduced the *mild curvature assumption*, which states that successive links have very nearly the same

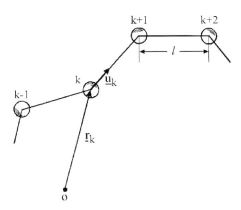

Figure 7.3-6 Three segments of a Kramers chain

orientation, although the whole chain may be greatly distorted, as illustrated in Fig. 7.3-7. The bead-rod system is then approximated by a smooth curve without links.

The rheological equation derived by Curtiss and Bird (1981) can be written as

$$\underline{\underline{\sigma}} = -Nnk_BT\left[\int_0^1 \left\langle \underline{uu} - \frac{1}{3}\underline{\underline{\delta}}\right\rangle d\sigma + \epsilon\lambda\underline{\underline{\kappa}} : \int_0^1 \langle \underline{uuuu}\rangle \sigma(1-\sigma)d\sigma\right], \quad (7.3\text{-}18)$$

where $\langle \ \rangle$ refers to the average with respect to the orientation distribution function given by the following diffusion equation:

$$\frac{\partial f}{\partial t} = \frac{1}{\lambda}\frac{\partial^2 f}{\partial \sigma^2} - \left(\frac{\partial}{\partial \underline{u}} \cdot G_t(\underline{u})f\right), \quad (7.3\text{-}19a)$$

with the operator G_t defined as

$$G_t(\underline{u}) = \underline{\underline{\kappa}} \cdot \underline{u} - \underline{\underline{\kappa}} : \underline{uuu}. \quad (7.3\text{-}19b)$$

The characteristic time, λ, is defined by

$$\lambda = \frac{N^{3+\beta}\zeta l^2}{2k_BT}. \quad (7.3\text{-}20)$$

In the limit of linear viscoelasticity, the Curtiss-Bird model reduces to

$$\underline{\underline{\sigma}} = -Nnk_BT\left[\frac{1.6}{\pi^2}\int_{-\infty}^t \sum_{k\text{ odd}} \frac{1}{k^2}\exp\left(-\frac{\pi^2 k^2(t-t')}{\lambda}\right)\underline{\dot{\gamma}}(t')dt' - \frac{1}{90}\epsilon\lambda\underline{\dot{\gamma}}(t)\right], \quad (7.3\text{-}21)$$

which leads to the following expression for the relaxation modulus:

$$G(t-t') = Nnk_BT\left[\frac{1.6}{\pi^2}\sum_{k\text{ odd}} \frac{1}{k^2}\exp\left(-\frac{\pi^2 k^2(t-t')}{\lambda}\right) + \frac{1}{45}\epsilon\lambda\delta(t-t')\right], \quad (7.3\text{-}22)$$

where $\delta(t-t')$ is the Dirac delta function. The complex viscosity is then

$$\eta^*(\omega) = \int_{-\infty}^t G(s)e^{-i\omega s}ds = Nnk_BT\left[\frac{1.6\lambda}{\pi^2}\sum_{k\text{ odd}}\frac{1}{k^2(\pi^2 k^2 + i\lambda\omega)} + \frac{1}{90}\epsilon\lambda\right]. \quad (7.3\text{-}23)$$

In the limit of zero frequency (or zero shear), this result gives

$$\eta_0 = Nnk_BT\lambda\left(\frac{1}{60} + \frac{\epsilon}{90}\right) = \frac{N\rho RT}{M}\lambda\left(\frac{1}{60} + \frac{\epsilon}{90}\right), \quad (7.3\text{-}24)$$

Figure 7.3-7 View of a polymeric chain that respects the mild curvature assumption

where ρ is the polymer melt density, or the mass concentration of the polymeric solution. Equation 7.3-23 predicts a nonzero value for the real part of the complex viscosity at high frequencies, $\eta'(\infty)$:

$$\frac{\eta'(\infty)}{\eta_0} = \left(\frac{2}{3}\right)\frac{\epsilon}{1+\frac{2}{3}\epsilon}. \tag{7.3-25}$$

A plateau modulus is predicted as $\omega \to \infty$:

$$G_N^0 = G'(\infty) = \frac{1}{5}Nnk_BT. \tag{7.3-26}$$

Hence, the Curtiss-Bird theory predicts a plateau modulus proportional to N or M (molecular weight). This and the constant value of G' at high frequencies are in contradiction with experiments. This failure is attributed to the fact that the motion of side groups is neglected in the theory.

For steady shear flow, the zero shear normal stress coefficients are

$$\psi_{10} = \frac{N\rho RT\lambda^2}{300M} \tag{7.3-27}$$

and

$$\psi_{20} = -\frac{2}{7}(1-\epsilon)\psi_{10}. \tag{7.3-28}$$

If we assume N to be proportional to the molecular weight, M, then $\eta_0 \propto \lambda \propto M^{3+\beta}$, and $\psi_{10} \propto M^{6+2\beta}$. For $\beta = 0.4$, the theory predicts the well-established relation for linear polymer melts of high M: $\eta_0 \propto M^{3.4}$. It is interesting to note that secondary normal stress differences are predicted only in the case of anisotropic motion, that is, for $\epsilon < 1$ (equivalent to the idea of reptation introduced by de Gennes (1971) and used by Doi and Edwards (1978, 1979)).

The original Curtiss-Bird theory has been assessed and compared to steady and transient shear flow data for concentrated polymer solutions (Bird et al., 1982; and Saab and Bird, 1982). The data for nearly monodisperse polymers are very well described by the theory for ϵ values ranging between 0.3 and 0.5. The Curtiss-Bird constitutive equation appears to be more flexible than the Doi-Edwards equation except possibly for high frequency linear viscoelasticity. The Doi-Edwards model can be approximately recovered by setting $\epsilon = 0$ and $\beta = 0$ in the Curtiss-Bird equation. The most serious limitations of the reptation theories are the high shear rate predictions. The Doi-Edwards theory predicts a viscosity proportional to $\dot{\gamma}^{-1.5}$, whereas for the Curtiss-Bird equation with $\epsilon > 0$, $\eta \propto \dot{\gamma}^{-1.0}$. This is unrealistic, and modifications to the theories have to be proposed to correct this undesired feature. Considerable improvements have been obtained by Schieber (1987) who modified the Curtiss-Bird model to account for polydispersity. The material functions are calculated in terms of ϵ, the polydispersity M_w/M_n, and a time constant λ_{z+j} defined by

$$\lambda_{z+j} = \frac{M_n^{1+\beta}M_{z+j}^2\zeta l^2}{2k_BTM_0^{3+\beta}}, \tag{7.3-29}$$

where M_n and M_{z+j} are the number and $z+j$ weight average molecular weights respectively. M_0 is the molecular weight of one bead in the Curtiss-Bird model. The main finding of Schieber is that polydispersity allows for better fits of viscosity and first normal stress coefficient (for polydisperse polymers). The results of Schieber are quite promising, but efforts should be devoted to simplify such complex theories to forms suitable to be used by non-specialists of the field.

7.4 Conformation Tensor Rheological Models

In this section, we summarize the results obtained by Grmela and collaborators (Grmela, 1985, 1986; Grmela and Carreau, 1987; Grmela and Chhon Ly, 1987; Ait-Kadi et al., 1988; Ghosh et al., 1995a, b), who used a thermodynamic approach based on the conformation tensor as an internal variable. The salient features of the model are presented, followed by a few specific examples.

7.4-1 Basic Description of the Conformation Model

To describe mathematically the time evolution of complex fluids, we have to choose a quantity which characterizes the internal structure (internal state variable). Thus, we introduce equations that govern the time evolution of the hydrodynamic fields and the chosen internal state variable. The choice of the internal variables is always a compromise between completeness (the ability to describe molecular details) and simplicity (the ease with which the governing equations are solved). For the sake of simplicity, we choose the conformation tensor $\underline{\underline{c}}$ as the internal state variable (Hand, 1964; Giesekus, 1982). The tensor $\underline{\underline{c}}$ is assumed to be symmetric and positive definite. From a physical point of view, we can regard it for example as a molecular chain deformation tensor. The conformation tensor can be viewed in reference to a distribution function $\psi(\underline{R}, t)$ as the second moment of the end-to-end vector $\underline{R}$ of the conformation (see Fig. 7.4-1), that is,

$$\underline{\underline{c}}(\underline{R}, t) = \langle \underline{R}\underline{R} \rangle = \int \psi(\underline{R}, t)\underline{R}\underline{R}d\underline{R}. \qquad (7.4\text{-}1)$$

Here it represents the average conformation of coil structure under any flow conditions.

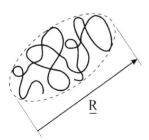

Figure 7.4-1 Polymeric chain in a coil conformation

We assume that the internal structure of the fluid is spatially (locally) homogeneous (i.e., $\underline{\underline{c}}$ is independent of the position vector $\underline{r}$) and that the fluid is isothermal and incompressible. This means that the only state variables considered are the velocity field $\underline{V}(\underline{r}, t)$ and the conformation tensor $\underline{\underline{c}}(t)$.

In formulating the time evolution equations, we require that the governing equations have a clear physical meaning. This implies that we should be able to clearly express the molecular nature of the fluid in the governing equations and thus consider them as a bridge between the molecular properties and observable macroscopic thermodynamic and flow properties. What we want to avoid in particular is the introduction of physically unjustified approximations, such as closure approximations, that are necessary if the governing equations are searched at a more microscopic level of description.

Following Grmela (1986), we propose the following equation governing the time evolution of the conformation tensor $\underline{\underline{c}}$:

$$\frac{D}{Dt}\underline{\underline{c}}(t) = \frac{1}{2}(2-\xi)[\underline{\underline{\kappa}} \cdot \underline{\underline{c}} + \underline{\underline{c}} \cdot \underline{\underline{\kappa}}^+] - \frac{\xi}{2}[\underline{\underline{\kappa}}^+ \cdot \underline{\underline{c}} + \underline{\underline{c}} \cdot \underline{\underline{\kappa}}] - \underline{\underline{\Lambda}}(\underline{\underline{c}}) \cdot \underline{\underline{c}} \cdot \frac{dA(\underline{\underline{c}})}{d\underline{\underline{c}}}, \qquad (7.4\text{-}2)$$

where κ is the transpose of the velocity gradient tensor, defined by equation 7.1-30; $A(\underline{\underline{c}})$ is the polymer contribution to the Helmholtz free energy functional; $\underline{\underline{\Lambda}}(\underline{\underline{c}})$ is the mobility tensor which in general is anisotropic and a function of $\underline{\underline{c}}$; and ξ is a slip parameter as used in the Phan-Thien-Tanner model (Section 6.3-5). For affine deformation, $\xi = 0$, and the first three terms of equation 7.4-2 correspond to the upper-convected derivative. The last term of equation 7.4-2 corresponds to the diffusive part, to make it compatible with the Clausius-Duhem inequality, which implies positive entropy production (Grmela, 1986). Another consequence of the requirement of compatibility of equation 7.4-2 with equilibrium thermodynamics is the relation between the conformation tensor $\underline{\underline{c}}$ and the polymer contribution to the extra stress tensor. This relation can be written as (Grmela, 1986)

$$\underline{\underline{\sigma}}^p = -2n(1-\xi)\frac{dA(\underline{\underline{c}})}{d\underline{\underline{c}}} \cdot \underline{\underline{c}}, \qquad (7.4\text{-}3)$$

where $\underline{\underline{\sigma}}^p$ is the polymer contribution to the extra stress tensor and n is the number density of macromolecules. Equations 7.4-2 and 7.4-3 constitute the governing equations of the model. In order to relate the actual model clearly to rheological equations introduced previously, we reduce equations 7.4-2 and 7.4-3 into a single tensorial equation for $\underline{\underline{\sigma}}^p$. This step requires solving for the internal strain tensor $\underline{\underline{c}}$. It should be noted, however, that solutions of equations 7.4-2 and 7.4-3 can be analyzed without reducing them to a single equation for $\underline{\underline{\sigma}}^p$.

Let a solution for $\underline{\underline{c}}$ be the following:

$$\underline{\underline{c}} = f(\underline{\underline{\sigma}}^p)[\underline{\underline{\sigma}}^p - (1-\xi)nk_BT\underline{\underline{\delta}}], \qquad (7.4\text{-}4)$$

where $f(\underline{\underline{\sigma}}^p)$ is a scalar function uniquely determined by $A(\underline{\underline{c}})$. Specific examples are given in Section 7.4-2. Examples of the Helmholtz free energy $A(\underline{\underline{c}})$ for which equation 7.4-4 is not a solution of equation 7.4-3 are discussed by Grmela and Carreau (1987). Inserting equation 7.4-4 into equation 7.4-2 yields for an *isotropic mobility* (see Problem 7.5-5):

$$\frac{\delta}{\delta t}\underline{\underline{\sigma}}^p + \frac{\xi}{2}(\underline{\underline{\sigma}}^p \cdot \underline{\underline{\dot{\gamma}}} + \underline{\underline{\dot{\gamma}}} \cdot \underline{\underline{\sigma}}^p) + [\underline{\underline{\sigma}}^p - (1-\xi)nk_BT\underline{\underline{\delta}}]\frac{1}{f(\underline{\underline{\sigma}}^p)}\frac{D}{Dt}f(\underline{\underline{\sigma}}^p)$$
$$-\frac{\Lambda(\underline{\underline{\sigma}}^p)}{2n(1-\xi)f(\underline{\underline{\sigma}}^p)}\underline{\underline{\sigma}}^p = -n(1-\xi)^2 k_B T \underline{\underline{\dot{\gamma}}},$$
(7.4-5)

where $\underline{\underline{\dot{\gamma}}} = \underline{\underline{\kappa}} + \underline{\underline{\kappa}}^+$. This result was originally obtained in a slightly different form by Ait-Kadi et al. (1988). As for the dumbbell model, the solvent contribution can be added as follows:

$$\underline{\underline{\sigma}} = \underline{\underline{\sigma}}^p - \eta_s \underline{\underline{\dot{\gamma}}}.$$
(7.4-6)

Various specific constitutive equations can be obtained by choosing different functions $f(\underline{\underline{\sigma}}^p)$ and $\Lambda(\underline{\underline{\sigma}}^p)$. One route is to specify the Helmholtz free energy as a function of the conformation tensor and the mobility. A complete example is presented in Section 7.4-2. An alternative method is to obtain the functions f and Λ from experimental results. For steady shear flow ($V_x = \dot{\gamma}y$, $V_y = V_z = 0$, where $\dot{\gamma}$ = constant), equations 7.4-5 and 7.4-6 simplify to

$$-\dot{\gamma}\begin{bmatrix} 2\sigma^p_{yx} & \sigma^p_{yy} & \sigma^p_{yz} \\ \sigma^p_{yy} & 0 & 0 \\ \sigma^p_{yz} & 0 & 0 \end{bmatrix} + \frac{\xi}{2}\dot{\gamma}\begin{bmatrix} 2\sigma^p_{yx} & \sigma^p_{xx}+\sigma^p_{yy} & \sigma^p_{yz} \\ \sigma^p_{xx}+\sigma^p_{yy} & 2\sigma^p_{yx} & \sigma^p_{zx} \\ \sigma^p_{yz} & \sigma^p_{zx} & 0 \end{bmatrix}$$
$$-\frac{\Lambda(\underline{\underline{\sigma}}^p)}{2n(1-\xi)f(\underline{\underline{\sigma}}^p)}\begin{bmatrix} \sigma^p_{xx} & \sigma^p_{xy} & \sigma^p_{xz} \\ \sigma^p_{yx} & \sigma^p_{yy} & \sigma^p_{yz} \\ \sigma^p_{zx} & \sigma^p_{zy} & \sigma^p_{zz} \end{bmatrix} = -[n(1-\xi)^2 k_B T]\dot{\gamma}\begin{bmatrix} 0 & 1 & 0 \\ 1 & 0 & 0 \\ 0 & 0 & 0 \end{bmatrix}.$$
(7.4-7)

The three material functions are then

$$\eta - \eta_s = -\frac{2n^2(1-\xi)^3 k_B T f(\underline{\underline{\sigma}}^p)}{\Lambda(\underline{\underline{\sigma}}^p)\left\{1 + (2\xi - \xi^2)\left[\frac{2n(1-\xi)\dot{\gamma}f(\underline{\underline{\sigma}}^p)}{\Lambda(\underline{\underline{\sigma}}^p)}\right]^2\right\}},$$
(7.4-8a)

$$\frac{\psi_1}{\eta - \eta_s} = -\frac{4n(1-\xi)f(\underline{\underline{\sigma}}^p)}{\Lambda(\underline{\underline{\sigma}}^p)},$$
(7.4-8b)

and $\quad \psi_2 = -\frac{\xi}{2}\psi_1.$
(7.4-8c)

Through appropriate choices of $f(\underline{\underline{\sigma}}^p)$ and $\Lambda(\underline{\underline{\sigma}}^p)$, using for example tr $\underline{\underline{\sigma}}^p$, we can in principle describe the simple shear behavior observed for most polymeric systems. The corresponding results for uniaxial elongational flow can be found in Ait-Kadi et al. (1988). For affine deformation, $\xi = 0$ and results 7.4-8 reduce to

$$\eta - \eta_s = -\frac{2n^2 k_B T f(\underline{\sigma}^p)}{\Lambda(\underline{\sigma}^p)}, \qquad (7.4\text{-}9a)$$

$$\psi_1 = \frac{2(\eta - \eta_s)^2}{nk_B T}, \qquad (7.4\text{-}9b)$$

and $\psi_2 = 0$. $\qquad (7.4\text{-}9c)$

The identical result to equation 7.4-9b was also obtained for the FENE-P model (equation 7.1-62). It implies that the primary normal stress coefficient is uniquely related to the shear viscosity. As illustrated in Fig. 6.4-9, several polymeric solutions seem to obey relation 7.4-9b. Without loss of generality, for affine deformation, we can use a constant mobility, $\Lambda = \Lambda_0$, in equation 7.4-9a. The relative viscosities can then be described by a simple equation such as the Ellis model (equation 2.2-3) or the Carreau viscosity model (equation 2.2-4). Note that as discussed below, the model in this formulation contains a single characteristic (or relaxation) time. Hence, the model reduces to a simple (one-mode) Maxwell model for small deformation, as discussed in Chapter 5. It is therefore inappropriate for describing linear viscoelasticity. A generalization to multimode models is briefly discussed in Section 7.4-4.

In obtaining the above results, we did not specify the Helmholtz free energy functional in which the physics is embedded. In Section 7.4-2, we present a detailed example in which the Helmholtz free energy is specified.

7.4-2 FENE-Charged Macromolecules

Charged polymers known as polyelectrolytes are widely used in various industrial processes encountered in the food, pharmaceutical, paint, and pulp and paper industries, and are also used for mobility control of fluids in porous media. A rheological model for polyelectrolytes has been proposed by Dunlap and Leal (1984), who used a Coulombic potential to account for electrostatic repulsive forces. The concept of an anisotropic but conformation-dependent friction coefficient (de Gennes, 1974; Hinch, 1974; Tanner, 1975) has also been introduced into the setting of the FENE-P model (Ait-Kadi et al., 1988) to represent the coil-stretch transition experienced by macromolecules during flow.

Ait-Kadi et al. (1988) used a FENE potential and a Coulombic potential to write the Helmholtz free energy as

$$A(\underline{c}) = -H_0 R_\infty^2 \ln\left(1 - \frac{\text{tr}\,\underline{c}}{R_\infty^2}\right) + E\,\text{tr}(\underline{c})^{-\frac{1}{2}} - \frac{k_B}{2} T \ln \det \underline{c}, \qquad (7.4\text{-}10)$$

where H_0 is the coil (spring) modulus, R_∞ is its maximum extension, E is a constant, and the second term on the right side implies that the electrostatic forces derived from the electrostatic potential U_c decrease as the inverse of the square of the extension of the macromolecule, as proposed by Dunlap and Leal (1984). This is also equivalent to the third term of equation 7.2-40, introduced in the network framework. The electrostatic charges are assumed to be concentrated at the ends of the polymer chain. This is equivalent to assuming that the distance between chain sub-elements is proportional to $\sqrt{\text{tr }\underline{c}}$. The parameter E is related to the amount of available electrostatic charges. The last term in equation 7.4-10 is due to the Brownian motion (entropic contribution). It has been justified by Grmela (1985, 1986).

Following de Gennes (1974), Hinch (1974) and Tanner (1975), we retain the following relation for the mobility:

$$\Lambda(\text{tr }\underline{c}) = \frac{\Lambda_0}{(1+\beta\chi)}. \tag{7.4-11}$$

This relation expresses the fact that as the coil is deformed and stretched under flow, its mobility is reduced. Λ_0 and β are two parameters, and $\chi (= \sqrt{\text{tr }\underline{c}/R_\infty^2})$ is the reduced extension.

For any flow kinematics, the set of governing equations can be solved to obtain material functions in terms of the following four parameters:

$$b = \frac{2H_0 R_\infty^2}{k_B T}, \quad \text{extensibility parameter,} \tag{7.4-12}$$

$$e = \frac{E}{k_B T}, \quad \text{electrostatic parameter,} \tag{7.4-13}$$

$$\lambda_\beta = \frac{2\chi_0^2 R_\infty^2}{3 k_B T \Lambda_0}(1+\beta\chi_0), \quad \text{time constant,} \tag{7.4-14}$$

and β, a conformation or friction parameter. Affine deformation is assumed, that is, $\xi = 0$. In equation 7.4-14, χ_0 is the equilibrium (no-flow) value of the reduced end-to-end distance or extension of the coil. It is given by the positive root of the following equation:

$$\chi_0^3 + \frac{e}{3+b}\chi_0^2 - \frac{3}{3+b}\chi_0 - \frac{e}{3+b} = 0. \tag{7.4-15}$$

λ_β reduces to λ_e, which is related to the zero shear viscosity by

$$\lambda_e = \left[\frac{\eta_o - \eta_s}{c}\right]\frac{M}{RT}, \tag{7.4-16}$$

where c is the polymer concentration, M its molecular weight, and R the gas constant.

Many examples of model predictions for steady shear and elongational flows have been presented by Ait-Kadi et al. (1988) considering affine deformation ($\xi = 0$ in equations 7.4-3 to 7.4-5). We summarize here the effects of the electrostatic and friction parameters. We note that if e, β, and ξ are taken to be equal to zero, the model is identical to the FENE-P model discussed in Section 7.1-2. We can show (see Problem 7.5-5) that equation 7.4-5 reduces to the Tanner equation 7.1-51. Figure 7.4-2 shows the influence of the parameter e on the viscosities and chain extension. Part (a) reports the steady shear and elongational viscosities as a function of a dimensionless deformation rate, which is the product of λ_e and the square root of the second invariant of the rate-of-deformation tensor (equal to $\lambda_e \dot{\gamma}$ for shear and $\sqrt{3}\lambda_e \dot{\varepsilon}$ for elongational flow). Part (b) illustrates the reduced extension as a function of the dimensionless deformation rate.

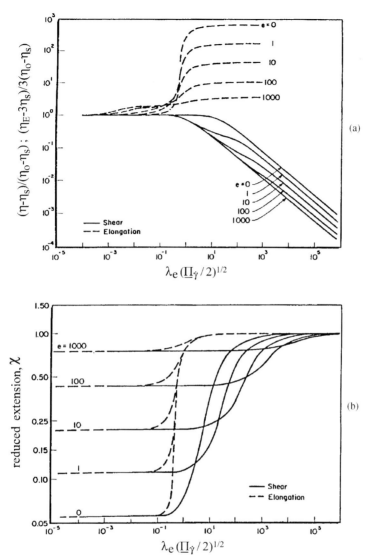

Figure 7.4-2 Effect of the electrostatic parameter e on
(Adapted from Ait-Kadi et al., 1988)
(a) the steady shear and elongational reduced viscosity;
(b) on the reduced extension in shear and elongational flows; $b = 1000$, $\xi = \beta = 0$

Model predictions are reported for $b = 1000$, $\beta = 0$, and e ranging from 0 to 1000. As the electrostatic parameter increases, an intermediate region appears between the low rate-of-deformation zone (constant shear and elongational viscosities) and the high rate-of-deformation zone (shear thinning viscosity with a constant power-law index equal to $-2/3$ and a constant elongational viscosity). The electrostatic parameter affects the onset of the decrease of viscosity in shear flow and shifts the onset of the increase of the elongational viscosity toward lower values of the deformation rate. Moreover, as e increases, the zero shear viscosity increases and the high rate-of-deformation elongational viscosity decreases.

Figure 7.4-2b shows that the equilibrium extension increases with increasing e. The equilibrium hydrodynamic volume is then larger, due to the repulsive forces acting on the macromolecular chain; the deformation due to the flow field is considerably restricted and the behavior is that of a rigid chain. We also note that e has no apparent effect on the transition from the equilibrium conformation to the fully extended conformation in elongational flow. In shear flow, large values of the electrostatic parameter result in a longer transition from the equilibrium to the stretched conformation. At the same time, the onset of the transition is shifted toward higher values of the velocity gradient. The results described above are comparable to those obtained by Dunlap and Leal (1984). Note that Fuller and Leal (1980) and Dunlap and Leal (1984) used an incorrect expression for the stress tensor. They used formula 7.4-3 without ξ, that is, with $\xi = 0$.

The influence of the friction parameter β is shown in Fig. 7.4-3. In shear flow, β has little influence on the onset of nonlinear effects. However, the transition from equilibrium to full extension is more rapid as β increases. At the onset of the deformation state, for $\beta > 0$, the shear viscosity starts to increase with increasing rate of deformation (shear thickening), up to a maximum corresponding to 50% of the full extension. Then the effect of the FENE connector becomes important, and the viscosity starts to decrease, following a power-law model with a constant exponent equal to $-2/3$. The amplitude of the shear thickening phenomenon increases with increasing value of β, but tends toward an asymptotic value for high values of the friction parameter. Tanner (1975), using a Dirac-delta function to approximate the configuration distribution function, obtained identical results. Equivalent results were also obtained by Fuller and Leal (1980) and Dunlap and Leal (1984) using the *Peterlin pre-averaging approximation*.

As shown in Fig. 7.4-3a and b, the effect of β is more pronounced in the case of elongational flow. The onset of the transition from undeformed to extended conformation is dramatically shifted to lower values as β increases. This corresponds to an increase in the elongational viscosity at lower rate of deformation. Moreover, the value of the second elongational viscosity plateau increases with β up to a limiting value for $\beta > 100$. The model also predicts S-shaped curves for both the elongational viscosity and the molecular extension for nonzero values of β. This phenomenon has been predicted by several authors: de Gennes (1974), Tanner (1975), Fuller and Leal (1980), Dunlap and Leal (1984), and others. Models predicting S-shaped viscosity functions are considered by many authors as inadmissible. S-shaped curves necessarily lead to unstable flow situations. We could consider that rheologically induced instabilities are similar to hydrodynamically induced instabilities, but the forces involved are different.

One objective in developing this model was to describe the shear-thickening behavior observed for polyacrylamide solutions in simple shear experiments. Figure 7.4-4 compares

7.4 Conformation Tensor Rheological Models

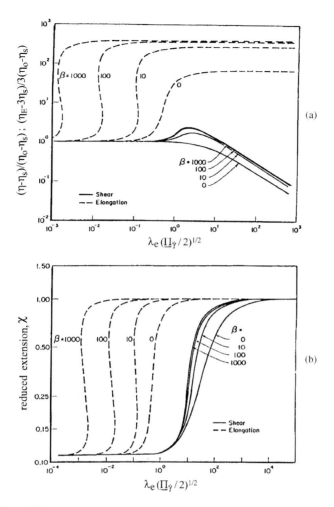

Figure 7.4-3 Effect of the friction parameter β on
(a) the steady shear and elongational-reduced viscosity
(b) on the reduced extension in shear and elongational flows; $b = 1000$, $e = 1$ and $\xi = 0$
(Adapted from Ait-Kadi et al., 1988)

the viscosity data for partially hydrolyzed PAA (polyacrylamide, Pusher 700 from Dow Chemicals) solutions in a mixed solvent containing 20 mass % distilled water and 80% glycerol. The solvent also contains 20 g/L of added sodium chloride (see Ait-Kadi et al., 1987). The shear-thickening behavior observed for dimensionless shear rates, $\lambda_e \dot{\gamma}$, greater than 10 is fairly well predicted by the model using a friction parameter β equal to 5. We stress that none of these polymer solutions except possibly the 170 ppm are dilute solutions, and λ_e is an increasing function of the polymer concentration. Hence, intermolecular interactions and possibly entanglements may play an important role in the observed phenomenon. Further discussion can be found in Ait-Kadi et al. (1988).

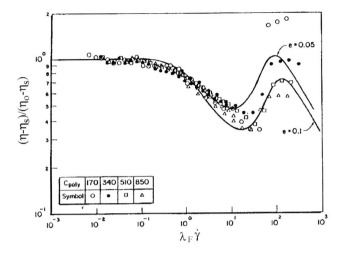

Figure 7.4-4 Reduced shear viscosity data for PAA Pusher-700 solutions $C_{NaCl} = 20$ g/L. Polymer concentrations are in ppm. The model parameters are $b = 10^5$, $\beta = 5$, and $\xi = 0$. The data are from Ait-Kadi et al. (1987) (Adapted from Ait-Kadi et al., 1988)

7.4-3 Rod-Like and Worm-Like Macromolecules

In this section, we focus on rod-like or semi-flexible worm-like macromolecules as encountered in the case of liquid crystalline polymers (LCP). We consider the polymer chains to be rigid in average, but we allow for some flexibility in the chains. We choose the worm-like model with Kuhn segments of length l, shown in Fig. 7.4-5.

The results of Khokhlov and Semenov (1984, 1985) obtained for the entropy of chains consisting of subsegments vibrating about an angle θ close to 2π can be approximated as

$$S(\underline{\underline{c}}) = -k_B S_L \mathrm{tr}\left(\frac{\underline{\underline{c}}}{R_0^2}\right)^{-1}. \tag{7.4-17}$$

This is referred to as the Lifshitz contribution to the Brownian motion due to the semi-flexibility of the chains (Khokhlov and Semenov, 1984). We assume that the parameter S_L is related to the chain flexibility

$$S_L = \frac{L}{l}, \tag{7.4-18}$$

where L is the contour length of the polymer chains. We note that if $S_L \ll 1$, the chains are rigid. On the contrary, if $S_L \gg 1$, the chains are quite flexible. The intramolecular energy will be replaced by a constraint $F(\underline{\underline{c}}) = \mathrm{tr}\, c = R_0^2$ (rigid-in-average chains).

7.4 Conformation Tensor Rheological Models

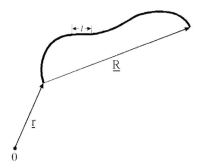

Figure 7.4-5   Sketch of a worm-like molecule

The expression for the free energy becomes

$$A(\underline{\underline{c}}) = U(\underline{\underline{c}}) - TS(\underline{\underline{c}}) = \frac{1}{2}k_B T\left[BF(\underline{\underline{c}}) - S_B \ln \det\left(\frac{\underline{\underline{c}}}{R_0^2}\right) + 2S_L \text{tr}\left(\left(\frac{\underline{\underline{c}}}{R_0^2}\right)^{-1}\right)\right]. \quad (7.4\text{-}19)$$

In equation 7.4-19, we have introduced the parameter S_B, which weighs the influence of the Boltzmann contribution of the Brownian motion. In qualitative terms, it can be expressed as follows. Brownian motion is of importance for small or molecular-sized particles. For larger particles such as crystallites, fillers, and fibers, the contribution of the Brownian motion to the free energy will gradually diminish as the particle size increases. The Lagrange multiplier B is obtained from the requirement that the constraint holds for all times, that is,

$$\frac{dF}{dt} = \frac{d}{dt}(\text{tr}\,\underline{\underline{c}} - R_0^2) = 0. \quad (7.4\text{-}20)$$

To obtain the components of the stress tensor and the material functions under various flow situations, we follow the procedure outlined in Example 7.4-1. The solutions are in general not analytical, and numerical schemes have to be used.

Example 7.4-1   Steady Shear Properties of Rod-Like Macromolecules

Determine the steady shear material functions described by the conformation model for rod-like macromolecules. Assume rigid-in-average chains with no flexibility ($S_L = 0$) and consider a constant mobility Λ_0 and affine deformation $\xi = 0$. Take the Brownian motion parameter S_B to be 1.

Solution

The rheological model is given by equations 7.4-2, 7.4-3, 7.4-19, and 7.4-20. Solving the set of equations in terms of B, we obtain

$$\underline{\underline{\sigma}}^p(\underline{\underline{c}}) = nk_B T(\underline{\underline{\delta}} - B\underline{\underline{c}}) \quad (7.4\text{-}21)$$

and

$$\frac{D}{Dt}(\underline{\underline{c}}) = \underline{\underline{\kappa}} \cdot \underline{\underline{c}} + \underline{\underline{c}} \cdot \underline{\underline{\kappa}}^+ + \frac{1}{2}k_B T\Lambda_0(\underline{\underline{\delta}} - B\underline{\underline{c}}), \quad (7.4\text{-}22)$$

where the Lagrange multiplier is obtained from equations 7.4-19 and 7.4-20 (for $S_B = 1$, $S_L = 0$):

$$B = \frac{4 \, \text{tr} \, (\underline{\underline{c}} \cdot \underline{\underline{\kappa}})}{k_B T \Lambda_0 \, \text{tr} \, \underline{\underline{c}}} + \frac{3}{\text{tr} \, \underline{\underline{c}}}. \qquad (7.4\text{-}23)$$

We consider here only simple shear flow:

$$\underline{\underline{\kappa}} = \begin{bmatrix} 0 & \dot{\gamma} & 0 \\ 0 & 0 & 0 \\ 0 & 0 & 0 \end{bmatrix}, \qquad (7.4\text{-}24)$$

and $\underline{\underline{c}}$ can be written as

$$\underline{\underline{c}}^* = \frac{\underline{\underline{c}}}{R_0^2} = \begin{bmatrix} \frac{1}{2}(C + N) & C_{12} & 0 \\ C_{12} & \frac{1}{2}(C - N) & 0 \\ 0 & 0 & 1 - C \end{bmatrix}, \qquad (7.4\text{-}25)$$

where C, N and C_{12} are the three dimensionless characteristic values of the tensor $\underline{\underline{c}}$. Equation 7.4-22 combined with equations 7.4-23 and 7.4-25 yields the following three equations:

$$\frac{\partial C_{12}}{\partial t} = \frac{1}{2}(C - N)\dot{\gamma} - \frac{\beta^*}{\lambda} C_{12}, \qquad (7.4\text{-}26)$$

$$\frac{\partial N}{\partial t} = 2C_{12}\dot{\gamma} - \frac{\beta^*}{\lambda} N, \qquad (7.4\text{-}27)$$

and

$$\frac{\partial C}{\partial t} = 2C_{12}\dot{\gamma} + \frac{1}{\lambda}(2 - \beta^* C), \qquad (7.4\text{-}28)$$

where β^* is given by

$$\beta^* = BR_0^2 = \frac{4\dot{\gamma} C_{12} R_0^2}{k_B T \Lambda_0} + 3 = 2\dot{\gamma}\lambda C_{12} + 3, \qquad (7.4\text{-}29)$$

and the time constant is defined by

$$\lambda = \frac{2R_0^2}{k_B T \Lambda_0}. \qquad (7.4\text{-}30)$$

The four unknowns in equations 7.4-26 to 7.4-29, C_{12}, N, C, and β^*, can be solved numerically by taking, for example, isotropic conditions for the initial values of the conformation tensor. The components of the stress tensor are then obtained from equation 7.4-21.

The governing equations of the model can be solved analytically for steady shear and elongational flows and for start-up and relaxation in elongational flows (Grmela and

Carreau, 1987; Grmela, 1985). For steady simple shear flow, including the contribution of the solvent, equation 7.4-21 and equations 7.4-26 to 7.4-30 lead to the following results:

$$\lambda\dot{\gamma} = \left[\frac{1 - \left(\frac{\eta - \eta_s}{\eta_0 - \eta_s}\right)}{\frac{2}{27}\left(\frac{\eta - \eta_s}{\eta_0 - \eta_s}\right)^3}\right]^{\frac{1}{2}}, \qquad (7.4\text{-}31)$$

where η is the shear viscosity and η_0 is the zero shear viscosity obtained by taking $\beta^* = 3$ as $\dot{\gamma} \to 0$, that is,

$$\eta_0 - \eta_s = \lim_{\dot{\gamma} \to 0}\left(-\frac{\sigma_{12}}{\dot{\gamma}}\right) = \frac{1}{3}nk_BT\lambda \qquad (7.4\text{-}32)$$
$$= \frac{1}{3}\frac{cRT\lambda}{M},$$

where c is the polymer mass concentration. The primary and secondary normal stress coefficients are respectively given by

$$\psi_1 = -\frac{(\sigma_{11} - \sigma_{22})}{\dot{\gamma}^2} \qquad (7.4\text{-}33)$$
$$= \frac{2(\eta - \eta_s)^2}{nk_BT} = \frac{2}{3}\left(\frac{\eta - \eta_s}{\eta_0 - \eta_s}\right)^2 (\eta_0 - \eta_s)\lambda$$

and

$$\psi_2 = 0. \qquad (7.4\text{-}34)$$

These results have been obtained by Grmela and Carreau (1987) in a more general framework.

For high shear rates, the model predicts unique power-law curves, that is,

$$\eta \propto (\lambda\dot{\gamma})^{-\frac{2}{3}}$$
$$\text{and} \quad \psi_1 \propto (\lambda\dot{\gamma})^{-\frac{4}{3}}. \qquad (7.4\text{-}35)$$

Note that this result is significantly different from that obtained by Bird et al. (1987) for rigid dumbbells in the context of the phase-space kinetic theory, which does not admit a closed form solution. The power-law exponents for the viscosity and the primary normal stress coefficient were found to be $-1/3$ and $-2/3$, respectively, that is, half of the values obtained for the rigid-in-average model. It is not clear why such differences in the high shear rate behavior are observed. Nevertheless, both models yield single master curves for the steady shear viscosity and primary normal stress coefficient. With one adjustable parameter, λ, these models are obviously not flexible enough.

Following the procedure established in Example 7.4-1, the effect of chain flexibility can be examined. The zero shear viscosity is not affected by the chain flexibility and is expressed by equation 7.4-32. On the other hand, the zero shear primary normal stress coefficient is now given by

$$\psi_{10} = \frac{2nk_BT\lambda^2}{9(S_B + 6S_L)} = \frac{2(\eta_0 - \eta_S)\lambda}{3(S_B + 6S_L)} \qquad (7.4\text{-}36)$$

As expected, contributions to the Brownian motion affects the material elasticity as well as the shear thinning properties. Figure 7.4-6 shows steady shear viscosity master curves for different values of the parameter S_L.

It is interesting to note that with increasing chain flexibility (increasing value of S_L), the onset of shear thinning appears at higher reduced shear rate, $\lambda\dot{\gamma}$. Also, with increasing flexibility, the fluid becomes less shear thinning: the slope in the power-law region increases from $-2/3$ for rigid chains to approximately $-1/2$ for semi-flexible chains. Similar master curves for ψ_1 have been obtained (see Carreau et al., 1990).

Figure 7.4-7 illustrates the conformation tensor for simple shear flow, using the principal values to generate ellipses. In Fig. 7.4-7a, the conformation in the case of rigid rods ($S_L = 0$) is shown to be more oriented in the flow direction as the dimensionless shear rate is increased. Figure 7.4-7b shows the influence of flexibility on the orientation of the conformation tensor. As expected, the more flexible chains are less oriented by the flow at the same dimensionless flow rate.

The transient behavior in shear flow, and the elongational properties predicted by the model are of considerable interest. Figure 7.4-8 compares the model predictions of the shear stress growth and relaxation functions for the rigid and semi-flexible chains. In Fig. 7.4-8a we show the effect of the dimensionless shear rate on the stress growth and stress relaxation functions for rigid rods. With increasing shear rate, the model predicts overshoots in the growth function (η^+) that increase with increasing shear rate, and the maximum of the overshoot occurs at shorter time. This is in agreement with experiments for typical viscoelastic fluids. The relaxation curve, however, is seen to be a unique function of

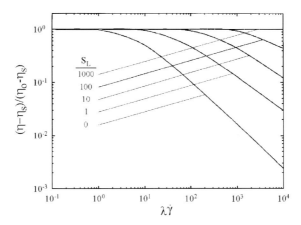

Figure 7.4-6 Effect of chain flexibility on the reduced shear viscosity: $S_B = 1.0$, $\Lambda = \Lambda_0$ (Adapted from Carreau et al., 1990)

7.4 Conformation Tensor Rheological Models

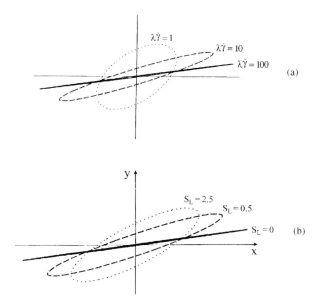

Figure 7.4-7 Conformation tensor in simple shear flow ($V_x = \dot{\gamma} y$)
(a) rigid rods $S_B = 1$ and $S_L = 0$
(b) effect of flexibility
(Adapted from Carreau et al., 1990)

the dimensionless time t/λ, which is at variance with the experimental observations showing faster relaxation at higher shear rates.

Figure 7.4-8b shows that the chain flexibility affects the transient behavior. With increasing values of S_L, the behavior becomes more Newtonian, that is, the shear stress grows more rapidly to the steady-state value but depicts no overshoot; the stress relaxes more rapidly after cessation of steady shear flow. This is in line with the predictions of the onset of shear thinning at higher shear rates (see Fig. 7.4-6).

As a last example, we show in Fig. 7.4-9 the effect of chain flexibility on the elongational viscosity. The elongational viscosity increases with the dimensionless elongational rate, as it was predicted for coil type chains (Section 7.4-2); however, the high strain rate plateau is equal to about twice the low deformation rate elongational viscosity, irrespective of the extent of the chain flexibility. Moreover, we observe that strain hardening occurs at a much smaller dimensionless strain rate for rigid rods ($S_L = 0$). For $S_L = 50$, the onset is observed at a value of $\lambda \dot{\varepsilon}$ almost 1000 times larger than that for rigid rods.

The predictions of the conformation model compared to the steady shear data of Baird and Ballman (1979) are shown in Fig. 7.4-10. Two types of polymers were investigated. The first one was a polyester (PPT) of molecular weight equal to 27 600 and 32 000 kg/kmol in sulphuric acid. The chains of this polymer system are believed to be rigid. The second system consisted of Nylon 6,6 ($M_w = 25\,100$, 35 200, and

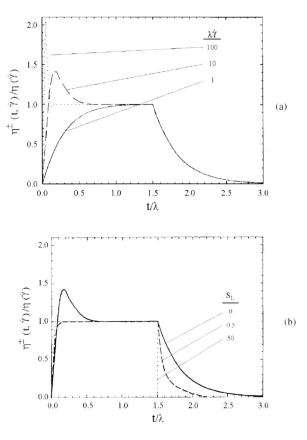

Figure 7.4-8 Stress growth and stress relaxation functions versus dimensionless time. The parameter S_B is taken to be 1.0 and the solvent viscosity, η_s, is set equal to zero.
(a) rigid rod case
(b) effect of chain flexibility
(Adapted from Carreau et al., 1990)

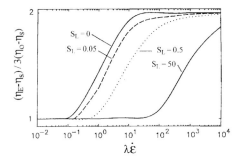

Figure 7.4-9 Effect of chain flexibility on the steady elongational viscosity as a function of the dimensionless elongational rate; $S_B = 1.0$ (Adapted from Carreau et al., 1990)

7.4 Conformation Tensor Rheological Models 319

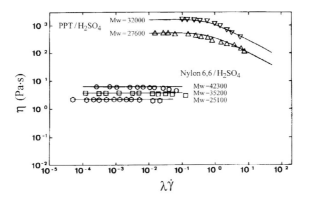

Figure 7.4-10   Viscosity of PPT and Nylon 6,6 solutions in H_2SO_4
Data from Baird and Ballman (1979) (Adapted from Carreau et al., 1990)
——— model predictions

42 300 kg/kmol) also in sulphuric acid. The polymeric chains of these Nylon solutions are known to be somewhat flexible. In all cases, the polymer concentration was kept at 0.117 g/mL, and the rheological measurements were made at 22 °C. The corresponding comparison for the primary normal stress coefficient is shown in Fig. 7.4-11. The parameters are reported in Table 7.4-1.

It is interesting to note that the viscosity, the primary normal stress coefficient, and hence the time constant, λ, for the rigid PPT polymeric chains are much higher than for the semi-flexible Nylon chains. Since the same concentration and the same range of molecular

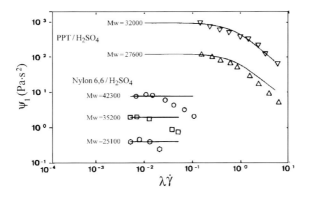

Figure 7.4-11   Primary normal stress coefficient of PPT and Nylon 6,6 solutions in H_2SO_4
Data from Baird and Ballman (1979) (Adapted from Carreau et al., 1990)
——— model predictions

Table 7.4-1 Parameters for Two Sets of Polymer Solutions in H_2SO_4 at 0.117 g/mL and 22 °C, $S_B = 0.2$, $S_L = 0$, $\eta_s = 0$ (Data from Baird and Ballman, 1979)

Polymer	M_w kg/kmol	η_0 Pa·s	λ ms
PPT	27 600	520	149
PPT	32 000	1725	573
Nylon 6,6	25 100	2.00	0.52
Nylon 6,6	35 200	3.40	1.24
Nylon 6,6	42 300	6.00	2.64

weight were used, it is obvious that chain rigidity is responsible for the high values. In the model, this is accounted for in the definition of the time constant (equation 7.4-30). It is reasonable to expect large end-to-end distance, R_0, and lower mobility, Λ_0, for the more rigid chains. It is also probable that particle-particle interactions and chain entanglements play an important role and reduce the mobility. Note from Table 7.4-1 that λ is strongly dependent on the molecular weight.

Overall, the model predictions for the PPT solutions are good. The parameter S_L was set equal to zero and the parameter S_B was found to be equal to 0.2. It is reasonable to expect a reduction of the effect of the Brownian motion since, as discussed earlier, particle-particle interactions are highly dominant. The onset of shear thinning in the case of the Nylon 6,6 solutions is observed at values of the reduced shear rate, $\lambda\dot\gamma$, less than 0.1. This is much lower than predicted by the model. Hence, no value for the parameter S_L could be obtained, and the low shear rate data were fitted with $S_B = 0.2$ and $S_L = 0$.

There are several ways to improve the model predictions. Two examples are discussed in Section 7.4-4.

7.4-4 Generalization of the Conformation Tensor Model

We generalize first the simple case of the rod-like macromolecule discussed in Example 7.4-1, to account for the molecular weight distribution of the polymer. The extra stress tensor given by equation 7.4-6 is now expressed by

$$\underline{\underline{\sigma}} = \sum_{i=1}^{N} \underline{\underline{\sigma}}_i - \eta_s \dot{\gamma}, \qquad (7.4\text{-}37)$$

where $\underline{\underline{\sigma}}_i(\underline{\underline{c}}_i)$ is the contribution of the polymeric chains of molecular weight M_i to the extra stress tensor. We consider here N different molecular weights (or relaxation modes), and

7.4 Conformation Tensor Rheological Models

for the sake of simplicity, we restrict the results to steady shear flow. The results of Example 7.4-1 can be readily extended. For each mode, we have

$$\frac{2}{27}\left(\frac{\eta_i - \eta_s}{\eta_{0i} - \eta_s}\right)^3 (\lambda_i \dot{\gamma})^2 + \left(\frac{\eta_i - \eta_s}{\eta_{0i} - \eta_s}\right) - 1 = 0, \tag{7.4-38a}$$

with

$$\eta_{0i} - \eta_s = \frac{1}{3} n_i k_B T \lambda_i, \qquad i = 1, 2 \ldots, N, \tag{7.4-38b}$$

where n_i is the chain density of molecular weight M_i, and the characteristic times, λ_i, are defined by

$$\lambda_i = \frac{2R_{0i}^2}{k_B T \Lambda_{0i}}. \tag{7.4-38c}$$

For rod-like macromolecules, R_0 should be proportional to the chain molecular weight M_i, and the mobility Λ_{0i} is expected to be inversely proportional to M_i; hence, as a first approximation, the λ_i are expected to be proportional to M_i^3.

The steady shear viscosity and primary normal stress coefficient are given by

$$\eta - \eta_s = \sum_{i=1}^{N} (\eta_i - \eta_s) \tag{7.4-39}$$

and

$$\psi_1 = \frac{2}{k_B T} \sum_{i=1}^{N} \frac{(\eta_i - \eta_s)^2}{n_i}. \tag{7.4-40}$$

For affine deformation ($\xi = 0$), the secondary normal stress coefficient ψ_2 vanishes. The zero shear viscosity is obtained from 7.4-37 and 7.4-38b as

$$\eta_0 - \eta_s = \frac{1}{3} k_B T \left(\sum_{i=1}^{N} n_i \lambda_i\right) = \frac{1}{3} RT \left(\sum_{i=1}^{N} \frac{c_i \lambda_i}{M_i}\right), \tag{7.4-41}$$

where c_i is the mass concentration of polymeric chains of molecular weight M_i. In principle, these results can be used to predict steady shear properties of rod-like macromolecules, knowing the molecular weight distribution and the characteristic times from linear viscoelasticity. Result 7.4-41 can be used to define an overall weight average characteristic time, λ_w, related to the zero shear viscosity by

$$\eta_0 - \eta_s = \frac{RT}{3M_v} \lambda_w, \tag{7.4-42}$$

where M_v is the viscosity average molecular weight (usually close to the weight average molecular weight). Note that for small deformation, this multimode conformation model reduces to the generalized Maxwell model (equations 5.2-28). It could describe most viscoelastic systems.

Using only two characteristic times, Ortiz (1992) and Ortiz et al. (1994) were able to fit the shear viscosity and primary normal stress differences of a series of polyethylene oxide solutions. Figure 7.4-12 reports viscosity master curves calculated from equations 7.4-37

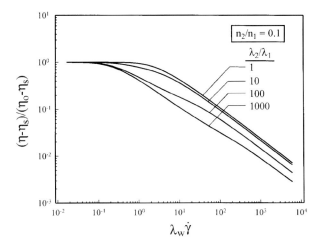

Figure 7.4-12 Viscosity master curves for rod-like macromolecules characterized by two relaxation times ($\xi = 0$ and constant mobilities) (Adapted from Ortiz, 1992)

and 7.4-38 in terms of λ_w defined by equation 7.4-42. Curve (1) represents the results of Example 7.4-1 (for a single characteristic time). For two characteristic times, the viscosity master curves are not sensitive to the ratio n_2/n_1 (density ratio); however, the shear-thinning properties change markedly with the characteristic time ratio (λ_2/λ_1). The high shear rate power-law slope is $-2/3$ in all cases.

The corresponding normal stress difference master curves are reported in Fig. 7.4-13. The function $\psi_1 \dot{\gamma}^2 \lambda_w/(\eta - \eta_S)$ is retained from previous work (Ait-Kadi et al., 1989).

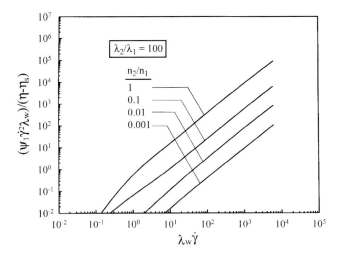

Figure 7.4-13 Normal stress master curves for rod-like macromolecules characterized by two relaxation times ($\xi = 0$ and constant mobilities) (Adapted from Ortiz et al., 1994)

Curve (1) represents the case of a single mode (result of Example 7.4-1). For two characteristic times, this function is not affected by the ratio of the characteristic times; however, the density ratio n_2/n_1 shows a very strong influence. Note that the high shear rate slope is not affected by the ratio.

These two figures illustrate the potential of multimode conformation models to describe the rheological behavior of polymeric systems. Figure 7.4-14 compares the model predictions with shear data obtained for polyethylene oxide solutions.

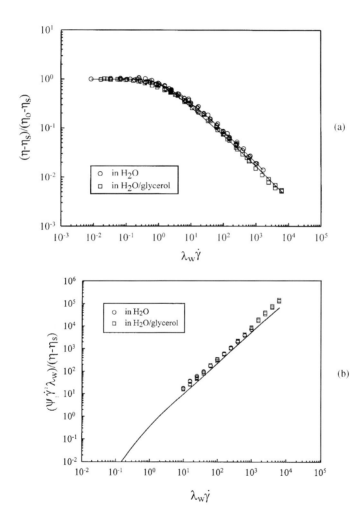

Figure 7.4-14 Comparison of predictions for the modified model (———) with the data for 3% PEO ($M_v = 1.8 \times 10^6$ kg/kmol) solutions (Adapted from Ortiz et al., 1994)
(a) Viscosity data, upper curve: $\lambda_2/\lambda_1 = 18.4$ and $n_2/n_1 = 0.66$; lower curve: $\lambda_2/\lambda_1 = 100$ and $n_2/n_1 = 0.65$
(b) Normal stress data

It is interesting to note that the reduced viscosity and normal stress data for these 3% PEO solutions in two different solvents are about equal, even though the zero shear viscosity and the overall relaxation time are considerably different: $\eta_0 = 16.5$ Pa·s and 117.5 Pa·s in water and water/glycerol respectively; and $\lambda_w = 1.2$ s and 7.2 s for water and water/glycerol respectively. Ortiz et al. (1994) showed that such dimensionless plots can lead to master curves which are almost independent of concentration, solvent, and molecular weight. Here, the solvent slightly affects the reduced viscosity curves. This is accounted for by changing the ratio of the relaxation times, λ_2/λ_1, from 18.4 to 100. This can be explained through the effect of the solvent on chain mobility. Obviously, these solutions are concentrated ones, and the mobility is controlled by entanglements. The fits obtained for the viscosity data (Fig. 7.4-14a) are excellent. Using the same parameters, the normal stress predictions (Fig. 7.4-14b) are not as good, but a better fit of the normal stress data could be obtained using a more sophisticated nonlinear regression technique. As mentioned earlier, the normal stress master curve is not sensitive to the λ_2/λ_1 ratio. Hence, the two sets of ratios lead to the same predictions.

In a second example, we present a generalization for semi-flexible macromolecules proposed by Ghosh et al. (1995). The result (equation 7.4-17) for the entropy is generalized to include topological interactions or excluded volume effects, due to Onsager (1949):

$$S(\underline{\underline{c}}) = k_B \left\{ S_B \frac{1}{2} \ln \det\left(\frac{\underline{\underline{c}}}{R_0^2}\right) - S_L \mathrm{tr}\left(\frac{\underline{\underline{c}}}{R_0^2}\right)^{-1} - S_0 \left[\left(\mathrm{tr}\frac{\underline{\underline{c}}}{R_0^2}\right)^2 - \mathrm{tr}\frac{(\underline{\underline{c}} \cdot \underline{\underline{c}})}{R_0^4} \right] \right\}. \tag{7.4-43}$$

S_0 is a parameter proportional to the excluded volume of the macromolecules. For dilute polymer solutions S_0 can be taken as the polymer concentration. The other parameters have the same meaning as in the original model of Section 7.4-3.

Following Giesekus (1985), Ghosh et al. (1995) introduced a form of the mobility tensor that is linearly dependent on the conformation tensor. Also introduced is the thinking that the more oriented, or less concentrated, the macromolecules are, the easier it is for them to move along the direction of their axis (Doi, 1981), and hence the larger is the mobility. Mathematically, this is represented as

$$\Lambda_{ij} = \Lambda_0 \frac{1}{\left[\left(\mathrm{tr}\frac{\underline{\underline{c}}}{R_0^2}\right)^2 - a\,\mathrm{tr}\frac{(\underline{\underline{c}} \cdot \underline{\underline{c}})}{R_0^4} \right]^2} \left[\delta_{ij} + l_G\left(\frac{c_{ij}}{R_0^2} - \delta_{ij}\right) \right], \tag{7.4-44}$$

where Λ_0 is a scalar constant and l_G is the *Giesekus parameter* ($0 \leq l_G < 1$). The factor in the denominator depends upon the orientation of the macromolecules; the more the macromolecules are oriented, the smaller is its magnitude and the higher is the mobility. The parameter a ($0 \leq a < 1$) serves as an empirical correction for this orientation effect. The Giesekus parameter, l_G, makes the mobility tensor anisotropic and accounts for the basic structure associated with the deformation of the macromolecules (Giesekus, 1985). Ghosh (1993) also considered the FENE potential, but we summarize here only the results for the model with the spatial constraint *inextensibility in average*, which is included in the model in Section 7.4-3. The model parameters are presented in Table 7.4-2. Details for the calculation of the rheological functions can be found in Ghosh et al. (1995).

7.4 Conformation Tensor Rheological Models

Table 7.4-2 Conformation Model Parameters (inextensible in Average Case)

Parameter	Physical significance	Acceptable range
Orientation parameter, a	Accounts for dependence of mobility on ordering	$0 \leq a \leq 1$
Giesekus parameter, l_G	Accounts for anisotropy in mobility	$0 \leq l_G < 1$
Boltzmann parameter, S_B	Weighs contribution to entropy due to Brownian motion of macromolecules	$0 \leq S_B \leq 1$
Semi-flexibility parameter, S_L	Weighs contribution to entropy due to persistent semi-flexibility of macromolecular chains	$S_L \geq 0$
Onsager parameter, S_0	Weighs contribution to entropy due to excluded volume effect of the macromolecular chains	$S_0 \geq 0$
Slip parameter*, ξ	Accounts for non-affine deformation	$0 \leq \xi \leq 1$

* Through Sections 7.4-3 and 7.4-4, only affine deformation is considered, i.e., $\xi = 1$

The effect of chain flexibility on the reduced shear viscosity has been presented in Fig. 7.4-6. Here, in Fig. 7.4-15, we present effects of the other three parameters on the reduced shear viscosity. The time constant, λ, used in the figure, is defined as

$$\lambda = \frac{2\left(1 - \frac{a}{3}\right)^2 R_0^2}{\left(1 - 2\frac{l_G}{3}\right)\Lambda_0 k_B T}. \quad (7.4\text{-}45)$$

The time constant is related to the zero shear viscosity by equation 7.4-32. For $a = l_G = 0$, equation 7.4-45 reduces to the expression for the time constant of rod-like macromolecules (equation 7.4-30). The effect of the Giesekus parameter, l_G, is shown in Fig. 7.4-15a to have the opposite effect of the semi-flexibility parameter S_L. Shear thinning is enhanced as l_G increases from 0 to 1. For $l_G = 1$, the slope of the reduced viscosity at high shear rate is equal to -2, which is inadmissible. However, for $l_G < 1$, the predictions are acceptable, with slopes between -0.67 and -1.0.

Figure 7.4-15b illustrates that the slope of the viscosity function is slightly affected by the orientation parameter, a. With increasing a, shear thinning is enhanced, with the slope in the power-law region decreasing from -0.67 to -0.86. This is expected since the mobility of the chains increases with increasing a. The most interesting effects are those of the Onsager parameter, as illustrated in part (c) of Fig. 7.4-15. This parameter, related to topological effects, may introduce yield stress behavior, observed for suspensions of colloidal particles or fibers (Ghosh, 1993). It may be used to describe effects of crowding or entanglements in polymer solutions or melts. Note that for semi-flexible chains ($S_L = 1$), topological effects lead to the appearance of a Newtonian plateau in the intermediate shear rate range, as observed with liquid crystalline polymers.

These examples illustrate how different physical ideas can be incorporated in conformation tensor models. Also, Ghosh (1993) has shown that by considering appropriate forms for the entropic, intramolecular energy and mobility terms, we can recover various rheological models proposed in the literature, in particular those of Leonov (1976) and Giesekus (1985).

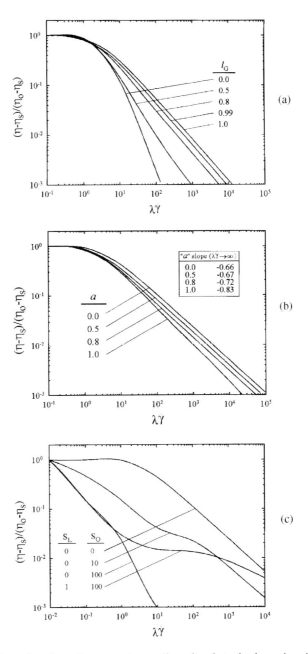

Figure 7.4-15 Effects of conformation parameters on the reduced steady shear viscosity
(a) effect of the Giesekus parameter ($S_L = S_O = 0$, $a = S_B = 1$)
(b) effect of the orientation parameter ($S_L = l_G = 0$, $S_B = 1$)
(c) effect of the Onsager parameter ($a = l_G = 0$, $S_B = 1$)
For $S_0 > 0$, the model does not predict a zero shear viscosity; the reduced viscosity was arbitrarily set equal to 1 at $\lambda\dot{\gamma} = 10^{-2}$ (Adapted from Ghosh, 1993)

Obviously, as discussed above, the linear viscoelastic behavior can be adequately described only if multimode conformation models are used. The generalization to multi-modes for rod-like molecules discussed here still needs to be fully assessed and extended. Ajji et al. (1989) have shown that the conformation tensor approach can be used with success to model entangled polymers and describe shear-refining effects. Other interesting aspects such as those involving irreversibility remain to be investigated.

7.5   Problems*

7.5-1$_b$ Hookean Dumbbell Model

Consider the Hookean dumbbell discussed in Section 7.1-1.

(a) Verify equation 7.1-19 by deriving all the intermediate results.
(b) Obtain result 7.1-35a from equation 7.1-32.
(c) Verify that for Hookean dumbbells, the Giesekus equation leads to the upper-convected Maxwell equation.
(d) Obtain the expression of the time constant λ_0 in terms of polymer concentration and molecular weight, the universal gas constant, and the relative viscosity $\eta_0 - \eta_s$.

7.5-2   Tanner Equation

Equation 7.1-51 is known as the Tanner equation. Referring to the original article of Tanner (1975), discuss the major differences in hypotheses used by Tanner compared to those on which the results of Section 7.1-2 are based.

7.5-3$_b$ Complex Viscosity of Rouse Fluid

Consider the Rouse model given by equation 7.1-92.

(a) Obtain the expressions in linear viscoelasticity for η^*, η', G', and G''.
(b) Illustrate typical results in terms of $\lambda_R \omega$ with $\eta_s/\eta_0 = 0.01$, 0.1, and 0.5. Discuss the applicability of the Rouse model to fit real linear viscoelastic data.

7.5-4$_b$ Network Model

(a) Show that the constitutive equation 7.2-27 obtained for elastic liquids using the network approach is equivalent to the Lodge equation 6.4-1. Relate the network parameters to those of the memory function 6.4-3.

---

* Problems are labelled by a subscript a or be indicating the level of difficulty.

(b) Show that the constitutive equation 7.2-27 can be generalized to obtain a BKZ type equation. Specify the functions $u(I_{c^{-1}}, II_{c^{-1}})$ that could be derived from the network theory.

7.5-5$_b$ Conformation Model

(a) Verify that equation 7.4-5 satisfies the evolution equation 7.4-2. (Note that $\delta\underline{\underline{\delta}}/\delta t = -\underline{\underline{\dot\gamma}}$, as shown by equation 6.1-11 for the upper-convected derivative.)
(b) Verify results 7.4-8a to 7.4-8c.

7.5-6$_b$ FENE Conformation Model

(a) Show that the FENE conformation model of Section 7.4-2 reduces to the Tanner equation 7.1-51 if we set the parameters e, β, and ξ equal to zero.
(b) Obtain the results for the steady shear viscosity, primary normal stress coefficients, and the steady elongational viscosity, in terms of $\lambda_e \dot\gamma$ or $\lambda_e \dot\epsilon$.
(c) Calculate the reduced shear viscosity function and plot the results for a few values of the extensibility parameter, b.

7.5-7$_b$ Rod-Like Macromolecules

Consider the conformation tensor model developed in Section 7.4-3 for rod-like macromolecules.

(a) Solve Example 7.4-1 in detail and verify results 7.4-31 and 7.4-34.
(b) Show that the viscosity obeys a power-law expression with exponent equal to $-2/3$.
(c) Obtain the corresponding result for the steady uniaxial elongational viscosity. Illustrate the behavior predicted by the model.

8 Multiphase Systems

8.1 Systems of Industrial Interest 330
8.2 Rheology of Suspensions 331
    8.2-1 Viscosity of Dilute Suspensions of Rigid Spheres 332
    8.2-2 Rheology of Emulsions 334
        8.2-2.1 Oldroyd's Emulsion Model 336
        8.2-2.2 The Palierne Model 338
    8.2-3 Rheology of Concentrated Suspensions of Non-Interactive Particles 343
        8.2-3.1 Elasticity of Suspensions of Spheres 346
    8.2-4 Rheology of Concentrated Suspensions of Interactive Particles 347
    8.2-5 Concluding Remarks 351
8.3 Flow About a Rigid Particle 352
    8.3-1 Flow of a Power-Law Fluid Past a Sphere 352
    8.3-2 Other Fluid Models 356
    8.3-3 Viscoplastic Fluids 356
    8.3-4 Viscoelastic Fluids 357
    8.3-5 Wall Effects 357
    8.3-6 Non-Spherical Particles 359
    8.3-7 Drag-Reducing Fluids 360
    8.3-8 Behavior in Confined Flows 361
8.4 Flow Around Fluid Spheres 362
    8.4-1 Creeping Flow of a Power-Law Fluid Past a Gas Bubble 362
    8.4-2 Experimental Results on Single Bubbles 363
8.5 Creeping Flow of a Power-Law Fluid Around a Newtonian Droplet 366
    8.5-1 Experimental Results on Falling Drops 367
8.6 Flow in Packed Beds 368
    8.6-1 Creeping Power-Law Flow in Beds of Spherical Particles: The Capillary Model . . . 368
    8.6-2 Other Fluid Models 373
    8.6-3 Viscoelastic Effects 373
    8.6-4 Wall Effects 374
    8.6-5 Effects of Particle Shape 375
    8.6-6 Submerged Objects' Approach to Fluid Flow in Packed Beds: Creeping Flow 376
8.7 Fluidized Beds 377
    8.7-1 Minimum Fluidization Velocity 378
    8.7-2 Bed Expansion Behavior 380
    8.7-3 Heat and Mass Transfer in Packed and Fluidized Beds 382
8.8 Problems 383
    8.8-1_b Einstein's Result 383
    8.8-2_b Oldroyd's Emulsion Model 383
    8.8-3_b Palierne's Emulsion Model 383
    8.8-4_b Flow About a Sphere 384
    8.8-5_b Friction Factor for a Packed Bed 384
    8.8-6_a Criterion for Flow in a Viscoplastic Fluid 385

The term "multiphase systems" is often used to describe a wide variety of phenomena involving more than one phase in the engineering literature. Broadly speaking, we can classify multiphase systems into several sub-classes. The phenomenon of suspension (emulsion, foam) viscosity or resistance to deformation is encountered when the dispersed and continuous phases (solid/droplets/bubbles) move and deform relative to each other. In spite of the tendency for separation, formation, and disintegration of internal structures within the multiphase mixture, it is customary to assign average physical properties (viscosity, thermal conductivity, and so on) to such systems, thereby tacitly assuming it to be a pseudo-homogeneous substance. On the other hand, particles may move together in bulk through a fluid, as in hindered settling or sedimentation of concentrated suspensions. In turn, the particles may be more or less stationary as in a packed bed. The relative particle-fluid motions may be more complex, as in liquid-solid fluidized beds. Clearly, it is beyond the scope of this chapter to cover the entire spectrum of multiphase systems. Consideration is primarily given to the rheological behavior of suspensions and filled polymeric systems. The behavior of single particles (rigid, drops, bubbles), and the phenomena of momentum, mass, and heat transfer in packed and fluidized beds with rheologically complex media are then discussed.

8.1 Systems of Industrial Interest

In many industrial operations, the fluids or materials handled are non-homogeneous, because they consist of more than one phase. Typical examples include foodstuffs, pharmaceutical products, paints, and greases, which are either emulsions of liquid droplets or suspensions of solid particles. Slurries such as mineral and paper pulps, fermentation broths, and biological fluids (such as blood) are other examples of suspensions that exhibit a complex rheology. Most of these systems are strongly shear-thinning and frequently show time-dependent effects. Many other examples of suspensions or multiphase systems can be found in the plastics industry. Glass fiber reinforced thermoplastics are currently used in the fabrication of a variety of automobile parts, and filled polymeric materials find their uses in many household as well as industrial articles.

Polymer blends attract much interest for technical as well as commercial reasons. For example, about 60% of all the linear low density polyethylene (LLDPE) produced goes into blends with other polymers. Polymers used in commercial blends are often immiscible. The properties of multiphase materials are strongly influenced by the morphology, which, in turn, depends on the interactions or chemical affinity of the polymeric components, their rheological behavior, and the processing conditions. Considerable research is going on worldwide to explore the properties of various blends of commodity thermoplastics, and to relate material structure, processing, morphology, and final properties.

At the other end of the spectrum of multiphase systems we have the flow of single particles in fluids: the motion of rigid (solid) and deformable (bubbles and drops) particles through rheologically complex media occurs in a variety of processes carried out in the chemical, polymer, food, and allied industries. The motion of oil droplets through polymer solutions in rocks is at the core of the enhanced oil recovery process, via polymer flooding. Degassing of polymer melts involves the movement of gas bubbles through a non-Newtonian viscous mass. The filtration of polymer melts using sand pack filters prior to processing, is yet another example of a complex interplay between particulates and non-Newtonian characteristics. Fixed and fluidized bed reactors used to carry out catalytic polymerization reactions as well as for leaching of some metals are other important examples involving relative motion between particles and a non-Newtonian liquid phase. Other examples involving such non-Newtonian fluid/particle interactions include the hydraulic transport of particulate matter (such as coal) using non-Newtonian carrier fluids, gravity separation, three-phase fluidized beds, and slurry reactors. There is no dearth of industrial applications involving non-Newtonian fluid/particle interactions.

Although in most real-life applications, we encounter ensembles or clusters of particles rather than single particles, experience has shown that the behavior of isolated spheres, bubbles, and drops serves as a precursor to the understanding of the more realistic multi-particle situations, as the latter are not amenable to theoretical modeling with any reasonable level of rigor. This chapter sets out to provide an overview of basic results on the rheology of suspensions, and of the developments in the field of particle motion in non-Newtonian fluids. In particular, consideration is given to macroscopic momentum, heat, and mass transfer between a particle (or a bed of particles) in relative motion with a non-Newtonian medium. More detailed accounts, as well as extensive bibliographies on these topics, are available in books and reviews (Schowalter, 1978; Metzner, 1985; Kamal and Mutel, 1986; Utracki, 1988; Shenoy, 1988; Adler et al., 1990; Chhabra, 1993a, 1993b, 1993c; De Kee et al., 1996; Ghosh et al., 1994).

8.2 Rheology of Suspensions

The rheology of suspensions is a very difficult topic, and rigorous analytical solutions have been obtained only for the special case of very dilute suspensions of rigid spheres (Einstein, 1906, 1911), of slightly deformable spheres (Taylor, 1932), and of ellipsoids (Jeffrey, 1922). Those results that are restricted to Newtonian suspending media are of low technological interest, since the suspensions are far too dilute to be representative of industrial systems. Nevertheless, these theories have been empirically or semi-empirically extended to concentrated systems with some degree of success. The rheology of concentrated suspensions in polymers and non-Newtonian fluids in general is a very complex subject of which the principles are not yet well defined. Appropriate constitutive relations for even the most simple model systems (e.g., suspensions of glass beads in a polymer matrix) still need to be developed, and many published rheological results are debatable.

The flow behavior of polymer composites and blends is hardly predictable from first principles, due to unknown factors associated with the interface, variations with time under processing, orientation of solid particles, agglomeration, breakage or coalescence, and so on, under flow conditions. Strictly speaking, the rheology of multiphase systems is nonsense! We should rather try to solve problems using equations of fluid mechanics for two-phase flow. For many systems, including viscoelastic polymeric materials, it is more practical to define rheological properties for the mixture, and eventually account for changes due to processing.

In this section, we present the fundamental results of Einstein and Taylor. Some of the important extensions to non-Newtonian fluids and results of practical interest are then discussed. Obviously, this section is far from being exhaustive. The interested reader should refer to the book of Schowalter (1978), to the reviews of Metzner (1985), Kamal and Mutel (1986), and Shenoy (1988) on the rheology of filled systems, and to the comprehensive review on the rheology of two-phase systems by Utracki (1988) for detailed treatments.

8.2-1 Viscosity of Dilute Suspensions of Rigid Spheres

As a first example, we summarize the analysis which leads to the viscosity expression initially obtained by Einstein (1906, 1911) for a dilute suspension of rigid spheres in a Newtonian fluid. We follow the approach of Landau and Lifshitz (1959), and consider a sphere of radius R, placed in a symmetrical flow of an incompressible fluid of viscosity μ_m. The origin of the coordinates is at the center of the sphere, as illustrated in Fig. 8.2-1 for a simple shear situation.

Far away from the sphere, or in the absence of a sphere, the velocity field is undisturbed and the velocity is given by

$$V_{0_i} = \frac{\dot{\gamma}_{0_{ik}}}{2} x_k, \tag{8.2-1}$$

where $\dot{\gamma}_{0_{ik}}$ are the components of the rate-of-deformation tensor.

We look for a solution of the form

$$\underline{V} = \underline{V}_0 + \underline{V}_1, \tag{8.2-2}$$

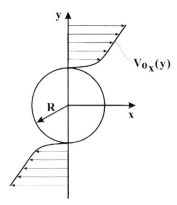

Figure 8.2-1 Flow about a sphere in simple shear

8.2 Rheology of Suspensions

where $\underline{V}_1$ is the perturbation due to the presence of the sphere. Obviously, $\underline{V}_1$ must vanish at infinity, but will be equal to $-\underline{V}_0$ at the sphere surface (the velocity has to be equal to zero at the sphere surface due to symmetry and no-slip condition). For steady-state conditions and negligible inertial and gravitational effects, the Navier-Stokes equation reduces to

$$0 = -\nabla P + \mu_m \nabla^2 \underline{V}. \tag{8.2-3}$$

The curl of this equation yields

$$0 = \nabla^2 \nabla \times \underline{V}, \tag{8.2-4}$$

since the curl of a gradient of a scalar vanishes (that is, $\nabla \times \nabla P = 0$). The solution has to satisfy equation 8.2-4 and the continuity equation ($\nabla \cdot \underline{V} = 0$). The solution for $\underline{V}_0$ is already a solution of the Navier-Stokes equations. $\underline{V}_1$ has to be proportional to the rate-of-strain tensor, and the solution that satisfies the boundary conditions at infinity and at the sphere surface is (see Landau and Lifshitz, 1959)

$$\underline{V}_1 = \nabla \times \nabla \times [\dot{\underline{\gamma}}_{=0} \cdot \nabla f(r)], \tag{8.2-5}$$

where $f(r)$ is expressed by

$$f(r) = ar + \frac{b}{r}. \tag{8.2-6}$$

The solution for the velocity profile is

$$V_{1_i} = \frac{5}{4}\left(\frac{R^5}{r^4} - \frac{R^3}{r^2}\right)\dot{\gamma}_{0_{kl}} n_i n_k n_l - \frac{R^5}{2r^4}\dot{\gamma}_{0_{ik}} n_k, \tag{8.2-7}$$

where n_i is the i-component of the unit vector normal to the sphere surface. The corresponding pressure profile is given by

$$P = -\frac{5}{2}\mu_m \frac{R^5}{r^3} \dot{\gamma}_{0_{ik}} n_i n_k. \tag{8.2-8}$$

The viscosity of the suspension is determined by averaging the momentum flux tensor per unit volume ($\rho \underline{V}\underline{V} + \underline{\underline{\pi}}$), which in the absence of inertial terms is the total stress tensor

$$\bar{\pi}_{ik} = \frac{1}{V}\int (\sigma_{ik} + P\delta_{ik})dV, \tag{8.2-9}$$

where the integration is carried over a large volume V compared to the solid sphere dimension. We write the following identity:

$$\bar{\pi}_{ik} = -\mu_m \bar{\dot{\gamma}}_{ik} + \bar{P}\delta_{ik} + \frac{1}{V}\int \{\sigma_{ik} + \mu_m \dot{\gamma}_{ik}\}dV, \tag{8.2-10}$$

where the overbar refers to the average over the volume integral. For a Newtonian fluid, the integral on the right side of equation 8.2-10 is zero except within the solid sphere. Since the suspension is assumed to be dilute, the calculation for a single sphere will be carried out, and then the result will be multiplied by c, the concentration of the suspension. To avoid calculating the stresses inside a solid sphere, we make use of the Gauss divergence theorem (see Appendix A-5 of Bird, Stewart, and Lightfoot, 1960). We note first that the equation of motion (neglecting inertial and gravitational effects) is

$$0 = -\frac{\partial P}{\partial x_i} - \frac{\partial \sigma_{ij}}{\partial x_i} = -\frac{\partial \pi_{ij}}{\partial x_i}, \qquad (8.2\text{-}11)$$

which can be written as

$$\pi_{ij} = \frac{\partial}{\partial x_l}(\pi_{il} x_j). \qquad (8.2\text{-}12)$$

Because of symmetry, the net pressure contribution is zero, and equation 8.2-10 can be written as

$$\bar{\sigma}_{ik} = -\mu_m \bar{\dot{\gamma}}_{ik} + c \int_0^{2\pi} \int_0^{\pi} \{n_l \sigma_{il} x_k + \mu_m (V_i n_k + V_k n_i)\} R_v^2 \sin\theta d\theta d\phi, \qquad (8.2\text{-}13)$$

where R_v is the radius of a very large spherical domain, which lies entirely in the fluid. For large R_v, only the terms in $1/r^2$ have to be considered in the expression for the velocity profile:

$$V_{1_i} \approx -\frac{5}{4}\frac{R^3}{r^2} \dot{\gamma}_{0_{km}} n_i n_k n_m, \qquad (8.2\text{-}14)$$

where $\dot{\gamma}_{0_{km}}$ are the components of the undisturbed rate-of-deformation tensor. The integral can be evaluated using $\sigma_{il} \approx -\mu_m \dot{\gamma}_{0_{il}}$ to obtain

$$\bar{\sigma}_{ik} = -\mu_m \bar{\dot{\gamma}}_{ik} - \frac{5}{2}\mu_m \dot{\gamma}_{0_{ik}}\left(\frac{4}{3}\pi R^3 c\right). \qquad (8.2\text{-}15)$$

For dilute suspensions, the average flow field is not highly disturbed, that is, $\bar{\dot{\gamma}}_{ik} \approx \dot{\gamma}_{0_{ik}}$ and

$$\bar{\sigma}_{ik} = -\mu_s \bar{\dot{\gamma}}_{ik} = -\mu_m \left[1 + \frac{5}{2}\phi\right]\bar{\dot{\gamma}}_{ik}, \qquad (8.2\text{-}16)$$

where ϕ is the volume fraction occupied by the spheres, $(4\pi R^3 c/3)$, and μ_s is the equivalent viscosity for the suspension, which can be expressed in terms of the relative or reduced viscosity as follows:

$$\mu_r = \frac{\mu_s}{\mu_m} = 1 + \frac{5}{2}\phi. \qquad (8.2\text{-}17)$$

The dashed line in Fig. 8.2-2 represents the Einstein relation (equation 8.2-17). The figure reports the relative viscosity as a function of volume fraction of rigid spheres of narrow size distribution in Newtonian fluids. Clearly, the Einstein theoretical result is valid for very low concentration ($\phi < 0.01$) only.

8.2-2 Rheology of Emulsions

Einstein's result has been extended to slightly deformable particles by Taylor (1932). For a dilute emulsion of Newtonian droplets in another Newtonian fluid of viscosity μ_m, under creeping shear flow, the viscosity of the emulsion is given by

$$\mu_s = \mu_m\left(1 + \frac{5k+2}{2k+2}\phi\right), \qquad (8.2\text{-}18)$$

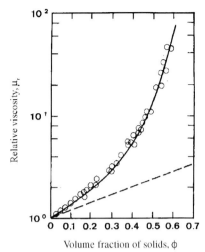

Figure 8.2-2 Relative viscosity vs concentration for suspensions of spheres of low interaction, narrow size distributions, and particle diameters in the range of 0.1 to 440 μm (Adapted from Thomas, 1965)
—— Equation 8.2-34
- - - Equation 8.2-17

where k is the ratio of the droplet and the matrix viscosity ($k = \mu_d/\mu_m$). For $k \to \infty$ (solid spheres) this result reduces to the Einstein expression. For $k \to 0$ (cases of gas bubbles), equation 8.2-18 reduces to

$$\mu_s = \mu_m(1 + \phi). \tag{8.2-19}$$

This result correctly predicts that the presence of tiny gas bubbles increases the viscosity of a liquid. Schowalter et al. (1968) included the effect of the interfacial tension to obtain the following expressions for the stress tensor in simple shear flow (Schowalter, 1978):

$$\begin{pmatrix} \sigma_{11} & \sigma_{12} & 0 \\ \sigma_{21} & \sigma_{22} & 0 \\ 0 & 0 & \sigma_{33} \end{pmatrix} = -\left[1 + \frac{5k+2}{2k+2}\phi\right]\mu_m \begin{pmatrix} 0 & \dot{\gamma} & 0 \\ \dot{\gamma} & 0 & 0 \\ 0 & 0 & 0 \end{pmatrix} - \frac{\mu_m^2 d_p \phi}{2\alpha}\left[\frac{19k+16}{20(k+1)^2}\right]$$

$$\times \left\{ \left[\frac{25k^2 + 41k + 4}{28(k+1)}\right] \begin{pmatrix} \dot{\gamma}^2 & 0 & 0 \\ 0 & \dot{\gamma}^2 & 0 \\ 0 & 0 & -2\dot{\gamma}^2 \end{pmatrix} \right.$$

$$\left. - \frac{(19k+16)}{4} \begin{pmatrix} -\dot{\gamma}^2 & 0 & 0 \\ 0 & \dot{\gamma}^2 & 0 \\ 0 & 0 & 0 \end{pmatrix} \right\}, \tag{8.2-20}$$

where d_p is the droplet diameter and α is the interfacial tension. This result is of considerable interest, because it predicts nonzero normal stress coefficients. The shear stress component leads to the result obtained by Taylor (1932). For low viscosity droplets (gas bubbles), $k \to 0$, and equation 8.2-20 reduces to

$$\psi_1 = -\frac{(\sigma_{11} - \sigma_{22})}{\dot{\gamma}^2} = \frac{32}{10} \frac{\mu_m^2 d_p \phi}{\alpha} \tag{8.2-21a}$$

and $\quad \psi_2 = -\frac{(\sigma_{22} - \sigma_{33})}{\dot{\gamma}^2} = -\frac{10}{7} \frac{\mu_m^2 d_p \phi}{\alpha}.$ \hfill (8.2-21b)

For highly viscous droplets, $k \to \infty$, and result 8.2-20 becomes

$$\psi_1 = \frac{361}{80} \frac{\mu_m^2 d_p \phi}{\alpha} \tag{8.2-21c}$$

and $\quad \psi_2 = -\frac{551}{560} \frac{\mu_m^2 d_p \phi}{\alpha}.$ \hfill (8.2-21d)

It is worthwhile to note here that elastic properties are predicted for emulsions of Newtonian droplets in a Newtonian matrix. The viscoelastic properties are functions of the ratio d_p/α, which characterizes the deformability of the droplets. Surprisingly, the results of equations 8.2-21 are very similar for the two extreme limits. Obviously, the assumption of slightly deformable spheres for low viscosity droplets at high shear rates is unrealistic.

8.2-2.1 Oldroyd's Emulsion Model

Following a different approach, Oldroyd (1953, 1955) included time-dependent effects to describe the rheology of dilute suspensions. Again the fluids are assumed to be Newtonian and the analysis is restricted to dilute suspensions and small droplet deformation. In the simple case of a thin interface and constant interfacial tension, the complex viscosity is given by

$$\eta^* = \mu_m \left[\frac{1 + 3\phi H^*(\omega)}{1 - 2\phi H^*(\omega)} \right], \tag{8.2-22}$$

where the operator $H^*(\omega)$ is defined by

$$H^*(\omega) = \frac{A + Bi\omega}{C + Di\omega} = \frac{AC + i\omega(BC - AD) + \omega^2 BD}{C^2 + D^2\omega^2}, \tag{8.2-23a}$$

with

$$A = \frac{8\alpha}{d_p}(2 + 5k), \tag{8.2-23b}$$

$$B = \mu_m(k - 1)(16 + 19k), \tag{8.2-23c}$$

$$C = \frac{80\alpha}{d_p}(1 + k), \tag{8.2-23d}$$

and $\quad D = \mu_m(3 + 2k)(16 + 19k).$ \hfill (8.2-23e)

As for the Schowalter result (equation 8.2-20), the Oldroyd emulsion model predicts viscoelastic behavior. This is illustrated is Example 8.2-1.

Example 8.2-1 Linear Viscoelastic Behavior Predicted by the Oldroyd Emulsion Model

Determine the linear functions η^*, G', and G'' for the Oldroyd emulsion model. Discuss the low frequency limits in terms of a simple Maxwell model.

Solution

Expanding equation 8.2-22 and retaining the linear terms in ϕ (the Oldroyd model is valid only for small ϕ), we obtain

$$\eta^* = \mu_m[1 + 5\phi H^*]$$
$$= \mu_m\left[1 + 5\phi\frac{AC + BD\omega^2}{C^2 + D^2\omega^2}\right] + i\omega 5\phi\mu_m\frac{BC - AD}{C^2 + D^2\omega^2}. \quad (8.2\text{-}24)$$

Hence,

$$\eta' = \frac{G''}{\omega} = \mu_m\left[1 + 5\phi\frac{AC + BD\omega^2}{C^2 + D^2\omega^2}\right] \quad (8.2\text{-}25a)$$

and

$$\eta'' = \frac{G'}{\omega} = 5\omega\mu_m\phi\frac{AD - BC}{C^2 + D^2\omega^2}. \quad (8.2\text{-}25b)$$

We note that at low frequencies the storage modulus, G', is quadratic in ω and that the loss modulus is linear with respect to ω. Also at low frequencies,

$$\lim_{\omega \to 0} \eta^* = \eta_0 = \mu_m\left[1 + 5\phi\frac{A}{C}\right] = \mu_m\left[1 + \frac{2 + 5k}{2(1 + k)}\phi\right], \quad (8.2\text{-}26)$$

which is identical to the result of Taylor (equation 8.2-18). To recover a characteristic elastic time, we compare the result of equation 8.2-25b to that of a simple Maxwell fluid (equation 5.2-15) at low frequencies. The terminal relaxation time of the emulsion is given by

$$\lambda_0 = \lim_{\omega \to 0}\frac{G'}{\eta'\omega^2} = \frac{\dfrac{\mu_m\phi}{160\left(\dfrac{\alpha}{d_p}\right)}\left[\dfrac{16 + 19k}{1 + k}\right]^2}{1 + \dfrac{2 + 5k}{2(1 + k)}\phi}. \quad (8.2\text{-}27)$$

That is, at low frequencies, the emulsion behaves as a typical viscoelastic fluid. Note that the elastic characteristic time is proportional to the concentration of droplets at low concentration and also to d_p/α (droplet diameter over the interfacial tension).

An assessment of the Oldroyd model with experimental data can be found in Graebling and Muller (1990).

8.2-2.2 The Palierne Model

The most interesting extension of the Oldroyd emulsion model is that proposed by Palierne (1990), who analyzed the linear viscoelastic behavior of non-dilute suspensions of viscoelastic droplets in a viscoelastic matrix. Dipole-type particle interactions are considered and effects of particle size distribution are included.

The Palierne model describes the complex modulus, G_s^*, of viscoelastic emulsions as a function of the complex moduli of both phases (G_m^* for the matrix and G_d^* for the droplets), and of the ratio of the interfacial tension and the droplet size and size distribution, as follows:

$$G_s^*(\omega) = G_m^*(\omega) \frac{1 + 3\sum_i \phi_i H_i^*(\omega)}{1 - 2\sum_i \phi_i H_i^*(\omega)}, \qquad (8.2\text{-}28)$$

with $H_i^*(\omega)$ given by

$$H_i^*(\omega) = \frac{8\left(\dfrac{\alpha}{d_i}\right)[2G_m^*(\omega) + 5G_d^*(\omega)] + [G_d^*(\omega) - G_m^*(\omega)][16G_m^*(\omega) + 19G_d^*(\omega)]}{80\left(\dfrac{\alpha}{d_i}\right)[G_m^*(\omega) + G_d^*(\omega)] + [2G_d^*(\omega) + 3G_m^*(\omega)][16G_m^*(\omega) + 19G_d^*(\omega)]}, \qquad (8.2\text{-}29)$$

where α is the interfacial tension and ϕ_i is the volume fraction of droplets with diameter d_i.

The general Palierne model formulation also contains parameters related to the deformability of the interface: $\beta'(\omega)$ due to the change of interfacial area and $\beta''(\omega)$ related to the local shear. Since these parameters are virtually impossible to determine experimentally, they are usually set equal to zero. It is interesting to note that the model in its simpler form described here contains no empirical parameters, and we can predict the linear viscoelastic properties of emulsions such as immiscible polymer blends from the knowledge of the complex moduli of both phases, the particle size and size distribution, the volume fraction of droplets, and the interfacial tension. All of the required data for using the model are experimentally accessible.

In the case of constant interfacial tension and uniform particle size, the Palierne model and the Oldroyd model expressions are somewhat similar, although the two approaches are quite different. Recall that the Oldroyd model is restricted to dilute emulsions of Newtonian fluids, whereas the Palierne model is formulated for a non-diluted suspension of viscoelastic droplets in a viscoelastic fluid. Since the volume average diameter, d_v, is known to take most of the particle size distribution effects into account (Bousmina and Muller, 1993), the model can be simplified, assuming a monodisperse size, using d_v as the characteristic dimension. Under this assumption, the storage and loss moduli of the blends can be expressed explicitly in terms of the moduli of both components (Bousmina et al., 1995):

$$G_s' = \frac{1}{D}[G_m'(B_1B_2 + B_3B_4) - G_m''(B_4B_1 - B_2B_3)] \qquad (8.2\text{-}30)$$

and

$$G_s'' = \frac{1}{D}[G_m'(B_1B_4 - B_2B_3) + G_m''(B_1B_2 + B_3B_4)], \qquad (8.2\text{-}31)$$

where the constants are expressed by

$$B_1 = C_1 - 2\phi C_3, \tag{8.2-32a}$$

$$B_2 = C_1 + 3\phi C_3, \tag{8.2-32b}$$

$$B_3 = C_2 - 2\phi C_4, \tag{8.2-32c}$$

$$B_4 = C_2 + 3\phi C_4, \tag{8.2-32d}$$

$$\text{and} \quad D = (C_2 - 2\phi C_4)^2 + (C_1 - 2\phi C_3)^2, \tag{8.2-32e}$$

with

$$C_1 = 80\frac{\alpha}{d_v}(G'_m + G'_d) + 38(G'^2_d - G''^2_d) + 48(G'^2_m - G''^2_m) + 89(G'_m G'_d - G''_m G''_d), \tag{8.2-32f}$$

$$C_2 = 80\frac{\alpha}{d_v}(G''_m + G''_d) + 96G'_m G''_m + 76G'_d G''_d + 89(G''_m G'_d + G'_m G''_d), \tag{8.2-32g}$$

$$C_3 = 8\frac{\alpha}{d_v}(2G'_m + 5G'_d) - 16(G'^2_m - G''^2_m) + 19(G'^2_d - G''^2_d) - 3(G'_m G'_d - G''_m G''_d), \tag{8.2-32h}$$

and $\quad C_4 = 8\frac{\alpha}{d_v}(2G''_m + 5G''_d) - 32G'_m G''_m + 38G'_d G''_d - 3(G''_m G'_d + G'_m G''_d). \tag{8.2-32i}$

Two limiting cases are worth mentioning: the first case with $\alpha = 0$ corresponds to a hypothetical two-phase miscible blend, and no increase of elasticity due to droplet deformation is expected; the second case with $\alpha = \infty$ represents a suspension of rigid (nondeformable) spheres. Surprisingly, both cases yield nearly the same result. If the complex modulus of the particles is much larger than that of the matrix, $H_i^* = H^*(\omega) = 0.5$ and the model simplifies to

$$G_s^*(\omega) = G_m^*(\omega)\left(\frac{1 + \frac{3}{2}\phi}{1 - \phi}\right). \tag{8.2-33}$$

The complex modulus of the emulsion is then independent of the drop size and interfacial tension. For a very dilute suspension in a Newtonian matrix, equation 8.2-33 reduces to Einstein's result. An approximate expression for the relaxation time of the deformed droplets can be used to explain the shift toward the low frequency region for the terminal zone of the emulsion with increasing concentration of the minor (dispersed) phase (Bousmina et al., 1995).

Figure 8.2-3 compares the linear viscoelastic data and the Palierne emulsion model predictions for two immiscible blends of an amorphous copolyethylene terephthalate glycol (PETG 6763 of Eastman Kodak) and an ethylene vinyl acetate copolymer (EVA AT 2803 M of A.T. Plastics Inc.). Rheological parameters and characteristics of the components and blends are reported in Table 8.2-1. All the curves show a very similar increase in elasticity in the terminal zone of the matrix, although the droplets are much larger at higher temperatures. The rheological behavior over the whole frequency range as well as the increase in elasticity are well predicted by the emulsion model. The agreement is shown to be excellent for both blends. The enhancement of elasticity is attributed to the deformability and the relaxation of the EVA droplets suspended in the PETG matrix. Measurements have to be carried out at low frequencies in order to observe phenomena associated with droplet

340   Multiphase Systems

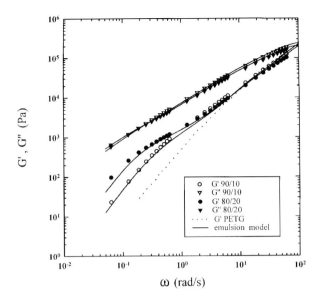

Figure 8.2-3 Comparison between experimental data and the Palierne emulsion model predictions for the 90/10 and 80/20% PETG/EVA blends; T = 210 °C (From Lacroix et al., 1996)

deformation. The G'' data are not sensitive to the emulsion concentration and the G' data at high frequencies are very close to those of the matrix.

Note from Table 8.2-1 that the zero shear viscosity of EVA at 210 °C is slightly less than that of PETG (k = 0.72). Also, the zero shear viscosities of the blends are considerably larger than those of both unblended components. This surprising effect is predicted by the Palierne model.

Figure 8.2-4a reports on the particle size variation with time for the 80/20 PETG/EVA blend at 210 °C. The increases in diameter occurred in the first five minutes. After preparation in the Brabender at 210 °C, the diameter of the droplets was approximately equal to 1.56 μm. During curing at 210 °C, the value of d_v increased to 2.46 μm, while the number average diameter was stable at about 0.95 μm. Hence, as shown also by the micrographs of Figs. 8.2-4b and c, polydispersity in diameter and coalescence are enhanced by curing at 210 °C. These results stress the necessity of determining the morphology of polymer blends as a function of time if we wish to make correct interpretations of the rheological measurements. The equilibrium d_v values were used to calculate the Palierne model predictions shown in Fig. 8.2-3. The interfacial tension for the PETG/EVA system was unknown and the model predictions were used to obtain the interfacial tension values reported in Table 8.2-1.

Figures 8.2-5 and 8.2-6 compare the Palierne model predictions to the G' and G'' data of two other blends of polystyrene and polymethyl methacrylate. The PMMA is a commercial polymer (Altulite 2773) of molecular weight $M_w = 133\,000$ kg/kmol and $M_w/M_n = 1.9$. The polystyrene is Gedex 1541 GA 100 of $M_w = 23\,000$ kg/kmol and $M_w/M_n = 3.6$. The average molecular weight M_w and the polydispersity (M_w/M_n) of the

Table 8.2-1 Rheological Parameters and Characteristics of the Components and PETG/EVA Blends at $T_0 = 210$ °C

	η_0 kPa·s	$\lambda_0^{(1)}$ ms	$d_v^{(2)}$ μm	α mN/m
PETG	7.753	17.6	–	–
EVA	5.55	562	–	–
90/10% PETG/EVA	10.07	83.1	1.10	3.9
80/20% PETG/EVA	14.45	1444	2.46	4.5

(1) Relaxation time in the terminal zone $\lambda_0 = \lim_{\omega \to 0} G'/\eta'\omega^2$
(2) Volume average diameter at equilibrium

polymers were determined by gel permeation chromatography (GPC) in tetrahydrofuran (THF). The first blend consists of 90 mass % of PMMA and 10 mass % of PS. The zero shear viscosity ratio, k, is equal to 0.035. The average droplet radius was estimated by digital image analysis to be around 0.22 μm, and the interfacial tension is 1.5 mN/m for this system. It should be mentioned that the morphology of the blend was found to be unchanged following rheological experiments carried out at 200 °C for about 2 h. It is clear from Fig. 8.2-5 that the Palierne emulsion model describes the dynamic moduli of the blend over the whole frequency range well. It also predicts the observed increase in elasticity. The low frequency shoulder in G' is a characteristic of a long time relaxation process of about 100 s, which is much longer than the terminal relaxation times of the phases (2 s for PMMA and 0.5 s for PS). The loss modulus for the blend is identical to that of the matrix, even at low frequencies. (The G'' data for the matrix are not shown in the figure.)

Figure 8.2-6 confirms that the low frequency elasticity of blends is related to the deformability of the minor phase particles. The figure reports the linear viscoelastic properties of the inverse composition (10/90 PMMA/PS) blend. In this case, the dispersed PMMA inclusions are much more viscous than the PS matrix ($\eta_{0(PMMA)}/\eta_{0(PS)} = 30$). As expected and predicted by the model, no significant enhancement of elasticity is observed at low frequencies, confirming that the deformability of the suspended droplets is responsible for the increase in blend elasticity in the low frequency range, when the dispersed phase is less viscous than the matrix.

In summary, the Palierne model predictions are in good agreement with the experimental results of viscoelastic emulsions (Graebling et al., 1993), but fail as expected in the case of strong particle-particle interactions or agglomerated particles (see Carreau et al., 1994b; Bousmina and Muller, 1993). Similar failures have also been reported for systems with a high concentration of a copolymer (Brahimi et al., 1991). This is not surprising, since those systems that contain a compatibilizing agent are no longer simple suspensions. Possibly substantial interphases of a totally different rheological behavior exist for such systems. We hope that the more general Palierne (1990) model will be able to cope with such interphase effects.

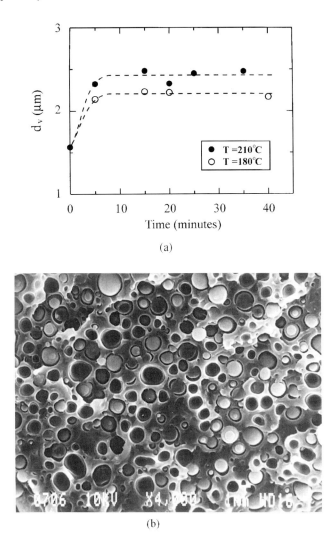

Figure 8.2-4 Variation of d_v with curing time, 80/20 PETG/EVA
(a) Scanning electron micrographs
(b) before curing
(c) after 20 min curing, T = 210 °C
(From Lacroix et al., 1996)

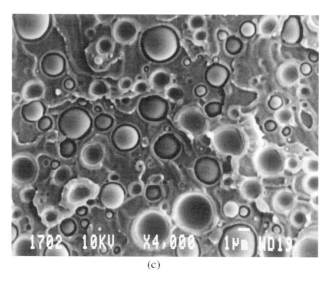

(c)

Figure 8.2-4 (*continued*)

8.2-3 Rheology of Concentrated Suspensions of Non-Interactive Particles

The viscosity of non-interactive spherical particles in Newtonian fluids is illustrated in Fig. 8.2-2 in terms of the relative or reduced viscosity, $\mu_r = \mu_s/\mu_m$. These data from Thomas

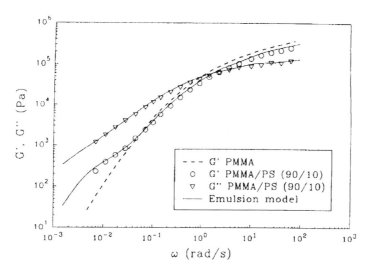

Figure 8.2-5 Comparison between experimental data and Palierne's model predictions for the 90/10 PMMA/PS blend at 200 °C (From Carreau et al., 1994a)

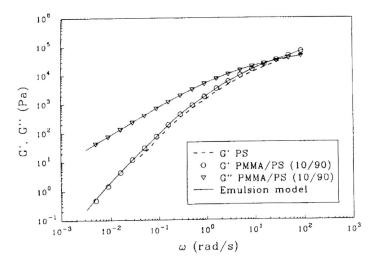

Figure 8.2-6 Comparison between experimental data and Palierne's model predictions for the 10/90 PMMA/PS blend at 200 °C (From Carreau et al., 1994a)

(1965) show that the relative viscosity is a unique function of the volumetric fraction of solids. The whole curve can be described by the Maron and Pierce (1956) empirical equation

$$\mu_r = \frac{\mu}{\mu_m} = \left[1 - \frac{\phi}{\phi_m}\right]^{-2}, \qquad (8.2\text{-}34)$$

where ϕ_m is the solids fraction at maximum packing, theoretically equal to 0.68 for solid spheres of narrow size distribution.

The non-Newtonian behavior of polymers filled with non-interactive spheres is very similar to that of non-filled polymers, at least up to a solid fractions close to maximum packing. Figure 8.2-7 shows the data obtained by Poslinski et al. (1988) for filled thermoplastics, and the description of the data using the Carreau (1972) equation:

$$\eta_s = \eta_{s_0}[1 + (t_s\dot{\gamma})^2]^{\frac{(n-1)}{2}}, \qquad (8.2\text{-}35)$$

where the subscript s refers to the suspension and η_{s_0} is the zero shear viscosity. We note that the data of Poslinski et al. (1988) were obtained by using a cone-and-plate rheometer (Rheometrics RMS) for the lower shear rates and an Instron capillary viscometer for high shear rates. The Mooney correction was found to be negligible, and the excellent agreement between both sets of data lends support to the idea that these filled systems behave like homogeneous fluids. The time constant for the composite is given by a Maron-Pierce equation as

$$t_r = \frac{t_s}{t_m} = \left[1 - \frac{\phi}{\phi_m}\right]^{-2}. \qquad (8.2\text{-}36)$$

8.2 Rheology of Suspensions 345

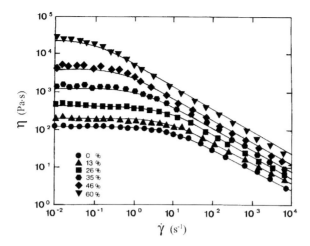

Figure 8.2-7 Shear viscosity of glass spheres of various volume fractions, dispersed in a thermoplastic polymer at 150 °C
The glass spheres' average diameter was 15 μm with a narrow size distribution. The solid lines are the predictions using the Carreau equation (Replotted from Poslinski et al., 1988)

Note the large increases of the reduced time constant, up to three decades, with increasing solid content. This is contrary to the conventional thinking that the addition of solid particles decreases the fluid's elasticity.

Another very important finding of Poslinski et al. (1988) is that equation 8.2-34 can be applied to non-Newtonian polymeric systems, provided that the viscosity values of the composite and of the matrix are compared at the same shear stress, that is, equation 8.2-34 has to be rewritten in the following form:

$$\eta_r = \left(\frac{\eta_s}{\eta_m}\right)\bigg|\sigma_{21} = \left[1 - \frac{\phi}{\phi_m}\right]^{-2}. \qquad (8.2\text{-}37)$$

It means that the viscosity of thermoplastics filled with narrow size distribution and low interacting spheres can be predicted from the viscosity of the matrix. In fact the findings of Poslinski et al. (1988) have much wider implications, as we will see next.

It has been observed that polydispersity reduces the viscosity of filled systems at fixed solids concentration (see Eveson, 1959). Poslinski et al. (1988) have shown that a formula for the maximum packing parameter proposed by Gupta and Seshadri (1986) can be used to calculate, with the help of equation 8.2-37, the relative viscosity of polymers filled with polydisperse spheres. Equation 8.2-37 was found to give very good predictions for the viscosity of bimodal systems. The results of Poslinski et al. (1988), reproduced in Fig. 8.2-8, confirm this expectation.

At very high solids fractions ($\phi = 0.60$), the reduced viscosity goes through a sharp minimum for systems containing approximately 15% of 15 μm glass beads, the rest consisting of 78 μm beads. This minimum corresponds to a maximum packing of 0.75 obtained for these bimodal suspensions. The viscosity reduction compared to the mono-

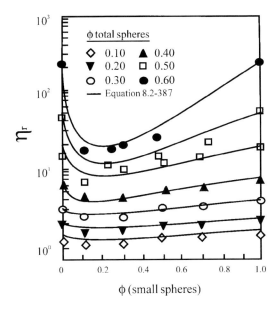

Figure 8.2-8 Relative shear viscosity of suspensions of glass beads in a polybutene at 22 °C as a function of the volume percent of 15 μm spheres in a mixture of 15 μm and 78 μm beads (Replotted from Poslinski et al., 1988)

dispersed systems is of the order of ten. This is not a negligible effect for engineering applications and indeed even larger viscosity reductions have been reported. Finally, these results suggest that ϕ_m can be taken as a fitting parameter for polymer systems filled with particles of more complex shapes.

8.2-3.1 Elasticity of Suspensions of Spheres

Only limited data on the elastic properties of suspensions (normal stresses in steady simple shear flow, storage modulus, and so on) are available in the literature. Both the Bagley correction term and the extrudate swell are known to decrease with solids content. On the other hand, the (elastic) storage modulus of filled polypropylene melts was found by Faulkner and Schmidt (1977) to increase with the solids fraction according to the formula (for solids concentration ϕ less than 0.26)

$$G'_r = \frac{G'_s}{G'_m} = 1.0 + 1.8\phi, \tag{8.2-38}$$

whereas the (viscous) loss modulus obeys the relation

$$G''_r = \frac{G''_s}{G''_m} = 1.0 + 2.0\phi + 3.3\phi^2. \tag{8.2-39}$$

The results are in agreement with the observations on extrudate swell and entrance effects, since the viscous forces are increasing more rapidly than the elastic forces. The more recent results of Poslinski et al. (1988) are somewhat at variance with those of Faulkner and Schmidt. They found that the storage modulus (for frequencies larger than 10 rad/s) can be correlated by a Maron-Pierce type equation, as follows:

$$G'_r = \left[1 - \frac{\phi}{\phi_m}\right]^{-2}, \qquad (8.2\text{-}40)$$

where the moduli are obtained at the same frequency. The loss modulus follows a different correlation:

$$G''_r = 1 + \left[\frac{0.75\dfrac{\phi}{\phi_m}}{1 - \dfrac{\phi}{\phi_m}}\right]^2. \qquad (8.2\text{-}41)$$

These results indicate that the elastic forces increase more rapidly than the viscous forces. They also measured the primary normal stress differences in steady simple shear flow. Again the Maron-Pierce type equation is successful in correlating the data:

$$\psi_{1r} = \frac{\psi_{1s}}{\psi_{1m}} = \left[1 - \frac{\phi}{\phi_m}\right]^{-2}, \qquad (8.2\text{-}42)$$

where the primary normal stress coefficient, $\psi_1 = -(\sigma_{11} - \sigma_{22})/\dot{\gamma}^2$, is compared at the same shear rate, $\dot{\gamma}$.

It is obvious that more research is needed to clarify how fillers affect the elastic properties of the polymeric matrix. Clearly, the general impression that elasticity is reduced as the solids concentration increases is not always true.

8.2-4 Rheology of Concentrated Suspensions of Interactive Particles

In this section, we present and discuss a few interesting results for suspensions of interacting particles. When interparticle interactions are important compared to viscous forces, many complications can arise. A major one is the agglomeration of particles, which results in substantial increases of the viscosity, mainly at low shear rates, as shown by Nielsen (1977) for glass beads in the presence of water in a hydrocarbon liquid. The behavior does not only become shear-thinning, but the suspension appears to show a yield stress. Another interesting example is reported by Minagawa and White (1975) for a high density polyethylene filled with rutile particles (reproduced in Fig. 8.2-9). The increase in viscosity at low shear stress for high solids loading is found to be quite spectacular. For the highly concentrated suspensions, the low shear rate or low shear stress viscosity appears to be unbounded. This is an indication of a yield stress. The viscosity of these systems depends strongly on the interfacial properties and sizes of the particle aggregates. Hence, the blending conditions may considerably alter the viscosity of the systems.

Carbon black particles are known to yield very different rheologies dependent on the properties and sizes of the particles. Figures 8.2-10 and 8.2-11 show two sets of viscosity data reported by Lakdawala and Salovey (1987) for a polystyrene melt filled with carbon

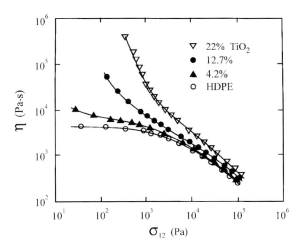

Figure 8.2-9 Steady shear viscosity of suspensions of rutile in a high density polyethylene melt (HDPE) The TiO$_2$ content is in volume percent (Adapted from Minagawa and White, 1975)

black. Note the apparently unbounded viscosity at low shear rate for high filler content. The behavior at 230 °C (Fig. 8.2-11) is considerably different from that at 180 °C (Fig. 8.2-10), mainly for dilute systems that exhibit viscosities even lower than the unfilled matrix. This is not uncommon and similar results have been obtained by others. It is possibly due to a reduction of polymeric chain entanglements in the presence of solid particles. Also, we

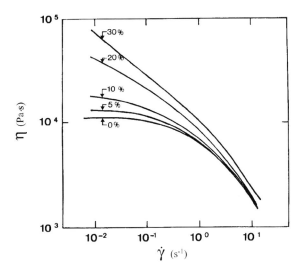

Figure 8.2-10 Viscosity of a high molecular weight polystyrene filled with carbon black (24 m^2/g) at 180 °C
The carbon black concentration is in mass percent (Adapted from Lakdawala and Salovey, 1987)

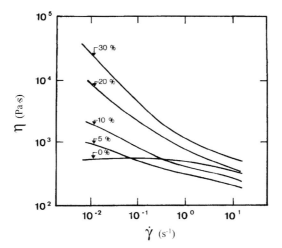

Figure 8.2-11 Viscosity of a high molecular weight polystyrene filled with carbon black (24 m²/g) at 230 °C
The carbon black concentration is in mass percent (Adapted from Lakdawala and Salovey, 1987)

cannot rule out the possibility of polymer degradation in preparing these composites at high temperature.

Figure 8.2-12 shows the relative viscosity of the polyethylene filled with five different types of carbon black. The viscosity data are compared at $\dot{\gamma} = 10$ s^{-1}, and the solids concentration is on a mass basis. The different carbon black samples were obtained from

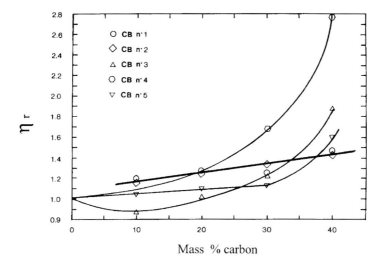

Figure 8.2-12 Relative viscosity of a high density polyethylene filled with carbon black particles at $\dot{\gamma} = 10$ s^{-1} (Data from Dufresne, 1989)

two suppliers and their main characteristics can be found in Dufresne (1989). Obviously, the viscosity is strongly dependent on the surface properties of the carbon black. Properties of these systems have been discussed by Malik et al. (1988).

In the case of very small particles suspended in polymer solutions, the reduction of the viscosity can be attributed to the polymer adsorption on the particle surface. This results in a somewhat depleted solution compared to the unfilled solution, as postulated by Otsubo (1986).

The hydroxyl groups at the surface of the fumed silica particles favor the adsorption of the polyacrylamide chains (Fig. 8.2-13). Otsubo (1986) has estimated that the adsorption layer is of the order of 8 to 10 nm, which is comparable to the size of the silica particles. At high solids concentration, the rheological behavior becomes quite complex. The fact that the viscosity increases at low shear rates is indicative of a gel-type behavior, and the shear thickening at high shear rates suggests flow-induced structure.

Concentrated colloidal particle suspensions are known to be highly thixotropic materials, with structure changes over long time periods. The behavior of a few of these materials is described in Section 2.3. Considerable research efforts are currently devoted to elucidating the flow mechanisms and to developing appropriate constitutive relations. A summary of the current work on these topics is outside the scope of this book.

Finally, we report viscosity data for mica-polyethylene composites. Figure 8.2-14 shows the viscosity as a function of shear rate for different suspensions of mica (mass percent) in a high density polyethylene (HDPE) at 220 °C. The viscosity for this polymer melt at various temperatures is reported in Figure 3.1-9. The viscosity data obtained on a Rheometrics RSR at low shear rate are consistent with those obtained on the capillary rheometer. Due to alignment and orientation of the mica flakes, the viscosity increases for

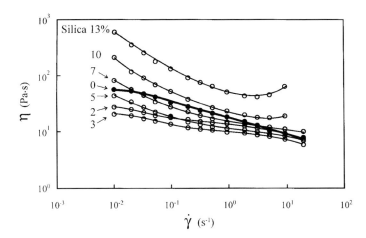

Figure 8.2-13 Shear viscosity of suspensions of fumed silica particles in a 1.5 mass % polyacrylamide (PAA) solution in glycerine
The average particle size is 20 nm and the weight average molecular weight of the PAA is 2.10×10^6 kg/kmol. The particle concentrations are in mass percent (Adapted from Otsubo, 1986)

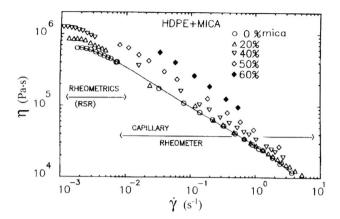

Figure 8.2-14   Viscosity vs shear rate for a high density polyethylene filled with mica at 220 °C
The concentrations are in mass percent (Adapted from Malik et al., 1988)

the highly concentrated suspensions are not large. At comparable loading, the viscosity increase for the PE with mica is considerably less than that with carbon black.

8.2-5   Concluding Remarks

The rheology of polymeric multiphase systems is still in its infancy. The only aspect that appears to be well established is the shear viscosity of suspensions of non-interacting (or weakly interacting) spheres. In that respect, equation 8.2-34 or 8.2-37 combined with viscosity expressions such as 8.2-35 is a very useful empiricism to predict the viscosity of filled thermoplastics. For large aspect ratio (slender) particles, it should be used with caution. Quite powerful theories are also available to predict the viscosity of dilute suspensions of rigid rods (see for example Ausias et al., 1992). However, the properties of concentrated fiber suspensions are hardly predictable. On the other hand, the viscosity of suspensions of interactive particles depends strongly on the surface properties of the particles, on the particle size, and on the geometry and blending methods. Results in the literature should be used with extreme care to predict the viscosity of such systems.

The elasticity of filled thermoplastics remains a subject of considerable controversy. The elastic forces may increase more slowly or more quickly than the viscous forces in loaded systems. The effects of fillers on the transient response and on the extensional viscosity of thermoplastics remain open questions. Definitely, more research is needed in order to clarify these crucial points.

8.3 Flow About a Rigid Particle

8.3-1 Flow of a Power-Law Fluid Past a Sphere

The schematic representation of this flow is shown in Fig. 8.3-1. We consider the creeping and isothermal flow of an incompressible fluid under steady-state conditions. The boundary conditions for this flow are those of no slip at the sphere and a free-stream velocity far away from the sphere; that is,

$$V_r = V_\theta = 0 \quad \text{at} \quad r = R$$
$$\text{and} \quad V_z = V_\infty \quad \text{at} \quad r \to \infty. \tag{8.3-1}$$

Owing to axisymmetry, $V_\phi = 0$, and there is no ϕ-dependence; thus, we postulate $V_r = V_r(r, \theta)$, $V_\theta = V_\theta(r, \theta)$, and $P = P(r, \theta)$.

The r- and θ-components of the equation of motion reduce to

$$0 = -\frac{\partial P}{\partial r} - \left[\frac{1}{r^2}\frac{\partial}{\partial r}(r^2 \sigma_{rr}) + \frac{1}{r \sin\theta}\frac{\partial}{\partial \theta}(\sigma_{r\theta} \sin\theta) - \frac{\sigma_{\theta\theta} + \sigma_{\phi\phi}}{r}\right] + \rho g_r \tag{8.3-2}$$

and

$$0 = -\frac{1}{r}\frac{\partial P}{\partial \theta} - \left[\frac{1}{r^2}\frac{\partial}{\partial r}(r^2 \sigma_{r\theta}) + \frac{1}{r \sin\theta}\frac{\partial}{\partial \theta}(\sigma_{\theta\theta} \sin\theta) + \frac{\sigma_{r\theta}}{r} - \frac{\cot\theta}{r}\sigma_{\phi\phi}\right] + \rho g_\theta. \tag{8.3-3}$$

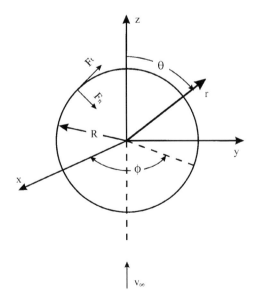

Figure 8.3-1 Flow about a rigid sphere

The equation of continuity yields

$$\frac{1}{r^2}\frac{\partial}{\partial r}(r^2 V_r) + \frac{1}{r\sin\theta}\frac{\partial}{\partial \theta}(V_\theta \sin\theta) = 0. \tag{8.3-4}$$

For this non-viscometric flow, the power-law fluid model is written as

$$\eta = m\left(\frac{1}{2}\underline{\dot{\gamma}}:\underline{\dot{\gamma}}\right)^{\frac{n-1}{2}}, \tag{8.3-5}$$

where

$$\left(\frac{1}{2}\underline{\dot{\gamma}}:\underline{\dot{\gamma}}\right) = 2\left[\left(\frac{\partial V_r}{\partial r}\right)^2 + \left(\frac{1}{r}\frac{\partial V_\theta}{\partial \theta} + \frac{V_r}{r}\right)^2 + \left(\frac{V_r}{r} + \frac{V_\theta \cot\theta}{r}\right)^2\right] + \left[r\frac{\partial}{\partial r}\left(\frac{V_\theta}{r}\right) + \frac{1}{r}\frac{\partial V_r}{\partial \theta}\right]^2. \tag{8.3-6}$$

Equations 8.3-1 to 8.3-6 define this problem completely. We are seeking the solution in terms of the r and θ velocity components and pressure. The latter can subsequently be manipulated to estimate the form and friction drag force acting on the sphere as follows:

$$F_n = 2\pi R^2 \int_0^\pi (-P|_{r=R}\cos\theta)\sin\theta d\theta \tag{8.3-7}$$

and

$$F_t = 2\pi R^2 \int_0^\pi (\sigma_{r\theta}|_{r=R}\sin\theta)\sin\theta d\theta. \tag{8.3-8}$$

It is customary to introduce a dimensionless drag coefficient, C_D, defined as follows:

$$C_D = \frac{F_D}{(\frac{1}{2}\rho V_\infty^2)(\pi R^2)}, \tag{8.3-9}$$

where F_d is the total drag force, that is, the sum of F_n and F_t.

The non-dimensionalization of the equations of motion, the equation of continuity, and the boundary conditions suggests that for the flow of power-law fluids,

$$C_D = C_D(Re, n). \tag{8.3-10}$$

Often the results are expressed in the form of a correction factor $X(n)$ defined as follows:

$$C_D = \frac{24}{Re}X(\vec{n}) \tag{8.3-11}$$

where the Reynolds number is defined as $Re = \rho V_\infty^{2-n}(2R)^n/m$.

However, owing to the nonlinear viscosity relation, an analytical solution for this boundary value problem is not possible. Early attempts (Tomita, 1959; Wasserman and Slattery, 1964) employed the so-called velocity and stress variational principles (Slattery, 1972) to obtain upper and lower bounds on the correction factor, X. However, more accurate values of this correction factor have now been established using numerical methods (Crochet et al., 1984; Gu and Tanner, 1985; Tripathi et al., 1994) and these are summarized in Table 8.3-1. A good correspondence exists between various predictions. Furthermore, it is also reasonably well-established that creeping flow is limited to

Table 8.3-1 Numerical Predictions of Drag on a Sphere ($Re \leq 0.1$)

	X		
n	Gu & Tanner (1985)	Cho & Hartnett (1983)	Tripathi et al. (1994)
1	1.002	1.002	1.003
0.8	1.24	1.239	1.230
0.6	1.382	1.37	1.381
0.4	1.442	1.366	1.440
0.2	1.413	1.346	1.398

$Re < \sim 1$, although a visible wake only appears at $Re \sim 20$ for power-law fluids. This is also consistent with the behavior for incompressible Newtonian flow (Tripathi et al., 1994).

Numerical predictions of sphere drag in power-law fluids are now available up to $Re = 170$, and these are shown in Figs. 8.3-2 and 8.3-3. With the increasing Reynolds number and/or degree of shear thinning, the relative contribution of the pressure drag increases (Fig. 8.3-2). Some experimental results are also included in Fig. 8.3-3 to substantiate the theoretical predictions (Tripathi et al., 1994; Graham and Jones, 1994). Extensive comparisons between theory and experiments suggest that the match is poor in the low Reynolds number region and is only fair for the $Re > 1$ regime. The flow behavior index is seen to be a much more significant variable in the creeping region, whereas at high Reynolds number, the value of the drag coefficient is largely determined by the inertial forces. Indeed, the Newtonian standard drag curve is about as accurate ($\pm 30\%$) to describe the power-law drag coefficient of spheres as are the numerical predictions for power-law fluids (Chhabra, 1993a). Similar numerical predictions of drag on spheres and oblate- and prolate-shaped particles in shear-thickening fluids ($n > 1$) have also been reported recently in the literature (Tripathi and Chhabra, 1995), and these show exactly the opposite type of behavior, namely, $X < 1$ in the creeping regime.

Example 8.3-1 Settling Velocity of a Glass Bead

Estimate the terminal settling velocity of a glass ball (density $= 2500$ kg/m^3) of diameter d_p equal to 1 mm in a quiescent power-law polymer melt ($m = 33.7$ Pa·s$^{0.6}$ and $n = 0.6$). The polymer melt has a density of 775 kg/m^3.

Solution

At equilibrium, the fluid drag is equal to the buoyant weight of the ball. Equation 8.3-11 can now be rearranged to obtain the following expression for the terminal settling velocity, V_∞:

$$V_\infty = \left[\frac{d_p^{n+1}(\rho_s - \rho)g}{18mX}\right]^{\frac{1}{n}}.$$

Substituting the numerical values, we obtain $V_\infty = 1.4$ μm/s.

8.3 Flow About a Rigid Particle 355

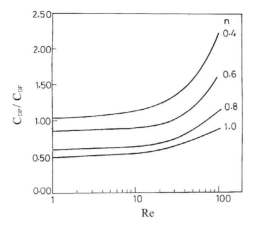

Figure 8.3-2 Variation of pressure-to-friction drag ratio with Reynolds number and power-law index for spheres (Reprinted with permission from A. Tripathi, R.P. Chhabra, and T. Sundararajan, Ind. Eng. Chem. Res., 33, 403 (1994) Copyright (1993) American Chemical Society)

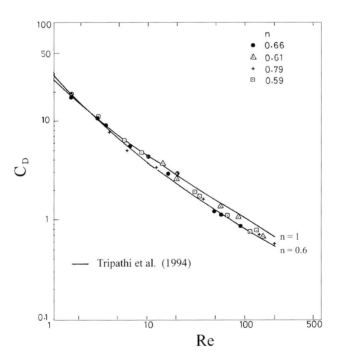

Figure 8.3-3 Comparison between theoretical and experimental drag on spheres in power-law fluids (Reprinted with permission from A. Tripathi, R.P. Chhabra, and T. Sundararajan, Ind. Eng. Chem. Res., 33, 403 (1994) Copyright (1993) American Chemical Society)

Check:

$$Re = \frac{\rho V_\infty^{2-n} d_p^n}{m} = \frac{775 \times (1.4 \times 10^{-6})^{1.4}(1 \times 10^{-3})^{0.6}}{33.7} = 2.32 \times 10^{-9} \ll 1$$

So it is quite small, and the flow is in the creeping flow regime.

8.3-2 Other Fluid Models

Numerous investigators have studied the creeping flow of a variety of generalized Newtonian fluids around a sphere using variational principles, albeit not as extensively as for power-law fluids. All such studies have been critically reviewed in a recent book (Chhabra, 1993a). The main implication of using a GNF model containing the zero shear viscosity is that the correction factor, X, now also becomes a function of a dynamic parameter such as the Carreau number or the Ellis number. Also, the fact that all such studies predict lower drag ($X < 1$) than the Stokesian value is simply an artifact of the definition of the Reynolds number, which is based on the zero shear viscosity in this case. There is no inconsistency as to whether the drag is increased or decreased above the Stokes value. Figure 8.3-4 shows the typical dependence of X upon the flow behavior index and the Carreau number ($Cu = t_1 V_\infty / R$). Note that an iterative scheme is now needed to estimate the free settling velocity of a particle, as the unknown velocity occurs in both X and Cu.

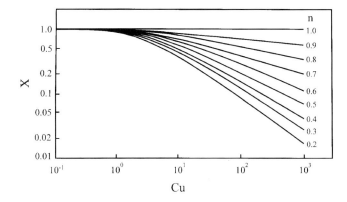

Figure 8.3-4 Approximate drag correction factor for Carreau model fluids (Replotted from Chhabra and Uhlherr, 1980)

8.3-3 Viscoplastic Fluids

The motion of spheres in viscoplastic media has generated considerable interest (Chhabra and Uhlherr, 1988) due to its potential applications in slurry pipelines. The main difficulty in this case is the identification of the sheared zone surrounding a moving sphere and the criterion for the initiation or cessation of sphere motion in such systems. However, once the motion is initiated in the creeping regime, the drag can be correlated using equation 8.3-11, where the correction factor X is given by the simple relation (Chhabra, 1993a)

$$X = 1 + \frac{\sigma_0}{m\left(\dfrac{V_\infty}{d_p}\right)^n}, \qquad (8.3\text{-}12)$$

and m, n, and σ_0 are the Herschel-Bulkley model parameters (equation 2.2-11). However, this approach must be regarded as tentative and has not been tested extensively. Numerical predictions for the creeping Bingham plastic flow over a sphere are also available (Beris et al., 1985).

8.3-4 Viscoelastic Fluids

The motion of a sphere in viscoelastic fluids has been analyzed for a wide range of fluid models (Walters and Tanner, 1992). Notwithstanding the subtle difficulties inherent in numerical simulations, most investigations predict a small shift in streamlines, but the drag can increase, decrease, or remain unchanged with reference to the Stokesian value, depending upon the magnitude of the viscoelastic effects, namely, the values of the Weissenberg number. In most practical situations when shear thinning and viscoelasticity are simultaneously encountered, the drag on a sphere is determined primarily by the shear-dependent viscosity. Other interesting time-dependent effects including velocity overshoot, rotation, and bouncing associated with sphere motion in viscoelastic media have been documented in the literature (Bird, Armstrong, and Hassager, 1977, 1987; Walters and Tanner, 1992). The scant experimental literature on the drag of spheres in viscoelastic fluids has been reviewed by Chhabra (1993a).

8.3-5 Wall Effects

While most theoretical treatments assume an infinite fluid extent, it is virtually impossible to meet this requirement in experimental work. Cylindrical tubes are often used to carry out such studies. From a theoretical standpoint, the presence of confining walls changes the boundary conditions, which must now be satisfied at definite boundaries rather than far away from the submerged body. In practical terms, the bounding walls exert an extra retardation effect on the free settling particles in both quiescent and moving fluids (Clift et al., 1978; Chhabra, 1993a). In other words, the free settling velocity of a particle measured in a cylindrical tube will always be smaller than that in the absence of walls. Thus, the knowledge of wall effects is often necessary to deduce the net fluid dynamic drag on the particle. Conversely, the measured terminal settling velocity must be corrected for wall effects prior to its use in process calculations.

While some theoretical analyses are available dealing with the extent of wall effects on the motion of spherical and axisymmetric particles in Newtonian fluids in the low Reynolds number regime (Happel and Brenner, 1973; Clift et al., 1978), no such results are available for non-Newtonian fluids. However, the effect of containing walls on the free settling velocity has been determined experimentally for pseudoplastic, viscoplastic, and viscoelastic types of fluids. For power-law fluids, the wall factor (terminal velocity in unbounded medium/terminal velocity in the presence of walls), f, is given by the following function of $(d_p/D_c) \leq 0.5$ and $Re \leq 1000$:

$$\frac{f - f_\infty}{f_0 - f_\infty} = (1 + 1.3\, Re^2)^{-0.33}, \tag{8.3-13a}$$

where

$$f_0 = \left(1 - 1.6\frac{d_p}{D_c}\right)^{-1} \tag{8.3-13b}$$

and

$$f_\infty = \frac{1}{1 - 3\left(\dfrac{d_p}{D_c}\right)^{3.5}}. \tag{8.3-13c}$$

Qualitatively, the extent of wall effects in non-Newtonian fluids is generally smaller than that in Newtonian media under otherwise identical conditions. The effect is further suppressed in viscoelastic media (Chhabra, 1993a).

Example 8.3-2 Wall Effects in a Viscoplastic Fluid
Discuss qualitatively the nature of wall effects on a sphere moving in a viscoplastic medium.

Solution
Recall that a viscoplastic material acts like a fluid only when the value of the existing stress exceeds the yield stress of the material. A solid sphere whose weight is sufficient to overcome the force due to the yield stress will sediment under the influence of gravity in a viscoplastic substance contained in a cylindrical tube. The deformation caused by the moving sphere in an otherwise quiescent medium is thus transmitted over a short distance in the fluid. The detailed velocity profile measurements reveal that the size and shape of the sheared cavity surrounding a sphere depend strongly on the size and velocity of the sphere as well as on the yield stress and the viscosity of the substance. Figure 8.3-5 qualitatively shows the shape of such a cavity.

Beyond the sheared region, the viscoplastic material behaves like an elastic solid. If the size of the sheared fluid cavity extends up to the cylindrical fall tube, that is, $\delta > D_c$ (see Fig. 8.3-5a), the moving sphere will experience the wall presence. On the other hand, if the local stresses drop below the yield stress and the material behaves like a solid even before reaching the cylindrical boundary, that is, $\delta < D_c$, the sphere motion will not be influenced by the container walls (see Fig. 8.3-5b). The sphere will experience wall effects only if the

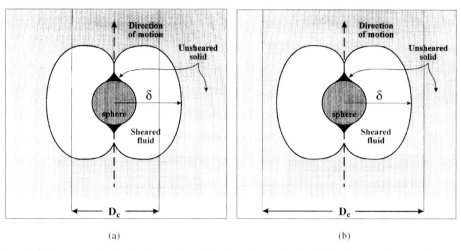

Figure 8.3-5 Sheared cavity for the motion of a sphere in a cylindrical fall tube containing a viscoplastic fluid
(a) wall effects, $\delta > D_c/2$
(b) no wall effects, $\delta < D_c/2$

sheared zone around the sphere extends to the confining walls. See Atapattu et al. (1990) for more details on this subject.

8.3-6 Non-Spherical Particles

There is not much information available on the sedimentation of non-spherical particles in rheologically complex fluids. Some analytical results on the slow flow of needles, long cylindrical and axisymmetric spheroidal particles in power-law fluids are available (Chhabra, 1993a; Tripathi et al., 1994). The latter authors have analyzed the flow of a power-law fluid over oblates and prolates up to values of $Re \approx 100$. While the flow structures remain virtually unaffected by the non-Newtonian behavior, the drag shows significant deviations from its corresponding Newtonian value. Some experimental results on cylinders, prisms, and irregular shaped gravel chips in pseudoplastic as well as viscoelastic media are also available (Rodrigue et al., 1994). It appears that the use of the volume equivalent sphere diameter together with a sphericity factor provides an adequate scheme for handling the drag on such particles. For instance, Rodrigue et al. (1994) presented the following expression for the hydrodynamic drag on non-spherical objects in free settling mode in Carreau model fluids:

$$C_D = \frac{24}{Re_c}[1 + 0.14\,Re_c^{0.7}], \qquad (8.3\text{-}14a)$$

where

$$Re_c = \frac{\rho V_\infty d_e}{\eta_0} \left[1 + \left(\frac{t_1 V_\infty}{d_e}\right)^{1-n} \right]. \quad (8.3\text{-}14b)$$

The Reynolds number associated with the Carreau viscosity equation involves the particle equivalent diameter d_e.

More recently, Liu and Joseph (1993) have reported an interesting experimental study on the sedimentation of non-spherical objects in viscoelastic polymer solutions. The equilibrium orientation achieved by the object was explained by an interplay between viscoelastic, shear-dependent viscosity, and geometry effects.

8.3-7 Drag-Reducing Fluids

The phenomenon of drag reduction in external flows has not been studied as extensively as in pipe flows. Drag reduction for freely falling spheres occurs in the subcritical Reynolds number region. Figure 8.3-6 illustrates representative data showing drag reduction in polyethylene oxide (PEO) solutions. Results reported by different workers seldom agree quantitatively. Although a range of drag reducing additives has been tried, most studies have been carried out using different grades of PEO. Despite the lack of quantitative agreement among various investigators, drag reduction for spheres occurs in the laminar boundary layer region ($Re \sim 2 \times 10^3$ to 3×10^5). This is in sharp contrast to the skin friction reduction in turbulent pipe flows. Several plausible mechanisms, including the modified boundary layer and wake characteristics due to viscoelasticity (James and Acosta, 1970), and the existence of a finite shear wave velocity (Denn and Porteous, 1971), have been postulated. However, none of these has proved to be completely satisfactory (Sellin et al., 1982).

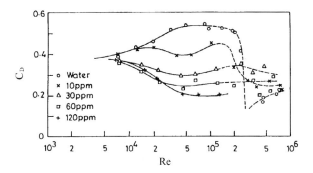

Figure 8.3-6 Typical results showing drag reduction for spheres in Polyox WSR-301 solutions (Replotted from White, 1967)

8.3-8 Behavior in Confined Flows

A sizeable body of information is now available that elucidates the interplay between particle shape, nonlinear liquid characteristics, geometry, and the type of flow field (shear, extensional) prevailing in the system. Extensive reviews are available in this area (Leal, 1980; Brunn, 1980). It is readily acknowledged that any departure, however small, from the Newtonian fluid behavior, from the creeping flow approximation, or from the spherical shape can result in dramatic changes in the orientation of particles that have serious implications in the rheometry of suspensions. Most of the research efforts have been directed toward the elucidation of one or more of the following aspects of particle motion.

(i) Both Leal (1975) and Brunn (1977) have predicted that rod-like axisymmetric particles rotate to attain a stable configuration and settle in viscoelastic media with their longest axis parallel to the direction of gravity. Preliminary experimental results seem to support this notion (Tiefenbruck and Leal, 1980).
(ii) Several attempts have also been made to predict the lateral position of particles relative to the confining walls in one-dimensional shearing motion. Both spheres and non-spheres have a tendency to rotate in Couette and Poiseuille flow of viscoelastic fluids (Gauthier et al., 1971; Karnis and Mason, 1967; Bartram et al., 1975).
(iii) Another interesting feature is the lateral translation of spherical particles across the streamlines as encountered in the viscometry of concentrated suspensions (Metzner, 1985). It is worthwhile to recall that similar effects in Newtonian media also occur but only due to either inertial effects or particle deformation. This is in sharp contrast to the aforementioned behavior in viscoelastic media, which is observed for spheres and at very small Reynolds numbers.
(iv) Numerous other interesting (and so far unexplained) effects have been reported in the literature. For instance, Riddle et al. (1977) studied the interaction between two spheres settling in-line in viscoelastic fluids. Depending upon the initial separation and the extent of viscoelasticity, the two spheres may converge or diverge. Michele et al. (1977) presented some evidence on the alignment and segregation of spherical particles when the suspension is subjected to a shearing motion. The phenomenon of hindered settling in power-law fluids has been investigated recently (Chhabra et al., 1992). For mildly shear-thinning polymer solutions, the degree of hindrance on the velocity of a particle can be estimated using the procedures available in the literature for Newtonian fluids (see Example 8.3-3). Viscoelasticity seems to result in segregation and formation of internal structures in such systems involving significant particle interactions (Allen and Uhlherr, 1989).

Example 8.3-3 Hindered Settling Velocity of Glass Beads in a Suspension
Estimate the hindered settling velocity of a 30 vol % suspension of 1 mm glass beads in a polymer solution ($n=0.8$, $m=2.8$ Pa·s^n). The densities of the glass beads and polymer solution are 2500 kg/m^3 and 1050 kg/m^3 respectively.

Solution

As shown by Chhabra et al. (1992), the sedimentation velocities of the suspension and of the single sphere are related via the expression of Richardson and Zaki (1954), that is,

$$\frac{V}{V_\infty} = (1 - \phi)^Z, \qquad (8.3\text{-}15a)$$

where

$$Z = f(Re_\infty), \qquad (8.3\text{-}15b)$$

and ϕ is the volume fraction of the suspension.

The velocity of a single particle is calculated as in Example 8.3-1 as

$$V_\infty = \left[\frac{(1 \times 10^{-3})^{1.8} \times 9.81 \times (2500 - 1050)}{18 \times 2.8 \times 1.24}\right]^{\frac{1}{0.8}} = 157\ \mu m/s$$

Check:

$$Re_\infty = \frac{\rho V_\infty^{2-n} d_p^n}{m} = \frac{1050 \times (157 \times 10^{-6})^{1.2} (1 \times 10^{-3})^{0.8}}{2.8} = 4.1 \times 10^{-6} \ll 1.$$

For these conditions, the method of Richardson and Zaki (1954) yields $Z = 4.65$ and

$$V = 157(1 - 0.3)^{4.65} = 30\ \mu m/s.$$

Thus, the settling velocity is hindered by a factor of more than 5. This value may appear to be rather small, but in the rheometry of suspensions, where the characteristic linear scale is also of the order of 500 to 1000 μm, separation or sedi-mentation may occur, thereby falsifying the rheometric measurements.

8.4 Flow Around Fluid Spheres

8.4-1 Creeping Flow of a Power-Law Fluid Past a Gas Bubble

The flow under consideration is similar to that shown in Fig. 8.3-1. However, additional complications arise due to the mobile surface and the likely presence of surface active agents on the bubble surface. Besides, a rising bubble also undergoes a change in volume due to the variation in hydrostatic pressure. Here only quasi-steady state is considered, that is, it is assumed that the rate of change of bubble size is much slower than its rise velocity and that it retains its spherical shape (large surface tension systems). The boundary conditions for this problem are

$$\text{at}\quad r = R, \quad V_r = 0, \qquad (8.4\text{-}1a)$$

$$\text{at}\quad r = R, \quad \sigma_{r\theta} = 0, \qquad (8.4\text{-}1b)$$

$$\text{and at}\quad r \to \infty, \quad V_z = V_\infty. \qquad (8.4\text{-}1c)$$

The condition outlined in equation 8.4-1b stems from the fact that the gas viscosity is much smaller than the liquid viscosity. Also, surface tension effects are assumed to be negligible.

Owing to axisymmetry, $V_\phi = 0$ and there is no ϕ-dependence; thus, we postulate $V_r = V_r(r, \theta)$, $V_\theta = V_\theta(r, \theta)$, and $P = P(r, \theta)$. Once again, the resulting equations are nonlinear due to the power-law viscosity relation. Only approximate solutions have been obtained (Hirose and Moo-Young, 1969; Bhavaraju et al., 1978). Hirose and Moo-Young (1969) reported the following approximate functional relationship between X and the flow behavior index, n (for $n \ll 1$):

$$X = 2^n 3^{\frac{n-3}{2}} \left[\frac{13 + 4n - 8n^2}{(2n+1)(n+2)} \right]. \tag{8.4-2}$$

Note that in the limit of $n = 1$, equation 8.4-2 predicts $X = 2/3$, which confirms the Hadamard-Rybczynski result.

Similar approximate theoretical results for the motion of single bubbles in other generalized Newtonian fluids such as the Ellis model, the Carreau model, and Bingham plastics, have been summarized by Chhabra (1993a). The viscoelastic effects on bubble motion have been considered by Tiefenbruck and Leal (1980, 1982), whereas the role of surfactants on bubble and drop motion in viscoelastic media has been investigated by Quintana (1992). The growth and collapse of gas bubbles in viscoelastic systems has been studied by Papanastasiou et al. (1984).

8.4-2 Experimental Results on Single Bubbles

The voluminous literature on this subject has recently been summarized elsewhere (Chhabra and De Kee, 1992; Chhabra, 1993a). Only the salient features are recapitulated here. Perhaps the most striking phenomenon in this field is the so-called discontinuity observed in the bubble size-rise velocity relationship. Note a tenfold increase in the rise velocity of air bubbles in a viscoelastic polymer solution at a critical bubble size shown in Fig. 8.4-1. It is also, however, appropriate to add here that more recent results show a gradual transition of slope as anticipated from the switchover to a mobile surface, as shown in Fig. 8.4-2 (De Kee et al., 1986b, 1990a). Thus, it is far from clear whether the abrupt discontinuity is an artifact or a real phenomenon. Current research, however, supports the existence of a discontinuity, as observed by Astarita and Appuzzo (1965) and others. This discontinuity appears to be a function of the bubble volume, fluid elasticity, and surfactant concentration (De Kee et al., 1996). Rising bubbles through rheologically complex media display a great variety of shapes (see Figs. 1.2-14 and 8.4-3), which differ significantly from those encountered in incompressible Newtonian fluids (Clift et al., 1978; Carreau et al., 1974). Indeed, rheological complexity coupled with the experimental uncertainty (nonspherical shape, level of cleanliness, impurities, etc.) have been the main impediments in the development of a generalized drag curve for single bubbles (De Kee and Carreau, 1993). Figure 8.4-3 shows that the bubble shape in a viscoelastic fluid can be a strong function of the injection period, that is, of the time between the injection of consecutive bubbles.

364   Multiphase Systems

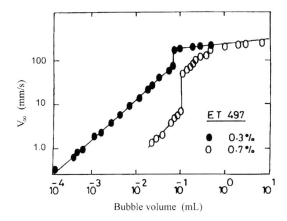

Figure 8.4-1 Terminal velocity-bubble volume data showing an abrupt increase in velocity for air bubbles rising in viscoelastic polymer solutions (Replotted from Astarita and Appuzzo, 1965)

The dependence on injection period can be responsible for a shift in the velocity-volume relation, as depicted in Fig. 8.4-4. This does not explain the discontinuity mentioned earlier, as data are normally collected at constant injection period. In addition, the reported discontinuities are all within a narrow volume range, around 0.1 mL. Note also in Fig. 8.4-4 that the data for the lower molecular weight polyacrylamide solution show no effect of injection period.

The longtime effects observed for the 0.1% PAA solution cannot be attributed to the elastic properties of this solution, for which the relaxation time is only of the order of seconds. This is more probably an indication of a strong orientation under flow of the

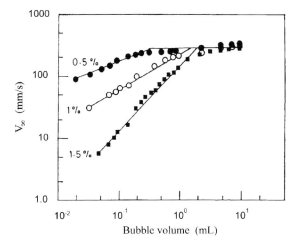

Figure 8.4-2 Terminal velocity-bubble volume data showing gradual change of slope in polyacrylamide solutions (Data from Dajan, 1985)

8.4 Flow Around Fluid Spheres   365

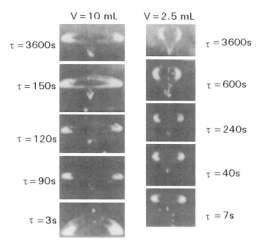

Figure 8.4-3 Constant volume bubble shapes in a 0.5 mass % Separan (MG-700) in a mixture of 20/80 mass % of glycerine/water, as a function of the injection period (Adapted from Carreau et al., 1974)

macromolecules. High molecular weight and hydrolyzed polyacrylamide have in an aqueous solvent a highly developed structure with a large aspect ratio compared to a non-ionic polymer. The chains can be easily oriented, yielding a lower resistance to the motion of subsequent bubbles. A return to the initial isotropic configuration would be achieved through molecular diffusion only if sufficient time between consecutive injections is allowed. Visual refractive effects in the liquid during the experiments were observed, indicating that the hypothesis of molecular orientation is quite reasonable. Even more intriguing are the phenomena of bubble interactions and coalescence in non-Newtonian media (see Fig. 1.2-14). De Kee and Chhabra (1988) presented a photographic study of

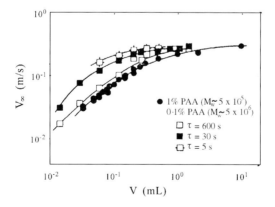

Figure 8.4-4 Effect of injection period on the terminal velocity for polyacrylamide solution (Adapted from De Kee et al., 1986b)

8.5 Creeping Flow of a Power-Law Fluid Around a Newtonian Droplet

Consider the flow of a power-law fluid over a Newtonian fluid sphere (see Fig. 8.3-1). This flow differs from that around a solid sphere in that we are now required to solve the field equations in the $0 \leq r \leq R$ region as well as in the outer field, $R \leq r \leq \infty$. In the absence of inertial forces, the pertinent equations are of the form of equations 8.3-2 to 8.3-4. The boundary conditions for droplets, however, merit special attention. Taking a reference frame attached to the drop with origin at its center, these are

at $r = 0$,
$$V_r^{(1)} \text{ and } V_\theta^{(1)} \text{ are finite;} \tag{8.5-1a}$$

at $r = R$,
$$V_r^{(1)} = V_r^{(2)} = 0 \quad \text{(no flow across the interface),} \tag{8.5-1b}$$
$$V_\theta^{(1)} = V_\theta^{(2)}, \tag{8.5-1c}$$
$$\sigma_{r\theta}^{(1)} = \sigma_{r\theta}^{(2)}, \tag{8.5-1d}$$
$$\left(P + \sigma_{rr} + \frac{2\alpha}{R}\right)^{(2)} = (P + \sigma_{rr})^{(1)}; \tag{8.5-1e}$$

at $r \to \infty$,
$$V_z^{(2)} = V_\infty; \tag{8.5-1f}$$

where the superscripts (1) and (2) refer to the dispersed (inner) and continuous (outer) phases respectively. In writing these conditions, it is tacitly assumed that there are no surface active impurities present at the interface and thus, the constant surface tension shows up as a fixed term in the normal stress balance. Thus, the continuity and momentum equations, written for both media, together with the boundary conditions, provide the theoretical framework for the calculation of inner and outer flow fields, which in turn can be manipulated to evaluate the drag force on a spherical droplet.

The non-dimensionalization of the aforementioned equations and boundary conditions suggests that it is still possible to express the results in the form of the correction factor, X. But in this case, it becomes a function of another dimensionless parameter, namely, the ratio of inner to outer phase viscosity (β_μ). That is,
$$X = X(n, \beta_\mu), \tag{8.5-2a}$$

8.5 Creeping Flow of a Power-Law Fluid Around a Newtonian Droplet

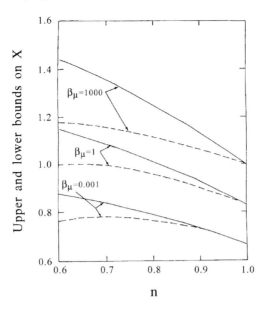

Figure 8.5-1 Drag correction factor for creeping fluid sphere motion in power-law media (Replotted from Mohan, 1974)

where

$$\beta_\mu = \frac{\mu^{(1)}}{m^{(2)}} \left(\frac{R}{V_\infty}\right)^{1-n^{(2)-1}}. \quad (8.5\text{-}2b)$$

Representative results on the variations of X in terms of upper and lower bounds on X with n and β_μ are depicted in Fig. 8.5-1. In the absence of any definitive information, the use of the arithmetic mean of the two bounds has been suggested (Mohan, 1974). At this juncture, two observations can be made. For $\beta_\mu \ll 1$, we recover the case of a gas bubble, whereas $\beta_\mu \gg 1$ denotes the case of a solid sphere. Note that for a Newtonian liquid, β_μ is equal to k used in Section 8.2-2.

Similar approximate results for other generalized Newtonian fluids have been summarized by Chhabra (1993a). The interplay between viscoelasticity and surface impurities on the dynamics of a single drop has been analyzed by Quintana (1992). Small deviations from a spherical shape have been simulated by Wagner and Slattery (1971). The intermediate Reynolds number flow of power-law fluids around spherical drops has been numerically analyzed by Nakano and Tien (1970).

8.5-1 Experimental Results on Falling Drops

Just as is the case with bubbles, drops also exhibit a variety of shapes under free falling conditions in non-Newtonian media. Broadly, for a drop size such that the surface forces are overcome by the viscous forces, the shape changes from spherical to oblate with increasing drop size. A further increase in size leads to tear drop with a tailing filament, and eventually, for very large drops, flattening occurs and the rear surface begins to fold in. Good photographs showing all such shapes have been presented by Acharya et al. (1978). Empirical correlations for drag on drops in generalized Newtonian fluids are also available in the literature but these have not been checked using independent data (Yamanaka and Mitsuishi, 1977; Acharya et al., 1978), and hence are too tentative to be included here. Aside from the aforementioned literature, the behavior of a single drop moving in confined flows in viscoelastic media has also been analyzed (Chan and Leal, 1979, 1981; Olbricht and Leal, 1982, 1983).

8.6 Flow in Packed Beds

8.6-1 Creeping Power-Law Flow in Beds of Spherical Particles: The Capillary Model

In the preceding two sections, we have discussed, in detail, very simple non-viscometric flows involving single particles. Multiparticle systems of considerable interest in chemical and polymer processing applications involve packed beds. Generally speaking, there have been two main approaches for studying pressure drops through packed beds. In the first approach, the packed bed is regarded as a bundle of tangled tubes of arbitrary cross-section; the method extends the previous results for single straight circular tubes to a collection of tubes. In the second approach, the bed is viewed as a collection of submerged objects and the pressure drop is simply the summation of the fluid dynamic drag on the objects. The capillary model is presented here in detail, while the second approach is touched upon only briefly and is left as an exercise.

It is assumed throughout the ensuing discussion that the packing is uniform everywhere in the column and that the diameter of the packing is much smaller than that of the column (i.e., no wall effects), and that there is no channeling. There are essentially three models available which fall within the general framework of capillary models: the Blake, the Blake-Kozeny, and the Kozeny-Carman model. In the Blake model, the bed is replaced by a bundle of straight tubes of complicated cross-section (characterized by a hydraulic radius, R_h). The mean interstitial velocity $\langle V \rangle$ is related to the superficial velocity V_0 via the well-known Dupuit equation, given by

$$V_0 = \varepsilon \langle V \rangle. \qquad (8.6\text{-}1)$$

V_0 is the average linear or superficial velocity of the fluid in the absence of packing, and ε refers to the void fraction. For a bed of uniform sized spheres, the hydraulic radius is given by (Bird et al., 1960)

$$R_h = \frac{\varepsilon d_p}{6(1 - \varepsilon)}, \qquad (8.6\text{-}2)$$

where d_p is the mean particle diameter.

8.6 Flow in Packed Beds

The Blake-Kozeny model, on the other hand, postulates that the effective length L_e of tangled capillaries is somewhat longer than the physical height of the bed, thereby introducing the so-called tortuosity factor, T, defined as (L_e/L). Lastly, the Kozeny-Carman model is similar to the Blake-Kozeny model, but it also corrects the average velocity for the tortuous nature of the flow path in the following manner:

$$\langle V \rangle = \left(\frac{V_o}{\varepsilon}\right)\left(\frac{L_e}{L}\right). \tag{8.6-3}$$

In equation 8.6-3, the term (L_e/L) reflects the fact that a fluid particle moving with the superficial velocity V_0 traverses the path length L in the same time as an actual fluid particle moving with velocity $\langle V \rangle$ covers an average effective length L_e. Early developments dealing with the flow of power-law fluids through packed beds employed the Blake-Kozeny model, whereas it has been argued in recent years that the Kozeny-Carman approach yields a somewhat better representation of pressure drop data (Kemblowski et al., 1987).

The starting point is the expression for the average velocity $\langle V \rangle$ for the laminar flow of a power-law fluid in a tube of diameter D, given by equation 4.2-8 (using eq. 4.2-2):

$$\langle V \rangle = \frac{4Q}{\pi D^2} = \frac{nD^{\frac{n+1}{n}}}{2(3n+1)}\left(-\frac{\Delta P}{4mL}\right)^{\frac{1}{n}}. \tag{4.2-8a}$$

Equation 4.2-8a is adapted for packed beds by noting that $D \equiv 4R_h$ and $L = L_e$. After substituting for these quantities from equations 8.6-1 to 8.6-3, we obtain the following relationship in terms of the superficial velocity:

$$V_0 = \frac{2n}{T(3n+1)}\left\{\frac{\varepsilon d_p}{6(1-\varepsilon)}\right\}^{\frac{n+1}{n}}\left(-\frac{\Delta P}{mLT}\right)^{\frac{1}{n}}. \tag{8.6-4}$$

Note the similarity between the form of equation 8.6-4 and the empirical modification of the Darcy law (Christopher and Middleman, 1965) given by

$$\nabla P = \frac{\mu_{eff}}{k} V_0^n. \tag{8.6-5}$$

It is convenient to introduce the dimensionless friction factor and Reynolds number as follows:

$$f = \left(-\frac{\Delta P}{L}\right)\left(\frac{d_p}{\rho V_0^2}\right)\left(\frac{\varepsilon^3}{1-\varepsilon}\right), \tag{8.6-6}$$

and

$$Re = \frac{\rho V_0^{2-n} d_p^n}{m}. \tag{8.6-7}$$

Now equation 8.6-4 can be rewritten in terms of f and Re as

$$f = \frac{A(n, \varepsilon, T)}{Re}, \tag{8.6-8}$$

where

$$A(n, \varepsilon, T) = \left(\frac{3n+1}{2n}\right)^n T^{n+1} \left[\frac{6(1-\varepsilon)}{\varepsilon}\right]^{n+1} \left(\frac{\varepsilon^3}{1-\varepsilon}\right). \quad (8.6\text{-}9)$$

Considerable confusion exists in the literature concerning the value of the tortuosity factor, T. For instance, based on the assumption that the streamlines on the average digress by 45° from the direction of flow, Carman (1956) suggested $T = \sqrt{2}$, whereas Christopher and Middleman (1965) employed a value of $25/12$. Based on heuristic arguments, Sheffield and Metzner (1976) proposed a value of $\pi/2$ for spherical particles. In most cases, however, the tortuosity factor is embodied into the experimentally determined constants and is not amenable to independent determination. Finally, the functional dependence of A on the flow behavior index n and on ε also depends on the version of the capillary model employed.

Perhaps the work of Kemblowski et al. (1987) is the most definitive and extensive study, and it is worthwhile to briefly mention their work here. Based on the analogy between the flow in circular pipes and in packed beds, Kemblowski et al. (1987) rearranged the Hagen-Poiseuille equation as follows:

$$R_h \left(\frac{-\Delta P}{L}\right) = \mu \left(\frac{K_0 V_0}{R_h}\right), \quad (8.6\text{-}10)$$

where K_0 is a constant and depends on the geometry; $K_0 = 2$ for circular tubes. The term on the left side is the shear stress at the wall, whereas the quantity within brackets on the right side is the shear rate at the wall for a Newtonian fluid. The latter may be regarded as the nominal shear rate at the wall for generalized Newtonian fluids, that is,

$$\langle \sigma_w \rangle = R_h \left(-\frac{\Delta P}{L}\right) \quad (8.6\text{-}11)$$

and

$$\langle \dot{\gamma}_{wn} \rangle = \frac{K_0 \langle V \rangle}{R_h}. \quad (8.6\text{-}12)$$

After substituting for R_h and $\langle V \rangle$ from equations 8.6-2 and 8.6-3, we obtains

$$\langle \sigma_w \rangle = \left(\frac{d_p}{6}\right) \frac{\varepsilon}{(1-\varepsilon)} \left(-\frac{\Delta P}{TL}\right) \quad (8.6\text{-}13)$$

and

$$\langle \dot{\gamma}_{wn} \rangle = 6 K_0 T \left(\frac{1-\varepsilon}{\varepsilon^2}\right) \left(\frac{V_0}{d_p}\right). \quad (8.6\text{-}14)$$

For generalized Newtonian fluids, we can postulate the following relation between the wall shear stress and the nominal shear rate:

$$\langle \sigma_w \rangle = m' \langle \dot{\gamma}_{wn} \rangle^{n'}, \quad (8.6\text{-}15)$$

where m' and n' are the apparent consistency and flow behavior indices respectively, akin to those introduced by Metzner and Reed (1955) for flow in circular tubes. For a true power-law fluid, for instance,

$$n' = n \tag{8.6-16}$$

and

$$m' = \left(\frac{3n+1}{4n}\right)^n m. \tag{8.6-17}$$

It should, however, be noted that the Rabinowitsch factor of $(3n+1/4n)^n$ is strictly applicable for circular tubes; but, fortunately, the calculations of Miller (1972) for non-circular conduits suggest that it is nearly independent of the conduit geometry. Kemblowski et al. (1987) further argued that the cross-section of capillaries in a packed bed is not necessarily circular. They recommended a value of $K_0 = 2.5$, which is the mean of the values for circular and slit geometries. They also employed $T = \sqrt{2}$, which is fairly close to $\pi/2$, and finally presented the following relation between the friction factor and a modified Reynolds number:

$$f = \frac{150}{Re*}, \tag{8.6-18}$$

where

$$Re* = Re\left(\frac{4n}{3n+1}\right)^n \left(\frac{15\sqrt{2}}{\varepsilon^2}\right)^{1-n} (1-\varepsilon)^{-n}. \tag{8.6-19}$$

Equation 8.6-18 seems to correlate the bulk of the literature data on the power-law fluid flow through packed beds of spheres up to $Re* \approx 1$. Subsequently, based on extensive experimental data, equation 8.6-18 has been generalized to encompass the flow at high Reynolds number. The resulting modified expression is given by

$$f = \frac{150}{Re*} + 1.75 \frac{KH^2}{\sqrt{K^2(H^2-1)^2 + H^2}}, \tag{8.6-20a}$$

where K and H are further correlated as

$$H = \zeta Re* \tag{8.6-20b}$$

$$\text{and} \quad \log K = \sum_{i=0}^{5} A_i n^i, \tag{8.6-20c}$$

with $A_0 = -1.7838$; $A_1 = 5.219$; $A_2 = -6.239$; $A_3 = 1.559$; $A_4 = 2.394$; and $A_5 = -1.12$; and

$$\log \zeta = \sum_{i=0}^{5} B_i n^i, \tag{8.6-20d}$$

with $B_0 = -4.9035$; $B_1 = 10.91$; $B_2 = -12.29$; $B_3 = 2.364$; $B_4 = 4.25$; and $B_5 = -1.896$.

Figure 8.6-1 shows the predictions of equations 8.6-20a-d.

372 Multiphase Systems

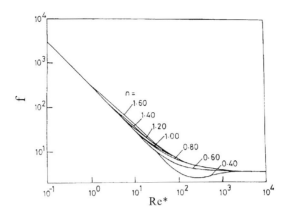

Figure 8.6-1 Predictions of equation 8.6-20
The modified Reynolds number, Re^*, is defined by equation 8.6-19 (Replotted from Kemblowski et al., 1987)

Example 8.6-1 Flow Through a Packed Bed

Estimate the pressure gradient required to maintain the flow of a Nylon 6 melt at 0.01 m³/s through a 150 mm diameter column packed with 0.5 mm glass beads. The physical properties of the Nylon 6 melt are: $m = 33.7$ kPa·s^n; $n = 0.6$, and $\rho = 770$ kg/m³. The packed column has a voidage, ε, of 0.39.

Solution

First, estimate the superficial velocity as

$$V_0 = \frac{0.01}{\frac{\pi}{4}(0.15)^2} = 0.57 \text{ m/s}.$$

Then, it follows that

$$Re^* = \frac{770(0.0005)^{0.6}(0.57)^{1.4}}{33.7 \times 1000 \times (1 - 0.39)^{0.6}} \left(\frac{4 \times 0.6}{3 \times 0.6 + 1}\right)^{0.6} \left(\frac{15\sqrt{2}}{0.39^2}\right)^{0.4} = 961 \times 10^{-6} \ll 1$$

and

$$f = \frac{150}{Re^*} = 156 \times 10^3.$$

Hence

$$\left(-\frac{\Delta P}{L}\right) = \frac{f \rho V_0^2}{d_p} \frac{1-\varepsilon}{\varepsilon^3} = \frac{156 \times 10^3 \times 770 \times 0.57^2}{5 \times 10^{-4}} \times \frac{(1-0.39)}{(0.39)^3} \text{ Pa/m}$$

$$= 8.02 \times 10^{11} \text{ Pa/m}$$

8.6-2 Other Fluid Models

The aforementioned analysis has also been extended to embrace the laminar flow of a wide range of generalized Newtonian fluids, including Ellis model fluids (Sadowski and Bird, 1966), Carreau fluids (Park et al., 1975), Herschel-Bulkley fluids (Al-Fariss and Pinder, 1987), and Bingham plastic fluids (Vradis and Protopapas, 1993). These have recently been critically evaluated by Chhabra (1993a).

8.6-3 Viscoelastic Effects

It is now generally agreed that the flow of viscoelastic fluids in packed beds results in a greater pressure drop than that which can be ascribed to the shear rate dependent viscosity. At low flow rates, the pressure drop is determined largely by shear viscosity. As the flow rate is gradually increased, viscoelastic effects begin to appear. When the pressure drop is plotted against a suitably defined Deborah or Weissenberg number, beyond a critical value of the Deborah number, the pressure drop increases rapidly. Figure 8.6-2 shows this behavior clearly for the flow of three different polymer solutions. While numerous workers have reported similar results, there is very little quantitative agreement on the magnitude of increase in pressure drop or the critical value of the Deborah number, De, marking the onset of viscoelastic effects (Kemblowski et al., 1987; Chhabra, 1993a). One possible reason for this lack of agreement is the diversity of methods employed to evaluate the fluid relaxation time used to define the Deborah number. A survey of the available correlations for pressure drop for viscoelastic fluid flow in packed beds reveals that the following generic form provides an adequate representation of the experimental data:

$$f = \frac{A}{Re^*}(1 + BDe^2). \tag{8.6-21}$$

The quadratic dependence on the Deborah number in equation 8.6-21 has some theoretical basis (Wissler, 1971; Zhu, 1990). Based on the available experimental data, Kemblowski and Dziubinski (1978) evaluated A and B to be given by 150 and 8 respectively. Note that for inelastic fluids or at very small values of the Deborah number, equation 8.6-21 reduces to equation 8.6-18.

In recent years, it has been argued that the flow through porous media involves a substantial extensional component, which is mainly responsible for the reported dramatic increases in pressure drop (James and McLaren, 1975; Durst and Haas, 1981; Durst et al., 1987). Other mechanisms, including slip effects, adsorption, pore blockage, and *in situ* pseudo-dilatant behavior, have also been dealt with in detail by others (e.g., Sorbie, 1991; Chhabra, 1993a). The role of extensional flow has been recently examined by Kozicki and Slegr (1994).

374   Multiphase Systems

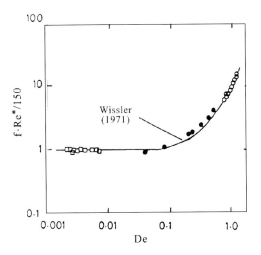

Figure 8.6-2 Typical results showing viscoelastic effects in porous media flow (Replotted from Kemblowski et al., 1987)

8.6-4   Wall Effects

In most applications, packed columns are of finite size in the radial direction and thus the confining walls influence the flow. Wall effects manifest in two ways. First, the wall of the tube provides an extra surface (and wetted perimeter), which is not taken into account while defining the hydraulic radius via equation 8.6-2. The second effect is that the bed voidage in the wall region is significantly different from that in the central part of the bed. Three approaches have evolved for taking wall effects into account while calculating the frictional pressure gradient. The first method involves correlating the numerical constants appearing in equations such as 8.6-20 explicitly with the particle-to-tube diameter ratio. This approach has been used for Newtonian as well as for power-law fluids (Fand and Thinakaran, 1990; Srinivas and Chhabra, 1992). There is, however, little quantitative agreement amongst various workers regarding the nature of the relationship between the constants and the particle-to-tube diameter ratio.

The second approach accounts for the contribution of the bounding walls to the wetted perimeter and flow area in the calculation of the hydraulic radius. Mehta and Hawley (1969) thus obtained the following expression for R_h:

$$R_h = \frac{\varepsilon d_p}{6(1-\varepsilon)M}, \qquad (8.6\text{-}22)$$

where

$$M = 1 + \frac{\left(\frac{4}{6}\right)\left(\frac{d_p}{D_c}\right)}{(1-\varepsilon)}. \qquad (8.6\text{-}23)$$

Hence, we replace d_p by (d_p/M) in the definitions of the Reynolds number and friction factor. Note that $M \to 1$ with decreasing (d_p/D_c). The utility of this approach in correcting the results for wall effects, especially in narrow columns, for power-law and other generalized Newtonian fluids, has been demonstrated by Park et al. (1975) and by Rao and Chhabra (1993).

The third method relies on the knowledge of the detailed radial voidage profiles prevailing in a packed column. The available experimental results clearly show that the voidage is almost unity in the vicinity of a wall, it oscillates about its mean value moving away from the wall, and finally, it attains its mean value. Such a three-region model was developed by Cohen and Metzner (1981). They fitted the bulk of the literature data to analytical expressions and suggested that though the wall region extends only up to 1/4 of the particle diameter, the voidage becomes independent of the radial location only at about eight particle diameters away from the wall. Based on extensive comparisons between experiments and predictions, Cohen and Metzner (1981) concluded that if the wall effects are to be avoided, packed beds with tube-to-particle diameter ratio greater than 30 should be used. Notwithstanding its sound basis, the three-region model is rather cumbersome to use, for it is neither always possible to measure detailed voidage profiles nor are such data readily available. From a process calculations standpoint, the second method of applying the wall correction represents a good compromise, and no additional information is required for its application.

8.6-5 Effects of Particle Shape

Both particle shape and surface roughness influence the frictional pressure loss for flow through packed beds. Based on a critical evaluation of the pertinent data for Newtonian fluids, MacDonald et al. (1979) concluded that while the effect of particle shape can reasonably be accounted for by using an effective size (volume equivalent sphere diameter multiplied by sphericity) in the estimation of the pressure loss, the surface roughness exerts a significant influence on the pressure drop only under highly turbulent conditions. The latter is thus of little consequence in the case of non-Newtonian fluids, which are rarely processed in the turbulent region. Not much is known about the effect of particle shape or the particle size distribution on pressure loss incurred by non-Newtonian fluids in packed beds. The limited available data seem to suggest that the use of the volume equivalent diameter multiplied by a sphericity factor in lieu of d_p provides an adequate scheme for dealing with non-spherical packing particles (Chhabra and Srinivas, 1991; Sharma and Chhabra, 1992). Likewise, the use of the mean of hydraulic radii for beds with particle size distribution has been proposed, at least in the streamline region, to estimate pressure losses in beds of binary-sized particles (Rao and Chhabra, 1993).

8.6-6 Submerged Objects' Approach to Fluid Flow in Packed Beds: Creeping Flow

As mentioned earlier, in this approach, a packed bed is envisioned to be a collection of objects immersed in a moving fluid, and the frictional pressure gradient across the packed bed is simply a summation of hydrodynamic drag on the objects. Thus, the central problem in this analysis is that of evaluating the fluid dynamic drag on a typical particle present in the assemblage. For instance, for the flow of Newtonian fluids, Brinkman (1947) calculated the force on a typical particle by postulating it to be immersed in a homogeneous and isotropic porous medium, and derived an expression for the effective permeability of the surrounding porous medium. In another approach, the drag on a representative particle in the ensemble has been evaluated by solving the Navien-Stokes equation over periodic arrays of known geometric configuration, such as body- and face-centered cubic arrays (Sangani and Acrivos, 1982). Within the framework of this approach, another method, albeit less rigorous than that involving periodic arrays, involves the use of cell models (Happel and Brenner, 1973). Here, the influence of neighboring particles is simulated by enclosing the particle in question in an artificial cell. One such formulation, namely the free surface cell model, is presented here to give the reader an idea of what is involved in this line of analysis. The free surface cell model description envisions each sphere of radius R to be surrounded by a spherical envelope of radius R_∞, as shown schematically in Fig. 8.6-3, with the sphere moving slowly in a spherical cavity. Owing to the ϕ-symmetry, the flow is two-dimensional and we seek a solution of equations 8.3-2 to 8.3-4 in terms of $V_r(r, \theta)$, $V_\theta(r, \theta)$, and $P(r, \theta)$. While the boundary conditions on the sphere surface are the standard no-slip conditions, these are not immediately obvious at $r = R_\infty$. Indeed, the latter is a matter of longtime debate (Happel and Brenner, 1973). The free surface cell model assumes no radial mass flux and no friction at $r = R_\infty$, thereby emphasizing the non-interactive nature of cells. Thus, we can write

$$\text{at} \quad r = R, \qquad V_r = V_0 \cos \theta, \qquad (8.6\text{-}24a)$$

$$\text{and} \quad V_\theta = -V_0 \sin \theta, \qquad (8.6\text{-}24b)$$

$$\text{at} \quad r = R_\infty, \qquad V_r = 0, \qquad (8.6\text{-}25a)$$

$$\text{and} \quad \sigma_{r\theta} = 0. \qquad (8.6\text{-}25b)$$

The simplest relation between R and R_∞ is obtained by assuming that the porosity of each cell is equal to the mean voidage for the overall assemblage. That is,

$$R_\infty = R(1 - \varepsilon)^{-\frac{1}{3}}, \qquad (8.6\text{-}26)$$

For the slow flow of power-law fluids, we can show via the standard dimensional arguments that the drag correction factor, X, now becomes a function of the bed voidage and the power-law index. This functional relationship has been established by numerically solving equations 8.3-2 and 8.3-4 in conjunction with equations 8.6-24 and 8.6-25 for power-law fluids (Jaiswal et al., 1993a).

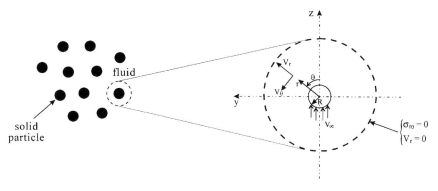

Figure 8.6-3 Schematic representation of the free surface cell description of flow in a packed bed

The relation between the drag coefficient and the friction factor for packed beds is simply obtained by equating the rate of energy dissipation per unit cell volume to the pressure drop per unit bed height, that is,

$$\left(-\frac{\Delta P}{L}\right) = \frac{F_D}{\frac{4}{3}\pi R_\infty^3}. \tag{8.6-27}$$

Now, substituting for the drag force ($F_D = C_D/(1/2)\rho V_0^2 \pi R^2$) and for the friction factor from equation 8.6-6, we obtain

$$f = \frac{3}{4} C_D \varepsilon^3. \tag{8.6-28}$$

Preliminary comparisons between predictions and experimental data are satisfactory for shear-thinning fluids up to about $Re^* \approx 20$ and for shear-thickening fluids in the creeping flow region, $Re^* < 1$ (Jaiswal et al., 1991, 1994). Similar results are also available for the Carreau model fluids (Jaiswal et al., 1993b).

8.7 Fluidized Beds

With the upward flow of a liquid through a bed of particles, we can discern three different flow regimes depending upon the flow rate of a liquid and the size and density of the particles. At sufficiently small flow rates, the bed behaves like a fixed or packed bed wherein the particles are in contact with each other; but at sufficiently high flow rates, the solid particles will be freely supported in the liquid to give rise to what is known as a fluidized bed. At very high flow rates, the particles will be conveyed out of the system. The bed in which the conditions cease to exist as a fixed bed is called the *incipiently fluidized bed*, and the corresponding velocity at this condition is known as the *minimum fluidization velocity*. In between this value and the transporting velocity, the bed continually expands with increasing liquid velocity. Behavior of this kind is known as *particulate fluidization*, and occurs in most liquid-solid systems except when the solids are too heavy. Fluidized

378   Multiphase Systems

beds are widely used as heat/mass exchangers and chemical reactors, since significantly enhanced heat and mass transfer rates can be achieved due to vigorous mixing between the liquid and the particulate phases. Potential applications of fluidized beds involving non-Newtonian media have been discussed by Joshi (1983) and by Tonini (1987).

It is readily recognized that the minimum fluidization velocity and the velocity-voidage relationship represent important design parameters for fluidized beds, and therefore the present discussion is limited to these two aspects.

8.7-1   Minimum Fluidization Velocity

For the upward flow of a liquid through a bed, the pressure drop, ΔP, will initially increase as the superficial velocity, V_0, of the liquid increases. The system behaves like a fixed bed (see Fig. 8.7-1). When the velocity has reached a value such that the frictional pressure drop is equal to the buoyant weight per unit area, any further increase in velocity results in the rearrangement of particles such that the pressure drop across the bed remains constant, as shown schematically in Fig. 8.7-1. The critical value of the velocity marking this transition from a fixed to a fluidized condition is called the minimum fluidization velocity, V_{mf}, and the corresponding bed voidage is designated by ε_{mf}. Some hysteresis is observed in the reverse direction, especially in the fixed bed regime due to repacking of the bed, as shown schematically in Fig. 8.7-1. The generally accepted method for the determination of V_{mf} from $\Delta P - V_0$ behavior is to draw two separate lines through the fixed and fluidized regions. Their point of intersection yields the value of V_{mf}.

From a theoretical standpoint, at the point of the incipient fluidization, the pressure drop across the bed is given by its buoyant weight, that is,

$$-\Delta P = (1 - \varepsilon_{mf})(\rho_s - \rho)Lg. \tag{8.7-1}$$

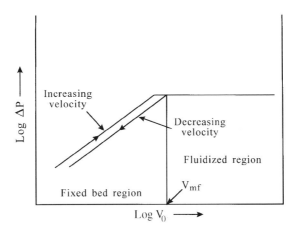

Figure 8.7-1   Ideal pressure drop-velocity curve showing fixed and fluidized bed regions

Though an incipiently fluidized bed represents a slightly loosened bed, it is reasonable to treat it as a fixed bed, since the particles are still in contact with each other, and hence we can estimate the pressure loss across the bed using equation 8.6-6:

$$-\Delta P = \frac{fL\rho V_{mf}^2}{d_p}\left(\frac{1-\varepsilon_{mf}}{\varepsilon_{mf}^3}\right). \tag{8.7-2}$$

By eliminating ΔP in these two expressions, we obtain

$$\frac{f\rho V_{mf}^2}{d_p} = (\rho_s - \rho)g\varepsilon_{mf}^3. \tag{8.7-3}$$

Thus, equation 8.7-3 provides a convenient method to estimate the value of V_{mf} in an envisaged application for the known size and density of particles (d_p, ρ_s) and the liquid density ρ. The corresponding value of f is estimated using expressions similar to equation 8.6-18 or 8.6-20. Clearly, in the so-called streamline flow, equation 8.7-3 is explicit in V_{mf}, whereas an iterative procedure is required outside the creeping flow regime. The main limitation of equation 8.7-3 is the fact that a fairly accurate value of ε_{mf} is required, which is generally not known *a priori* in most situations. Most of the available methods for the prediction of V_{mf} have been critically evaluated in a recent paper (Chhabra, 1993d).

Example 8.7-1 Minimum Fluidization Velocity
Estimate the value of the minimum fluidization velocity for a bed of 3 mm glass beads of density 2500 kg/m^3 to be fluidized by a polymer solution with $m = 13.7$ Pa·s^n, $n = 0.67$, and $\rho = 1000$ kg/m^3. The value of voidage at the incipient fluidized condition is 0.38.

Solution
As a first approximation, assume the Reynolds number to be small so that equation 8.6-18 can be used to estimate the value of the friction factor, f. Substituting for f in equation 8.7-3, we get

$$\frac{150\, \rho V_{mf}^2}{Re^* \, d_p} = (\rho_s - \rho)g\varepsilon_{mf}^3,$$

which upon rearrangement and substitution for Re^* yields

$$V_{mf}^n = \frac{(\rho_s - \rho)g\varepsilon_{mf}^3 d_p^{n+1}}{150m}\left(\frac{4n}{3n+1}\right)^n \left(\frac{15\sqrt{2}}{\varepsilon^2}\right)^{1-n}(1-\varepsilon)^{-n}.$$

Now, substituting numerical values, we get

$$V_{mf}^{0.67} = \frac{(2500 - 1000)(9.81)(0.38)^3(3 \times 10^{-3})^{1.67}}{150 \times 13.7}\left(\frac{15\sqrt{2}}{0.38^2}\right)^{0.33}\left(\frac{4 \times 0.67}{3 \times 0.67 + 1}\right)^{0.67}(1 - 0.38)^{-0.67}$$

or

$$V_{mf} = 0.780\ \mu\text{m/s}.$$

380 Multiphase Systems

We check the value of the Reynolds number:

$$Re^*_{mf} = \frac{(1000)(0.78 \times 10^{-6})^{2-0.67}(3 \times 10^{-3})^{0.67}}{13.7} \left(\frac{4 \times 0.67}{3 \times 0.67 + 1}\right)^{0.67}$$

$$\times \left(\frac{15\sqrt{2}}{0.38^2}\right)^{1-0.67} (1-0.38)^{-0.67}$$

$$= 74 \times 10^{-9},$$

which is quite small. Note that

$$V_{mf} \propto [\varepsilon_{mf}^{2n+1}(1-\varepsilon_{mf})^{-n}]^{\frac{1}{n}}$$

$$= \frac{\varepsilon_{mf}^{2+\frac{1}{n}}}{1-\varepsilon_{mf}}.$$

So a small error in the determination of ε_{mf} will yield a substantial change in the prediction of V_{mf}. For instance, the use of a value of 0.4 for ε_{mf} leads to a 25% increase in the value of V_{mf}. In order to circumvent this difficulty, some investigators have correlated the minimum fluidization velocity, V_{mf}, with the free settling velocity, V_∞, of a single particle in the same fluid, but this approach often entails large errors. For instance, Machac et al. (1986) presented the following simple relationship:

$$V_{mf} = 0.019 V_\infty.$$

For the present example, the value of V_∞ is estimated to be 151.6 μm/s via equation 8.3-11, and this in turn yields a value of V_{mf} = 2.88 μm/s, which is about 3.5 times the value calculated previously!

8.7-2 Bed Expansion Behavior

Once the superficial velocity of the liquid exceeds the minimum fluidization velocity, the mean voidage of the bed gradually increases and the frictional pressure drop across the bed remains constant at a value equal to its buoyant weight. Figure 8.7-2 confirms this kind of behavior. It is customary to depict bed expansion behavior of fluidized beds by plotting dimensionless velocity ratio (V_0/V_∞) versus bed voidage, as shown in Fig. 8.7-3 for 3.57 mm glass spheres fluidized by a carboxy methyl cellulose solution. The type of dependence is seen to be similar to that observed with Newtonian fluids. Often, such data have been represented and correlated using expressions of the following form:

$$\frac{V_0}{V_\infty} = \varepsilon^Z. \qquad (8.7\text{-}4)$$

In turn, the index Z is correlated with the single particle Reynolds number Re_∞, defined via equation 8.3-11. Khan and Richardson (1989) have reviewed the extensive literature on the prediction of Z for Newtonian liquids in the absence of wall effects. The available limited data for power-law fluids suggest that the same values of Z can be used to predict the bed expansion behavior for power-law fluids, provided the modified Reynolds number

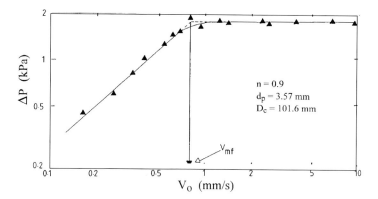

Figure 8.7-2 Experimental pressure drop-velocity curve and determination of V_{mf} (Replotted from Srinivas and Chhabra, 1991)

$(Re = \rho V_\infty^{2-n} d_p^n / m)$ is used instead of the usual Reynolds number (Srinivas and Chhabra, 1991).

The effect of particle shape on the fluidization behavior has been investigated recently, and the use of the volume equivalent sphere diameter was found to be adequate for design calculations (Sharma and Chhabra, 1992). Very little is known concerning the importance of viscoelastic effects in this flow configuration. The currently available information is of conflicting and inconclusive nature (Chhabra, 1993a).

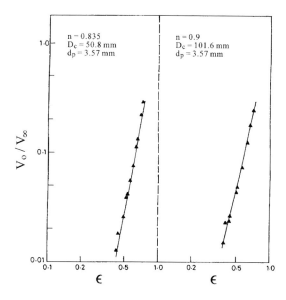

Figure 8.7-3 Typical bed expansion data for $n = 0.835$ and $n = 0.9$ (Replotted from Srinivas and Chhabra, 1991)

8.7-3 Heat and Mass Transfer in Packed and Fluidized Beds

Little work is available on the interphase or the wall-to-bed heat and mass transfer processes in packed and fluidized beds (Chhabra, 1993a, b). Kawase and Ulbrecht (1981) extended the analysis of Pfeffer (1964) to study particle-liquid mass transfer in packed beds. They developed the following expression for the Sherwood number

$$Sh = \frac{k_L d_p}{D_A} = A_1(n) \varepsilon^{-\frac{1}{n+1}} Re^{\frac{n+2}{3(n+1)}} Sc^{\frac{1}{3}}, \tag{8.7-5}$$

where $A_1(n)$ is a weak function of n, k_L is the mass transfer coefficient, and Sc is the Schmidt number defined for power-law fluids by

$$Sc = \left(\frac{m}{\rho D_A}\right)\left(\frac{d_p}{V_0}\right)^{1-n}. \tag{8.7-6}$$

Some experimental results on the interphase particle-to-liquid mass transfer in packed and fluidized beds are available in the literature. Kumar and Upadhyay (1981) proposed the following empirical correlation in terms of the j-factor:

$$\varepsilon j = \frac{0.765}{Re_1^{0.82}} + \frac{0.365}{Re_1^{0.386}}, \tag{8.7-7}$$

where the mass transfer factor j and the modified Reynolds number Re_1 are defined respectively as

$$j = \left(\frac{k_L}{V_0}\right)\left[Sc\left(\frac{3n+1}{4n}\right)^n\left\{\frac{12(1-\varepsilon)}{\varepsilon^2}\right\}^{n-1}\right]^{\frac{2}{3}} \tag{8.7-8}$$

and

$$Re_1 = Re\left(\frac{4n}{3n+1}\right)^n\left\{\frac{12(1-\varepsilon)}{\varepsilon^2}\right\}^{1-n}. \tag{8.7-9}$$

The independent experimental data available in the literature (Wronski and Szembek-Stoeger, 1988; Hilal et al., 1991; Hwang et al., 1993) also correlate satisfactorily with equation 8.7-7, as shown in Fig. 8.7-4. Combined, these studies encompass wide ranges of conditions such as $10^{-4} \le Re_1 \le 40$; $800 \le Sc \le 71\,000$; and $1 > n > 0.63$. Virtually no results are available for heat transfer and for viscoelastic fluids in packed and fluidized beds.

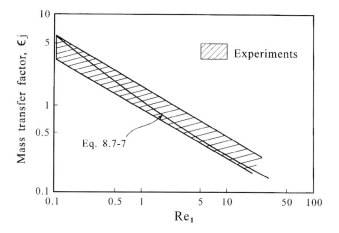

Figure 8.7-4 Particle-liquid mass transfer in packed and fluidized beds (Reprinted with permission R.P. Chhabra, *Bubbles, Drops and Particle in Non-Newtonian Fluids*, 1993. Copyright CRC Press, Boca Raton, Florida)

8.8 Problems*

8.8-1$_b$ Einstein's Result

(a) Show that the velocity profile (equation 8.2-7) and the pressure profile (equation 8.2-8) satisfy the Navier-Stokes equation for low inertial flow.
(b) Integrate equation 8.2-13 to obtain equation 8.2-15.

8.8-2$_b$ Oldroyd's Emulsion Model

(a) Following the procedure outlined in Example 8.2-1, compare the Oldroyd emulsion model predictions for $G'(\omega)$ and $G''(\omega)$ to those of the Jeffreys model (Example 5.2-7), and obtain the expressions for λ_1 and λ_2.
(b) Compute λ_1 and λ_2 for $\phi = 0.05$, $\mu_m = 1$ Pa·s, $\alpha/d_p = 2$ mPa, and for $k = 0.1$, 1.0, and 10. Illustrate the result in terms of $G'(\omega)$ and $G''(\omega)$ on log-log scales.

8.8-3$_b$ Palierne's Emulsion Model

Show that the Palierne emulsion model reduces to the Oldroyd model for Newtonian droplets of uniform diameter in a Newtonian matrix. (In the simpler formulation of the

*Problems are labelled by subscript a or b indicating the level of difficulty

Palierne model, presented in Section 8.2-2, the effect of particle interactions averages to zero (Palierne, 1990).)

8.8-4$_b$ Flow About a Sphere

For mildly shear-thinning fluids, Hirose and Moo-Young (1969) argued that we can use the Newtonian flow field to evaluate the power-law viscosity expressed by equation 2.2-18, where the second invariant of the rate-of-deformation tensor, II_{ij}, is evaluated using the Stokes streamfunction obtained for the creeping flow of a Newtonian fluid about a solid sphere or about a spherical bubble.

(a) Show that this approximation yields the following expressions for X defined in equation 8.3-11 in the case of a sphere and a bubble respectively:
For a solid sphere:

$$X = 3^{\frac{3n-3}{2}} \left(\frac{-22n^2 + 29n + 2}{n(n+2)(2n+1)} \right); \qquad (8.8\text{-}1)$$

For a spherical bubble:

$$X = 2^n 3^{\frac{n-3}{2}} \left(\frac{-8n^2 + 4n + 13}{n(2n+1)(n+2)} \right). \qquad (8.8\text{-}2)$$

(b) Compare your results critically with the values reported in Table 8.3-1.
(c) Alternatively, we can argue that the flow variables can be expanded in terms of a series as

$$\psi = \psi_0 + (n-1)\psi_1 + (n-1)^2 \psi_2 + \cdots$$
$$\text{and} \quad P = P_0 + (n-1)P_1 + (n-1)^2 P_2 + \cdots.$$

Show that this approximation leads to the following expression based on first-order terms for the flow about a spherical bubble:

$$X = 2^n 3^{\frac{n-3}{2}} [1 - 3.83(n-1)].$$

Compare these predictions with those obtained in (a).

8.8-5$_b$ Friction Factor for a Packed Bed

(a) Obtain the friction factor-Reynolds number relationship for the laminar flow of a Bingham plastic fluid through a packed bed via the Kozeny-Carman model, that is, the equivalent of equation 8.6-18.
(b) Derive the corresponding expression to equation 8.6-22, including wall effects.

(c) Obtain an approximate expression for $X(n, \varepsilon)$ for the flow of a power-law fluid through a packed bed via the free surface cell model employing the approximation of problem 8.8-4.

8.8-6a Criterion for Flow in a Viscoplastic Fluid

A sphere of radius R and density ρ_s is embedded in a viscoplastic fluid (yield stress σ_0 and density $\rho < \rho_s$). Obtain the criterion for the commencement of the sedimentation of the sphere on account of its own weight.

$$\text{Answer:} \quad \frac{3\pi\sigma_0}{4R(\rho_s - \rho)g} \leq 1$$

9 Liquid Mixing

9.1 Introduction	387
9.2 Mechanisms of Mixing	388
9.2-1 Laminar Mixing	389
9.2-2 Turbulent Mixing	391
9.3 Scale-Up and Similarity Criteria	391
9.4 Power Consumption in Agitated Tanks	396
9.4-1 Low Viscosity Systems	396
9.4-2 High Viscosity Inelastic Fluids	397
9.4-3 Viscoelastic Systems	412
9.5 Flow Patterns	414
9.5-1 Class I Agitators	415
9.5-2 Class II Agitators	416
9.5-3 Class III Agitators	418
9.6 Mixing and Circulation Times	420
9.7 Gas Dispersion	423
9.7-1 Gas Dispersion Mechanisms	423
9.7-2 Power Consumption in Gas Dispersed Systems	425
9.7-3 Bubble Size and Holdup	428
9.7-4 Mass Transfer Coefficient	429
9.8 Heat Transfer	430
9.8-1 Class I Agitators	431
9.8-2 Class II Agitators	432
9.8-3 Class III Agitators	434
9.9 Mixing Equipment and its Selection	436
9.9-1 Mechanical Agitation	436
9.1-1.1 Tanks	436
9.1-1.2 Baffles	436
9.1-1.3 Impellers	437
9.9-2 Extruders	437
9.10 Problems	439
9.10-1_a Power Requirement for Shear-Thinning Fluids	439
9.10-2_a Effective Deformation Rate	440
9.10-3_a Bottom Effects on the Metzner-Otto Constant	440
9.10-4_b Effective Deformation Rate in the Transition Regime	440

In many industrial mixing operations, the materials, products, and intermediates are rheologically complex. Numerous examples can be found in the polymer-based industries (synthetic rubbers, plastics, fibers, resins, paints, coatings, and adhesives), food-processing industries, biochemical operations, and in the manufacturing of detergent, fertilizers, explosives, and propellants. This chapter is devoted to mixing concepts applied to non-Newtonian fluids. As most non-Newtonian fluids are rather viscous materials, mixing then proceeds in the so-called laminar regime. Hence, very little of the chapter is on turbulent mixing.

9.1   Introduction

Mixing is perhaps one of the most common processes encountered in the chemical and processing industries. The term *mixing* is used to denote the operation aimed at reducing the degree of non-uniformity, or gradient of a property in a system. Examples of such properties are temperature, concentration, and viscosity. Mixing is accomplished by moving material from one region to another. The process of mixing may be of interest simply as a means of producing a desired degree of homogenization, but it is also frequently used to enhance heat and mass transfer processes, especially in chemically reactive systems.

At the outset, it is instructive and useful to consider some common examples of industrial mixing operations, as this will not only elucidate the ubiquitous nature of mixing, but will also provide an appreciation of some of the associated difficulties. We can classify mixing problems based on a variety of factors such as the uniformity and the degree of homogenization of the final product. Perhaps it is more meaningful to use a classification scheme based on the phases present in a mixing application, for example, liquid-liquid or liquid-solid, since this strategy of classification also allows the development of a unified approach to the problems encountered in a range of process industries. Table 9.1-1 lists some common examples involving different types of mixing processes. An examination of this table clearly shows that the mixing process cuts across the boundaries among a range of industries, and it may be required to mix virtually anything, be it a gas, a liquid, or a solid. It is thus not possible to consider the complete spectrum of mixing problems here. In this chapter consideration is given primarily to the batch mixing of Newtonian and non-Newtonian liquids, followed by terse descriptions of gas-liquid systems and heat/mass transfer in agitated systems. However, readers interested in the other types of mixing applications are referred to excellent books and review articles available on this subject (Harnby et al., 1992; Ulbrecht and Patterson, 1985; Ottino, 1990; Tatterson 1991).

Broadly, there are two kinds of problems to be considered—how to design and select mixing equipment for a potential application, and how to ascertain whether a mixer is suitable for a particular duty. In either case, a thorough understanding of the following aspects of mixing is required (covered in the corresponding sections of the chapter):

9.2 Mechanisms of Mixing
9.3 Scale-Up or Similarity Criteria
9.4 Power Consumption in Agitated Tanks

388   Liquid Mixing

Table 9.1-1   Types and Examples of Mixing

Type of mixing	Examples	Remarks
Single liquid	Blending of miscible petroleum products and silicone oils. Agitation to promote heat and mass transfer processes.	More difficult to mix/agitate highly viscous Newtonian and non-Newtonian systems.
Liquid-liquid	Immiscible liquids as encountered in liquid-liquid extraction; emulsions in food, brewing, and pharmaceutical processes; polymeric alloys.	Main objective to produce large interfacial area.
Liquid-solid	Suspension of particles in low viscosity systems by mechanical agitation; incorporation of carbon black powder and other fillers into a viscous non-Newtonian matrix to produce composites.	In case of fine particles surface forces play an important role.
Gas-liquid-solid	Slurry reactors; three-phase fluidized beds.	Good mixing is required for effective operation.
Solid-solid	Formation of concrete by blending sand, cement, and aggregates; production of gun powder.	Strongly dependent upon size and shape as well as surface properties of the components.

9.5 Flow Patterns
9.6 Mixing and Circulation Times
9.7 Gas Dispersion
9.8 Heat Transfer
9.9 Mixing Equipment and its Selection

9.2   Mechanisms of Mixing

For an efficient accomplishment of mixing to produce a uniform blend, it is necessary to fulfill two requirements in the case of liquid mixing devices. First, there must be a bulk or convective flow so that there are no stagnant zones. Second, there must be a zone of high shear in which the inhomogeneities are progressively broken down into smaller and smaller units. Both these steps are energy-consuming, and eventually the mechanical energy is dissipated as heat; the fraction of energy consumed by each of these steps varies from one application to another. Depending upon the fluid properties (viscosity in particular) the flow in mixing vessels may be laminar or turbulent, with a substantial transition zone in between these two regimes, and generally both laminar and turbulent conditions prevail in different parts of the vessel. It is, however, convenient to consider laminar and turbulent mixing separately simply because of the inherently different underlying mechanisms responsible for the mixing process.

9.2.1 Laminar Mixing

This mode of mixing is usually associated with highly viscous Newtonian or non-Newtonian liquids ($\eta > 10$ Pa·s). In such systems, the inertial forces die out quickly and the impeller must cover a significant proportion of the cross-section of the mixing vessel to cause sufficient bulk motion. Owing to the high velocity gradients in the vicinity of the rotating impeller, the fluid elements in this region deform and stretch continually and gradually become thinner and with each pass of the fluid element through the high shear zone. Figure 9.2-1 shows this mode of mixing schematically.

Elongational flow occurs simultaneously. This also leads to the thinning or flattening of the fluid elements of constant volume (as shown in Fig. 9.2-2). Both shear and elongation mechanisms give rise to stresses that lead to a reduction in droplet size and an increase in interfacial area, thereby resulting in a homogeneous mixture.

Finally, it must be recognized that the ultimate homogenization is brought about only by molecular diffusion, which unfortunately is a slow process in highly viscous systems.

A similar mixing process also occurs when a liquid is sheared between two rotating cylinders. During each turn, the fluid element becomes thinner, and the molecular diffusion takes over when the elements become sufficiently thin. This mode of mixing is shown schematically in Fig. 9.2-3, wherein the second component is shown as being introduced perpendicular to the direction of motion.

Finally, in highly viscous systems, mixing can also be effected by physically splicing the fluid elements into smaller units and redistributing them as shown in Fig. 9.2-4. In-line mixers rely largely on this mechanism.

Thus, mixing in viscous liquids is accomplished by several mechanisms, which gradually reduce the size or scale of fluid elements and then redistribute them in the bulk. If there are initial differences in concentration of a soluble material, homogenization is gradually achieved, with molecular diffusion becoming progressively more important as the element size is reduced. Ottino (1989) has presented excellent color photographs illustrating the different stages of mixing in highly viscous liquids.

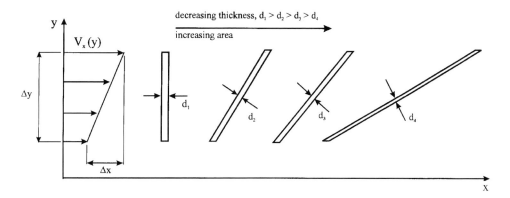

Figure 9.2-1 Thinning of fluid elements due to laminar shear flow

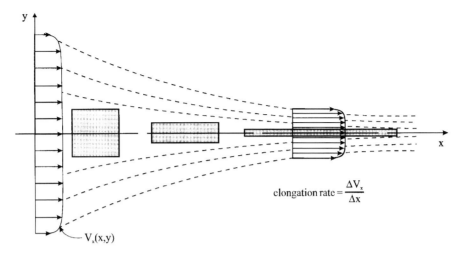

Figure 9.2-2　Thinning of fluid elements due to extensional flow

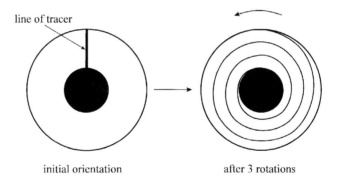

Figure 9.2-3　Laminar shear mixing in a coaxial cylinder arrangement

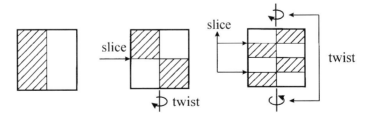

Figure 9.2-4　Schematic representation of distributive mixing due to cutting and folding

9.2-2 Turbulent Mixing

For low viscosity systems ($\eta < 10$ mPa·s), the bulk flow pattern in mixing vessels with rotating impellers is invariably turbulent. Hence, the inertia imparted by the rotating impeller is sufficient to cause the liquid to circulate throughout the vessel and return to the impeller. Turbulent eddy diffusion occurs throughout the vessel with a maximum in the impeller region. There is no question that eddy diffusion is inherently much faster than molecular diffusion (and, consequently, turbulent mixing is less time-consuming than laminar mixing), but the ultimate homogenization takes place only via molecular diffusion, which is also enhanced in low viscosity systems.

Turbulent flow is hopelessly complex, and the calculation of the prevailing flow fields in mixing vessels is not amenable to a theoretical treatment. However, when the Reynolds number of the main flow is sufficiently high, some useful but qualitative insights into the mixing process can be developed via the theory of local isotropic turbulence (Levich, 1965). Traditionally, turbulent flow is considered as a spectrum of velocity fluctuations and eddies of different sizes superimposed on an overall time- averaged mean flow. In a mixing tank, it is fair to postulate that the primary eddies (of a size corresponding approximately to the impeller diameter) would give rise to large velocity fluctuations, but would be of low frequency. These eddies are anisotropic and account largely for the kinetic energy present in the system. Interaction between these primary eddies and the slow moving liquid elements yield smaller eddies of higher frequency, which undergo further disintegration until they finally dissipate their energy as heat. During this disintegration process, kinetic energy is transferred from the fast moving large eddies to the smaller ones. Admittedly, the aforementioned description is oversimplified, but nevertheless, it does provide a qualitative picture of the salient features of turbulent mixing. Some quantitative results on the scale of eddies and structure of turbulence in mixing vessels are available in the literature (Banerjee, 1992), but at present it is not at all clear how this information can be integrated into the various design methodologies.

9.3 Scale-Up and Similarity Criteria

One of the main problems confronting the designers of mixing equipment is that of deducing the power consumption for a large unit from experiments with small units. In order to achieve the same kind of flow pattern in two units, geometrical, kinematic, and dynamic similarity as well as identical boundary conditions must be maintained between the two units. Since detailed discussions of scale-up procedures for mixing are available in a number of sources (Harnby et al., 1992; Skelland, 1967, 1983; Oldshue, 1983; Tatterson, 1991, 1994), only their salient features are recapitulated here. It has been found convenient to relate the power consumption to the geometrical (various dimensions, clearances, etc.) and mechanical arrangement (type of vessel, impeller, baffles, their number and size, etc.). Figure 9.3-1 shows the basic configuration of a typical-batch mixing unit.

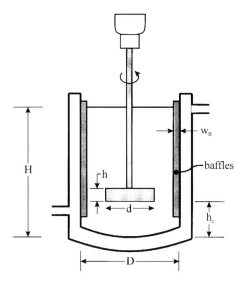

Figure 9.3-1 Typical batch mixing configuration

For reliable scale-up of laboratory data, it is necessary to ensure geometric, kinematic, and dynamic similarity between two mixing systems. Geometric similarity prevails between two systems of different sizes if all corresponding linear dimensions have a constant ratio. Thus, with reference to Fig. 9.3-1, the following ratios must be the same in the two systems:

$$\frac{d}{D}, \quad \frac{H}{D}, \quad \frac{w_B}{D}, \quad \frac{h}{D}, \quad \frac{h_c}{D}, \quad \text{etc.} \tag{9.3-1}$$

Kinematic similarity exists in two geometrically similar units when the velocities at corresponding points have a constant ratio. Also, the flow pattern (streamlines) in the two systems must be alike. Two geometrically similar systems are said to be dynamically similar if all the corresponding forces at congruous points in the flow field have a constant ratio. It is thus necessary here to distinguish between the various types of forces pertinent to the flow in mixing tanks: inertial, gravitational, viscous, and surface tension, and other forces such as those due to viscoelasticity and yield stress. Some or all of these forces may be significant in an envisaged mixing application. Considering congruous positions in systems (1) and (2), which refer to the laboratory and large-scale systems respectively, when the different types of forces occurring are F_a, F_b, F_c, and so on, dynamic similarity requires

$$\frac{F_{a_1}}{F_{a_2}} = \frac{F_{b_1}}{F_{b_2}} = \frac{F_{c_1}}{F_{c_2}} = a \quad \text{(constant)} \tag{9.3-2}$$

or

$$\frac{F_{a_1}}{F_{b_1}} = \frac{F_{a_2}}{F_{b_2}}; \quad \frac{F_{a_1}}{F_{c_1}} = \frac{F_{a_2}}{F_{c_2}}; \quad \text{etc.} \tag{9.3-3}$$

Some of the commonly encountered forces in mixing applications will now be formulated.

9.3 Scale-Up and Similarity Criteria

Inertial forces are associated with the reluctance of a body of fluid to change from its current state. Consider a mass m of fluid flowing with velocity V through an area A during the time interval dt; then $dm = \rho V A\, dt$, where ρ is the fluid density. The inertial force $F_i = (\text{mass} \times \text{acceleration})$ or

$$dF_i = dm \times \frac{dV}{dt} = \rho V A\, dV \tag{9.3-4}$$

and

$$F_i = \int_0^{F_i} dF_i = \int_0^V \rho V A\, dV = \frac{\rho A V^2}{2}. \tag{9.3-5}$$

The area for flow is, however, $A \propto L^2$, where L is the characteristic linear dimension of the system. It is customary to use the impeller diameter d for L in mixing applications, and likewise, the representative velocity V is taken to be the velocity at the tip of the impeller, that is, Nd where N is in revolutions per unit time. Therefore, the inertial force may be written as

$$F_i \propto \rho d^4 N^2. \tag{9.3-6}$$

The rate of change, dV/dt due to F_i, has to be balanced by viscous forces F_V, which for a Newtonian fluid are given by

$$F_V = \mu A' \left(\frac{dV}{dy}\right). \tag{9.3-7}$$

For the purpose of dimensional considerations, dV/dy, the velocity gradient, may be taken to be proportional to V/L, and A' (a characteristic drag surface) is again proportional to L^2. The viscous force is thus given by (again using d for L, and Nd for V)

$$F_V \propto \mu d^2 N. \tag{9.3-8}$$

Note that the physical significance of the viscous force remains the same regardless of the type of fluid: viscous (GNF) or viscoelastic fluid. Note, however, that equation 9.3-8 is restricted to Newtonian fluids.

The force due to gravity, F_g, is simply given by the weight of the fluid, that is,

$$F_g \propto \rho d^3 g. \tag{9.3-9}$$

Similarly, the surface tension force F_s can be formulated as

$$F_s \propto \alpha d, \tag{9.3-10}$$

where α is the interfacial tension.

Finally, the forces arising due to yield stress, F_y, and viscoelasticity, F_{N_1} (measured in terms of the primary normal stress difference), are given by

$$F_y \propto \sigma_0 d^2 \tag{9.3-11}$$

and

$$F_{N_1} \propto \psi_1 N^2 d^2, \tag{9.3-12}$$

where σ_0 is the yield stress and ψ_1 is the primary normal stress coefficient.

Now, identifying F_a, F_b, F_c, and so on in equation 9.3-2 with F_i, F_V, F_g, and so on, the dynamic similarity of the two systems requires that

$$\left(\frac{F_i}{F_V}\right)_1 = \left(\frac{F_i}{F_V}\right)_2, \tag{9.3-13}$$

which upon substitution of the respective expressions for F_i and F_V leads to

$$\left(\frac{\rho d^2 N}{\mu}\right)_1 = \left(\frac{\rho d^2 N}{\mu}\right)_2. \tag{9.3-14}$$

This is, of course, the familiar Reynolds number, which determines the nature of the flow to be laminar, transitional, or turbulent. In a similar manner, the requirement of constant ratios between other forces leads to the Froude (Fr) and Weber (We) numbers defined by

$$Fr = \frac{N^2 d}{g} \propto \frac{F_i}{F_g} \tag{9.3-15}$$

and

$$We = \frac{N^2 d^3 \rho}{\alpha} \propto \frac{F_i}{F_s}. \tag{9.3-16}$$

In the case of viscoplastic materials, another dimensionless group, the Bingham number (Bm) arises by keeping the ratio of viscous to yield stress forces constant as

$$Bm = \frac{\sigma_0}{\mu N} \propto \frac{F_y}{F_V}. \tag{9.3-17}$$

Similarly, the Weissenberg number emerges for viscoelastic fluids as

$$Wi = \frac{\psi_1 N}{\mu} \propto \frac{F_N}{F_V}, \tag{9.3-18}$$

where (ψ_1/η) can be regarded as a fluid relaxation time. We use here the non-Newtonian viscosity, η, instead of μ. (See Section 4.1 for a general definition of the Weissenberg number.)

Thus, the ratios of various forces acting on a fluid element in mixing vessels can be represented in terms of the aforementioned dimensionless groups, which, in turn, serve as similarity parameters for scale-up of mixing equipment of similar geometry. Additional dimensionless groups arise in the case of multiphase mixing problems such as the Froude number associated with gas-liquid or fluid-solid systems, and the Prandtl and Schmidt numbers in heat and mass transfer processes. From the theory of similitude it can be shown that the complete similarity between the two systems (1) and (2) requires all the pertinent dimensionless numbers (Re, Fr, Bm, etc.) to be equal in both systems. In practice, however, this is not always possible, owing to conflicting requirements on the physical properties of materials and operating conditions. It is generally possible to identify one or two key features that need to be satisfied. Obviously, in the case of substances without a yield stress, the Bingham number is redundant, as is the Weissenberg number for purely viscous (or GNF) fluids. Similarly, the Froude number is generally important only under conditions when gross vortexing occurs, or in the case involving turbulent mixing of low viscosity liquids. It is a common practice when dealing with low viscosity fluids to use mixing vessels fitted with baffles, thereby minimizing vortex formation (the use of baffles is not recommended for viscous and non-Newtonian fluids). Therefore, the Froude number (N^2d/g) is rarely important. However, additional problems can arise for viscoelastic media due to the Weissenberg effect, but we rarely employ high speed impellers for mixing such liquids. In applications involving gas dispersions and solid suspensions, the gas or particle Froude numbers may still be important. Finally, the Weber number usually exerts little influence on the power requirement and is often neglected.

Aside from these theoretical considerations, further difficulties can arise depending upon the choice of scale-up criteria. This choice is strongly dependent on the type of mixing (liquid or gas-liquid, for instance), as well as on the objectives of the mixing process itself. For geometrically similar systems, the size of the equipment has already been fixed by the scale-up factor. For power consumption, one commonly used criterion is to maintain the value of the power input per unit volume equal for both systems. This criterion leads to the following relationships for power:

For laminar flow:

$$\frac{\left(\frac{P}{V}\right)_1}{\left(\frac{P}{V}\right)_2} = \left(\frac{N_1}{N_2}\right)^2 \left(\frac{\mu_1}{\mu_2}\right); \qquad (9.3\text{-}19)$$

For turbulent flow:

$$\frac{\left(\frac{P}{V}\right)_1}{\left(\frac{P}{V}\right)_2} = \left(\frac{N_1}{N_2}\right)^3 \left(\frac{d_1}{d_2}\right)^2 \left(\frac{\rho_1}{\rho_2}\right), \qquad (9.3\text{-}20)$$

where P and V are the power consumption and the fluid volume respectively. In deducing equations 9.3-19 and 9.3-20, use has been made of the fact that the power number ($P/\rho N^3 d^5$) is a unique function of the Reynolds number and is inversely proportional to the Reynolds number in the laminar regime. The power number is constant in the fully turbulent region (see Section 9.4). In essence, the equality of the Reynolds numbers ensures the dynamic similarity between the two systems. The volume of the fluids, $V(\propto d^3)$, is fixed by the scale-up factor. Further simplifications result when the same fluid is being used in both systems, thereby leaving only the impeller velocity N_2 as unknown. Similarly, processes involving heat transfer in agitated systems are often scaled up either on the basis of equal heat transferred per unit volume, or by maintaining the same value of the heat transfer coefficient. Other commonly used criteria include equal mixing times, equal bubble/drop sizes, and equal mass transfer coefficients, and most of these have been described in detail by Tatterson (1994). We conclude this section by noting that most of the aforementioned information has been discussed in terms of Newtonian fluids or purely viscous non-Newtonian media, and therefore extrapolation to viscoelastic fluids must be carried out with caution.

9.4 Power Consumption in Agitated Tanks

From a practical point of view, power consumption is perhaps the most important design parameter. Owing to the inherently different flow patterns and mixing mechanisms involved, it is convenient to consider power consumption in low and high viscosity systems separately.

9.4-1 Low Viscosity Systems

Typical equipment for low viscosity liquids consists of a vertical cylindrical tank, with a height-to-diameter ratio of 1.5 to 2, fitted with an agitator. For such systems, high speed propellers with a diameter of about one third of that of the vessel are suitable, running at 500 to 1500 rpm. Admittedly, work on single-phase mixing of low viscosity liquids is of limited utility in industrial applications, but it does provide a useful starting point for the subsequent treatment of highly viscous liquids. Using dimensional analysis, we can easily show that, in the absence of surface tension effects, the power consumption is related to the other system variables via the following functional relation:

$$\underbrace{\frac{P}{\rho N^3 d^5}}_{\text{(power number)}} = f\bigg(\underbrace{\frac{\rho N d^2}{\mu}}_{\text{(Reynolds number)}}, \underbrace{\frac{N^2 d}{g}}_{\text{(Froude number)}}, \text{geometric ratios}\bigg). \qquad (9.4\text{-}1)$$

Thus, for geometrically similar systems, equation 9.4-1 reduces to

$$Po = \frac{P}{\rho N^3 d^5} = f\bigg(\frac{\rho N d^2}{\mu}, \frac{N^2 d}{g}\bigg). \qquad (9.4\text{-}2)$$

In equation 9.4-2, the Froude number is generally important only when severe vortexing occurs, and its influence can be neglected if the Reynolds number is less than about 300. In view of the detrimental effect of vortexing on the process of mixing, the tanks are usually fitted with baffles, hence, in most situations involving low viscosity fluids, the power number is a function of the Reynolds number only. Figure 9.4-1 shows this functional relationship for a range of impellers used to mix Newtonian liquids.

The vast amount of work reported on the mixing of low viscosity liquids suggests that, for a given geometrical design and configuration of impeller and vessel, all single-phase experimental data can be represented by a single power curve. We can generally discern three distinct zones in a power curve. At small values of the Reynolds number ($Re < \sim 10$), a laminar zone exists, and the slope of the power curve on log-log coordinates is -1, which is typical of most viscous flows. This region, which is characterized by slow mixing at both micro- and macroscales, is where the majority of the highly viscous Newtonian and non-Newtonian liquids are processed.

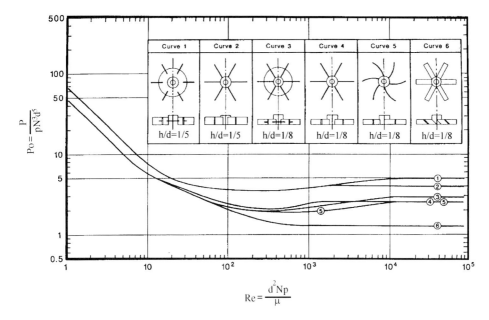

Figure 9.4-1 Power number-Reynolds number correlation in Newtonian fluids for various turbine impeller designs (Adapted from Bates et al., 1963)

At very high Reynolds number ($Re > 10^4$), the flow is fully turbulent, resulting in rapid mixing. In this regime, the power number is virtually constant and independent of the Reynolds number, but depends upon the impeller/vessel configuration. Often gas-liquid, solid-liquid, and liquid-liquid contacting operations are carried out in this region. Though the mixing itself is quite rapid, the controlling step may be mass transfer.

In between the laminar and turbulent regimes, there exists a substantial transition zone in which the viscous and inertial forces are of comparable magnitudes. No simple analytical relationship can be developed between power number and Reynolds number in this regime, and we must resort to a graphical form of representation.

Power curves for several impeller geometries, baffle arrangements, and so on are available in the literature (Nagata, 1975; Oldshue, 1983), but it must always be remembered that each power curve is system-geometry specific. Suffice it to add here that adequate information is now available on low viscosity systems for the estimation of the power requirements for a given duty under most conditions of practical interest.

9.4-2 High Viscosity Inelastic Fluids

As noted earlier, mixing in highly viscous liquids is slow both at the macroscale, due to poor bulk flow, as well as at the microscale, on account of the small values of molecular diffusion coefficients. In contrast to the low viscosity systems where momentum is transferred from a rotating impeller through a relatively large body of fluid, in highly

Liquid Mixing

viscous systems, only a small fraction of the fluid in the immediate vicinity of the impeller is influenced by the agitator, and the flow is generally laminar.

For highly viscous Newtonian and non-Newtonian systems, it is necessary to use specially designed impellers involving close clearances with the vessel walls (as shown in Figs. 9.4-3 and 9.4-4). The power curve approach is usually applicable in this case also. Most of the highly viscous fluids of practical interest also exhibit non-Newtonian characteristics, though viscous Newtonian fluids such as glycerol and many lubricating oils are also encountered in processing applications.

A simple relationship has been shown to exist between much of the data on power consumption for purely viscous non-Newtonian fluids, and Newtonian systems in the laminar region. This link, which was first suggested by Metzner and Otto (1957) for shear-thinning (pseudoplastic) liquids, hinges on the fact that there appears to be an average or effective deformation rate $\dot{\gamma}_e$ for a mixer, which adequately characterizes the power consumption, and which is directly proportional to the impeller speed

$$\dot{\gamma}_e = k_s N, \qquad (9.4\text{-}3)$$

where k_s is a function of the type of impeller and other geometrical factors. If the apparent viscosity corresponding to the effective deformation rate defined by equation 9.4-3 is used in the equation for Newtonian fluids, the power consumption in the laminar regime is satisfactorily predicted for shear-thinning liquids. For a given impeller/vessel configuration, the experimental evaluation of k_s proceeds as follows:

(i) The power number is determined for a particular value of N;
(ii) The corresponding value of the Reynolds number is obtained from the appropriate power curve for Newtonian fluids in a geometrically similar system; the effective viscosity is thus deduced from the value of Re;
(iii) The corresponding shear rate is obtained, either directly from viscometric shear stress-shear rate data or via an appropriate fluid model such as the power-law model; and
(iv) k_s evaluated using equation 9.4-3 for a specific impeller configuration.

Typical power data for shear-thinning fluids involving different impellers are shown in Fig. 9.4-2. The values of k_s and experimental details are reported in Table 9.4-1.

Example 9.4-1 The Metzner-Otto Constant for a Turbine Impeller

The following power data have been obtained for a power-law polymer solution ($n = 0.7$ and $m = 28$ Pa·s^n) using a six-blade flat turbine of 0.6 m diameter in a 1.8 m diameter tank fitted with 4 equispaced baffles (width = 180 mm). Estimate the value of k_s for $N = 10, 25,$ and 75 rpm. The corresponding values for P are 10, 50.5, and 410 W. The polymer solution has a density of 1000 kg/m^3.

Solution
For $N = 10$ rpm,

$$Po = \frac{P}{\rho N^3 d^5} = \frac{10}{(1000)\left(\frac{10}{60}\right)^3 (0.6)^5} = 28.$$

9.4 Power Consumption in Agitated Tanks

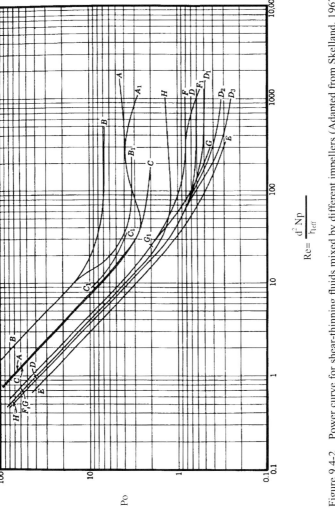

Figure 9.4-2 Power curve for shear-thinning fluids mixed by different impellers (Adapted from Skelland, 1967)

Table 9.4-1 Values of k_s for Various Types of Impellers and Key to Figure 9.4-2 (Adapted from Skelland, 1967)

Curve	Impeller	Baffles	d (m)	D/d	N (Hz)	$k_s(n<1)$
AA	Single turbine with 6 flat blades	4, $w_B/D=0.1$	0.05–0.20	1.3–5.5	0.05–1.5	11.5 ± 1.5
AA$_1$	Single turbine with 6 flat blades	None	0.05–0.20	1.3–5.5	0.18–0.54	11.5 ± 1.4
BB	Two turbines, each with 6 flat blades and $D/2$ apart	4, $w_B/D=0.1$	–	3.5	0.14–0.72	11.5 ± 1.4
BB$_1$	Two turbines, each with 6 flat blades and $D/2$ apart	4, $w_B/D=0.1$, or none	–	1.02–1.18	0.14–0.72	11.5 ± 1.4
CC	Fan turbine with 6 blades at 45 °C	4, $w_B/D=0.1$, or none	0.10–0.20	1.33–3.0	0.21–0.26	13 ± 2
CC$_1$	Fan turbine with 6 blades at 45 °C	4, $w_B/D=0.1$, or none	0.10–0.20	1.33–3.0	1.0–1.42	13 ± 2
DD	Square-pitch marine propellers with 3 blades (downthrusting)	None, (i) shaft vertical at vessel axis, (ii) shaft 10° from vertical, displaced $R/3$ from center	0.13	2.2–4.8	0.16–0.40	10 ± 0.9
DD$_1$	Same as for DD but upthrusting	None, (i) shaft vertical at vessel axis, (ii) shaft 10° from vertical, displaced $R/3$ from center	0.13	2.2–4.8	0.16–0.40	10 ± 0.9
DD$_2$	Same as for DD	None, position (ii)	0.30	1.9–2.0	0.16–0.40	10 ± 0.9
DD$_3$	Same as for DD	None, position (i)	0.30	1.9–2.0	0.16–0.40	10 ± 0.9
EE	Square-pitch marine propeller with 3 blades	4, $w_B/D=0.1$	0.15	1.67	0.16–0.60	10
FF	Double-pitch marine propeller with 3 blades (downthrusting)	None, position (ii)	–	1.4–3.0	0.16–0.40	10 ± 0.9
FF$_1$	Double-pitch marine propeller with 3 blades (downthrusting)	None, position (i)	–	1.4–3.0	0.16–0.40	10 ± 0.9
GG	Square-pitch marine propeller with 4 blades	4, $w_B/D=0.1$	0.12	2.13	0.05–0.61	10
GG$_1$	Square-pitch marine propeller with 4 blades	4, $w_B/D=0.1$	0.12	2.13	1.28–1.68	–
HH	2-bladed paddle	4, $w_B/D=0.1$	0.09–0.13	2–3	0.16–1.68	10
–	Anchor (see Fig. 9.9-1)	None	0.28	1.02	0.34–1.0	11 ± 5
–	Cone impellers	0 or 4, $w_B/d_T=0.08$	0.10–0.15	1.92–2.88	0.34–1.0	11 ± 5

From curve AA in Fig. 9.4-2 we obtain

$$Re = \frac{\rho N d^2}{\eta_{eff}} = 2.6, \quad \text{which yields } \eta_{eff} = 23.1 \text{ Pa} \cdot \text{s}.$$

Thus,

$$\eta_{eff} = 28(k_s N)^{n-1} = 28\left(k_s \frac{10}{60}\right)^{0.7-1}$$

and

$$k_s = 11.4$$

Similarly, the remaining two data points yield $k_s = 11.6$ and $k_s = 11.23$ respectively. Therefore, the mean value of k_s is 11.41, which is quite close to the mean value listed in Table 9.4-1.

A compilation of the experimental values of k_s for a variety of impellers has been presented by Skelland (1983) and is reproduced here in Table 9.4-1. For shear-thinning systems, the value of k_s is seen to lie approximately in the range of 10 to 13 for the range of configurations (mainly for the so-called class I agitators, see Section 9.5-1) studied so far.

Skelland (1983) has also reconciled most of the data on the agitation of purely viscous non-Newtonian fluids using the so-called high speed agitators (see Fig. 9.4-2). Although the aforementioned approach (Metzner and Otto, 1957) has gained wide acceptance, it has also come under some criticism, especially with regard to its failure in generating a unique power curve for highly shear-thinning fluids (Skelland, 1967; Mitsuishi and Hirai, 1969). Despite this uncertainty, it is perhaps safe to conclude that this method predicts power consumption with an accuracy of ± 25 to 30%. Furthermore, Godfrey (1985) has asserted that the constant k_s is independent of equipment size when there are no scale-up problems. It is well-known, however, that k_s depends upon the non-Newtonian characteristics. For example, for anchors and other close clearance impellers, Calderbank and Moo-Young (1959) and Beckner and Smith (1966) have correlated the coefficient, k_s, with the impeller/vessel configuration and the power-law flow behavior index. The effect of n was found to be rather weak. For helical ribbon impellers, k_s has been found to be a stronger function of the power-law index. This is discussed below (see also Carreau et al., 1993; Brito et al., 1991).

This approach has also been successfully extended to the agitation of viscoplastic (yield stress) media (Nagata et al., 1970; Tran et al., 1992), shear-thickening fluids (Nagata et al, 1970; Jomha and Edwards, 1990), and thixotropic fluids (Edwards et al., 1976).

In recent years, there has been recognition of the fact that high speed agitators and close clearance anchor or gate impellers are not very effective in mixing highly viscous liquids. Impellers with high pumping capacity are preferred. The two geometries which have received considerable attention are helical screw and helical ribbon impellers. A single-blade helical ribbon impeller with key notation is shown in Fig. 9.4-3. Figure 9.4-4 compares the helical screw configuration with the double helical ribbon impeller. D_t, in Fig. 9.4-4a, refers to the diameter of a draft tube placed around the screw to improve its pumping capacity.

402   Liquid Mixing

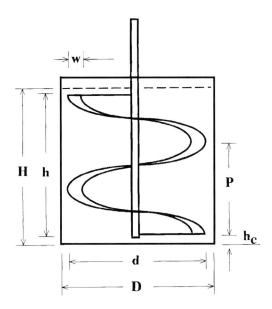

Figure 9.4-3 Schematic representation of a single-blade helical ribbon impeller (From Carreau et al., 1993a, reproduced with permission of the American Institute of Chemical Engineers. © 1993 AIChE. All rights reserved.)

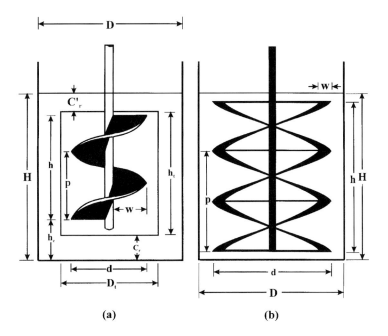

(a)   (b)

Figure 9.4-4 (a) Helical screw impeller; (b) double helical ribbon impeller (Adapted from Ulbrecht and Carreau, 1985)

9.4 Power Consumption in Agitated Tanks

Based on the idea of Bourne and Butler (1969), Chavan and Ulbrecht (1972a, b, 1973 a, b) simulated the flow between the rotating helical surface and the vessel wall by a Couette flow, and they also replaced the rotating surface by a cylinder of equivalent diameter experiencing the same torque as the actual impeller. For power-law fluids, they proposed the following expressions for power consumption with helical screw and helical ribbon impellers respectively. The mixing systems are illustrated in Fig. 9.4-4.

For *helical screw impellers*,

$$Po = \frac{\pi A}{2}\left(\frac{d_e}{d}\right)\left(\frac{D_t}{d_e}\right)^2 \left\{\frac{4\pi}{n\left[\left(\frac{D_t}{d_e}\right)^{\frac{2}{n}}-1\right]}\right\}^n \left(1+\frac{d}{h_c}\right)^{0.37}$$

$$\times \left(\frac{D-D_t}{h_t}\right)^{-0.046}\left(\frac{C_r}{d}\right)^{-0.036}\left(\frac{d^2 N^{2-n}\rho}{m}\right)^{-1}, \qquad (9.4\text{-}4)$$

where the equivalent diameter is given by

$$\frac{d_e}{d} = \frac{D_t}{d} - \frac{\frac{2w}{d}}{\ln\left\{\frac{\left(\frac{D_t}{d}\right)-1+2\left(\frac{w}{d}\right)}{\left(\frac{D_t}{d}-1\right)}\right\}}. \qquad (9.4\text{-}5)$$

The dimensionless blade surface area is

$$A = \frac{\left(\frac{p}{d}\right)\left(\frac{h}{d}\right)}{3\pi}\left[\frac{\pi\sqrt{\left(\frac{p}{d}\right)^2+\pi^2}}{\left(\frac{p}{d}\right)^2}+\ln\left(\frac{\pi}{\left(\frac{p}{d}\right)}+\frac{\sqrt{\left(\frac{p}{d}\right)^2+\pi^2}}{\left(\frac{p}{d}\right)}\right)\right]$$

$$\times \left\{1-\left(1-\left[2\left(\frac{w}{d}\right)\right]\right)^2\right\} + \pi\left[1-2\left(\frac{w}{d}\right)\right]\left(\frac{h}{d}\right). \qquad (9.4\text{-}6)$$

D_t is the diameter of the draft tube, h_t is the height of the draft tube, and C_r is the clearance between the draft tube and the vessel bottom.

For *helical ribbon impellers*,

$$Po = 2.5A\pi\left(\frac{d_e}{d}\right)\left(\frac{D}{d_e}\right)^2\left\{\frac{4\pi}{n\left[\left(\frac{D}{d_e}\right)^{\frac{2}{n}}-1\right]}\right\}^n\left(\frac{d^2N^{2-n}\rho}{m}\right)^{-1}, \qquad (9.4\text{-}7)$$

where

$$\frac{d_e}{d} = \frac{D}{d} - \frac{\frac{2w}{d}}{\ln\left\{\frac{\left(\frac{D}{d}\right) - 1 + 2\left(\frac{w}{d}\right)}{\left(\frac{D}{d} - 1\right)}\right\}},\qquad(9.4\text{-}8)$$

and

$$A = \frac{\left(\frac{h}{d}\right)\left(\frac{p}{d}\right)}{3\pi}\left[\frac{\pi\sqrt{\left(\frac{p}{d}\right)^2 + \pi^2}}{\left(\frac{p}{d}\right)^2} + \ln\left(\frac{\pi}{\left(\frac{p}{d}\right)} + \frac{\sqrt{\left(\frac{p}{d}\right)^2 + \pi^2}}{\left(\frac{p}{d}\right)}\right)\right]$$
$$\times\left\{1 - \left[1 - 2\left(\frac{w}{d}\right)\right]^2\right\}. \qquad(9.4\text{-}9)$$

The power consumption is doubled for two ribbons if mounted on the same shaft. These theoretically derived equations were stated to predict experimental data with an average error of about 10% in the viscous flow regime. In another interesting study, Patterson et al. (1979) and Yap et al. (1979) employed the drag flow analogy and developed expressions for the agitation of viscous Newtonian and power-law fluids by helical ribbons. For instance, for power-law and mildly viscoelastic fluids, Yap et al. (1979) presented their results in terms of k_s as

$$k_s = 4^{\frac{1}{1-n}}\left(\frac{d}{D}\right)^2\left(\frac{l}{d}\right), \qquad(9.4\text{-}10)$$

where l is the total length of a ribbon. Equation 9.4-10 predicts a strong dependence of k_s on the flow behavior index (n) and it shows an order of magnitude variation ranging from 20 to 200, eventually becoming indeterminate at $n = 1$.

In order to elucidate the influence of shear thinning on the effective deformation rate in a mixing vessel, Ulbrecht and Carreau (1985) developed an idealized framework using the Couette flow analogy. In this approach, the impeller is simulated by a cylinder of equivalent diameter d_e rotating inside a coaxial cylinder, but the rheological properties and the torque are evaluated at the wall of the vessel, and the contribution to the torque from the bottom of the tank is neglected. For power-law fluids in the laminar regime, the effective shear rate is found to be proportional to the speed of rotation. The coefficient k_s is now a function of the geometry and the power-law index as given below:

$$k_s = \left[\frac{K_p d^3}{\pi^2 D^2 H}\right]^{\frac{1}{1-n}}\left[\frac{n\left\{\left(\frac{D}{d_e}\right)^{\frac{2}{n}} - 1\right\}}{4\pi}\right]^{\frac{n}{1-n}}, \qquad(9.4\text{-}11)$$

where $K_p = PoRe$ = constant for a given impeller/vessel geometry. The equivalent diameter d_e is evaluated from the corresponding expression for Newtonian fluids, namely

$$K_p = \frac{4\pi^3 D^2 H}{d^3 \left[\left(\dfrac{D}{d_e}\right)^2 - 1\right]}. \tag{9.4-12}$$

It is instructive to mention here that, for a turbine impeller with $K_p = 70$, the value of k_s changes from 6 to 11 as the power-law index drops from 0.9 to 0.1, which is consistent with the findings of others (Skelland, 1983). On the other hand, for a helical ribbon impeller with $K_p = 400$, the value of k_s shows a weak dependence on the flow behavior index, and the resulting mean value of 35 is fairly close to the value of 30 suggested by Nagata (1975). More recently, Carreau et al. (1993) have improved upon the analysis of Ulbrecht and Carreau (1985) and their final expression for the coefficient k_s is given by

$$k_s = \left[\frac{\pi^{n+2}}{K_p d^3}\right]^{\frac{1}{n-1}} \left\{\left[D^2 H \frac{4}{n\left[\left(\dfrac{D}{d_e}\right)^{\frac{2}{n}} - 1\right]}\right]^n + \frac{D^3}{2(n+3)} \left(\frac{H}{h_c}\right)^n \left(\frac{D}{H}\right)^n \left(\frac{d_e}{D}\right)^{n+3}\right\}^{\frac{1}{n-1}}, \tag{9.4-13}$$

where h_c is the clearance between the impeller and the vessel bottom. The second term on the right side is the correction due to the bottom (Cheng et al., 1996). The experimentally

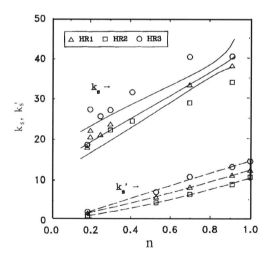

Figure 9.4-5 Variation of k_s with the flow index n
See Carreau et al. (1993) for details on fluids and helical ribbon geometries used. The rheological parameters are reported in Table 9.4-2 (Adapted from Carreau et al., 1993, reproduced with permission of the American Institute of Chemical Engineers. © 1993 AIChE. All rights reserved.)
---- best linear fits
—— predictions of equation 9.4-14

determined values of k_s as a function of the power-law index n for three geometrically different systems are shown in Fig. 9.4-5. The fluids used and their rheological parameters are listed in Table 9.4-2. The 0.7% gellan (polysaccharide) and 800 ppm PAA (polyacrylamide) solutions in corn syrup have nearly constant viscosity ($n = 0.910$ and 0.940 respectively) and exhibit non-negligible normal stress differences. The 0.35% polyisobutylene (PIB) in a mixture of polybutene (PB) and kerosene is a Boger type fluid with constant viscosity and quadratic normal stress differences ($N_1 \propto \dot{\gamma}^2$). The xanthan (XTN) and carboxy methyl cellulose (CMC) solutions are mildly elastic or inelastic shear-thinning fluids.

The increase of k_s with n is not predicted by equation 9.4-11, using K_p and the equivalent diameter, d_e, from power consumption data obtained for Newtonian fluids. The correction due to the bottom clearance between the impeller and the vessel was found to be negligible (Cheng et al., 1996). The experimental k_s and K_p values for shear-thinning fluids can be used to calculate the equivalent diameter, d_e, via equation 9.4-11. The results are reported in Fig. 9.4-6.

Figure 9.4-6 shows a clear dependence of the ratio D/d_e on the power-law index. The dependence is quite similar for the six different impeller geometries. Although the reasons for the dependence of D/d_e on n are not immediately obvious, they can be attributed in part to the changes in flow patterns for highly shear-thinning fluids as observed by Carreau et al. (1976), and more recently computed by Tanguy et al. (1992). However, in spite of the

Table 9.4-2 Rheological Parameters of the Fluids used by Carreau et al. (1993)

Fluids	$n^{*\ddagger}$	m^*	$t_1^{\ddagger}$	$\eta_0^{\ddagger}$	η_s	n'	m'	ρ
	—	Pa·s^n	s	Pa·s	Pa·s	—	Pa·s^n	kg/m^3
Dilute corn syrup #1	1	12.0						1440
Dilute corn syrup #2	1	4.16						1360
Dilute glycerol #1	1	0.470						1140
Dilute glycerol #2	1	0.067						1100
2.5% XTN	0.183	22.4						1080
0.5% XTN (gly/H$_2$O)	0.199	4.13			0.19	0.782	7.85	1200
1.8% XTN	0.200	11.8						1080
0.8% XTN	0.240	2.31						1050
0.5% XTN	0.250	1.84						1030
3% CMC	0.299		7.83	469				1060
1% CMC	0.409		0.110	1.57				1040
0.4% CMC (gly/H$_2$O)	0.530	9.75				0.740	18.0	1200
0.1% CMC (gly/H$_2$O)	0.701	1.20				1.12	0.140	1200
0.7% gellan (corn syrup)	0.910	0.750						1300
800 ppm PAA (corn syrup)	0.940	1.03				1.67	0.150	1350
0.35% PIB (PB+kerosene)	1	8.19				2.00	1.29	1100

* n, m are the power-law parameters (equation 2.2-2)
‡ n, t_1, and η_0 are the parameters in the Carreau viscosity equation 2.2-4
n' and m' and the power-law parameters for the primary normal stress differences, $N_1 = m'|\dot{\gamma}|^{n'}$

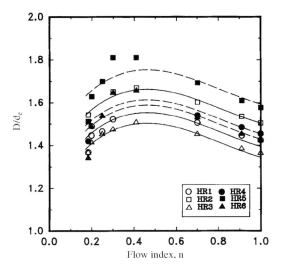

Figure 9.4-6 Dependence of D/d_e on the power-law index n for the shear-thinning fluids listed in Table 9.4-2
For the details of the six different geometries and the properties of the fluids, see Carreau et al. (1993). The lines represent smooth curves through the calculated points. (From Carreau et al., 1993, reproduced with permission of the American Institute of Chemical Engineers. © 1993 AIChE. All rights reserved.)

slight dependence of D/d_e on n, it can be argued that the calculation of power is not very sensitive to the variation of D/d_e with n. For engineering calculations, it is quite adequate to use the value of (D/d_e) from the analysis involving Newtonian fluids, together with the k_s values determined from equation 9.4-11 (see Carreau et al., 1993).

To avoid using an equivalent diameter that depends on the fluid properties, Carreau et al. (1993) suggested a relation between the effective rate of deformation and the wall shear rate. This leads to

$$k_s = \left[\frac{\pi^2}{K_p}\left(\frac{D}{d}\right)^2\left(\frac{H}{d}\right)\right]^{\frac{1}{n-1}} (k'_s)^{\frac{n}{n-1}}, \qquad (9.4\text{-}14)$$

where k'_s is the proportionality constant for the wall shear rate ($\dot{\gamma}_w$) and is determined from experimental torque (T) data via the following equation:

$$\dot{\gamma}_w = k'_s N = \left[\frac{2T}{\pi m D^2 H}\right]^{\frac{1}{n}}. \qquad (9.4\text{-}15)$$

Values of experimentally determined k'_s for three impeller geometries are reported in Fig. 9.4-5. The broken lines are the best linear fits. The k_s predictions using k'_s and equation 9.4-14 are indicated by the solid curves in the figure. The predictions are reasonable, considering the sensitivity of k_s to the inaccuracy of the power consumption data.

In the transition zone, however, the effective deformation rate shows a stronger dependence on the impeller speed. Recently, Cheng and Carreau (1994b) have attempted to

408   Liquid Mixing

establish the interconnection between the effective deformation rate and the impeller speed by recognizing that $K_p = PoRe^a$ where $a < 1$ in the transition zone:

$$\dot{\gamma}_e = \left[\frac{\pi^{n+2}}{K_p d^3}\right]^{\frac{1}{r(n-1)}} A^{\frac{1}{r(n-1)}} \left[\frac{d^2 \rho}{m}\right]^{\frac{1-a}{r}} N^{\frac{2-a}{r}}, \quad (9.4\text{-}16)$$

where $r = n(1-a) + a$, and

$$A = \left\{ D^2 H \left[\frac{64}{n\left\{\left(\frac{D}{d_e}\right)^{\frac{2}{n}} - 1\right\}}\right]^n + \frac{D^3}{2(n+3)} \left(\frac{H}{h_c}\right)^n \left(\frac{D}{H}\right)^n \left(\frac{d_e}{D}\right)^{n+3}\right\}. \quad (9.4\text{-}17)$$

A simplified model for the transition regime has also been proposed by Cheng and Carreau (1994b). It can be written as

$$\dot{\gamma}_e = \left[\frac{K'_p d^3}{\pi^2 d_e^2 H}\right]^{\frac{1}{r}} \left[\frac{d^2 \rho}{m}\right]^{\frac{(1-a)}{r}} N^{\frac{(2-a)}{r}}, \quad (9.4\text{-}18)$$

where K'_p and a are local values for the power consumption of Newtonian fluids in the transition regime.

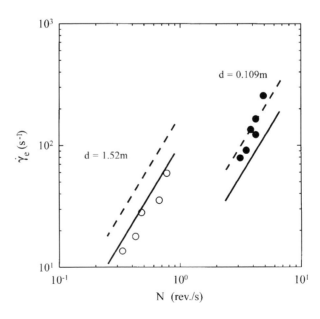

Figure 9.4-7 Predictions of the effective deformation rate for the data of Bourne et al. (1981) with anchor impellers (Adapted from Cheng and Carreau, 1994b)
- - - -   Predictions of equation 9.4-16
———   Predictions of equation 9.4-18

9.4 Power Consumption in Agitated Tanks   409

Cheng and Carreau (1994b) have shown that these two equations can predict reasonably well the effective deformation rate in the transition regimes for a variety of impellers and fluid properties. As a first example, the predictions of equation 9.4-16 or equation 9.4-18 for two geometrically similar but different sized anchor agitators are shown in Fig. 9.4-7. For the larger mixing system, equation 9.4-18 yields better predictions. For the smaller mixing system, the best results are obtained with equation 9.4-16. In both cases, the models predict that the effective rate of deformation is quadratic with respect to the impeller rotational speed.

As a second illustration, we reproduce in Fig. 9.4-8 the comparison between the model predictions and the data of Forschner et al. (1991). The data were obtained for the mixing of a 2.2% aqueous CMC solution ($n = 0.45$) using an intermig turbine impeller. The agreement between the experimental data and the predictions of equation 9.4-16 or equation 9.4-18 is quite good, even surprisingly good in the case of equation 9.4-18. Note that over the transition regime ($2 < N < 20$ rev/s), the effective rate of deformation increases by more than two decades.

In the turbulent flow regime ($Re > 10^4$), the power number is independent of the Reynolds number and the notion of an effective deformation rate is thus of little con-

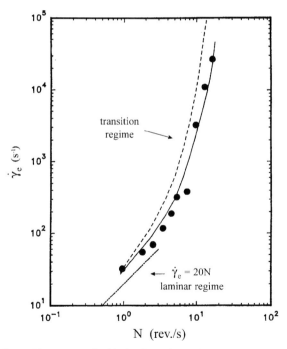

Figure 9.4-8 Predictions of $\dot{\gamma}_e$, compared with the data of Forschner et al. (1991) for an intermig turbine impeller
(Adapted from Cheng and Carreau, 1994b)
- - - - Predictions of equation 9.4-16
──── Predictions of equation 9.4-18

Liquid Mixing

sequence, at least for inelastic fluids. We must use the experimentally determined power curves to estimate the power consumption.

For purely viscous fluids, the estimation of power consumption proceeds as follows:

(i) Use the Metzner-Otto concept (equation 9.4-3) and calculate the effective deformation rate via equation 9.4-13, 9.4-16, or 9.4-18, for a given geometry and speed of rotation.
(ii) Evaluate the value of the effective viscosity at this effective deformation rate and calculate the value of the Reynolds number.
(iii) Obtain the corresponding value of the power number (and hence power) from an appropriate power curve or correlation.

Example 9.4-2 Power Requirement for a Turbine Impeller

A power-law fluid with $n = 0.84$ and $m = 15.5$ Pa·s^n is to be agitated at 20 °C in a cylindrical, flat-bottomed vessel which is 0.9 m in diameter and is filled to a height of 0.9 m. Four baffles, each 90 mm wide, are located radially at 90° around the vessel wall. Agitation is provided by a turbine with six flat blades, mounted along the vertical axis of the vessel and at a height of 300 mm above the vessel bottom. The turbine rotates at a speed of 60 rpm, is 300 mm in diameter, and 75 mm from the bottom. The density of the fluid is 1000 kg/m^3. Estimate the power requirement. What is the effect on the power requirement if the rotation speed is doubled?

Solution

All geometric ratios are within the range of the first entry of Table 9.4-1 for a turbine with six blades. Hence, we take

$$k_s = 11.5,$$

$$\dot{\gamma}_e = 11.5 \left(\frac{60}{60}\right) = 11.5 \text{ s}^{-1},$$

$$\eta_{\text{eff}} = 15.5(11.5)^{0.84-1} = 10.42 \text{ Pa·s},$$

$$\text{and} \quad Re = \frac{\rho N d^2}{\eta_{\text{eff}}} = \frac{1000 \times \left(\frac{60}{60}\right) \times (0.3)^2}{10.42} = 8.4.$$

From curve AA in Fig. 9.4-2,

$$Po = \frac{P}{\rho N^3 d^5} \approx 9,$$

$$\text{and} \quad P = 9 \times 1000 \times \left(\frac{60}{60}\right)^3 (0.3)^5 = 21.9 \text{ W}.$$

When N is doubled,

$$\eta_{\text{eff}} = 5.5 \left(11.5 \times \frac{120}{60}\right)^{0.84-1} = 9.38 \text{ Pa·s}$$

$$\text{and} \quad Re = \frac{\rho N d^2}{\eta_{\text{eff}}} = \frac{1000 \times (120/60)(0.3)^2}{9.38} = 19.2.$$

Again, from curve AA in Fig. 9.4-2,

$$\frac{P}{\rho N^3 d^5} = 3,$$

and $P = 3 \times 1000 \times \left(\frac{120}{60}\right)^3 (0.3)^5 = 58.32$ W.

Note that the power required has increased by more than a factor of 2! However, it must be borne in mind that the flow may no longer be laminar, hence, the effective deformation rate might show stronger dependence on N (see Cheng and Carreau, 1994b and equation 9.4-16 or 9.4-18).

Example 9.4-3   Scale-Up Criteria for Power-Law Fluids
Obtain the functional relationship for the power required, P, to mix a power-law fluid (m, n) of density ρ as a function of the rotational speed, N, for an impeller of diameter d. Assume that no vortex formation occurs and that surface tension effects are unimportant. Establish the scale-up criteria for geometrically similar systems.

Solution
Referring to Fig. 9.3-1, the functional relationship between the power, P, and the other variables can be expressed as $P = f(m, n, \rho, d, N,$ other geometrical parameters). For geometrically similar systems, this reduces to

$$P = f(m, n, \rho, d, N). \quad (9.4\text{-}19)$$

Any functional relationship can be expressed in terms of a power function as $P = K_P m^a n^b \rho^c d^d N^e$, where K_P is a dimensionless constant. Furthermore, since the power-law flow behavior index is already a dimensionless quantity, we can rewrite the above relation for a fixed value of n as $P = K_P m^a \rho^c d^d N^e$.

Using mass, length, and time as the base dimensions, we can write

$$[ML^2 T^{-3}] \equiv [M^0 L^0 T^0][ML^{-1} T^{n-2}]^a [ML^{-3}]^c [L]^d [T^{-1}]^e$$

The application of the law of dimensional homogeneity yields

$$M: \quad 1 = a + c,$$
$$L: \quad 2 = -a - 3c + d,$$
$$\text{and} \quad T: \quad -3 = a(n-2) - e.$$

We can solve for c, d, and e in terms of a as follows:

$$c = 1 - a; \quad d = 5 - 2a; \quad e = a(n-2) + 3.$$

Substituting these values in the equation for P, we obtain

$$P = K_P m^a \rho^{1-a} d^{5-2a} N^{a(n-2)+3}.$$

Rearrangement yields

$$Po \equiv \underbrace{\frac{P}{\rho N^3 d^5}}_{\text{(Power number)}} = K_P \underbrace{\left[\frac{\rho (Nd)^{2-n} d^n}{m}\right]^{-a}}_{\text{Reynolds number)]}}. \qquad (9.4\text{-}20)$$

Note that this relationship has been obtained for geometrically similar systems. In general, thus,

$$K_P = f(Re, n, \text{geometrical ratios}) \qquad (9.4\text{-}21)$$

By analogy with the Newtonian definition, we can rearrange the power-law Reynolds number as

$$Re = \frac{\rho (Nd) d}{\eta_{\mathit{eff}}}, \qquad (9.4\text{-}22)$$

where

$$\eta_{\mathit{eff}} = m \left(\frac{Nd}{d}\right)^{n-1} = m(N)^{n-1}. \qquad (9.4\text{-}23)$$

Thus, η_{eff} is the viscosity evaluated at an effective deformation rate N, which is also in agreement with equation 9.4-3, except for the empirical constant k_s. Experience has shown that the flow behavior index n is not an independent variable, and its influence is embodied in the definition of Re, provided that the effective viscosity based on the Metzner-Otto concept is used. Note that the Metzner-Otto parameter k_s is, in general, a function of n. However, for engineering calculations, it can be assumed to be constant for a given mixer geometry (Carreau et al., 1993). Thus, for geometrically similar systems and in the absence of surface tension and vortexing effects, the only scale-up parameter with respect to power consumption is the Reynolds number.

In concluding this section, it is appropriate to mention here that some of the more recent experimental work (Carreau et al., 1993; Cheng and Carreau, 1994b) suggests that the laminar flow regime extends to larger values of Reynolds number ($Re > 10$) in highly shear-thinning systems. Typical results shown in Fig. 9.4-9 illustrate this effect very clearly for a helical ribbon impeller. Similar trends have been observed by Hocker et al. (1981). To recapitulate, it is thus possible to estimate the power consumption for the agitation of purely viscous non-Newtonian systems with reasonable levels of confidence.

9.4-3 Viscoelastic Systems

Considerable confusion exists regarding the influence of viscoelasticity on power consumption. Some authors have argued that since the main flow pattern in mixing vessels is indistinguishable for inelastic and viscoelastic fluids, the fluid viscoelasticity is not expected to influence the power consumption. Indeed, the early experimental results of Chavan and Ulbrecht (1972a, b; 1973a, b) and Yap et al. (1979) provide support to this assertion. On the other hand, Nienow et al. (1983) reported a slight increase in power

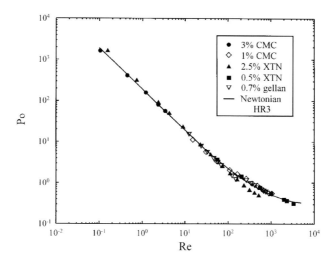

Figure 9.4-9 Power data for shear-thinning inelastic fluids mixed with a helical ribbon impeller
See Table 9.4-2 for the fluid properties. The Reynolds number is based on the Metzner-Otto concept.
(Adapted from Carreau et al., 1993)

consumption for viscoelastic fluids, whereas Ducla et al. (1983) observed a slight decrease in power consumption when using turbine impellers.

However, it must be realized that most polymer solutions also exhibit shear-dependent viscosity in addition to viscoelastic behavior, and it is therefore difficult to isolate the contributions of these two rheological features to power consumption. However, the use of constant viscosity, highly elastic fluids offers some hope in alleviating this difficulty. The experimental results obtained with these fluids suggest that the extent of viscoelastic effects is strongly dependent on the impeller geometry and on the operating conditions. For Rushton-type turbine impellers, Oliver et al. (1984) and Prud'homme and Shaqfeh (1984) have indicated that the power consumption may increase or decrease (above its Newtonian value) depending upon the values of the Reynolds and Weissenberg numbers. More recent results (Carreau et al., 1993) obtained with a helical ribbon type impeller, shown in Fig. 9.4-10, clearly suggest an early departure from the Newtonian behavior, thereby implying an increase in the value of the power number up to a factor of 5 with respect to the Newtonian case. The latter behavior is in sharp contrast to that for highly shear-thinning fluids that show an extended laminar regime (see Fig. 9.4-9). At high Reynolds number, viscoelasticity is believed to suppress the secondary motion, thereby resulting in a reduction in power consumption (Kale et al., 1974).

Before concluding this section on power consumption, it is of interest to refer to the scant work available on the effect of drag-reducing additives. It is generally believed that a torque reduction occurs in turbulent conditions for the agitation of drag-reducing polymer solutions (Quraishi et al., 1976). This saving in energy consumption is accompanied by a concomitant decrease in convective interphase transfer coefficients (Quiraishi et al., 1976; Ranade and Ulbrecht, 1978).

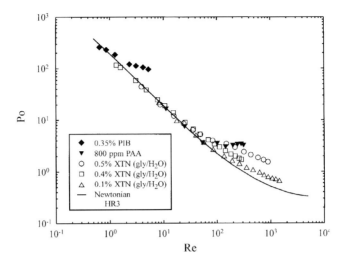

Figure 9.4-10 Power data for shear-thinning and Boger type viscoelastic fluids mixed with a helical ribbon impeller
The rheological parameters of the fluids are listed in Table 9.4-2. The Reynolds number is based on the Metzner-Otto concept (Adapted from Carreau et al., 1993)

9.5 Flow Patterns

A qualitative picture of the flow field created by an impeller in a mixing vessel is useful in ascertaining whether there are stagnant or dead zones in the vessel. Flow patterns are very much dependent upon the type of impeller and the geometry of the system. It is convenient to classify the agitators used with non-Newtonian fluids into three types:

(i) The first type operates at relatively high speeds, producing high deformation rates in the vicinity of the impeller, as well as relying on the favorable momentum transport to the whole liquid. Typical examples include turbine impellers and propellors.
(ii) The second type is characterized by the so-called close clearance impellers (such as gates and anchors), which are able to reach the far corners directly rather than relying on the turbulent momentum transport in highly viscous systems.
(iii) Finally, there are slowly rotating impellers which do not produce high deformation rates, but rely on their excellent pumping capacity to reach every corner of the vessel. Typical examples include the helical ribbon and the helical screw agitators sketched in Figs. 9.4-3 and 9.4-4.

9.5-1 Class I Agitators

Several workers, including Metzner and Otto (1957), Godleski and Smith (1962), Wichterle and Wein (1981), and Elson (1990), have provided good accounts of the flow patterns for Newtonian, shear-thinning, and shear-thickening fluids in agitated tanks. It is readily agreed that shear-thinning fluids experience high deformation rates in the impeller region, which diminish rapidly moving away from the impeller tip. The fluid motion decreases much faster in shear- thinning fluids than in Newtonian media of comparable viscosity. Shear- thickening fluids behave in exactly the opposite way. Using square cross- section vessels, Wichterle and Wein (1981) identified the region of motion/no motion for shear-thinning fluids agitated by turbine and propellor-type impellers. Typical results showing the stagnant regions are plotted in Fig. 9.5-1. Their results clearly show that the well-mixed region (D_c) is only about the size of the impeller diameter in the viscous flow regime, but it covers an increasing portion of the tank contents with increasing Reynolds number. They expressed their results as follows:

$$\frac{D_c}{d} = 1, \qquad Re < \frac{1}{a^2}, \qquad (9.5\text{-}1)$$

and

$$\frac{D_c}{d} = a(Re)^{0.5}, \qquad Re > \frac{1}{a^2}, \qquad (9.5\text{-}2)$$

where a was found to be 0.3 for propellers, 0.6 for turbines, and approximately equal to 0.375 $(Po_t)^{0.333}$ for any agitator, where Po_t is the value of the power number in the turbulent region.

In an interesting study dealing with the agitation of viscoplastic media, Nagata et al. (1970) reported a cyclic increase and decrease in power consumption which can be explained as follows. Initially, the power consumption is high due to the high viscosity of

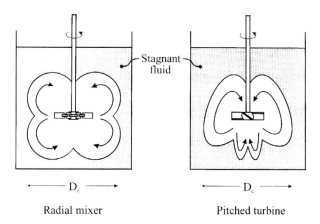

Figure 9.5-1 Shape of the mixing cavity in a shear-thinning suspension (Adapted from Wichterle and Wein, 1981)

the solid-like structure. However, once the yield stress is overcome in the vicinity of the impeller, and the material begins to behave like a fluid, the power consumption decreases, accompanied by the drop in the stress-level, which is conducive to structure buildup. Also, as fresh, solid-like material finds its way to the vicinity of the rotating impeller, the power consumption increases. The cycle repeats itself. A surface vortex was also observed during this cyclic torque behavior. Nagata et al. (1970) also noted that this behavior was drastically reduced or not observed at all with class II type agitators. In a recent study, Elson (1990) has compared the flow patterns produced by a six-bladed Rushton turbine for shear-thickening fluids. Their X-ray pictures clearly show that while the bulk motion is more intense in shear-thickening fluids than in a comparable Newtonian medium, shear thickening seems to suppress the extensional component of flow.

The influence of viscoelasticity on the flow pattern is much more striking and more difficult to assess. Based on the photographs of rotating turbine and propellor-type impellers in viscoelastic media (Giesekus, 1965), we can discern two distinct flow patterns. In a small region in the vicinity of the impeller, the flow is outward, whereas in the rest of the vessel the flow pattern is opposite, with the liquid flowing toward the impellers in the equatorial plane and outward from the impellers along the axis of rotation. A closed streamline separates the two regions, thereby suggesting no convective transport between these regions. Another more quantitative study of Kelkar et al. (1973) suggests that irrespective of the nature of the secondary flow pattern, the primary flow pattern (i.e., tangential velocity) around a rotating body is virtually unaltered by the fluid viscoelasticity in the viscous regime ($Re \leq 5$). Depending upon the relative magnitudes of the elastic, inertial, and viscous forces (Wi/Re), we may observe different types of flow patterns (Ulbrecht and Carreau, 1985). It is also instructive and useful to add here that the instability of the secondary flow is brought about not only by the variation in the impeller speed but, more significantly, by to the variation in the liquid rheological properties, namely, viscosity and the first normal stress differences (Ide and White, 1974).

9.5-2   Class II Agitators

While anchors and gate-type impellers are known to produce poor axial circulation of the liquid in the vessel, the only study available on this subject, by Peters and Smith (1967), suggests that viscoelasticity actually promotes axial flow. For instance, they reported an almost 15-fold increase in axial flow in the case of a viscoelastic solution as compared to a Newtonian medium. The shear rate profiles reported by Peters and Smith (1967) are shown in Fig. 9.5-2. The liquid axial velocity in the tank was virtually uninfluenced by the blade passage.

Broadly speaking, both gate and anchor agitators promote fluid motion close to the wall, but leave the region in the vicinity of the shaft relatively stagnant. Due to the modest top to bottom turnover, vertical concentration gradients usually exist, but these may be minimized by means of a helical ribbon or a screw added to the shaft. Such combined impellers would have a ribbon pumping upward near the wall, with the screw, twisted in the opposite sense, pumping the fluids downward near the shaft region. Typical flow

9.5 Flow Patterns

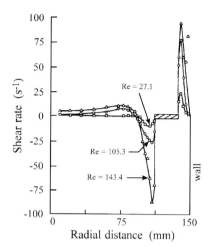

Figure 9.5-2 Shear rate profiles for an anchor impeller (Adapted from Peters and Smith, 1967)

patterns for an anchor-type impeller are shown in Fig. 9.5-3. Any rotational motion induced within the tank will produce a secondary flow in the vertical direction. The liquid in contact with the tank bottom is essentially stationary, while that at higher levels is rotating and will experience centrifugal forces. Consequently, the prevailing unbalanced forces within the fluid lead to the formation of a toroidal vortex. Depending upon the viscosity and type of fluid, single-cell or double-cell secondary flows may appear, as shown schematically in Fig. 9.5-4. Qualitatively, similar observations have been made by Murakami et al. (1980), and such flow patterns are also in qualitative agreement with a numerical study for Newtonian fluids (Abid et al., 1992).

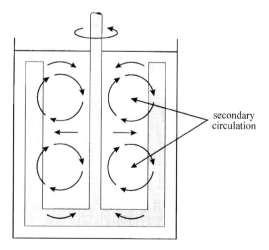

Figure 9.5-3 Secondary circulation in an anchor agitated tank (Adapted from Peters and Smith, 1967)

418   Liquid Mixing

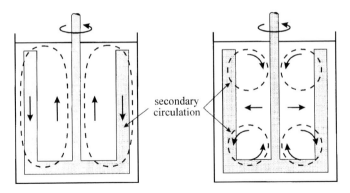

Figure 9.5-4 Schematic representation of single-cell and double-cell secondary flow with an anchor agitator

9.5-3   Class III Agitators

The salient features of the flow pattern produced by a helical ribbon impeller are shown in Fig. 9.5-5. The primary top to bottom circulation, mainly responsible for mixing, is due to the axial pumping action of the ribbons. The shear produced by the helical ribbon is localized to the regions inside and outside the blade, whereas the shear between the wall and the bulk liquid is cyclic in nature. Notwithstanding the considerable scatter present in the data of Bourne and Butler (1969), the axial velocity data shown in Fig. 9.5-6 seem to be independent of the scale of equipment and the fluid rheology. There is virtually no radial flow except in the top and bottom regions of the tank. The vertical velocity inside the ribbon helix varies roughly from 7% to 14% of the ribbon tangential speed. A simple but tentative correlation between the tangential velocity and the ribbon pitch for Newtonian media is also available in the literature (Chavan, 1983).

In addition to the above noted primary flow pattern, secondary flow cells develop with increasing impeller speed, similar to that observed by Peters and Smith (1967). Carreau et al. (1976) have also recorded flow patterns for the helical ribbon impeller. The fluid viscoelasticity seems to considerably reduce the axial circulation, as can be seen in Fig. 9.5-7, where the dimensionless axial velocity is plotted for an inelastic (2% CMC) and an elastic (1% polyacrylamide (PAA)) solution. The axial velocities for the inelastic solution were observed to be of the same order as that for a Newtonian solution.

On the other hand, for viscoelastic liquids, the tangential velocities were increased considerably, to the extent that the whole content of the vessel, except for a thin layer adhering to the wall, rotated as a solid body with an angular velocity equal to that of the impeller. More definitive conclusions regarding the role of non-Newtonian rheology must await additional work in this area.

Little is known about the flow patterns generated by helical screw impellers. Chapman and Holland (1965) have presented preliminary pictures of dye flow patterns of an off-centered, helical screw impeller pumping upward without a draft tube. There seems to exist a dispersive flow between the flights of the screw and the dispersion being completed at the

9.5 Flow Patterns   419

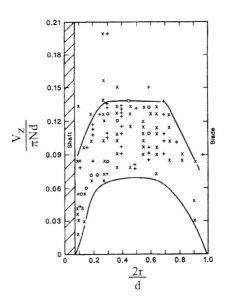

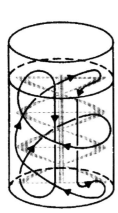

Figure 9.5-5 Flow pattern produced by a helical ribbon impeller (Adapted from Nagata et al., 1956)

Figure 9.5-6 Distribution of axial fluid velocities in the core region of helical ribbon impellers pumping downward in 6 and 160 gallon tanks
The curves represent the lower and upper bounds of the experimental data:
+ $d/D = 0.889$
× $d/D = 0.952$ in small vessel
○ $d/D = 0.954$ in large vessel (Adapted from Bourne and Butler, 1969)

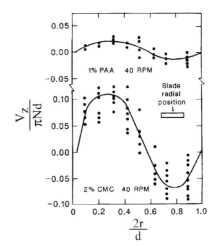

Figure 9.5-7 Axial velocity distribution in a vessel agitated by a helical ribbon impeller (Adapted from Carreau et al., 1976)

top of the screw. The flow into the screw impeller was from the other side of the tank, whereas the remaining parts of the tank appeared to be virtually stagnant. With a draft tube, the screw impeller has large axial pumping capacities, and its mixing performances for viscous and viscoelastic fluids are comparable to, if not better than the helical ribbon (see for example Chavan and Ulbrecht, 1973b; Carreau et al., 1992).

In recent years, several attempts have been made at numerical simulations of the flow patterns generated by simple geometry agitators such as anchor and paddle impellers (Hiraoka et al., 1978, 1979). Most of the simulations to date are two-dimensional, but nonetheless do seem to provide a theoretical justification for equation 9.4-3. Also, the resulting deformation rate distribution seems to be in qualitative agreement with experimental observations. More recently, Tanguy et al. (1992) have reported a 3-D numerical simulation of flow patterns in a vessel with a combined helical ribbon-screw impeller. They obtained good agreement between the calculated and measured values of integrated quantities such as torque and mixing time for Newtonian fluids. Their study thus offers some hope for the simulation of mixing of non-Newtonian fluids.

Aside from these three types of agitators, it is also useful to provide a brief account of flow patterns encountered in some other devices employed to mix thick pastes exhibiting complex rheological behavior. One common geometry used to mix thick pastes is the sigma blade mixer (see Fig. 9.9-1). Such devices have thick S- or Z-shaped blades that look like high-pitch helical ribbon impellors. Generally, two units are placed horizontally in separate troughs inside a mixing chamber, and these rotate in opposite directions at different speeds. Limited qualitative information on the flow patterns in the sigma blade mixer, as well as in minor variations thereof, is available in the literature (Earle, 1959; Hall and Godfrey, 1968; Kappel, 1979). Likewise, preliminary results are also available in the literature on a positive displacement mixer that seems to offer some advantages, such as easy mixing for thick pastes and extremely viscous materials, over helical ribbon and sigma mixers (Cheng and Schofield, 1974).

9.6 Mixing and Circulation Times

Before considering the question of circulation and mixing times, we must have methods of assessing the quality of a mixture. Due to the wide scope and spectrum of mixing problems, it is not possible to develop a single criterion for all applications. Criteria such as scale and intensity of segregation are appropriate to ascertain the quality of turbulent mixing (Harnby et al., 1985). For batch mixing of viscous fluids, a useful criterion is the mixing time, defined as the time needed to produce a mixture of predetermined quality. The rate of mixing is the rate at which the mixing proceeds toward the final state.

For single-phase liquids in a mixing vessel to which a small volume of tracer material (dye, electrolyte, acid, etc.) is added, the mixing time is measured from the instant the tracer is added, to the time when the contents of the tank have reached the required degree of uniformity. For a known amount of tracer, the equilibrium concentration may be calculated, and this value will be approached asymptotically. The mixing time is thus defined as the time required for the mixture composition to come within the prespecified limits of the equilibrium value. Clearly, the measurement of mixing time will also depend upon the quantity and the way the tracer is added, the location of the detector, and so on. The effect of the latter is often minimized by taking a mean of several observations at different places in the vessel. Despite these drawbacks, the notion of a single mixing time is a convenient one in practice, albeit extrapolations from one system to another must be treated with reserve. Irrespective of the technique used to measure the mixing time, the response curve will show periodic behavior. This is due simply to the repeated passages of a fluid element with a locally high concentration of the tracer. The time period between any two successive peaks is known as the circulation time. For a specific impeller/vessel configuration, mixing and circulation times are usually expressed in dimensionless form as functions of the Reynolds number and other pertinent parameters. Broadly speaking, both dimensionless mixing and circulation times are independent of the Reynolds number in the laminar and fully turbulent regimes, showing a substantial transition zone in between these two asymptotic values.

Not much is known about the mixing times for class I agitators in non-Newtonian systems. The limited available work (Norwood and Metzner, 1960) suggests that the correlations developed for Newtonian systems can also be used for non-Newtonian media, simply by using a generalized Reynolds number based on the effective viscosity (equation 9.4-3). The results of Godleski and Smith (1962), point to much larger mixing times than those predicted by Norwood and Metzner (1960), thereby implying severe segregation between the high shear (low viscosity) and low shear (high viscosity) zones. Furthermore, in highly shear-thinning systems with a yield stress, a cavern of turbulent flow surrounds the fast moving agitator, whereas the rest of the liquid may be at rest. Obviously, the use of mixing and circulation times is highly questionable to describe these systems. Intuitively, we would expect a similar deterioration in mixing for viscoelastic liquids, especially with the presence of secondary flows and flow reversal. The only mixing time study relating to the use of class II agitators for non-Newtonian media is that of Peters and Smith (1967) who reported a reduction in mixing and circulation times for viscoelastic systems agitated by an anchor-type impeller. The decrease in mixing time can be safely ascribed to the enhanced axial circulation.

In contrast, class III agitators have generated much more interest. It is generally believed that shear thinning does not exert an appreciable influence on the pumping capacity of helical impellers, hence the circulation times are not much affected by a shear-dependent viscosity (Chavan and Ulbrecht, 1973b; Chavan et al., 1975a, b; Carreau et al., 1976; Carreau et al., 1992; Cheng and Carreau, 1994b). Thus, the circulation time is constant in the viscous regime ($Re < 10$), whereas in the intermediate zone, it decreases with both increasing Reynolds number and decreasing degree of shear-thinning behavior, eventually becoming independent of the rheology (Cheng and Carreau, 1994b). Another study, with a helical ribbon impeller (Guérin et al., 1984), shows that even though the average circulation times are not influenced significantly by the shear-thinning characteristics, their distribution becomes progressively narrower with increasing shear thinning, thereby suggesting poor mixing between the high and low shear zones.

Similar observations can also be made regarding the mixing times in purely viscous systems; namely, the dimensionless mixing time, Nt_m, is independent of the Reynolds number in the low Reynolds number regime ($Re < 10$). However, for helical screw impellers in the intermediate zone ($10^1 < Re < 10^3$), the dimensionless mixing time decreases with increasing Reynolds number. The shear-thinning effects are completely embodied in the definition of a generalized Reynolds number. On the other hand, for helical ribbon impellers, Carreau et al. (1976) reported considerably longer mixing times as compared to Newtonian results, even though the circulation times were comparable. Dimensionless mixing times for five helical ribbons of different geometries are presented in Table 9.6-1, where a strong interaction between geometry and nonlinear flow properties is evident. The dimensionless mixing times are much longer in the viscoelastic fluids (1% Separan polyacrylamide and 3% CMC solutions) than in the Newtonian glycerol. The exceptionally large mixing time observed with the smaller diameter impeller IV (large clearance between the blades and the vessel wall) was attributed to the presence of a large stagnant zone, as shown in Fig. 1.2-13. The stagnant zone identified by the dark cone at the bottom of the tank was observed to persist for hours with the clockwise rotation of the impeller (i.e., pumping upward at the blades). No stagnant zones were present with the anticlockwise rotation of the helical ribbon impeller. Under these conditions, the mixing times were influenced by the shear-thinning characteristics. In a related study, dealing with the effect of mixers on the formation of aerated gels, Keirstead et al. (1980) observed, under certain conditions, a one order of magnitude improvement in mixing time, depending on the rotation direction. That is, a helical ribbon mixer rotating in the helix direction achieved a desired product homogeneity about ten times faster than if it were rotating in the counterhelix direction. This directional dependence continues to be elusive.

Viscoelasticity seems to suppress the axial circulation but augments the angular circulation for a helical ribbon and a helical ribbon-screw combination (Chavan and Ulbrecht, 1973b; Chavan et al., 1975a, b; Carreau et al., 1976), whereas it enhances the axial circulation for anchor-type impellers. These two sharply contradictory observations illustrate the interplay between viscoelasticity and geometry very well. Extrapolations from one geometry to another could lead to totally erroneous conclusions. For helical screw and ribbon agitators, both mixing and circulations times are believed to increase by differing

Table 9.6-1 Dimensionless Mixing Time for Helical Ribbon Agitators*

Impeller	n_b	D/d	p/d	w/D	Nt_m Glycerol	2% CMC	1% Separan
I	2	1.11	0.72	0.10	45	105	125
II	2	1.11	1.05	0.10	51	107	142
III	2	1.11	0.71	0.20	25	51	108
IV	2	1.37	0.85	0.12	55	189	137
V	1	1.11	0.70	0.10	61	120	163

* In all cases, D/H = 1. For details of the geometries, see Patterson et al. (1979). The main dimensions are defined in Fig. 9.4-3

amounts, on account of the viscoelasticity. Carreau et al. (1976) also concluded that viscoelastic liquids are inherently more difficult to mix than purely shear-thinning systems.

9.7 Gas Dispersion

In numerous instances, it is required to disperse a gas into a liquid. The main parameters describing the phenomenon of gas dispersion in agitated tanks are bubble size, gas holdup, interfacial area, mass transfer characteristics, and energy requirements. While a considerable body of information is available on various aspects of the dispersion of a gas into low viscosity liquids (Smith, 1985; Tatterson, 1991), little is known about the analogous processes in highly viscous Newtonian and non-Newtonian media, such as those encountered in polymer processing and fermentation applications.

9.7-1 Gas Dispersion Mechanisms

It is generally recognized that good dispersion of a gas into a liquid can only be achieved by using high speed (class I) agitators, which unfortunately, are not very useful for mixing high viscosity liquids. Hence, the process of gas dispersion into highly viscous media places two inherently conflicting requirements generally met in practice by using a combination of two impellers. For instance, a Rushton disc turbine combined with a 45° pitched blade agitator offers combined advantages, such as low flow-high shear, due to the disc turbine, and high flow-low shear produced by the other component of the dual impeller. Hence, such composite impellers may be regarded as a good compromise between those agitators which cause mixing by their size and shape on the one hand and those which rely on the mechanism of good momentum transport on the other hand. The intermig agitator of Fig. 9.7-1 is a good example of such a compromise. Note that two or more agitators can be mounted on the same shaft at 90° to each other, or in a staggered manner and, if designed suitably, can be rotated at moderately high speeds without excessive power consumption. In some applications two independent agitators are also employed. One of these, the positive displacement type, rotating slowly, results in liquid mixing, whereas the second one, operating at high rotational speed, facilitates gas dispersion. To date, however, most of the work has been carried out with Rushton disc turbines and a few dual impellers. The gas is invariably introduced through a sparger placed beneath the impeller.

Basically, the process of gas dispersion involves two competing phenomena: bubble breakage, which occurs predominantly in the high shear impeller region, and coalescence which takes place in quiet regions away from the impeller. Depending upon the impeller speed and the physical properties of the liquid (mainly viscosity and surface tension), gas bubbles may recirculate through the impeller region, may escape the system from the free surface, or may coalesce. The dispersion of gases into highly viscous liquids further differs from the corresponding process in low viscosity systems in regard to the tendency of gas

424   Liquid Mixing

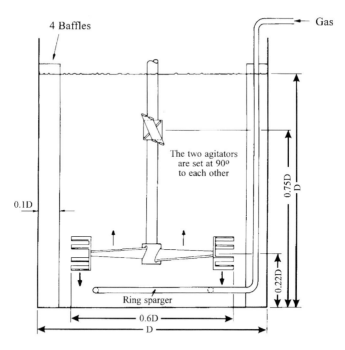

Figure 9.7-1 A pair of intermig agitators used for gas dispersion (Adapted from Nienow and Ulbrecht, 1985)

bubbles to follow the liquid motion. Admittedly, the exact mechanism of gas dispersion in liquids is not fully known, even for the much studied low viscosity systems, but qualitatively similar phenomena seem to be responsible for dispersion in highly viscous Newtonian and non-Newtonian systems. Essentially, the sparged gas gets sucked into the low pressure region (called cavities) behind the impeller. The shape and stability of these cavities are strongly influenced by the liquid rheology and the Froude and Reynolds numbers. There is, however, a minimum impeller speed required for the formation of such cavities. For instance, Van't Riet (1975) suggested that a Froude number larger than 0.1 is required for these cavities to form in liquids agitated by a disc turbine. High speed agitators are, however, not at all able to disperse gas in extremely viscous systems ($Re < 10$) as shown by Nienow et al. (1983). For $Re > 10^3$, the minimum impeller speed required for gas dispersion rises slowly with increasing liquid viscosity. That is, a 50% increase in impeller speed is associated with a ten-fold increase in viscosity. In the intermediate zone, $10 < Re < 10^3$, the process of cavity formation and the stability of the cavities are determined by complex interactions between Reynolds number, Froude number, and gas flow rate. Preliminary results seem to suggest that shear-thinning effects are negligible. The presence of a yield stress does not appear to play any role in this process, except for gas entrapped in cavities (Nienow et al., 1983). On the other hand, much bigger cavities, and cavities of different shapes (split-anvil) have been observed in viscoelastic systems (Solomon et al., 1981). Furthermore, the dispersion in viscoelastic liquids is more difficult, since the gas entrapped in a cavity does not disperse even if the gas flow is turned off. The complete release of gas occurs only when the agitation is stopped. The effect of viscoelasticity is even more pronounced with a twin-bladed intermig agitator. In this case, the gas-filled cavities can extend behind the outer twin split blades, almost all the way back to the following blade. Qualitatively, these differences have been attributed to the rather large extensional viscosities of the viscoelastic media.

9.7-2 Power Consumption in Gas-Dispersed Systems

The available limited work with Rushton turbines suggests that at low Reynolds number ($Re < 10$), the power number for gassed systems is almost the same as that without the introduction of a gas. This is possibly due to the fact that no gas cavities are formed at such low impeller speeds. In the intermediate region, $10 < Re < 10^3$, with the formation of gas cavities, the power number decreases below its ungassed value with the increasing impeller Reynolds number. It then goes through a minimum value for $300 < Re < 500$ and begins to increase again with impeller speed or Re. The gas flow rate does not seem to influence the value of the power number at a given impeller speed. Figure 9.7-2 shows this effect clearly for a variety of non-Newtonian and moderately viscous systems. The decrease in power number is a combined effect of streamlining, reduced pumping capacity of the impeller, and rheological properties of the fluid. At the point of the minimum power number, the cavities are of maximum size, and the impeller rotates in a pocket of gas without any dispersion. Undoubtedly, the reduction in power consumption as well as its minimum value are manifestations of the complex interplay between the size and structure of the cavities, the rheological characteristics, and the kinematic conditions. The nature of these interactions is far from being clear (Nienow and Ulbrecht, 1985; Tatterson, 1991).

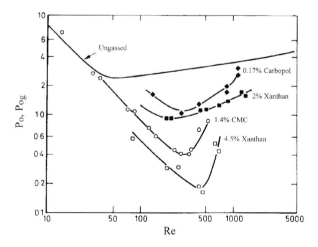

Figure 9.7-2 Gassed power number vs Reynolds number for various polymer solutions at sparging rates of 0.5 and 1.0 volume of gas over volume of fluid per min (Adapted from Nienow et al., 1983)

Another experimental work with helical ribbon impellers (Carreau et al., 1992; Cheng and Carreau, 1994a) suggests that the power consumption under gassed conditions may decrease or increase depending on the rheological characteristics of the liquid or flow regime. For instance, they reported reduced power consumptions in highly shear-thinning media (possibly due to the enhanced levels of deformation), whereas the power always increased in aerated viscoelastic systems. The latter was tentatively ascribed in part to the increase of the effective viscosity of the aerated fluids, due to the fact that small bubbles behave like solid spheres. Also, the elastic nature of the fluid is enhanced by the presence of deformable bubbles (see Section 8.2-2). Figures 9.7-3 and 9.7-4 illustrate the effects of aeration on power consumption in inelastic and viscoelastic systems respectively, mixed with a helical ribbon impeller. These results clearly reinforce the existence of strong interactions between viscoelasticity and impeller geometry. Furthermore, Cheng and Carreau (1994a) argued that the effective deformation rates prevailing under gassed conditions are likely to be larger than those estimated via equation 9.4-3. Thus, they introduced the concept of an aerated effective deformation rate, $\dot{\gamma}_a$, as

$$\dot{\gamma}_a = \sqrt{\dot{\gamma}_e^2 + \dot{\gamma}_b^2}, \qquad (9.7\text{-}1)$$

where $\dot{\gamma}_b$ is the extra shearing contribution due to aeration. Cheng and Carreau (1994a) found it adequate to estimate the magnitude of $\dot{\gamma}_b$ using the correlation of Nishikawa et al. (1977) modified by Henzler (1980) as

$$\dot{\gamma}_b = 1500 V_g, \qquad (9.7\text{-}2)$$

where V_g is the superficial gas velocity. Nishikawa et al. (1977) used 5000 instead of 1500 m^{-1}. Though the value of the Reynolds number under aerated conditions will increase with the increasing gassing rate, Cheng and Carreau (1994a) compared the gassed and ungassed power consumption data using a Reynolds number based on $\dot{\gamma}_a$ (Figs. 9.7-3 and 9.7-4). Figure 9.7-3 illustrates that the power requirement for inelastic shear-thinning fluids can be cut by half by sparging gas in the mixing vessel. In contrast, Fig. 9.7-4 shows that the power requirement may be increased by a factor up to 3.5 when gas is sparged in a constant viscosity elastic fluid.

Cheng and Carreau (1994a) have proposed the following expressions for the estimation of power (under aerated conditions):

Laminar flow regime: $0.28 \leq Re_a \leq 70$ $(0.028 \leq Na \leq 0.87;\ 0.0044 \leq Wi \leq 0.06)$:

$$Po_g = 1030 Re_a^{-0.942} Na^{0.604}(1 + 724 Wi^{2.15}); \qquad (9.7\text{-}3)$$

Transition flow regime: $70 \leq Re_a \leq 2600$ $(0.0087 \leq Na \leq 0.63;\ 0.013 \leq Wi \leq 0.96)$:

$$Po_g = 40.3 Re_a^{-0.623} Na^{0.06}(1 + 3.79 Wi^{0.057}), \qquad (9.7\text{-}4)$$

where $Na (= Q/Nd^3)$ is the aeration number, and $Wi (= \psi_1 N/\eta_a)$ is the Weissenberg number. ψ_1 is the primary normal stress difference coefficient and η_a is the aerated effective viscosity evaluated from the aerated effective deformation rate $\dot{\gamma}_a$. Note the diminishing significance of the aeration rate and the viscoelastic effects in the transition region, as evidenced by much smaller exponents of Na and Wi in equation 9.7-4.

9.7 Gas Dispersion   427

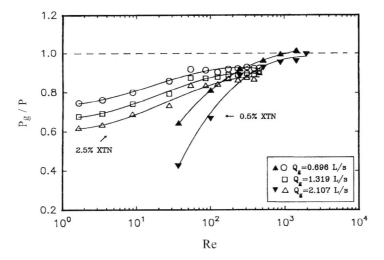

Figure 9.7-3 Gassed over ungassed power ratio for inelastic shear-thinning xanthan aqueous solutions for three gas sparging rates
The rheological parameters are reported in Table 9.4-2 (Reprinted from Cheng and Carreau, 1994a © 1994, with permission from Elsevier Science Ltd, The Boulevard, Langford Lane, Kidlington 0X5 1GB, UK)

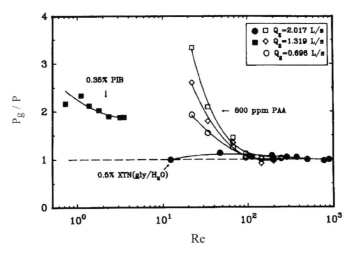

Figure 9.7-4 Gassed over ungassed power ratio for elastic Boger and shear-thinning fluids
The fluids are defined in Table 9.4-2. The 0.5% XTN (xanthan) solution is highly shear-thinning and moderately elastic; the 800 ppm PPA (polyacrylamide) solution in corn syrup has a nearly constant viscosity and moderate elastic properties; the 0.35% polyisobutylene (PIB) in a mixture of polybutene and kerosene is a Boger fluid with a constant viscosity and quadratic normal stress differences (Reprinted from Cheng and Carreau, 1994a © 1994, with permission from Elsevier Science Ltd, The Boulevard, Langford Lane, Kidlington 0X5 1GB, UK)

428   Liquid Mixing

Virtually no information is available on the behavior of more realistic dual impeller configurations, which are regarded to be more suitable for gas dispersion applications. As far as the scale-up is concerned, a preliminary study (Solomon et al., 1981) suggests that the power requirement for complete mixing decreases rapidly with the increasing size of impeller under both gassed and ungassed conditions.

9.7-3   Bubble Size and Holdup

These two parameters together with the interfacial area and volumetric mass transfer coefficients are used to characterize the effectiveness of gas dispersions into liquids. It is important to emphasize that all these parameters are extremely sensitive to the presence of surface active agents, as is the phenomenon of bubble coalescence. Undoubtedly, all these variables show spatial variation, and often only globally-averaged values are reported, which are found to be adequate for engineering design calculations. For a given fluid, the mean bubble size does not show a strong dependence on the level of agitation. Ranade and Ulbrecht (1978) investigated the influence of polymer addition on holdup and gas-liquid mass transfer in agitated vessels. Even small amounts of polymer addition were shown to result in substantial reductions in holdup and in mass transfer, albeit the extent of decrease showed a strong dependence on the type and concentration of polymer. Qualitatively similar results have been reported for markedly shear-thinning fluids (Machon et al., 1980), though the viscoelasticity seems to result in substantially larger holdup values with a helical ribbon impeller (Cheng and Carreau, 1994a). The reduction usually observed for polymer solutions has generally been ascribed to the formation of large bubbles, which in turn have a shorter residence time. The average holdup (ϕ) was found to vary with gassed power P_g and gas flow rate Q as

$$\phi \propto P_g^{0.3} Q^{0.7}, \tag{9.7-5}$$

or with power consumption per unit mass and superficial gas velocity V_g as

$$\phi \propto \left(\frac{P_g}{\rho V}\right)^{0.3} V_g^{0.7}. \tag{9.7-6}$$

For a fixed gassing rate, the power consumption decreases with increasing polymer concentration. Aside from these macroscopic observations, bubbles were found to be predominantly of two sizes, namely, a large population of small bubbles and very few large bubbles. Figure 9.7-5 shows typical results on the gas holdup in viscous shear-thinning liquids, obtained by Nienow and Ulbrecht (1985) with a single disc turbine and a composite impeller (45° downward pumping disc turbine). The point of minimum power consumption (largest amount of gas present in cavities) will obviously correspond to the point of the maximum gas holdup (Hickman, 1985).

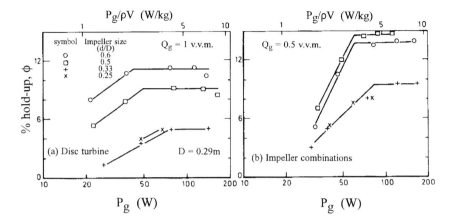

Figure 9.7-5 Examples of the relationship between holdup and power input when dispersing gas in a high viscosity, shear-thinning fluid (Adapted from Nienow and Ulbrecht, 1985)
(a) single disc turbine
(b) combination of 45° downward pumping impeller and disc turbine

9.7-4 Mass Transfer Coefficient

In view of what has been said so far regarding the mixing of viscous Newtonian and non-Newtonian systems, a deterioration in the mass transfer characteristics, namely a reduction in the value of $k_L a$, is inevitable. This conjecture is well borne out by the few studies available on this subject (Perez and Sandall, 1974; Yagi and Yoshida, 1975; Ranade and Ulbrecht, 1978; Nishikawa et al., 1981).

Dimensionless empirical expressions relating $k_L a$ to the system, as well as kinematic and physical variables, are available in the literature (Tatterson, 1991). The effective liquid viscosity has been evaluated at $\dot{\gamma}_e = k_s N$, except in the work of Nishikawa et al. (1981), who used the corresponding shear rate $\dot{\gamma}_e = 5000 V_g$, where V_g is the superficial gas velocity (in contrast to their earlier work, Nishikawa et al., 1977). By way of example, the following correlation due to Kawase and Moo-Young (1988) perhaps embraces the widest range of power-law parameters:

$$k_L a = 0.675 \frac{\rho^{0.6} \left(\frac{P_T}{V}\right)^{\frac{9+4n}{10(1+n)}}}{\left(\frac{m}{\rho}\right)^{0.5(1+n)} \alpha^{0.6}} (D_L)^{0.5} \left(\frac{V_g}{V_t}\right)^{0.5} \left(\frac{\eta_{eff}}{\mu_w}\right)^{-0.25}, \quad (9.7\text{-}7)$$

where P_T is the total power input, V is the volume of the dispersion, α is the surface tension, D_L is the gas diffusivity in the liquid, V_g and V_t are the gas superficial and the bubble velocities respectively, and μ_w is the viscosity of water. Equation 9.7-7 encompasses the following conditions: $0.59 < n < 0.95$; $0.00355 < m < 10.8 \text{ Pa} \cdot \text{s}^n$, and $0.15 < D < 0.6$ m.

9.8 Heat Transfer

It is an established practice to enhance heat transfer to process materials with externally applied motion, both within the bulk of the material and at the proximity of heat transfer surfaces. The imposed fluid motion enhances the rate of heat transfer beyond that caused by free convection in the case of mobile liquids, and that provided by conduction in the case of solids and highly viscous materials. In most process applications, fluid motion is promoted by either pumping or mechanical agitation. While a considerable body of information is available on heat transfer to low viscosity Newtonian fluids in agitated vessels (see Uhl and Gray, 1966; Hewitt et al., 1994), very little is known about the rates of heat transfer and the resulting heat transfer coefficient values in such systems for highly viscous Newtonian and non-Newtonian materials. It is somewhat surprising that despite their widespread occurrence in chemical and process applications, heat transfer to and from non-Newtonian fluids in mixing tanks through coils or jackets has received much less attention than that accorded to the fluid mechanical and mass transfer aspects, as described in the preceding sections. As in the case with power requirement and mixing time, the rate of heat transfer in mixing vessels is strongly dependent on the geometrical features of the tank-impeller combination, on the internals of the tanks (e.g., type and number of baffles), and on the operating variables (rotational speed, thermophysical properties of the fluid). Further complications arise from the type of heat transfer surface (e.g., coils, jacketed walls). Most of the progress in this area has therefore been made through the use of dimensional considerations aided by experimental observations. It is often not justifiable to make cross-comparisons between different studies unless the two systems exhibit complete similarity, namely geometrical, kinematic, and thermal.

It is customary to express the results of heat transfer studies in terms of the usual dimensionless groups as

$$Nu = f(Re, Pr, Gr, \quad \text{geometrical ratios}). \tag{9.8-1}$$

In equation 9.8-1, the new dimensionless groups are defined as

Nusselt number:

$$Nu = \frac{hL_c}{k}, \tag{9.8-2a}$$

Prandtl number:

$$Pr = \frac{\hat{C}_p \mu}{k}, \tag{9.8-2b}$$

and Grashof number:

$$Gr = \frac{g\beta \Delta T L_c^3 \rho^2}{\mu^2}. \tag{9.8-2c}$$

β is the coefficient of volume expansion and L_c is a characteristic linear dimension for heat transfer, which may be different from that used in the calculation of power consumption. L_c

is usually the impeller diameter. The Grashof number is a measure of the importance of free convection effects, which are generally insignificant in the case of low viscosity systems dominated by forced convection, when the characteristic temperature difference, ΔT, is small. It is of increasing importance in the case of highly viscous materials agitated by low rotational speed anchors and screw-type impellers (Carreau et al., 1994b). For geometrically similar systems, the relation between dimensionless groups, equation 9.8-1, reduces to

$$Nu = f(Re, Pr, Gr). \qquad (9.8\text{-}3)$$

No attempt is made to present an encyclopedic coverage of the pertinent studies in this area. Instead, only representative correlations for heat transfer for which effects of free convection are negligible are presented in the ensuing sections. It is convenient to present the information separately for each class of agitators.

9.8-1 Class I Agitators

As seen in Section 9.5-1, these agitators operate at relatively high rotational speeds, but are effective only in low to medium viscosity fluids. In most cases, the flow tends to be transitional or turbulent. The bulk of the literature related to heat transfer to Newtonian fluids, shear-thinning polymer solutions, and particulate slurries (cement, chalk) has been critically reviewed by Edwards and Wilkinson (1972), Poggermann et al. (1980), and Desplanches et al. (1980). In most cases involving paddle-, turbine-, and propeller-type impellers, the results have been correlated by the following generic expression:

$$Nu = A Re^b Pr^c (Vi)^d, \qquad (9.8\text{-}4)$$

where Vi is the viscosity number, that is, the ratio of the fluid viscosity evaluated at the wall over that at the bulk temperature. The viscosity number corrects for the effect of the temperature on the viscosity. Another method is to evaluate the fluid properties at the film temperature, that is, the average temperature between the wall and bulk temperatures.

For jacketed vessels, it is a well-accepted practice to use either the impeller diameter d or tank diameter D as the characteristic linear dimension in the Nusselt number, Nu. In most cases, the effective viscosity has been evaluated via equation 9.4-3, with $k_s = 4\pi N$, as suggested by Gluz and Pavlushenko (1966). The last term, namely $(Vi)^d$ in equation 9.8-4 has been introduced to account for the viscosity variation in the mixing tank, by analogy with that in the Dittus-Boelter equation for pipe flows. By way of illustration, Gluz and Pavlushenko (1966) presented the following correlation for a 300 mm diameter tank fitted with an electric heater and no baffles, agitated by a single-blade turbine impeller:

$$Nu = \frac{hd}{k} = 0.215 Re^{0.67} Pr^{0.33} \left(\frac{m_w}{m_b}\right)^{-0.18}, \qquad (9.8\text{-}5)$$

where m_w and m_b are the power-law consistency indices evaluated at the wall and bulk temperatures respectively. This equation is based on the experimental conditions $0.6 \leq n \leq 1$, $5 \leq Re \leq 2 \times 10^5$, and $Pr \leq 2.5 \times 10^4$, where $k_s = 4\pi N$ has been used. Suffice it to mention that reference should be made to the compilation presented by Edwards and Wilkinson (1972) for the other similar correlations involving additional geometric or rheological parameters.

On the other hand, the corresponding correlations for heat transfer to and from coils in vessels fitted with class I agitators tend to be more complex, and involve additional geometric parameters such as coil diameter/tank diameter ratio. The effective viscosity is still estimated via the Metzner-Otto concept with an appropriate value of k_s for the type of impeller in question. The following equation due to Edney and Edwards (1976) is a representative of this family of correlations:

$$Nu = \frac{hd_c}{k} = 0.036 Re^{0.64} Pr^{0.35} \left(\frac{D}{D_c}\right)^{0.375} \left(\frac{\eta_{eff}}{\eta_w}\right)^{0.2}, \qquad (9.8\text{-}6)$$

where d_c is the coil tube diameter and D_c is the mean helix diameter. The effective viscosities of carboxy-methyl cellulose and polyacrylamide solutions were estimated by using $k_s = 11.5$ for a six-blade turbine impeller. They concluded that this approach reconciled their data for Newtonian as well as non-Newtonian media encompassing the following ranges of variables: $400 \leq Re \leq 9.2 \times 10^5$; $4 \leq Pr \leq 1.9 \times 10^3$, and $0.65 \leq n_{eff} \leq 283$ mPa·s. Their preliminary results also seem to indicate that moderate levels of aeration did not influence the values of the heat transfer coefficient, though the power was seen to drop below the unaerated values.

9.8-2 Class II Agitators

As noted earlier, the performance of class I agitators deteriorates rapidly with increasing material viscosity. These impellers, namely gates and anchors, reach the far corners directly rather than relying on momentum transport, and operate at relatively low speeds. For heat transfer operations, it is even more important to induce fluid motion next to the heat transfer surface, such as the wall of the vessel. The bulk of the heat transfer literature involving anchors in viscous Newtonian and inelastic fluids has also been reviewed by Uhl and Gray (1966), Edwards and Wilkinson (1972), and Ayazi Shamlou and Edwards (1986). In the case of jacketed vessels, the thermal resistance seems to lie in the thin liquid film between the vessel wall and the impeller. The heat transfer modeling of such systems has been generally based on phenomenological considerations. The simplest model assumes that the heat transfer occurs primarily by conduction across the thin liquid film (Coyle et al., 1970). As expected, this oversimplified approach underestimates the value of the heat transfer coefficient by up to a factor of 4 (Ayazi Shamlou and Edwards, 1986). Heim (1980) developed a general model for anchor and helical agitators based on boundary layer considerations. He obtained an expression for the wall Nusselt number in terms of the Reynolds and Prandtl numbers and the ratio D/d. The latter appears to be a more significant variable in the laminar regime than under turbulent conditions. Subsequently, he has substantiated theoretical predictions by performing heat transfer experiments with carboxy-methyl cellulose solutions agitated by an anchor and a screw.

Other models of wall heat transfer include the penetration model due to Harriott (1959), which essentially treats it as unsteady, one-dimensional heat conduction with a semi-infinite medium, in between each impeller passage. This simple approach, however, seems to overestimate the value of the heat transfer coefficient by up to a factor of 7 (Rautenbach and Bollenrath, 1979). The latter authors attributed the observed discrepancies to the fact that the impeller does not completely wipe the liquid off the vessel wall, leaving a static liquid film. They thus modified the analysis due to Harriott (1959) by incorporating the effect of this static film, and presented the following expression for the Nusselt number:

$$Nu = 0.568 \left[\frac{\alpha}{(D-d)Nn_b^2} \right]^{-0.23} \left(\frac{D}{D-d} \right), \tag{9.8-7}$$

where α is the thermal diffusivity and n_b is the number of impeller blades. This equation, which is valid under laminar conditions, was stated to be applicable to both Newtonian and non-Newtonian fluids, since it is independent of the liquid viscosity. The limited data involving helical ribbons, reported by Ayazi Shamlou and Edwards (1986), are about twice as large as those predicted by equation 9.8-7, but nonetheless it is a considerable improvement over the previous predictions.

Several workers have correlated their experimental results on heat transfer through the use of dimensionless groups; most of these have been listed by Edwards and Wilkinson (1972). This section is concluded by citing one such correlation due to Sandall and Patel (1970) for jacketed wall heat transfer with anchors:

$$Nu = 0.315 Re^{\frac{2}{3}} Pr^{\frac{1}{3}} \left(\frac{\eta_{eff}}{\eta_w} \right)^{0.12}, \tag{9.8-8}$$

where the effective viscosity used in the Re and Pr numbers and in the viscosity ratio term has been evaluated via equation 9.4-3, with the value of k_s calculated by the method of Calderbank and Moo-Young (1961), that is,

$$k_s = \left(9.5 + \frac{9 \left(\frac{d}{d} \right)^2}{\left(\frac{d}{d} \right)^2 - 1} \right) \left(\frac{3n+1}{4n} \right)^{\frac{n}{n-1}}. \tag{9.8-9}$$

equation 9.8-8 is based on the results obtained with carbopol solutions ($0.3 \leq n \leq 1.0$) in a 180 mm diameter tank being heated/cooled for $320 \leq Re \leq 9 \times 10^4$ and $2 \leq Pr \leq 650$.

Similarly, Pollard and Kantyka (1969) carried out a comprehensive experimental study on coil heat transfer in tanks fitted with anchor agitators with chalk/water slurries ($0.3 \leq n \leq 1$) in vessels up to 1.1 m in diameter. They proposed the following correlation:

$$Nu = \frac{hD}{k} = 0.077 Re^{\frac{2}{3}} Pr^{\frac{1}{3}} \left(\frac{\eta_{eff}}{\eta_w} \right)^{0.14} \left(\frac{d}{d_c} \right)^{0.48} \left(\frac{d}{D_c} \right)^{0.27} \tag{9.8-10}$$

where d_c is the coil helix diameter. Equation 9.8-10 encompasses the Reynolds number in the range 200 to 6×10^5. The effective viscosity was estimated using equation 9.8-9 together with equation 9.4-3. Other correlations can be found in Edwards and Wilkinson (1972).

9.8-3 Class III Agitators

This class of agitators is characterized by relatively low shear rates, excellent pumping capacity for viscous fluids, and considerably improved mixing efficiency for highly viscous Newtonian as well as non-Newtonian media. Consequently, this category of agitators has generated much more interest in heat transfer studies than class II agitators. While the semi-theoretical framework mentioned in the preceding section is equally applicable here, most of the progress in this field has been made through dimensional considerations aided by experimental observations. Many excellent experimental studies on vessel wall heat transfer (Coyle et al., 1970; Heim, 1980; Ayazi Shamlou and Edwards, 1986; Kai and Shengyao, 1989; Kuriyama et al., 1983; Nagata et al., 1972b) and coil heat transfer (Nagata et al., 1972b; Carreau et al., 1994b) have been reported in the literature. However, since the systems employed by different investigators are not geometrically similar, it is virtually impossible to make meaningful comparisons between these studies. Instead, representative correlations are presented here to give the reader an idea of what is involved in attempting to develop universal correlations in this area.

In a recent study, Carreau et al. (1994b) have investigated the heat transfer between a coil (acting as a draft tube) and viscous Newtonian, shear-thinning, and viscoelastic polymer solutions agitated by a screw in a flat bottomed vessel. Experiments were performed both in the heating and the cooling mode to eliminate any spurious effects. The water flow rate inside the coil was sufficiently high to ensure the thermal resistance to be negligible on the inside of the coil. The effective viscosity was estimated via equation 9.4-3 with $k_s = 16$, a value deduced for their agitator in a previous study (Carreau et al., 1992). They were able to correlate their results for Newtonian, shear-thinning, and viscoelastic fluids by a single correlation as

$$Nu = \frac{h_{co} d_{co}}{k} = 0.387 Re^{0.51} Pr^{\frac{1}{3}} \left(\frac{d_{co}}{d}\right)^{0.594}, \qquad (9.8\text{-}11)$$

where the subscript "o" refers to the outside of the coil. All the properties are evaluated at the film temperature, $T_f = (T_w + T_b)/2$. Carreau et al. (1994b) noted that equation 9.8-11 entails somewhat greater uncertainty for viscoelastic fluids as compared to that for purely viscous fluids. Note the strong dependence on the ratio (d_{co}/d) in equation 9.8-11, which is qualitatively consistent with the analysis of Heim (1980).

Following Desplanches et al. (1983), Carreau et al. (1994) also proposed alternative correlations in terms of the liquid circulation velocity rather than the impeller tip velocity used in the conventional definition of the Reynolds number. The mean circulation velocity of the liquid, V_c, is defined by

$$V_c = \frac{l_c}{t_c}, \qquad (9.8\text{-}12)$$

where l_c is the mean circulation length experimentally found to be ≈ 1.08 m in their study (it will generally be strongly influenced by the actual circulation pattern); and t_c is the circulation time, which can be deduced from the correlation due to Carreau et al. (1992) as

$$\frac{V}{d^3Nt_c} = [0.124 + 0.265\{1 - \exp(0.00836Re)\}] \times (1 - 0.811Wi^{0.25}), \qquad (9.8\text{-}13)$$

where V is the vessel volume, Re is the Reynolds number based on η_{eff} evaluated at $\dot{\gamma}_e = k_s N$, and Wi is the Weissenberg number $(= N_1/2\eta_{eff}\dot{\gamma}_e)$. For the heat transfer, the effective deformation rate is taken as

$$\dot{\gamma}_e = \frac{V_c}{d_{c_0}}. \qquad (9.8\text{-}14)$$

Thus, Carreau et al. (1994b) introduced the following generalized definitions for the Reynolds, Prandtl, and Nusselt numbers:

$$Re_c = \frac{\rho V_c(n_s d_{c_0})}{\eta_{eff}} = \frac{\rho V_c^{2-n}(n_s d_{c_0}^n)}{m}, \qquad (9.8\text{-}15)$$

$$Pr_c = \frac{\hat{C}_p \eta_{eff}}{k} = \frac{\hat{C}_p m V_c^{n-1}}{k d_{c_0}^{n-1}}, \qquad (9.8\text{-}16)$$

and

$$Nu_c = \frac{h_{c_0}(n_s d_{c_0})}{k}. \qquad (9.8\text{-}17)$$

Note that $(n_s d_{c_0})$ has been used as the characteristic linear dimension in these definitions. This incorporates n_s, the number of loops in the coil. In terms of these definitions, Carreau et al. (1994b) proposed the following single correlation for all fluid/coil combinations:

$$Nu_c = 2.82 Re_c^{0.385} Pr_c^{1/3}. \qquad (9.8\text{-}18)$$

All the physical properties are evaluated at the film temperature. While both equations 9.8-11 and 9.8-18 are of similar form, the latter is preferable, since it incorporates some features of the flow patterns (in terms of l_c and t_c), which reflects the importance of the geometrical aspects of the mixer configuration. With the notable exception of Carreau et al. (1994b), little is known about the importance of viscoelastic effects on heat transfer in mixing vessels.

It is worthwhile to reiterate here that most of the currently available information pertains to specific geometrical arrangements, for a well-defined range of conditions. Extrapolation to larger scale and slightly different systems should be done with caution. The main intent here has been to provide a bird's-eye view of the activity in this area, and the correlations presented here must be regarded as being somewhat tentative. For purely viscous (inelastic) fluids, it is possible to use the corresponding correlation for Newtonian fluids together with an appropriately defined effective viscosity. This approach, however, may not work satisfactorily for viscoelastic systems.

9.9 Mixing Equipment and its Selection

The wide range of mixing equipment available commercially reflects the enormous range of mixing problems encountered in industry. It is thus reasonable to expect that no single piece of equipment will be able to carry out such a range of mixing duties in a satisfactory manner. This has led to the development of a number of distinct types of mixers over the years, but unfortunately very little has been done by way of the standardization of this equipment, and virtually no design codes are available. Consequently, the choice of a mixer type and its design is essentially governed by experience. In the following sections, the salient mechanical aspects of commonly used mixing equipment, together with their range of applications, is briefly described. Oldshue (1983) has presented a much more detailed account of mixing equipment.

9.9-1 Mechanical Agitation

This is easily the most commonly used method of mixing liquids, including gas- liquid and solid-liquid systems of low to moderately high viscosity. Basically, there are three elements in such devices.

9.9-1.1 Tanks

These are often vertically mounted cylindrical tanks (up to 10 m in diameter), which are typically filled to a depth equal to about one diameter, except in some gas-liquid contacting operations where liquid depths may be as large as three tank diameters. The bottom of the tank may be flat, dished, conical, or contoured depending upon the envisaged application and other factors such as the ease of discharge of solids.

For the batch mixing of viscous pastes and doughs using ribbon impellers, Z- or S-blades, the mixing tank may be mounted horizontally. The working volumes are often small and the mixing blades are massive in construction.

9.9-1.2 Baffles

In order to minimize gross vortexing, which is detrimental to mixing, particularly in low viscosity systems, baffles are fitted to the walls of the vessel. These take the form of thin strips, typically one tenth of the tank diameter in width. Usually four equally-spaced baffles are employed. In some cases, the baffles are flush-mounted with the wall, although occasionally a small clearance is left between the wall and the baffle to promote fluid motion near the wall region. Baffles are, however, usually not required for high viscosity liquids because the viscous forces are sufficient to counteract the tendency of vortex formation. Their presence may be detrimental to the mixing efficiency for shear-thinning and viscoplastic fluids.

9.9-1.3 Impellers

The impeller represents the most important element of mechanically agitated tank systems. Figure 9.9-1 depicts some of the frequently used impellers for relatively low viscosity systems. Turbines, paddles, propellors, gates, anchors, helical screws, and helical ribbons are generally mounted on a central vertical shaft in a cylindrical vessel, and they are chosen for a particular duty largely on the basis of the liquid viscosity. Broadly speaking, as the liquid viscosity increases, the choice of an impeller changes from a propellor, to a turbine, to a paddle, to an anchor/gate, to a helical ribbon, and finally to a screw for highly viscous systems. In so doing, the speed of rotation decreases. Further minor variations in the impeller designs are possible, such as retreating-blade turbines, angled-blade turbines, and four- to twenty-blade turbines.

Anchors, helical ribbons, and screws are generally used for high viscosity liquids. Anchors and ribbons operate with a very small clearance between the impeller and the tank, whereas helical screws often have a smaller diameter, but are used inside a draft tube to enhance fluid motion throughout the vessel. Finally, kneaders, Z- and sigma-blade mixers, and Banbury mixers, used for mixing high viscosity pastes, doughs, and rubbers, are usually mounted in a horizontal position and are used in pairs. These impellers are massive and the clearances between the blades as well as between the wall and the blade are very small, thereby ensuring that the entire mass of material is sheared.

9.9-2 Extruders

In plastics industries, the mixing duties are generally carried out in either single- or twin-screw extruders. The feed usually contains the base polymer in powder or in granular form along with additives such as stabilizers, pigments, and plasticizers. During processing in the extruder, the polymer is melted and the additives are mixed. The extrudate is delivered at high pressure and at a controlled rate from the extruder for shaping either by means of a die or a mold. In recent years, considerable advances, especially in terms of numerical simulations, have been made in the design of extruders. One complicating feature of extruders is the enormous viscous dissipation generated, leading to a strong coupling between heat and momentum transfer.

In a typical single screw extruder shown in Fig. 9.9-2, the shearing action is not intense enough to induce good mixing. This difficulty is obviated by using co- or counter-rotating twin screws, as shown schematically in Fig. 9.9-3. Detailed descriptions concerning the design and performance of extruders are available in the literature (see for example Agassant et al., 1991). In addition to the devices described in the foregoing, many other types of mixers based on altogether different principles, such as jet mixers, and in-line static mixers, and in-line dynamic mixers, have also been developed for specialized applications (Harnby et al., 1992).

438 Liquid Mixing

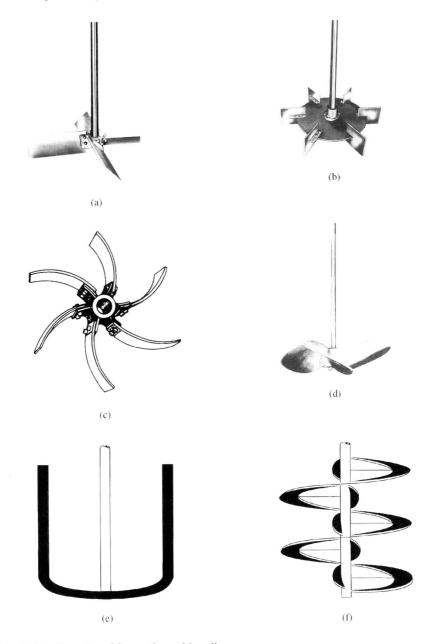

Figure 9.9-1 Examples of frequently used impellers
(a) 45° inclined blade turbine
(b) Rushton turbine
(c) curved-blade turbine
(d) propeller
(e) anchor or gate impeller
(f) helical ribbon with central screw

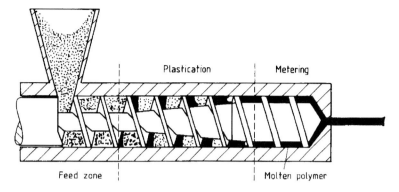

Figure 9.9-2 Single-screw extruder (From Agassant et al., 1991)

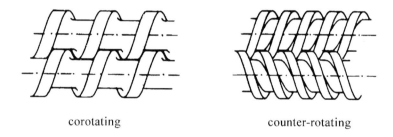

corotating   counter-rotating

Figure 9.9-3 Corotating twin-screw extruder in corotating and counter-rotating arrangements

9.10 Problems*

9.10-1$_a$ Power Requirement for Shear-Thinning Fluids

A cylindrical mixing vessel (fitted with four baffles) has a vertically located agitator turning at a speed of 1 Hz. A geometrically similar vessel (twice as large) operates with an agitator speed of 0.25 Hz. If the same liquid is used in both vessels, estimate the ratio of the power input per unit volume of fluid in the two cases, as a function of the power-law flow behavior index, $1 > n > 0.3$. Assume that the mixing always takes place in the laminar region and that the values of power-law parameters m and n remain constant in the shear rate range encountered in both vessels.

$$\text{Answer}: \quad \frac{P}{D^3} \propto N^{n+1}$$

---

* Problems are labelled by a subscript a or b indicating the level of difficulty.

440   Liquid Mixing

9.10-2$_a$ Effective Deformation Rate

Consider the schematic representation of a helical ribbon agitator system shown in Fig. 9.4-3. As mentioned in the text, one common approximation to evaluate the effective deformation rate in this configuration involves the replacement of the impeller by an equivalent cylinder (diameter d_e) rotating at the angular velocity Ω. Obtain an expression for the torque acting on the inner cylinder for a power-law fluid and thus derive the expression for k_s, given by equation 9.4-11. Note that the bottom effects have been neglected.

9.10-3$_a$ Bottom Effects on the Metzner-Otto Constant

It is readily seen that an additional contribution to the torque stems from the bottom of the vessel. This is likely to be a strong function of the clearance between the impeller (equivalent cylinder) and the bottom of the tank. For moderate clearances, we can argue that the $r\theta$-component of the extra stress would be negligible in comparison with the $z\theta$-component. Based on this hypothesis, derive an expression for the bottom correction to the torque on the inner cylinder. Verify the second term on the right side of equation 9.4-13.

The experimental results of Carreau et al. (1993) show that this correction is indeed small.

9.10-4$_b$ Effective Deformation Rate in the Transition Regime

With increasing impeller speed, the flow is no longer laminar and the axial flow is no longer negligible compared to the tangential velocity (transition regime). Thus, the effective viscosity should be determined by the overall deformation rate. One highly idealized flow configuration which captures both these flow features is the so-called helical flow, shown schematically in Fig. 4.2-5. Assume that both θ- and z-components of the velocity are functions of r.

(a) Obtain an expression for the effective deformation rate for power-law fluids to illustrate that it varies as $N^{\left(\frac{2-a}{a(1-n)}\right)}$, where

$$a = \frac{d\log Po}{d\log Re}(<1).$$

(b) In a recent publication (Cheng and Carreau, 1994b), two semi-empirical approaches have been presented to predict the effective deformation rates for power-law fluids agitated in the transition regime. Discuss and compare the predictions of these two approaches with that of the two-dimensional helical flow approximation.

Appendix A   General Curvilinear Coordinate Systems and Higher Order Tensors*

A.1 Cartesian Vectors and Summation Convention 442
A.2 General Curvilinear Coordinate Systems 445
    A.2-1 Generalized Base Vectors 445
    A.2-2 Transformation Rules for Vectors 449
        A.2-2.1 Contravariant Transformation 449
        A.2-2.2 Covariant Transformation 451
    A.2-3 Tensors of Arbitrary Order 452
    A.2-4 Metric and Permutation Tensors 454
    A.2-5 Physical Components 458
A.3 Covariant Differentiation 462
    A.3-1 Definitions 462
    A.3-2 Properties of Christoffel Symbols 464
    A.3-3 Rules of Covariant Differentiation 465
    A.3-4 Grad, Div, and Curl 468
A.4 Integral Transforms 474
A.5 Isotropic Tensors, Objective Tensors, and Tensor-Valued Functions 476
    A.5-1 Isotropic Tensors 476
    A.5-2 Objective Tensors 478
    A.5-3 Tensor-Valued Functions 480
A.6 Problems 483
    A.6-1$_a$ Rotation of Axes 483
    A.6-2$_a$ Contraction 484
    A.6-3$_a$ Quotient Law 484
    A.6-4$_a$ Transformation Rule for the Contravariant Components of a Second-Order Tensor 484
    A.6-5$_a$ Christoffel Symbols 484
    A.6-6$_a$ Cylindrical Coordinates 484
    A.6-7$_a$ Covariant Derivative 485
    A.6-8$_a$ Physical Components 485
    A.6-9$_b$ Divergence Theorem 485
    A.6-10$_b$ Isotropic Tensor 485
    A.6-11$_b$ Objectivity 485
    A.6-12$_a$ Invariants 486
    A.6-13$_b$ Tensor-Valued Function 486
    A.6-14$_b$ Elongational Flow 486

---

* Problems are labelled by a subscript a or b indicating the level of difficulty.

A.1 Cartesian Vectors and Summation Convention

A vector $\underline{v}$ is represented by its components, which depend on the choice of the coordinate system. If we change the coordinate system, the components will generally change. The vector $\underline{v}$ however is independent of the coordinate system. Thus, there is a relationship between the components of $\underline{v}$ in one coordinate system and the components of the same vector $\underline{v}$ in another coordinate system. It is the object of this section to establish such relationships.

We start by considering a transformation from one Cartesian coordinate system (x, y, z) to another Cartesian coordinate system $(\bar{x}, \bar{y}, \bar{z})$ obtained by rotating the (x, y, z) system as shown in Fig. A.1-1.

Vector $\underline{v}$ has components v_x, v_y, v_z in the (x, y, z) system and components $\bar{v}_x, \bar{v}_y, \bar{v}_z$ in the $(\bar{x}, \bar{y}, \bar{z})$ system. We wish to relate those components, and this can be achieved as follows:

$$\underline{v} = v_x \underline{\delta}_x + v_y \underline{\delta}_y + v_z \underline{\delta}_z = \bar{v}_x \underline{\bar{\delta}}_x + \bar{v}_y \underline{\bar{\delta}}_y + \bar{v}_z \underline{\bar{\delta}}_z = \sum_{i=1}^{3} \bar{v}_i \underline{\bar{\delta}}_i = \bar{v}_i \underline{\bar{\delta}}_i. \tag{A.1-1}$$

where the summation convention is used for the last term of equation A.1-1. The summation sign occurs so frequently that it is useful to adopt a convention, known as the Einstein summation convention. According to this convention, whenever an index occurs twice in an expression it implies summation over all the possible values of that index, unless stated otherwise. Thus in equation A.1-1 the index i is a dummy index, and the expression on the right side of equation A.1-1 implies summation over all possible values of i.

The component v_x is obtained by forming the dot product of equation A.1-1 with the unit vector $\underline{\delta}_x$:

$$v_x = \bar{v}_i \underline{\bar{\delta}}_i \cdot \underline{\delta}_x = \bar{v}_x \underline{\bar{\delta}}_x \cdot \underline{\delta}_x + \bar{v}_y \underline{\bar{\delta}}_y \cdot \underline{\delta}_x + \bar{v}_z \underline{\bar{\delta}}_z \cdot \underline{\delta}_x. \tag{A.1-2a}$$

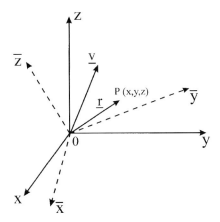

Figure A.1-1 Vectors $\underline{v}$ and $\underline{r}$ in two coordinate systems

A.1 Cartesian Vectors and Summation Convention

Similarly, v_y and v_z are obtained through forming dot products with unit vectors $\underline{\delta}_y$ and $\underline{\delta}_z$ respectively:

$$v_y = \bar{v}_i \bar{\underline{\delta}}_i \cdot \underline{\delta}_y \tag{A.1-2b}$$

$$\text{and} \quad v_z = \bar{v}_i \bar{\underline{\delta}}_i \cdot \underline{\delta}_z. \tag{A.1-2c}$$

Note that in the rectangular Cartesian system,

$$\underline{\delta}_i \cdot \underline{\delta}_j = \delta_{ij}, \tag{A.1-3a}$$

where

$$\delta_{ij} = \begin{cases} 0 & \text{if } i \neq j \\ 1 & \text{if } i = j \end{cases}. \tag{A.1-3b}$$

δ_{ij} is the *Kronecker delta*.

Inverting the set of equations A.1-2a to c, we obtain

$$\begin{pmatrix} \bar{v}_x \\ \bar{v}_y \\ \bar{v}_z \end{pmatrix} = \begin{pmatrix} l_{11} & l_{12} & l_{13} \\ l_{21} & l_{22} & l_{23} \\ l_{31} & l_{32} & l_{33} \end{pmatrix} \begin{pmatrix} v_x \\ v_y \\ v_z \end{pmatrix}, \tag{A.1-4}$$

where

$$l_{11} = \bar{\underline{\delta}}_x \cdot \underline{\delta}_x = \cos(\bar{\underline{\delta}}_x, \underline{\delta}_x), \tag{A.1-5a}$$

$$l_{21} = \bar{\underline{\delta}}_y \cdot \underline{\delta}_x = \cos(\bar{\underline{\delta}}_y, \underline{\delta}_x), \tag{A.1-5b}$$

etc.

$\cos(\bar{\underline{\delta}}_x, \underline{\delta}_x)$ is the cosine of the angle between the unit vectors $\bar{\underline{\delta}}_x$ and $\underline{\delta}_x$. The nine quantities $l_{11}, l_{21}, \ldots$ can be represented in a more compact form by l_{mn}, where both indices m and n take the values 1, 2, and 3.

In equation A.1-4 the indices in l_{mn} are written as numbers instead of as x, y, z, and this notation has the following advantages:

(i) In the case of an extension to an n-dimensional space, where n can be greater than 26, we would run out of letters. Indeed, we can extend it to an infinite dimensional space.
(ii) The notation is more compact.

The coordinates (x, y, z) and $(\bar{x}, \bar{y}, \bar{z})$ will be relabeled as (x^1, x^2, x^3) and $(\bar{x}^1, \bar{x}^2, \bar{x}^3)$ respectively. Note that the indices in (x^1, x^2, x^3) are written as superscripts. The reason for this notation will be explained later. Thus x^2 is not x squared and we shall denote x^2 squared as $(x^2)^2$.

The three equations given in matrix form A.1-4 can now be written as

$$v_m = \sum_{n=1}^{3} \bar{v}_n l_{nm} = \bar{v}_n l_{nm}, \tag{A.1-6}$$

where the summation convention is used for n; m can take the values of 1, 2, or 3. Once a value of m is chosen, we must apply the same value to m wherever m occurs. Thus the three equations represented by equation A.1-6 are obtained by assigning the value of $m = 1, 2,$ and 3 in turn.

The components $(\bar{v}_1, \bar{v}_2, \bar{v}_3)$ can be expressed in terms of (v_1, v_2, v_3) by inverting equation A.1-6, making use of the properties of l_{nm}, or repeating the process used in obtaining equation A.1-6. We shall adopt the second procedure to demonstrate the elegance and conciseness of the summation convention. Equation A.1-1 can be written as

$$\underline{v} = v_m \underline{\delta}_m = \bar{v}_n \bar{\underline{\delta}}_n. \tag{A.1-1}$$

Forming the dot product with $\bar{\underline{\delta}}_r$ (we cannot use $\bar{\underline{\delta}}_n$ or $\underline{\delta}_m$ due to the summation convention), we obtain

$$v_m \underline{\delta}_m \cdot \bar{\underline{\delta}}_r = \bar{v}_n \bar{\underline{\delta}}_n \cdot \bar{\underline{\delta}}_r = \bar{v}_n \delta_{nr} \tag{A.1-7}$$

From the definition of l_{mn}, and equations A.1-5a and b, we have

$$\underline{\delta}_m \cdot \bar{\underline{\delta}}_r = l_{rm}. \tag{A.1-8}$$

Note that the first index in l_{rm} (r) is associated with $\bar{\underline{\delta}}_r$, and the second index ($m$) is associated with $\underline{\delta}_m$, of the unbarred coordinate system. Combining equations A.1-7 and A.1-8, we obtain

$$l_{rm} v_m = \bar{v}_n \delta_{nr} = \bar{v}_r. \tag{A.1-9}$$

On summing the right side of equation A.1-9 for all possible values of n we find, due to the definition of δ_{nr}, that the only nonzero term is $\bar{v}_r$.

If P is any point in space, and the coordinates of P are (x^1, x^2, x^3) and $(\bar{x}^1, \bar{x}^2, \bar{x}^3)$ relative to the $Ox^1x^2x^3$ and $O\bar{x}^1\bar{x}^2\bar{x}^3$ coordinate systems respectively, then the components of the position vector $\underline{OP}(=\underline{r})$ will transform from the barred to the unbarred system or vice versa, according to equations A.1-6 and A.1-9. That is,

$$x^m = \bar{x}^p l_{pm} \tag{A.1-10}$$

and

$$l_{rm} x^m = \bar{x}^r. \tag{A.1-11}$$

Here we have freely written the indices as superscripts and subscripts. Although this is permissible in our present coordinate transformation, it is not generally permissible to do so in a general coordinate transformation, as we shall discuss in the next section.

Because the direction cosines l_{rm} are constants, as defined by equations A.1-10, the transformation from $Ox^1x^2x^3$ to $O\bar{x}^1\bar{x}^2\bar{x}^3$ is a linear transformation. Further, we note from equation A.1-11 that

$$l_{rm} = \frac{\partial \bar{x}^r}{\partial x^m}. \tag{A.1-11a}$$

Note that l_{rm} can also be obtained via equation A.1-10 by replacing the dummy index p by r, to yield

$$l_{rm} = \frac{\partial x^m}{\partial \bar{x}^r}. \tag{A.1-11b}$$

Equations A.1-11a and b imply that

$$\frac{\partial \bar{x}^r}{\partial x^m} = \frac{\partial x^m}{\partial \bar{x}^r}. \tag{A.1-11c}$$

This is only true in the case of a Cartesian transformation.

Substituting equation A.1-11 into equation A.1-9, we obtain

$$\bar{v}_r = \begin{cases} \dfrac{\partial x^m}{\partial \bar{x}^r} v_m & \text{(A.1-12a)} \\ \dfrac{\partial \bar{x}^r}{\partial x^m} v_m. & \text{(A.1-12b)} \end{cases}$$

Equations A.1-12a and b describe two types of transformation laws. Components that transform according to equation A.1-12a are called covariant components (v_m) and the indices are written as subscripts; components that transform according to equation A.1-12b are known as contravariant components and the indices are written as superscripts. Thus, in proper tensor notation, equation A.1-12b should be written as

$$\bar{v}^r = \dfrac{\partial \bar{x}^r}{\partial x^m} v^m. \qquad \text{(A.1-13)}$$

Note the symmetry in the notation. In equation A.1-12a, m is a dummy index: it occurs once as a superscript in $\partial x^m/\partial \bar{x}^r$ (in the expression $\partial x^m/\partial \bar{x}^r$ we regard m as a superscript and r as a subscript) and once as a subscript in v_m. We can consider them as cancelling each other. On the right side, we are left with a subscript r, and on the left side we also have only a subscript r. The superscript m is associated with $Ox^1 x^2 x^3$ and the subscript r with $O\bar{x}^1 \bar{x}^2 \bar{x}^3$. Similarly in equation A.1-13, the index m cancels and we have the superscript r on both sides of the equation. Thus, in a general coordinate transformation we need to distinguish between covariant and contravariant components. But for Cartesian components, both laws of transformation are equivalent (equations A.1-12a and b) and thus there is no need to make a distinction between subscripts (covariant components) and superscripts (contravariant components).

A.2 General Curvilinear Coordinate Systems

The rectangular Cartesian coordinate system is not always the most convenient coordinate system to use in solving problems. The laminar flow of a fluid in a circular pipe is solved using a cylindrical polar coordinate system. The velocity is then a function of the radial position r only and not of two variables x^1 and x^2. Similarly the spherical polar coordinate system is chosen for solving flow past a sphere. The choice of coordinate system depends on the geometry of the problem.

A.2-1 Generalized Base Vectors

We now consider a general curvilinear coordinate system (q^1, q^2, q^3) as shown in Fig. A.2-1. This coordinate system is related to the rectangular Cartesian coordinate system

(x^1, x^2, x^3) by

$$x^1 = x^1(q^1, q^2, q^3), \quad \text{(A.2-1a)}$$
$$x^2 = x^2(q^1, q^2, q^3), \quad \text{(A.2-1b)}$$
$$\text{and} \quad x^3 = x^3(q^1, q^2, q^3). \quad \text{(A.2-1c)}$$

Or in concise notation,

$$x^m = x^m(q^n). \quad \text{(A.2-2)}$$

We assume that the transformation from (x^1, x^2, x^3) to (q^1, q^2, q^3) is one-to-one and can be inverted as

$$q^n = q^n(x^m). \quad \text{(A.2-3)}$$

The base vectors $\underline{\delta}_m$ of the orthonormal coordinate system can be defined as being tangent to the coordinate axes x^m. Likewise, we define a set of base vectors $\underline{g}_m$ of the generalized curvilinear coordinate system as being tangent to the coordinate axes q^m, as illustrated in Fig. A.2-2. The vector connecting the origin to point P is $\underline{r}$.

The base vectors $\underline{g}_m$ are given by

$$\underline{g}_m = \frac{\partial \underline{r}}{\partial q^m} = \frac{\partial x^p}{\partial q^m} \underline{\delta}_p. \quad \text{(A.2-4)}$$

Base vectors defined as tangents to coordinate curves are covariant base vectors.

The base vectors $\underline{\delta}_m$ can also be considered as being normal to the planes $x^m = $ constant. For example, the base vector $\underline{\delta}_3$ is normal to the surface $x^3 = $ constant, that is, it is parallel to the $x^1 x^2$ plane. In the particular case of the Cartesian coordinate system, the tangent and the normal coincide, which explains why equation A.1-11c is valid.

This is not generally the case. Base vectors defined as normal to coordinate surfaces $q^n = $ constant are denoted by $\underline{g}^n$, as illustrated in Fig. A.2-2. These base vectors $\underline{g}^n$ are normal to the planes formed by the other two coordinate curves. They are contravariant base vectors defined by

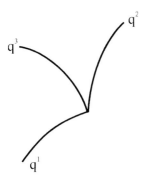

Figure A.2-1 Curvilinear coordinate system

A.2 General Curvilinear Coordinate Systems

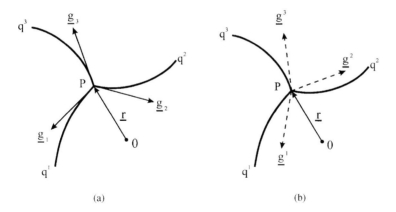

Figure A.2-2 Generalized coordinate system and base vectors: (a) covariant (tangential) base vectors; (b) contravariant (normal) base vectors

$$\underline{g}^n = \underline{\nabla} q^n = \text{grad } q^n = \frac{\partial q^n}{\partial x^l} \underline{\delta}_l, \quad \text{(A.2-5a)}$$

where the operator $\underline{\nabla}$ is defined by

$$\underline{\nabla} \equiv \underline{g}^n \frac{\partial}{\partial q^n}. \quad \text{(A.2-5b)}$$

Note that neither $\underline{g}_m$ nor $\underline{g}^n$ are necessarily unit vectors, but from orthogonality it follows that

$$\underline{g}_m \cdot \underline{g}^n = \delta_m^n \quad \text{(A.2-6a)}$$

and

$$\underline{g}^n \cdot \underline{g}_m = \delta_m^n, \quad \text{(A.2-6b)}$$

where δ_m^n and δ_m^n are the Kronecker deltas written in the notation convention for curvilinear coordinates.

Example A.2-1 Covariant and Contravariant Components of Base Vectors
Obtain the covariant and contravariant base vectors for the spherical polar coordinate system.

Solution
The transformation from the rectangular Cartesian coordinate system (x^1, x^2, x^3) to the spherical polar coordinate system (r, θ, ϕ) is

$$x^1 = r \sin\theta \cos\phi, \quad x^2 = r \sin\theta \sin\phi, \quad \text{and} \quad x^3 = r \cos\theta. \quad \text{(A.2-7)}$$

(In our notation $q^1 = r$, $q^2 = \theta$, $q^3 = \phi$.)

The covariant base vectors are given by equation A.2-4b:

$$\underline{g}_r = \frac{\partial x^1}{\partial r}\underline{\delta}_1 + \frac{\partial x^2}{\partial r}\underline{\delta}_2 + \frac{\partial x^3}{\partial r}\underline{\delta}_3 \tag{A.2-8a}$$

$$= \sin\theta\cos\phi\,\underline{\delta}_1 + \sin\theta\sin\phi\,\underline{\delta}_2 + \cos\theta\,\underline{\delta}_3, \tag{A.2-8b}$$

$$\underline{g}_\theta = \frac{\partial x^1}{\partial \theta}\underline{\delta}_1 + \frac{\partial x^2}{\partial \theta}\underline{\delta}_2 + \frac{\partial x^3}{\partial \theta}\underline{\delta}_3 \tag{A.2-8c}$$

$$= r\cos\theta\cos\phi\,\underline{\delta}_1 + r\cos\theta\sin\phi\,\underline{\delta}_2 - r\sin\theta\,\underline{\delta}_3, \tag{A.2-8d}$$

and $\quad \underline{g}_\phi = \dfrac{\partial x^1}{\partial \phi}\underline{\delta}_1 + \dfrac{\partial x^2}{\partial \phi}\underline{\delta}_2 + \dfrac{\partial x^3}{\partial \phi}\underline{\delta}_3 \tag{A.2-8e}$

$$= -r\sin\theta\sin\phi\,\underline{\delta}_1 + r\sin\theta\cos\phi\,\underline{\delta}_2. \tag{A.2-8f}$$

To obtain the contravariant base vectors $\underline{g}^n$ we have to invert equations A.2-7a, b, and c. We then obtain

$$r = \sqrt{(x^1)^2 + (x^2)^2 + (x^3)^2}, \tag{A.2-9a}$$

$$\theta = \arctan\left(\sqrt{\frac{(x^1)^2 + (x^2)^2}{(x^3)^2}}\right), \tag{A.2-9b}$$

and $\quad \phi = \arctan\left(\dfrac{x^2}{x^1}\right). \tag{A.2-9c}$

We calculate $\underline{g}^n$ from equation A.2-5:

$$\underline{g}^r = \frac{\partial r}{\partial x^1}\underline{\delta}_1 + \frac{\partial r}{\partial x^2}\underline{\delta}_2 + \frac{\partial r}{\partial x^3}\underline{\delta}_3 \tag{A.2-10a}$$

$$= \sin\theta\cos\phi\,\underline{\delta}_1 + \sin\theta\sin\phi\,\underline{\delta}_2 + \cos\theta\,\underline{\delta}_3, \tag{A.2-10b}$$

$$\underline{g}^\theta = \frac{\partial \theta}{\partial x^1}\underline{\delta}_1 + \frac{\partial \theta}{\partial x^2}\underline{\delta}_2 + \frac{\partial \theta}{\partial x^3}\underline{\delta}_3 \tag{A.2-10c}$$

$$= \frac{\cos\phi\cos\theta}{r}\underline{\delta}_1 + \frac{\sin\phi\cos\theta}{r}\underline{\delta}_2 - \frac{\sin\theta}{r}\underline{\delta}_3, \tag{A.2-10d}$$

and $\quad \underline{g}^\phi = -\dfrac{\sin\phi}{r\sin\theta}\underline{\delta}_1 + \dfrac{\cos\phi}{r\sin\theta}\underline{\delta}_2. \tag{A.2-10e}$

Note that

$$\underline{\delta}_r = \sin\theta\cos\phi\,\underline{\delta}_1 + \sin\theta\sin\phi\,\underline{\delta}_2 + \cos\theta\,\underline{\delta}_3, \tag{A.2-11a}$$

$$\underline{\delta}_\theta = \cos\theta\cos\phi\,\underline{\delta}_1 + \cos\theta\sin\phi\,\underline{\delta}_2 + \sin\theta\,\underline{\delta}_3, \tag{A.2-11b}$$

and $\quad \underline{\delta}_\phi = -\sin\phi\,\underline{\delta}_1 + \cos\phi\,\underline{\delta}_2. \tag{A.2-11c}$

It follows that

$$\underline{g}^r = \underline{g}_r = \underline{\delta}_r, \qquad \text{(A.2-11d)}$$

$$\underline{g}^\theta = \frac{1}{r^2}\underline{g}_\theta = \frac{\underline{\delta}_\theta}{r}, \qquad \text{(A.2-11e)}$$

$$\text{and} \quad \underline{g}^\phi = \frac{1}{r^2 \sin^2\theta}\underline{g}_\phi = \frac{\underline{\delta}_\phi}{r\sin\theta}, \qquad \text{(A.2-11f)}$$

where $\underline{\delta}_r$, $\underline{\delta}_\theta$, and $\underline{\delta}_\phi$ are the conventional unit vectors used in spherical coordinates (see Appendix A.6 of Bird, Stewart, and Lightfoot, 1960). Clearly $\underline{g}_\theta$ and $\underline{g}_\phi$ are parallel to $\underline{g}^\theta$ and $\underline{g}^\phi$ respectively, but they are not generally equal in magnitude.

A.2-2 Transformation Rules for Vectors

We now consider the transformation from one generalized coordinate system q^m to another generalized coordinate system $\bar{q}^n$, where, as usual, the indices m and n can take the values 1, 2, and 3.

The relation between these coordinate systems is

$$q^m = q^m(\bar{q}^n) \qquad \text{(A.2-12)}$$

or

$$\bar{q}^n = \bar{q}^n(q^m). \qquad \text{(A.2-13)}$$

The base vectors will be given by

$$\underline{\bar{g}}_m = \frac{\partial \underline{r}}{\partial \bar{q}^m} \qquad \text{(A.2-14)}$$

and

$$\underline{\bar{g}}^n = \frac{\partial \bar{q}^n}{\partial x^l}\underline{\delta}_l = \underline{\nabla}\bar{q}^n. \qquad \text{(A.2-15)}$$

We will now establish the relationship between the components of a vector $\underline{v}$ in two generalized curvilinear coordinate systems q^m and $\bar{q}^n$. Any vector $\underline{v}$ may be written as

$$\underline{v} = v^m \underline{g}_m = \bar{v}^n \underline{\bar{g}}_n. \qquad \text{(A.2-16)}$$

Since in general, covariant and contravariant base vectors are different, we have to pay special attention to the position of the indices. As far as the summation convention is concerned, repeated indices should appear as subscript–superscript pairs.

A.2-2.1 Contravariant Transformation

In the general case, we can no longer take dot products, as in equation A.1-7, to establish the relation between components v^m and $\bar{v}^m$. General coordinate systems are not necessarily orthogonal, and $\underline{g}_1 \cdot \underline{g}_2$ is not necessarily zero. These relationships are obtained via the definitions (equations A.2-4, A.2-5) and the chain rule, as follows

$$\underline{g}_m = \frac{\partial \underline{r}}{\partial \bar{q}^n} \frac{\partial \bar{q}^n}{\partial q^m} \tag{A.2-17a}$$

$$= \underline{\bar{g}}_n \frac{\partial \bar{q}^n}{\partial q^m}. \tag{A.2-17b}$$

Substitution in equations A.2-16 yields

$$\underline{v} = v^m \frac{\partial \bar{q}^n}{\partial q^m} \underline{\bar{g}}_n = \bar{v}^n \underline{\bar{g}}_n. \tag{A.2-18}$$

It now follows that

$$\bar{v}^n = \frac{\partial \bar{q}^n}{\partial q^m} v^m. \tag{A.2-19}$$

Components of vectors which transform according to equation A.2-19 are *contravariant components*. See also equation A.1-13.

Example A.2-2 Transformation of a Vector
Obtain the law of transformation of the velocity components V^i.

Solution
The velocity components V^m, in the q^m coordinate system, are defined as

$$V^m = \lim_{\Delta t \to 0} \frac{\Delta q^m}{\Delta t} \tag{A.2-20a}$$

$$= \frac{\partial q^m}{\partial t}. \tag{A.2-20b}$$

On transforming to the $\bar{q}^s$ coordinate system, the components $\bar{V}^s$ are defined as

$$\bar{V}^s = \lim_{\Delta t \to 0} \frac{\Delta \bar{q}^s}{\Delta t} \tag{A.2-21a}$$

$$= \frac{\partial \bar{q}^s}{\partial t}. \tag{A.2-21b}$$

According to the chain rule

$$\Delta \bar{q}^s = \frac{\partial \bar{q}^s}{\partial q^m} \Delta q^m. \tag{A.2-22}$$

Substituting equation A.2-22 into A.2-21a we obtain

$$\bar{V}^s = \lim_{\Delta t \to 0} \frac{\partial \bar{q}^s}{\partial q^m} \frac{\Delta q^m}{\Delta t} \tag{A.2-23a}$$

$$= \frac{\partial \bar{q}^s}{\partial q^m} V^m. \tag{A.2-23b}$$

Thus the components $\bar{V}^s$ transform as contravariant components.

A.2 General Curvilinear Coordinate Systems

We further note from equation A.2-22 that the components Δq^m also transform as contravariant components. Although the coordinates q^m are not components of a vector, Δq^m are contravariant components, and the indices are written as superscripts. The transformation from q^m to $\bar{q}^n$ is arbitrary, as can be seen from equations A.2-22 and A.2-23, but we still write the indices as superscripts because Δq^m are contravariant components.

A.2-2.2 Covariant Transformation

In equation A.2-12 we have expressed the vector $\underline{v}$ in terms of the covariant base vectors $\underline{g}_m$. We could equally have expressed $\underline{v}$ in terms of the contravariant base vectors $\underline{g}^n$. That is,

$$\underline{v} = v_m \underline{g}^m = \bar{v}_n \underline{\bar{g}}^n \tag{A.2-24}$$

The relation between $\underline{\bar{g}}^n$ and $\underline{g}^m$ can be deduced, using the chain rule, as follows:

$$\underline{g}^m = \underline{\nabla} q^m \tag{A.2-25}$$

$$= \frac{\partial q^m}{\partial \bar{q}^n} \underline{\nabla} \bar{q}^n \tag{A.2-26a}$$

$$= \frac{\partial q^m}{\partial \bar{q}^n} \underline{\bar{g}}^n. \tag{A.2-26b}$$

Substituting equation A.2-26b into A.2-24 we obtain

$$\frac{\partial q^m}{\partial \bar{q}^n} v_m \underline{\bar{g}}^n = \bar{v}_n \underline{\bar{g}}^n. \tag{A.2-27}$$

We then deduce

$$\bar{v}_n = \frac{\partial q^m}{\partial \bar{q}^n} v_m. \tag{A.2-28}$$

Components of vectors which transform according to equation A.2-28 are known as *covariant components* (see also equation A.1-12a).

Example A.2-3 Gradient of a Scalar

Show that the components of $\underline{\nabla}\varphi$, where φ is a scalar function of q^1, q^2, and q^3, transform as covariant components of a vector.

Solution
Let

$$\underline{u} = \underline{\nabla}\varphi. \tag{A.2-29a}$$

Then we define u_m as

$$u_m = \frac{\partial \varphi}{\partial q^m}. \tag{A.2-29b}$$

In the (q^1, q^2, q^3) coordinate system, the components $\bar{u}_n$ are given by

$$\bar{u}_n = \frac{\partial \varphi}{\partial \bar{q}^n}. \tag{A.2-30}$$

Applying the chain rule to equation A.2-30 we have

$$\bar{u}_n = \frac{\partial \varphi}{\partial q^m} \frac{\partial q^m}{\partial \bar{q}^n} \tag{A.2-31a}$$

$$= \frac{\partial q^m}{\partial \bar{q}^n} u_m. \tag{A.2-31b}$$

Thus the components $\bar{u}_n$ transform as covariant components.

A.2-3   Tensors of Arbitrary Order

So far we have considered only scalars and vectors. A scalar is a tensor of order zero. Its numerical value at a point remains invariant when the coordinate system is transformed. A vector has both a magnitude and a direction and its components transform according to equations A.2-19 and A.2-28 when the coordinate system is transformed. One index is sufficient to specify its components. It is a tensor of order one. In Example A.2-3 we have seen that $\underline{\nabla}\varphi$ is a vector but φ is a scalar. Thus the quantity $\underline{\nabla}\underline{V}$ is a tensor of order two and is known as the velocity gradient tensor. In fluid mechanics the rate-of-deformation tensor, which is a tensor of order two, is equal to the sum of the velocity gradient and its transpose. The stress tensor, which is another tensor of order two, linearly maps the surface force on the surface of a deformable continuum to the unit normal on the surface. Thus, a tensor of order two transforms a vector $\underline{u}$ linearly to another vector $\underline{v}$. This can be written as

$$u_m = T_{mn} v^n. \tag{A.2-32}$$

In equation A.2-32 we have used as before the summation convention by writing the dummy index as a subscript-superscript pair. The components of $\underline{u}$ are written as covariant components and the components of $\underline{v}$ are written in contravariant form. The quantities T_{mn} are the components of a second order tensor which we denote as $\underline{T}$. In a three-dimensional space, m and n can take the values 1, 2, and 3 and $\underline{T}$ has nine components. We need two indices to specify the components of a second order tensor.

In the $(\bar{q}^1, \bar{q}^2, \bar{q}^3)$ coordinate system, equation A.2-32 becomes

$$\bar{u}_m = \bar{T}_{mn} \bar{v}^n. \tag{A.2-33}$$

Using equations A.2-19 and A.2-28, equation A.2-33 becomes

$$\frac{\partial q^r}{\partial \bar{q}^m} u_r = \bar{T}_{mn} \frac{\partial \bar{q}^n}{\partial q^s} v^s. \tag{A.2-34}$$

Multiplying both sides of equation A.2-34 by $\partial \bar{q}^m / \partial q^p$, we obtain

$$\frac{\partial \bar{q}^m}{\partial q^p} \frac{\partial q^r}{\partial \bar{q}^m} u_r = \bar{T}_{mn} \frac{\partial \bar{q}^m}{\partial q^p} \frac{\partial \bar{q}^n}{\partial q^s} v^s. \tag{A.2-35}$$

Note that $\partial q^r/\partial q^p$ can, according to the chain rule, be written as $(\partial q^r/\partial \bar{q}^m)(\partial \bar{q}^m/\partial q^p)$. We recognize the quantity $\partial q^r/\partial q^p$ (in the unbarred coordinate system) as representing the Kronecker delta. Similarly, $\partial \bar{q}^s/\partial \bar{q}^t$ would represent δ^s_t, while a quantity such as $\partial q^i/\partial \bar{q}^l$ is not in general δ^i_l. It thus follows that

$$\bar{T}_{mn} \frac{\partial \bar{q}^m}{\partial q^p} \frac{\partial \bar{q}^n}{\partial q^s} v^s = \delta^r_p u_r \quad \text{(A.2-36a)}$$

$$= u_p. \quad \text{(A.2-36b)}$$

In equation A.2-32, we are at liberty to change the free index m to p and the dummy index n to s. We then obtain

$$u_p = T_{ps} v^s \quad \text{(A.2-37)}$$

Comparing equations A.2-36 and A.2-37, we deduce

$$T_{ps} = \frac{\partial \bar{q}^m}{\partial q^p} \frac{\partial \bar{q}^n}{\partial q^s} \bar{T}_{mn}. \quad \text{(A.2-38)}$$

Interchanging (q^1, q^2, q^3) and $(\bar{q}^1, \bar{q}^2, \bar{q}^3)$ in equation A.2-38, it becomes

$$\bar{T}_{ps} = \frac{\partial q^m}{\partial \bar{q}^p} \frac{\partial q^n}{\partial \bar{q}^s} T_{mn}. \quad \text{(A.2-39)}$$

Components that transform according to equation A.2-38 or A.2-39 are known as *covariant components*. Note the similarity to equation A.2-28.

Contravariant components $\bar{T}^{mn}$ transform according to

$$\bar{T}^{mn} = \frac{\partial \bar{q}^m}{\partial q^r} \frac{\partial \bar{q}^n}{\partial q^s} T^{rs}. \quad \text{(A.2-40)}$$

For second-order tensors, in addition to covariant and contravariant components, we can have mixed components, $\bar{T}^m_n$, which transform according to

$$\bar{T}^m_n = \frac{\partial \bar{q}^m}{\partial q^r} \frac{\partial q^s}{\partial \bar{q}^n} T^r_s. \quad \text{(A.2-41)}$$

Tensors of order higher than two are also frequently used. An example of a tensor of order three is the permutation tensor, which will be defined in the next section. A tensor that linearly maps a second-order tensor to another second order tensor is a tensor of order four. The constitutive equation of a linear elastic material may be written as

$$\tau_{ij} = c_{ijkl} \gamma^{kl}, \quad \text{(A.2-42)}$$

where τ_{ij}, γ^{kl} and c_{ijkl} are the stress tensor, the infinitesimal strain tensor, and the elastic tensor respectively. The components c_{ijkl} of the fourth-order tensor require four indices to be specified.

454 General Curvilinear Coordinate Systems and Higher Order Tensors

The law of transformation for a fourth-order tensor can be obtained by generalizing equations A.2-39 to A.2-41 as follows

$$\bar{T}_{prst} = \frac{\partial \bar{q}^i}{\partial q^p} \frac{\partial \bar{q}^j}{\partial q^r} \frac{\partial \bar{q}^k}{\partial q^s} \frac{\partial \bar{q}^l}{\partial q^t} T_{ijkl}, \tag{A.2-43a}$$

$$T^{prst} = \frac{\partial q^p}{\partial \bar{q}^i} \frac{\partial q^r}{\partial \bar{q}^j} \frac{\partial q^s}{\partial \bar{q}^k} \frac{\partial q^t}{\partial \bar{q}^l} \bar{T}^{ijkl}, \tag{A.2-43b}$$

and $$T^p_{rst} = \frac{\partial q^p}{\partial \bar{q}^i} \frac{\partial \bar{q}^j}{\partial q^r} \frac{\partial \bar{q}^k}{\partial q^s} \frac{\partial \bar{q}^l}{\partial q^t} \bar{T}^i_{jkl}. \tag{A.2-43c}$$

A second-order tensor may also be defined as the juxtaposition of two vectors $\underline{u}$ and $\underline{v}$. The second-order tensor $\underline{\underline{T}}$ may be defined as the dyadic product of vectors $\underline{u}$ and $\underline{v}$:

$$\underline{\underline{T}} = \underline{u}\,\underline{v} = \underline{u} \otimes \underline{v} \tag{A.2-44}$$

Equation A.2-44 can also be written as

$$\underline{\underline{T}} = u^i \underline{g}_i v^j \underline{g}_j = u_m \underline{g}^m v_n \underline{g}^n \tag{A.2-45a}$$

$$= u^i v^j \underline{g}_i \underline{g}_j = u_m v_n \underline{g}^m \underline{g}^n \tag{A.2-45b}$$

$$= T^{ij} \underline{g}_i \underline{g}_j = T_{mn} \underline{g}^m \underline{g}^n. \tag{A.2-45c}$$

A second order tensor $\underline{\underline{T}}$ is also known as a dyad, and the notation adopted in equations A.2-45a to c is called the dyadic notation. The juxtaposition of the two vectors $\underline{u}$ and $\underline{v}$ is also known as the outer product of the two vectors. The commutative law does not hold, and $\underline{g}_i \underline{g}_j$ is in general not equal to $\underline{g}_j \underline{g}_i$. The component T_{ij} is not necessarily equal to the component T_{ji}. If they are equal, $\underline{\underline{T}}$ is a symmetric tensor. Tensors of higher order can likewise be defined.

A.2-4 Metric and Permutation Tensors

Equations A.2-4a and b define the covariant base vector $\underline{g}_m$. If P and Q are two neighboring points (as shown in Fig. A.2-3) with vector positions $\underline{r}$ and $\underline{r} + d\underline{r}$ and coordinates (q^i) and $(q^i + dq^i)$ respectively, then the square of the distance, ds^2, between P and Q is

$$ds^2 = d\underline{r} \cdot d\underline{r} \tag{A.2-46a}$$

$$= \frac{\partial \underline{r}}{\partial q^m} \cdot \frac{\partial \underline{r}}{\partial q^n} dq^m dq^n \tag{A.2-46b}$$

$$= \underline{g}_m \cdot \underline{g}_n dq^m dq^n \tag{A.2-46c}$$

$$= g_{mn} dq^m dq^n. \tag{A.2-46d}$$

where g_{mn} are known as the covariant components of the metric tensor $\underline{\underline{g}}$ and in general are functions of (q^i).

A.2 General Curvilinear Coordinate Systems 455

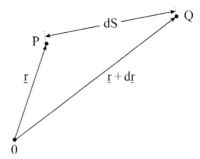

Figure A.2-3 Distance between two points

In an orthonormal coordinate system, the metric tensor simplifies to the Kronecker delta. In an orthogonal coordinate system

$$g_{mn} = \begin{cases} = 0 & \text{if } m \neq n \\ \neq 0 & \text{if } m = n \end{cases}. \quad (A.2\text{-}47)$$

Since the dot product is commutative, g_{mn} is symmetric.

The dual (conjugate or associate) of g_{mn}, which is denoted by g^{rs}, is defined as

$$g^{rs} = \underline{g}^r \cdot \underline{g}^s \quad (A.2\text{-}48a)$$

$$= \underline{\nabla} q^r \cdot \underline{\nabla} q^s. \quad (A.2\text{-}48b)$$

From equation A.2-4 we have

$$\underline{g}_m \cdot \underline{g}_j = \frac{\partial x^p}{\partial q^m} \delta_p \cdot \frac{\partial x^l}{\partial q^j} \delta_l \quad (A.2\text{-}49a)$$

$$= \frac{\partial x^p}{\partial q^m} \frac{\partial x^l}{\partial q^j} \delta_{pl} \quad (A.2\text{-}49b)$$

$$= \frac{\partial x^p}{\partial q^m} \frac{\partial x^p}{\partial q^j} = g_{mj}. \quad (A.2\text{-}49c)$$

Equation A.2-49c shows that the components of the metric tensor g_{mj} transform as covariant components, hence the subscript notation.

Starting from equation A.2-5, we can deduce an expression for g^{jn} in terms of (x^i) as follows:

$$\underline{g}^j \cdot \underline{g}^n = \frac{\partial q^j}{\partial x^l} \delta_l \cdot \frac{\partial q^n}{\partial x^r} \delta_r \quad (A.2\text{-}50a)$$

$$= \frac{\partial q^j}{\partial x^l} \frac{\partial q^n}{\partial x^r} \delta_{lr} \quad (A.2\text{-}50b)$$

$$= \frac{\partial q^j}{\partial x^r} \frac{\partial q^n}{\partial x^r} = g^{jn}. \quad (A.2\text{-}50c)$$

It can be seen from equation A.2-50c that the g^{jn} components transform as contravariant components. The relationship between g_{mj} and g^{jn} can be established by combining equations A.2-49c and A.2-50c, and the result is

$$g_{mj}g^{jn} = \frac{\partial x^p}{\partial q^m}\frac{\partial x^p}{\partial q^j}\frac{\partial q^j}{\partial x^r}\frac{\partial q^n}{\partial x^r} \tag{A.2-51a}$$

$$= \frac{\partial x^p}{\partial q^m}\frac{\partial x^p}{\partial x^r}\frac{\partial q^n}{\partial x^r} \tag{A.2-51b}$$

$$= \frac{\partial x^p}{\partial q^m}\frac{\partial q^n}{\partial x^r}\delta_{pr} \tag{A.2-51c}$$

$$= \frac{\partial x^r}{\partial q^m}\frac{\partial q^n}{\partial x^r} \tag{A.2-51d}$$

$$= \delta_m^n. \tag{A.2-51e}$$

In equation A.2-51e, we have written the Kronecker delta as a mixed tensor so as to conform to the convention that an index appearing as a subscript (superscript) on one side of the equation should also appear as a subscript (superscript) on the other side of the equation. The fundamental reason for writing the Kronecker delta as a mixed tensor is because it transforms as a mixed second-order tensor. The convention is framed so as to be compatible with the rules of transformation.

The permutation tensor is an example of a third order tensor, which in an orthonormal coordinate system is denoted by e_{ijk}, and is defined as

$$e_{ijk} = \begin{cases} 0, & \text{if any two indices are equal} \\ 1, & \text{if the indices 1, 2, 3 appear in the clockwise direction} \\ -1, & \text{if the indices 1, 2, 3 appear in the anticlockwise direction.} \end{cases} \tag{A.2-52}$$

Thus, for example,

$$e_{112} = e_{122} = 0, \tag{A.2-53a}$$

$$e_{123} = e_{312} = 1, \tag{A.2-53b}$$

$$\text{and} \quad e_{321} = e_{213} = -1. \tag{A.2-53c}$$

Let $\underline{w}$ be the vector product of two vectors $\underline{u}$ and $\underline{v}$. Then in an orthonormal coordinate system, w_i is given by

$$w_i = e_{ijk}u_j v_k. \tag{A.2-54}$$

If $\underline{\underline{D}}$ is a (3 × 3) matrix with elements d_{ij}, then the determinant of $\underline{\underline{D}}$ can be written as

$$|\underline{\underline{D}}| = e_{ijk}d_{i1}d_{j2}d_{k3}. \tag{A.2-55}$$

We extend the definition of the permutation tensor to a general curvilinear coordinate system and denote it by ε_{lmn}. We put a bar over the components of the vectors in the present coordinate system, so as to avoid confusion with the Cartesian components. Thus $\bar{w}_l$ is expressed in a general barred coordinate system as

$$\bar{w}_l = \varepsilon_{lmn}\bar{u}^m\bar{v}^n. \tag{A.2-56}$$

Transforming $\bar{w}_l$, $\bar{u}^m$, and $\bar{v}^n$ to the orthonormal coordinate system using equations A.2-19 and A.2-28, identifying the coordinates q^m as x^m and $\bar{q}^n$ as q^n, equation A.2-56 becomes

$$\frac{\partial x^r}{\partial q^l} w_r = \varepsilon_{lmn} \frac{\partial q^m}{\partial x^s} u^s \frac{\partial q^n}{\partial x^t} v^t \tag{A.2-57a}$$

$$= \varepsilon_{lmn} \frac{\partial q^m}{\partial x^s} \frac{\partial q^n}{\partial x^t} u^s v^t. \tag{A.2-57b}$$

Multiplying both sides of equation A.2-57b by $\partial q^l / \partial x^p$ we obtain

$$\frac{\partial q^l}{\partial x^p} \frac{\partial x^r}{\partial q^l} w_r = \varepsilon_{lmn} \frac{\partial q^l}{\partial x^p} \frac{\partial q^m}{\partial x^s} \frac{\partial q^n}{\partial x^t} u^s v^t. \tag{A.2-58}$$

Using the chain rule on the left side of the equation, and noting that in an orthonormal coordinate system we do not distinguish between covariant and contravariant components, equation A.2-58 becomes

$$w_p = \varepsilon_{lmn} \frac{\partial q^l}{\partial x^p} \frac{\partial q^m}{\partial x^s} \frac{\partial q^n}{\partial x^t} u_s v_t \tag{A.2-59a}$$

$$= e_{pst} u_s v_t. \tag{A.2-59b}$$

From equation A.2-59 we deduce that

$$e_{pst} = \varepsilon_{lmn} \frac{\partial q^l}{\partial x^p} \frac{\partial q^m}{\partial x^s} \frac{\partial q^n}{\partial x^t}, \tag{A.2-60}$$

which is the law of transformation of covariant components. Alternatively, ε_{lmn} may be written as

$$\varepsilon_{lmn} = \sqrt{g} \, e_{lmn}, \tag{A.2-61}$$

where g is the determinant of the metric tensor $\underline{\underline{g}}$. The contravariant form of the permutation tensor is

$$\varepsilon^{ijk} = \frac{1}{\sqrt{g}} e_{ijk}. \tag{A.2-62}$$

Example A.2-4 The Metric Tensor in Spherical Coordinates

Obtain the components of the metric tensor g_{ij}, its dual g^{rs}, and its determinant g for the spherical polar coordinate system.

Solution

The base vectors have been obtained in Example A.2-1. The components of the metric tensor are given by

$$g_{rr} = \underline{g}_r \cdot \underline{g}_r = 1, \tag{A.2-63a}$$

$$g_{\theta\theta} = \underline{g}_\theta \cdot \underline{g}_\theta = r^2, \tag{A.2-63b}$$

and $\quad g_{\phi\phi} = \underline{g}_\phi \cdot \underline{g}_\phi = r^2 \sin^2 \theta. \tag{A.2-63c}$

All the other g_{ij} are zero since the spherical polar coordinate system is orthogonal.

$$g^{rr} = \underline{g}^r \cdot \underline{g}^r = 1, \tag{A.2-64a}$$

$$g^{\theta\theta} = \underline{g}^\theta \cdot \underline{g}^\theta = \frac{1}{r^2}, \tag{A.2-64b}$$

$$\text{and} \quad g^{\phi\phi} = \underline{g}_\phi \cdot \underline{g}_\phi = \frac{1}{r^2 \sin^2 \theta}. \tag{A.2-64c}$$

The determinant g is given by

$$g = r^4 \sin^2 \theta. \tag{A.2-65}$$

A.2-5 Physical Components

It has been noted that the covariant and contravariant base vectors do not in general have the same dimensions, and it is not surprising that the covariant and contravariant components of a vector also do not have the same dimensions. Via the metric tensor it is possible to establish a relationship between these two types of components as follows:

$$\underline{v} = v_m \underline{g}^m = v^n \underline{g}_n. \tag{A.2-66}$$

Forming the dot product with $\underline{g}_r$ we have

$$v_m \underline{g}^m \cdot \underline{g}_r = v^n \underline{g}_n \cdot \underline{g}_r, \tag{A.2-67a}$$

$$v_m \delta^m_r = v^n g_{nr}, \tag{A.2-67b}$$

$$\text{and} \quad v_r = g_{nr} v^n. \tag{A.2-67c}$$

Equation A.2-67c shows the transformation from contravariant components to covariant components. Similarly, for higher order tensors we have

$$T_{rs} = g_{nr} g_{ms} T^{nm}, \tag{A.2-68a}$$

$$T^{nm} = g^{nr} g^{ms} T_{rs}, \tag{A.2-68b}$$

$$\text{and} \quad T^{nm}_{kl} = g^{nr} g^{ms} g_{pk} g_{ql} T^{pq}_{rs}. \tag{A.2-68c}$$

In a space in which a metric tensor is defined, it is possible to transform covariant components to contravariant components and vice versa. This process is known as lowering and raising indices. It was pointed out in Example A.2-1 that $\underline{g}_r$ and $\underline{g}_\theta$ do not have the same magnitude. In addition, we also note that they have different dimensions. This makes it necessary to define the so-called *physical components*. The physical components of a

A.2 General Curvilinear Coordinate Systems

vector are expressed in terms of normalized base vectors. The normalized covariant base vectors are given by

$$\underline{g}_{(n)} = \frac{\underline{g}_n}{|\underline{g}_n|} \tag{A.2-69a}$$

$$= \frac{\underline{g}_n}{\sqrt{g_{nn}}}. \tag{A.2-69b}$$

Note that in equations A.2-69a and b, there is no summation, and the index n is contained in brackets to designate physical components. A vector $\underline{v}$ may be written as

$$\underline{v} = v^m \underline{g}_m \tag{A.2-70a}$$

$$= v_{(n)} \underline{g}_{(n)} \tag{A.2-70b}$$

$$= \frac{v_{(n)} \underline{g}_n}{\sqrt{g_{nn}}}. \tag{A.2-70c}$$

Combining equations A.2-70a and c, noting that m is a dummy index, yields

$$\left(v^n - \frac{v_{(n)}}{\sqrt{g_{nn}}}\right) \underline{g}_n = \underline{0}. \tag{A.2-71}$$

Since the vectors $\underline{g}_n$ are base vectors and are linearly independent, equation A.2-71 implies that for each n,

$$v^n = \frac{v_{(n)}}{\sqrt{g_{nn}}} \tag{A.2-72a}$$

or

$$v_{(n)} = \sqrt{g_{nn}}\, v^n. \tag{A.2-72b}$$

That is, the physical components $v_{(n)}$ can be obtained via the metric tensor. Note that in equations A.2-72a and b there is no summation over n. The index n occurs three times!

The contravariant component v^n can be written as

$$v^n = g^{nm} v_m. \tag{A.2-73}$$

Substituting equation A.2-73 into A.2-72b we obtain

$$v_{(n)} = \sqrt{g_{nn}}\, g^{nm} v_m. \tag{A.2-74}$$

In an orthogonal coordinate system, combining equations A.2-47 and A.2-74, we obtain

$$\sqrt{g^{nn}}\, v_n = v_{(n)}. \tag{A.2-75}$$

Similarly, the physical components of higher order tensors are defined. For an orthogonal coordinate system we have

$$T_{(mr)} = \sqrt{g_{mm}}\sqrt{g_{rr}}T^{mr} \tag{A.2-76a}$$

$$= \sqrt{g^{mm}}\sqrt{g^{rr}}T_{mr} \tag{A.2-76b}$$

$$= \sqrt{g^{mm}}\sqrt{g_{rr}}T^r_m$$

So far, we have defined the physical components in terms of normalized covariant base vectors. We could equally have chosen to define physical components via normalized contravariant components. In the framework of an orthogonal coordinate system, both are identical. In the case of a non-orthogonal coordinate system, they may not be identical, since equation A.2-75 could not be obtained from equation A.2-74. Indeed, for a non-orthogonal system, g^{nn} in equation A.2-75 represents three nonzero components for each value of n. In equation A.2-75 there is no summation over n.

Example A.2-5 Physical Components of a Vector

Obtain the contravariant, covariant, and physical components of the velocity vector $\underline{V}$ of a particle in the spherical polar coordinate system.

Solution

In Example A.2-2, we have shown that $\partial q^m / \partial t$ transforms as contravariant components. The contravariant components of $\underline{V}$ are

$$V^r = \frac{dr}{dt} = \dot{r}, \tag{A.2-77a}$$

$$V^\theta = \frac{d\theta}{dt} = \dot{\theta}, \tag{A.2-77b}$$

and $\quad V^\phi = \dfrac{d\phi}{dt} = \dot{\phi}. \tag{A.2-77c}$

Using equation A.2-67c, we have

$$V_r = g_{rr}V^r = \dot{r}, \tag{A.2-78a}$$

$$V_\theta = g_{\theta\theta}V^\theta = r^2\dot{\theta}, \tag{A.2-78b}$$

and $\quad V_\phi = g_{\phi\phi}V^\phi = r^2(\sin^2\theta)\dot{\phi}. \tag{A.2-78c}$

The physical components are given by equation A.2-72b:

$$V_{(r)} = \sqrt{g_{rr}}V^r = \dot{r}, \tag{A.2-79a}$$

$$V_{(\theta)} = \sqrt{g_{\theta\theta}}V^\theta = r\dot{\theta}, \tag{A.2-79b}$$

and $\quad V_{(\phi)} = \sqrt{g_{\phi\phi}}V^\phi = r(\sin\theta)\dot{\phi}. \tag{A.2-79c}$

Example A.2-6 Physical Components in Spherical Coordinates
Calculate the covariant, contravariant, and physical components of $\underline{\nabla} f$ in the spherical polar coordinate system.

Solution
In Example A.2-3, we have shown that $\partial \varphi / \partial q^m$ is a covariant component. If we denote $\underline{\nabla} f$ by $\underline{u}$, then

$$u_r = \frac{\partial f}{\partial r}, \tag{A.2-80a}$$

$$u_\theta = \frac{\partial f}{\partial \theta}, \tag{A.2-80b}$$

$$\text{and} \quad u_\phi = \frac{\partial f}{\partial \phi}. \tag{A.2-80c}$$

The contravariant components are

$$u^r = g^{rr} u_r = \frac{\partial f}{\partial r}, \tag{A.2-81a}$$

$$u^\theta = g^{\theta\theta} u_\theta = \frac{1}{r^2} \frac{\partial f}{\partial \theta}, \tag{A.2-81b}$$

$$\text{and} \quad u^\phi = g^{\phi\phi} u_\phi = \frac{1}{r^2 \sin^2 \theta} \frac{\partial f}{\partial \phi}. \tag{A.2-81c}$$

The physical components are

$$u_{(r)} = \sqrt{g^{rr}}\, u_r = \frac{\partial f}{\partial r}, \tag{A.2-82a}$$

$$u_{(\theta)} = \sqrt{g^{\theta\theta}}\, u_\theta = \frac{1}{r} \frac{\partial f}{\partial \theta}, \tag{A.2-82b}$$

$$\text{and} \quad u_{(\phi)} = \sqrt{g^{\phi\phi}}\, u_\phi = \frac{1}{r \sin \theta} \frac{\partial f}{\partial \phi}. \tag{A.2-82c}$$

In Examples A.2-5 and A.2-6, we note that the r-covariant and contravariant components do not have the same dimensions as the θ- and ϕ-components, whereas the r, θ, and ϕ physical components all have the same dimensions. In a space in which a metric tensor exists, a tensor can be represented in terms of covariant, contravariant, or physical components. They can be transformed from one to the other by the process of raising or lowering the indices, as shown in Examples A.2-5 and A.2-6.

The laws of physics are independent of the coordinate system, and they should be written in tensorial form. The quantities that enter into the equations expressing these laws should be in covariant or contravariant components. Each expression in the equation should be a tensor component of the same kind and order. That is, if on the right side of the equation we have a mixed component that is covariant of order one and contravariant of order two, then on the left side we must also have a mixed component, covariant of order one and contravariant of order two.

Finally, the components of a tensor have to be measured in terms of certain units, and it is desirable to express all the components in terms of the same physical dimensions.

In the rectangular Cartesian coordinate system, there is no distinction between covariant, contravariant, and physical components. The metric tensor is the Kronecker delta. Many authors omit the word physical when referring to physical components of a tensor. Thus it is safe to assume that unless otherwise stated, the components of a tensor refer to physical components.

A.3 Covariant Differentiation

A.3-1 Definitions

We have shown in Example A.2-3 that if φ is a scalar function, $\partial\varphi/\partial q^i$ is a covariant component of a vector. We shall show that if v^i is the contravariant component of a vector $\underline{v}$, $\partial v^i/\partial q^j$ does not transform as a tensor.

Consider the transformation given by equations A.2-12 and A.2-13. On transforming from (q^1, q^2, q^3) to $(\bar{q}^1, \bar{q}^2, \bar{q}^3)$, we have

$$\frac{\partial v^i}{\partial q^j} = \frac{\partial}{\partial \bar{q}^s}\left(\frac{\partial q^i}{\partial \bar{q}^l}\bar{v}^l\right)\frac{\partial \bar{q}^s}{\partial q^j} \tag{A.3-1a}$$

$$= \frac{\partial \bar{v}^l}{\partial \bar{q}^s}\frac{\partial q^i}{\partial \bar{q}^l}\frac{\partial \bar{q}^s}{\partial q^j} + \bar{v}^l\frac{\partial^2 q^i}{\partial \bar{q}^s \partial \bar{q}^l}\frac{\partial \bar{q}^s}{\partial q^j}. \tag{A.3-1b}$$

Comparing equations A.2-41 and A.3-1b we note that $\partial v^i/\partial q^j$ transforms as a mixed component if $\partial^2 q^i/\partial \bar{q}^s \partial \bar{q}^l = 0$. That is, $\partial v^i/\partial q^j$ transforms as a component of a tensor only if the transformation given by equation A.2-12 is a linear transformation. In general the partial derivative of the components of a vector is not a tensor. This is because the base vectors are in general not constants, but functions of (q^1, q^2, q^3). On taking the partial derivative of a vector, contribution from the base vectors has to be taken into account.

From equation A.2-16a, we obtain

$$\frac{\partial \underline{v}}{\partial q^j} = \frac{\partial v^m}{\partial q^j}\underline{g}_m + v^m\frac{\partial \underline{g}_m}{\partial q^j}. \tag{A.3-2}$$

Using equation A.2-4b, we find

$$\frac{\partial \underline{g}_m}{\partial q^j} = \frac{\partial^2 x^p}{\partial q^m \partial q^j}\underline{\delta}_p. \tag{A.3-3}$$

Note that $\underline{\delta}_p$, the base vector in the rectangular Cartesian coordinate system, is a constant. The transformation of the covariant base vector $\underline{\delta}_p$ to $\underline{g}_l$ is given by equation A.2-17b:

$$\frac{\partial \underline{v}}{\partial q^j} = \frac{\partial v^m}{\partial q^j}\underline{g}_m + v^m \frac{\partial^2 x^p}{\partial q^m \partial q^j}\frac{\partial q^l}{\partial x^p}\underline{g}_l \tag{A.3-4a}$$

$$= \left[\frac{\partial v^l}{\partial q^j} + v^m \frac{\partial^2 x^p}{\partial q^m \partial q^j}\frac{\partial q^l}{\partial x^p}\right]\underline{g}_l \tag{A.3-4b}$$

$$= \left[\frac{\partial v^l}{\partial q^j} + \begin{Bmatrix} l \\ m\ j \end{Bmatrix} v^m \right]\underline{g}_l \tag{A.3-4c}$$

$$= [v^l{,}_j]\underline{g}_l. \tag{A.3-4d}$$

In going from equation A.3-4a to A.3-4b, we have replaced the dummy index m in the first term on the right side of the equation by l, allowing us to factor out $\underline{g}_l$. The quantity

$$\begin{Bmatrix} l \\ m\ j \end{Bmatrix} = \frac{\partial^2 x^p}{\partial q^m \partial q^j}\frac{\partial q^l}{\partial x^p} \tag{A.3-5}$$

is known as the Christoffel symbol of the second kind, and is also denoted by Γ^l_{mj}. The *covariant derivative* of v^l with respect to q^j is denoted by $v^l{,}_j$ or $v^l|_j$. From equations A.3-4c and d, it is given by

$$v^l{,}_j = \frac{\partial v^l}{\partial q^j} + \begin{Bmatrix} l \\ m\ j \end{Bmatrix} v^m. \tag{A.3-6}$$

The notation correctly suggests that $v^l{,}_j$ represents a mixed component of a second order tensor. $\partial v^l/\partial q^j$ and $\begin{Bmatrix} l \\ m\ j \end{Bmatrix}$ do not transform as tensors.

The Christoffel symbol of the first kind, denoted by $[rs, t]$ (or Γ_{rst}), is defined as

$$[rs, t] = g_{lt}\begin{Bmatrix} l \\ r\ s \end{Bmatrix}. \tag{A.3-7}$$

The *covariant derivative* of the covariant component v_l is given by

$$v_{l,j} = \frac{\partial v_l}{\partial q^j} - \begin{Bmatrix} m \\ l\ j \end{Bmatrix} v_m. \tag{A.3-8}$$

The covariant derivatives of higher order tensors are

$$T^{lm}_{,j} = \frac{\partial T^{lm}}{\partial q^j} + \begin{Bmatrix} l \\ s\ j \end{Bmatrix} T^{sm} + \begin{Bmatrix} m \\ t\ j \end{Bmatrix} T^{lt}, \tag{A.3-9a}$$

$$T_{lm,j} = \frac{\partial T_{lm}}{\partial q^j} - \begin{Bmatrix} s \\ l\ j \end{Bmatrix} T_{sm} - \begin{Bmatrix} t \\ m\ j \end{Bmatrix} T_{lt}, \tag{A.3-9b}$$

$$T^{l}_{m,j} = \frac{\partial T^{l}_{m}}{\partial q^j} + \begin{Bmatrix} l \\ s\ j \end{Bmatrix} T^{s}_{m} - \begin{Bmatrix} t \\ m\ j \end{Bmatrix} T^{l}_{t}, \tag{A.3-9c}$$

$$T^{lm}_{n,j} = \frac{\partial T^{lm}_{n}}{\partial q^j} - \begin{Bmatrix} l \\ s\ j \end{Bmatrix} T^{sm}_{n} + \begin{Bmatrix} m \\ t\ j \end{Bmatrix} T^{lt}_{n} - \begin{Bmatrix} u \\ n\ j \end{Bmatrix} T^{lm}_{u}, \tag{A.3-9d}$$

and $\quad T^{l}_{mn,j} = \dfrac{\partial T^{l}_{mn}}{\partial q^j} + \begin{Bmatrix} l \\ s\ j \end{Bmatrix} T^{s}_{mn} - \begin{Bmatrix} t \\ m\ j \end{Bmatrix} T^{l}_{tn} - \begin{Bmatrix} u \\ n\ j \end{Bmatrix} T^{l}_{mu}.$ (A.3-9e)

The covariant derivative of tensors of arbitrary order can be written by observing the pattern shown in equations A.3-9a to e.

A.3-2 Properties of Christoffel Symbols

(i) The Christoffel symbol of the second kind can be calculated in terms of the metric tensor, and is given by

$$\begin{Bmatrix} l \\ m\ j \end{Bmatrix} = \frac{1}{2} g^{lk} \left[\frac{\partial g_{jk}}{\partial q^m} + \frac{\partial g_{mk}}{\partial q^j} - \frac{\partial g_{mj}}{\partial q^k} \right]. \tag{A.3-10}$$

(ii)

$$\begin{Bmatrix} l \\ m\ j \end{Bmatrix} = \begin{Bmatrix} l \\ j\ m \end{Bmatrix} \quad \text{(symmetry)} \tag{A.3-11}$$

(iii) If the coordinate system is orthogonal,

$$\begin{Bmatrix} l \\ m\ j \end{Bmatrix} = 0, \quad \text{if } l, m, \text{ and } j \text{ are all different,} \tag{A.3-12a}$$

$$\begin{Bmatrix} l \\ l\ j \end{Bmatrix} = \frac{1}{2} g^{ll} \frac{\partial g_{ll}}{\partial q^j}, \tag{A.3-12b}$$

$$\begin{Bmatrix} l \\ j\ j \end{Bmatrix} = -\frac{1}{2} g^{ll} \frac{\partial g_{jj}}{\partial q^l}, \tag{A.3-12c}$$

and $\quad \begin{Bmatrix} l \\ l\ l \end{Bmatrix} = \dfrac{1}{2} g^{ll} \dfrac{\partial g_{ll}}{\partial q^l}.$ (A.3-12d)

In equations A.3-12a to d, no summation is implied.

(iv) On performing the transformation given by equations A.2-12 and A.2-13, the Christoffel symbol transforms as

$$\left\{\overline{\begin{matrix} l \\ m\ j \end{matrix}}\right\} = \left\{\begin{matrix} r \\ s\ t \end{matrix}\right\} \frac{\partial \bar{q}^l}{\partial q^r} \frac{\partial q^s}{\partial \bar{q}^m} \frac{\partial q^t}{\partial \bar{q}^j} + \frac{\partial \bar{q}^l}{\partial q^t} \frac{\partial^2 q^t}{\partial \bar{q}^m \partial \bar{q}^j}. \tag{A.3-13}$$

Note that for the linear transformation ($\partial^2 q^t / \partial \bar{q}^m \partial \bar{q}^j = 0$), the Christoffel symbol transforms as a mixed third-order tensor. In a rectangular Cartesian coordinate system all the Christoffel symbols are zero, since the metric tensors are constants.

A.3-3 Rules of Covariant Differentiation

(i) The covariant derivative of the sum (or difference) of two tensors is the sum (or difference) of their covariant derivatives:

$$(T_{ij} + S_{ij})_{,k} = T_{ij,k} + S_{ij,k}. \tag{A.3-14}$$

(ii) The covariant derivative of a dot (or outer) product of two tensors is equal to the sum of the two terms obtained by the dot (or outer) product of each tensor with the covariant derivative of the other tensor:

$$(T_{il}S_j^l)_{,k} = T_{il}S_{j,k}^l + (T_{il,k})S_j^l \tag{A.3-15a}$$

$$\text{and} \quad (T_{il}S_{mj})_{,k} = T_{il}S_{mj,k} + (T_{il,k})S_{mj}. \tag{A.3-15b}$$

Note that in equation A.3-15a, l is a dummy index, and the dot product of the two second-order tensors is another second-order tensor. In equation A.3-15b we have formed the outer product of two second-order tensors resulting in a fourth-order tensor.

(iii) The metric tensors g_{ij}, and g^{lm}, and the Kronecker delta δ_s^r are constants with respect to covariant differentiation:

$$g_{ij,k} = g_{,k}^{lm} = \delta_{s,k}^r = 0. \tag{A.3-16}$$

A consequence of this result is that the metric and Kronecker tensors can be put outside the covariant differentiation sign. That is,

$$(g_{ij}T^j_{lm})_{,k} = (T^j_{lm,k})g_{ij}. \tag{A.3-17}$$

(iv) The covariant derivative of v_l with respect to q^j is a second-order tensor, and we can take its covariant derivative with respect to q^k. Combining equations A.3-8 and A.3-9b we obtain

$$(v_{l,j})_{,k} = \frac{\partial}{\partial q^k}(v_{l,j}) - \begin{Bmatrix} s \\ l\ k \end{Bmatrix} v_{s,j} - \begin{Bmatrix} s \\ j\ k \end{Bmatrix} v_{l,s} \quad \text{(A.3-18a)}$$

$$= \frac{\partial}{\partial q^k}\left[\frac{\partial v_l}{\partial q^j} - \begin{Bmatrix} m \\ l\ j \end{Bmatrix} v_m\right] - \begin{Bmatrix} s \\ l\ k \end{Bmatrix}\left[\frac{\partial v_s}{\partial q^j} - \begin{Bmatrix} m \\ s\ j \end{Bmatrix} v_m\right]$$

$$- \begin{Bmatrix} s \\ j\ k \end{Bmatrix}\left[\frac{\partial v_l}{\partial q^s} - \begin{Bmatrix} m \\ l\ s \end{Bmatrix} v_m\right] \quad \text{(A.3-18b)}$$

$$= \frac{\partial^2 v_l}{\partial q^k \partial q^j} - \left[\frac{\partial}{\partial q^k}\begin{Bmatrix} m \\ l\ j \end{Bmatrix}\right] v_m - \begin{Bmatrix} m \\ l\ j \end{Bmatrix}\frac{\partial v_m}{\partial q^k} - \begin{Bmatrix} s \\ l\ k \end{Bmatrix}\frac{\partial v_s}{\partial q^j}$$

$$- \begin{Bmatrix} s \\ j\ k \end{Bmatrix}\frac{\partial v_l}{\partial q^s} + \begin{Bmatrix} s \\ l\ k \end{Bmatrix}\begin{Bmatrix} m \\ s\ j \end{Bmatrix} v_m + \begin{Bmatrix} s \\ j\ k \end{Bmatrix}\begin{Bmatrix} m \\ l\ s \end{Bmatrix} v_m. \quad \text{(A.3-18c)}$$

If we interchange the order of differentiation, equation A.3-18c becomes

$$v_{l,kj} = \frac{\partial^2 v_l}{\partial q^j \partial q^k} - \left[\frac{\partial}{\partial q^j}\begin{Bmatrix} m \\ l\ k \end{Bmatrix}\right] v_m - \begin{Bmatrix} m \\ l\ k \end{Bmatrix}\frac{\partial v_m}{\partial q^j} - \begin{Bmatrix} s \\ l\ j \end{Bmatrix}\frac{\partial v_s}{\partial q^k}$$

$$- \begin{Bmatrix} s \\ k\ j \end{Bmatrix}\frac{\partial v_l}{\partial q^s} + \begin{Bmatrix} s \\ l\ j \end{Bmatrix}\begin{Bmatrix} m \\ s\ k \end{Bmatrix} v_m + \begin{Bmatrix} s \\ k\ j \end{Bmatrix}\begin{Bmatrix} m \\ l\ s \end{Bmatrix} v_m. \quad \text{(A.3-18d)}$$

On the right side of equation A.3-18d, m is a dummy index in the third term and it can be replaced by s; likewise in the fourth term s can be replaced by m. Noting that the Christoffel symbol is symmetric (equation A.3-11), and assuming that v_l is continuous with continuous second partial derivatives, it follows from equations A.3-18c and d that

$$v_{l,jk} - v_{l,kj} = \left[\frac{\partial}{\partial q^j}\begin{Bmatrix} m \\ l\ k \end{Bmatrix} - \frac{\partial}{\partial q^k}\begin{Bmatrix} m \\ l\ j \end{Bmatrix} + \begin{Bmatrix} s \\ l\ k \end{Bmatrix}\begin{Bmatrix} m \\ s\ j \end{Bmatrix} - \begin{Bmatrix} s \\ l\ j \end{Bmatrix}\begin{Bmatrix} m \\ s\ k \end{Bmatrix}\right] v_m \quad \text{(A.3-19a)}$$

$$= R^m_{ljk} v_m. \quad \text{(A.3-19b)}$$

The left side of equation A.3-19b is a covariant component of a third-order tensor. R^m_{ljk} is a mixed fourth-order tensor, contravariant of order one and covariant of order three. The dot product $R^m_{ljk} v_m$ is a covariant component of a third-order tensor. Thus the terms on both sides of equation A.3-19b are covariant components of third-order tensors. The fourth-order tensor R^m_{ljk} is the Riemann-Christoffel tensor.

We can interchange the order of covariant differentiation if $R^m_{ljk} = 0$. The Riemann-Christoffel tensor is a property of the space and is independent of the vector $\underline{v}$. In a Euclidean space we can always set up a rectangular Cartesian coordinate system where the Christoffel symbols are zero, which implies that the Riemann-Christoffel tensor is zero. R^m_{ljk} is a fourth-order tensor and transforms according to equation A.2-43c. Thus it is zero in all coordinate systems which can be transformed to a rectangular Cartesian coordinate system. We conclude that in a Euclidean space, R^m_{ljk} is zero and the order of covariant differentiation is not important as long as the components of the tensor have continuous second partial derivatives.

Example A.3-1   Christoffel Symbols in Spherical Coordinates
Calculate the Christoffel symbols of the second kind for the spherical polar coordinate system (r, θ, ϕ).

A.3 Covariant Differentiation

Solution
Since the coordinate system is orthogonal, we make use of equations A.3-12a to d. The metric tensors are given by equations A.2-63 and A.2-64. The only nonzero Christoffel symbols are

$$\left\{\begin{matrix} r \\ \theta\ \theta \end{matrix}\right\} = -\frac{1}{2}g^{rr}\frac{\partial}{\partial r}(g_{\theta\theta}) = -r, \tag{A.3-20a}$$

$$\left\{\begin{matrix} r \\ \phi\ \phi \end{matrix}\right\} = -\frac{1}{2}g^{rr}\frac{\partial}{\partial r}(g_{\phi\phi}) = -r\sin^2\theta, \tag{A.3-20b}$$

$$\left\{\begin{matrix} \theta \\ \theta\ r \end{matrix}\right\} = \frac{1}{2}g^{\theta\theta}\frac{\partial}{\partial r}(g_{\theta\theta}) = \frac{1}{r}, \tag{A.3-20c}$$

$$\left\{\begin{matrix} \theta \\ \phi\ \phi \end{matrix}\right\} = -\frac{1}{2}g^{\theta\theta}\frac{\partial}{\partial \theta}(g_{\phi\phi}) = -\sin\theta\cos\theta, \tag{A.3-20d}$$

$$\left\{\begin{matrix} \phi \\ \phi\ r \end{matrix}\right\} = \frac{1}{2}g^{\phi\phi}\frac{\partial}{\partial r}(g_{\phi\phi}) = \frac{1}{r}, \tag{A.3-20e}$$

and $\left\{\begin{matrix} \phi \\ \phi\ \theta \end{matrix}\right\} = \frac{1}{2}g^{\phi\phi}\frac{\partial}{\partial \theta}(g_{\phi\phi}) = \cot\theta. \tag{A.3-20f}$

Example A.3-2 Covariant Derivative in Spherical Coordinates
Let V^i be the contravariant components of the velocity vector $\underline{V}$. Obtain $V^i_{,i}$ in the spherical polar coordinate system. That is, calculate $V^r_{,r} + V^\theta_{,\theta} + V^\phi_{,\phi}$. Rewrite this expression in physical components.

Solution
From equations A.3-6 and A.3-20a, we have

$$V^r_{,r} = \frac{\partial V^r}{\partial r} + \left\{\begin{matrix} r \\ m\ r \end{matrix}\right\} V^m \tag{A.3-21a}$$

$$= \frac{\partial V^r}{\partial r} + \left\{\begin{matrix} r \\ r\ r \end{matrix}\right\} V^r + \left\{\begin{matrix} r \\ \theta\ r \end{matrix}\right\} V^\theta + \left\{\begin{matrix} r \\ \phi\ r \end{matrix}\right\} V^\phi \tag{A.3-21b}$$

$$= \frac{\partial V^r}{\partial r}, \tag{A.3-21c}$$

$$V^\theta_{,\theta} = \frac{\partial V^\theta}{\partial \theta} + \left\{\begin{matrix} \theta \\ r\ \theta \end{matrix}\right\} V^r + \left\{\begin{matrix} \theta \\ \theta\ \theta \end{matrix}\right\} V^\theta + \left\{\begin{matrix} \phi \\ \phi\ \phi \end{matrix}\right\} V^\phi \tag{A.3-22a}$$

$$= \frac{\partial V^\theta}{\partial \theta} + \frac{1}{r}V^r, \tag{A.3-22b}$$

$$V^\phi_{,\phi} = \frac{\partial V^\phi}{\partial \phi} + \left\{\begin{matrix} \phi \\ r\ \phi \end{matrix}\right\} V^r + \left\{\begin{matrix} \phi \\ \theta\ \phi \end{matrix}\right\} V^\theta + \left\{\begin{matrix} \phi \\ \phi\ \phi \end{matrix}\right\} V^\phi \tag{A.3-23a}$$

$$= \frac{\partial V^\phi}{\partial \phi} + \frac{1}{r}V^r + \cot\theta V^\theta, \tag{A.3-23b}$$

and $V^r_{,r} + V^\theta_{,\theta} + V^\phi_{,\phi} = \frac{\partial V^r}{\partial r} + \frac{\partial V^\theta}{\partial \theta} + \frac{\partial V^\phi}{\partial \phi} + \frac{2}{r}V^r + \cot\theta V^\theta. \tag{A.3-24}$

Rewriting in physical components, using equation A.2-72b, we have

$$V^r = V_{(r)}, \quad V^\theta = \frac{1}{r}V_{(\theta)}, \quad V^\phi = \frac{1}{r\sin\theta}V_{(\phi)}. \tag{A.3-25}$$

Substituting equations A.3-25a to c into equation A.3-24, we obtain

$$V^r_{,r} + V^\theta_{,\theta} + V^\phi_{,\phi} = \frac{\partial}{\partial r}(V_{(r)}) + \frac{\partial}{\partial \theta}\left(\frac{1}{r}V_{(\theta)}\right) + \frac{\partial}{\partial \phi}\left(\frac{1}{r\sin\theta}V_{(\phi)}\right) + \frac{2V_{(r)}}{r} + \frac{\cot\theta}{r}V_{(\theta)}. \quad \text{(A.3-26)}$$

We note that on the right side of equation A.3-26, every term has the same physical dimension, namely the reciprocal of time. Equation A.3-26 expresses the divergence of the velocity vector $\underline{V}$ and is a scalar, which is shown in the next example.

Example A.3-3   Divergence of a Vector
Show that $V^i_{,i}$ is a scalar.

Solution
The component $V^i_{,j}$ is a mixed tensor and will transform according to equation A.2-41:

$$\bar{V}^i_{,j} = \frac{\partial \bar{q}^i}{\partial q^r}\frac{\partial q^s}{\partial \bar{q}^j} V^r_{,s}. \quad \text{(A.3-27)}$$

Setting $j = i$ and summing

$$\bar{V}^i_{,i} = \frac{\partial \bar{q}^i}{\partial q^r}\frac{\partial q^s}{\partial \bar{q}^i} V^r_{,s} \quad \text{(A.3-28a)}$$

$$= \delta^s_r V^r_{,s} \quad \text{(A.3-28b)}$$

$$= V^s_{,s} \quad \text{(A.3-28c)}$$

$$= V^i_{,i}. \quad \text{(A.3-28d)}$$

Equations A.3-28b and c are obtained by using the usual chain rule and the property of the Kronecker delta respectively. The indices i and s are dummy and can be interchanged freely. The sum $V^i_{,i}$ is independent of the coordinate system and is a scalar or an invariant. That is, we obtain the same value in the barred and in the unbarred coordinate system.

A.3-4   Grad, Div, and Curl

The operation $\underline{\nabla}$ in a general curvilinear coordinate system can be defined as

$$\underline{\nabla} = \underline{g}^r \frac{\partial}{\partial q^r}. \quad \text{(A.3-29)}$$

If φ is a scalar function,

$$\text{grad } \varphi = \underline{\nabla}\varphi = \underline{g}^r \frac{\partial \varphi}{\partial q^r}. \quad \text{(A.3-30)}$$

In Example A.2-3 we have shown that $\partial\varphi/\partial q^r$ transforms as a covariant tensor. If $\underline{v}$ is a vector

$$\underline{\nabla}\underline{v} = \underline{g}^r \frac{\partial}{\partial q^r} v_s \underline{g}^s = \underline{g}^r \frac{\partial}{\partial q^r} v^t \underline{g}_t \quad \text{(A.3-31a)}$$

$$= \underline{g}^r \underline{g}^s v_{s,r} = \underline{g}^r \underline{g}_t v^t_{,r}. \quad \text{(A.3-31b)}$$

The components $v_{s,r}$ and $v^t_{,r}$ are the covariant and mixed components respectively of a second-order tensor. This confirms the statement made earlier that the operation of taking the grad of a tensor raises the order of the tensor.

In Example A.3-3 we have defined the divergence of a vector, which in the $\underline{\nabla}$ notation is written as

$$\underline{\nabla} \cdot \underline{v} = \underline{g}^r \frac{\partial}{\partial q^r} \cdot v^s \underline{g}_s \qquad \text{(A.3-32a)}$$

$$= v^s_{,r} \underline{g}^r \cdot \underline{g}_s \qquad \text{(A.3-32b)}$$

$$= v^s_{,r} \delta^r_s \qquad \text{(A.3-32c)}$$

$$= v^s_{,s}, \qquad \text{(A.3-32d)}$$

which is the covariant derivative.

The divergence of a second-order tensor is defined as

$$\underline{\nabla} \cdot \underline{\underline{T}} = \underline{g}^r \frac{\partial}{\partial q^r} \cdot T^{st} \underline{g}_s \underline{g}_t \qquad \text{(A.3-33a)}$$

$$= T^{st}_{,r} \underline{g}^r \cdot \underline{g}_s \underline{g}_t \qquad \text{(A.3-33b)}$$

$$= T^{st}_{,r} \delta^r_s \underline{g}_t \qquad \text{(A.3-33c)}$$

$$= T^{st}_{,s} \underline{g}_t. \qquad \text{(A.3-33d)}$$

The component $T^{st}_{,s}$ is the contravariant component of a vector, since t is the only free index.

The divergence of a tensor of order n is a tensor of order $n - 1$. Thus, the divergence of a vector is a scalar and the divergence of a tensor of order two is a vector.

The Laplacian of a scalar function φ is given by

$$\nabla^2 \varphi = \text{div (grad } \varphi) = \underline{\nabla} \cdot (\underline{\nabla} \varphi). \qquad \text{(A.3-34)}$$

If we denote grad φ by $\underline{v}$, then

$$v_i = \frac{\partial \varphi}{\partial q^i}. \qquad \text{(A.3-35)}$$

$\nabla^2 \varphi$ can be written as

$$\nabla^2 \varphi = \underline{\nabla} \cdot \underline{v} = v^s_{,s} \qquad \text{(A.3-36a)}$$

$$= \left(g^{si} \frac{\partial \varphi}{\partial q^i} \right)_{,s} \qquad \text{(A.3-36b)}$$

$$= g^{si} \left(\frac{\partial \varphi}{\partial q^i} \right)_{,s} \qquad \text{(A.3-36c)}$$

$$= g^{si} \left[\frac{\partial^2 \varphi}{\partial q^i \partial q^s} - \begin{Bmatrix} j \\ i\ s \end{Bmatrix} \frac{\partial \varphi}{\partial q^j} \right]. \qquad \text{(A.3-36d)}$$

In equation A.3-36c we have used equation A.2-67c to transform the covariant components v_i to contravariant components v^s. The metric tensor is a constant with respect to the covariant derivative, and equation A.3-36c follows from equation A.3-36b. If the coordinate system is orthogonal, equation A.3-36d simplifies to

$$\nabla^2 \varphi = g^{ii}\left[\frac{\partial^2 \varphi}{\partial q^i \partial q^i} - \begin{Bmatrix} j \\ i\ i \end{Bmatrix}\frac{\partial \varphi}{\partial q^j}\right]. \qquad (A.3\text{-}37)$$

If $\underline{u}$ is curl $\underline{v}$,

$$\underline{u} = \underline{g}^r \frac{\partial}{\partial q^r} \times v_s \underline{g}^s \qquad (A.3\text{-}38a)$$

$$= v_{s,r}\underline{g}^r \times \underline{g}^s \qquad (A.3\text{-}38b)$$

$$= v_{s,r}\varepsilon^{rst}\underline{g}_t \qquad (A.3\text{-}38c)$$

$$= \left[\frac{\partial v_s}{\partial q^r} - \begin{Bmatrix} p \\ s\ r \end{Bmatrix} v_p\right]\frac{e_{rst}}{\sqrt{g}}\underline{g}_t. \qquad (A.3\text{-}38d)$$

Note that $\underline{u}$ is expressed in terms of contravariant components, and equation A.3-38d may be written as

$$u^t = \left[\frac{\partial v_s}{\partial q^r} - \begin{Bmatrix} p \\ s\ r \end{Bmatrix} v_p\right]\frac{e_{rst}}{\sqrt{g}}. \qquad (A.3\text{-}39)$$

If the coordinate system is orthogonal, then $\begin{Bmatrix} p \\ s\ r \end{Bmatrix}$ is zero if all three suffixes are different, but if $r=s$, the permutation tensor e_{rst} is zero. The only possibility of nonzero contributions from the Christoffel symbol arises when $p=s$ and $p=r$. The Christoffel symbol is symmetric (equation A.3-11), $e_{rst} = -e_{srt}$, and the contributions from the Christoffel symbol cancel out. To clarify this point further, consider the case $t=1$. The only nonzero contribution from e_{rst} arises when $r=2$, $s=3$ and when $r=3$, $s=2$. Thus equation A.3-39 becomes

$$u^1 = \frac{e_{231}}{\sqrt{g}}\left[\frac{\partial v_3}{\partial q^2} - \begin{Bmatrix} 3 \\ 3\ 2 \end{Bmatrix} v_3 - \begin{Bmatrix} 2 \\ 3\ 2 \end{Bmatrix} v_2\right] + \frac{e_{321}}{\sqrt{g}}\left[\frac{\partial v_2}{\partial q^3} - \begin{Bmatrix} 2 \\ 2\ 3 \end{Bmatrix} v_2 - \begin{Bmatrix} 3 \\ 2\ 3 \end{Bmatrix} v_3\right]. \qquad (A.3\text{-}40)$$

From the definition of the permutation tensor,

$$e_{231} = 1 \qquad (A.3\text{-}41a)$$
$$\text{and}\quad e_{321} = -1. \qquad (A.3\text{-}41b)$$

Using equations A.3-41a and b, we find that the contributions from the Christoffel symbols cancel out and Equation A.3-40 reduces to

$$u^1 = \frac{1}{\sqrt{g}}\left[\frac{\partial v_3}{\partial q^2} - \frac{\partial v_2}{\partial q^3}\right]. \qquad (A.3\text{-}42)$$

In an orthogonal coordinate system equation A.3-39 simplifies to

$$u^t = \frac{e_{rst}}{\sqrt{g}}\frac{\partial v_s}{\partial q^r}. \qquad (A.3\text{-}43)$$

Example A.3-4 Laplacian of a Scalar
If f is a scalar function of position, write $\nabla^2 f$ in the spherical polar coordinate system (r, θ, ϕ).

Solution
Since the spherical polar coordinate system is orthogonal, we can use equation A.3-37, and we have

$$\nabla^2 f = g^{rr}\left[\frac{\partial^2 f}{\partial r^2} - \left\{\begin{matrix}j\\r\ r\end{matrix}\right\}\frac{\partial f}{\partial q^j}\right] + g^{\theta\theta}\left[\frac{\partial^2 f}{\partial \theta^2} - \left\{\begin{matrix}j\\\theta\ \theta\end{matrix}\right\}\frac{\partial f}{\partial q^j}\right]$$
$$+ g^{\phi\phi}\left[\frac{\partial^2 f}{\partial \phi^2} - \left\{\begin{matrix}j\\\phi\ \phi\end{matrix}\right\}\frac{\partial f}{\partial q^j}\right]. \qquad (A.3\text{-}44)$$

Using equations A.2-64a to f and A.3-20a to f, equation A.3-44 becomes

$$\nabla^2 f = \frac{\partial^2 f}{\partial r^2} - \frac{1}{r^2}\left[\frac{\partial^2 f}{\partial \theta^2} + r\frac{\partial f}{\partial r}\right] + \frac{1}{r^2\sin^2\theta}\left[\frac{\partial^2 f}{\partial \phi^2} + r\sin^2\theta\frac{\partial f}{\partial r} + \sin\theta\cos\theta\frac{\partial f}{\partial \theta}\right] \qquad (A.3\text{-}45a)$$

$$= \frac{\partial^2 f}{\partial r^2} + \frac{1}{r^2}\frac{\partial^2 f}{\partial \theta^2} + \frac{1}{r^2\sin^2\theta}\frac{\partial^2 f}{\partial \phi^2} + \frac{2}{r}\frac{\partial f}{\partial r} + \frac{\cot\theta}{r^2}\frac{\partial f}{\partial \theta}. \qquad (A.3\text{-}45b)$$

Example A.3-5 Physical Components of Curl $\underline{v}$
Calculate the physical components of curl $\underline{v}$ in the spherical polar coordinate system.

Solution
If we denote curl $\underline{v}$ by $\underline{u}$, we have from equations A.2-65 and A.3-43,

$$u^r = \frac{1}{r^2\sin\theta}\left(\frac{\partial v_\phi}{\partial \theta} - \frac{\partial v_\theta}{\partial \phi}\right), \qquad (A.3\text{-}46a)$$

$$u^\theta = \frac{1}{r^2\sin\theta}\left(\frac{\partial v_r}{\partial \phi} - \frac{\partial v_\phi}{\partial r}\right), \qquad (A.3\text{-}46b)$$

$$\text{and}\quad u^\phi = \frac{1}{r^2\sin\theta}\left(\frac{\partial v_\theta}{\partial r} - \frac{\partial v_r}{\partial \theta}\right). \qquad (A.3\text{-}46c)$$

Transforming all the components to physical components via equations A.2-72b and A.2-75, we obtain

$$u_{(r)} = \frac{1}{r^2 \sin\theta}\left[\frac{\partial}{\partial\theta}(r\sin\theta v_{(\phi)}) - \frac{\partial}{\partial\phi}(rv_{(\theta)})\right] \quad \text{(A.3-47a)}$$

$$= \frac{1}{r\sin\theta}\left[\sin\theta\frac{\partial v_{(\phi)}}{\partial\theta} + \cos\theta v_{(\phi)} - \frac{\partial v_{(\theta)}}{\partial\phi}\right], \quad \text{(A.3-47b)}$$

$$u_{(\theta)} = \frac{r}{r^2\sin\theta}\left[\frac{\partial}{\partial\phi}v_{(r)} - \frac{\partial}{\partial r}(r\sin\theta v_{(\phi)})\right] \quad \text{(A.3-47c)}$$

$$= \frac{1}{r\sin\theta}\left[\frac{\partial v_{(r)}}{\partial\phi} - r\sin\theta\frac{\partial v_{(\phi)}}{\partial r} - \sin\theta v_{(\phi)}\right], \quad \text{(A.3-47d)}$$

and $$u_{(\phi)} = \frac{r\sin\theta}{r^2\sin\theta}\left[\frac{\partial}{\partial r}(rv_{(\theta)}) - \frac{\partial}{\partial\theta}(v_{(r)})\right] \quad \text{(A.3-47e)}$$

$$= \frac{1}{r}\left[r\frac{\partial v_{(\theta)}}{\partial r} + v_{(\theta)} - \frac{\partial v_{(r)}}{\partial\theta}\right]. \quad \text{(A.3-47f)}$$

Example A.3-6 Equation of Motion
The equation of motion for slow flows may be written as

$$\underline{\nabla}P = -\underline{\nabla}\cdot\underline{\underline{\sigma}}, \quad \text{(A.3-48)}$$

where P is a scalar and is $\underline{\underline{\sigma}}$ the stress tensor. Write equation A.3-48 in component form for the spherical polar coordinate system.

Solution
Assume that P and $\underline{\underline{\sigma}}$ are functions of r and θ only, and that $\underline{\underline{\sigma}}$ is symmetric. Equation A.3-48 is written in the so-called coordinate-free form. For a general curvilinear coordinate system (q^1, q^2, q^3), the equation can be written in component form as

$$\frac{\partial P}{\partial q^i} = -\sigma^j_{i,j}. \quad \text{(A.3-49)}$$

From Example A.2-2 it is known that $\partial P/\partial q^i$ is a covariant component of order one, as is the right side of equation A.3-49. Expanding equation A.3-49, we have

$$-\frac{\partial P}{\partial q^1} = \sigma^1_{1,1} + \sigma^2_{1,2} + \sigma^3_{1,3} \quad \text{(A.3-50a)}$$

$$= \frac{\partial \sigma^1_1}{\partial q^1} + \begin{Bmatrix}1\\s\ 1\end{Bmatrix}\sigma^s_1 - \begin{Bmatrix}t\\1\ 1\end{Bmatrix}\sigma^1_t + \frac{\partial \sigma^2_1}{\partial q^2} + \begin{Bmatrix}2\\s\ 2\end{Bmatrix}\sigma^s_1 - \begin{Bmatrix}t\\1\ 2\end{Bmatrix}\sigma^2_t$$

$$+ \frac{\partial \sigma^3_1}{\partial q^3} + \begin{Bmatrix}3\\s\ 3\end{Bmatrix}\sigma^s_1 - \begin{Bmatrix}t\\1\ 3\end{Bmatrix}\sigma^3_t, \quad \text{(A.3-50b)}$$

$$-\frac{\partial P}{\partial q^2} = \sigma^1_{2,1} + \sigma^2_{2,2} + \sigma^3_{2,3} \tag{A.3-50c}$$

$$= \frac{\partial \sigma^1_2}{\partial q^1} + \begin{Bmatrix} 1 \\ s\ 1 \end{Bmatrix} \sigma^s_2 - \begin{Bmatrix} t \\ 2\ 1 \end{Bmatrix} \sigma^1_t + \frac{\partial \sigma^2_2}{\partial q^2} + \begin{Bmatrix} 2 \\ s\ 2 \end{Bmatrix} \sigma^s_s - \begin{Bmatrix} t \\ 2\ 2 \end{Bmatrix} \sigma^2_t$$

$$+ \frac{\partial \sigma^3_2}{\partial q^3} + \begin{Bmatrix} 3 \\ s\ 3 \end{Bmatrix} \sigma^s_2 - \begin{Bmatrix} t \\ 2\ 3 \end{Bmatrix} \sigma^3_t, \tag{A.3-50d}$$

and $\quad -\dfrac{\partial P}{\partial q^3} = \sigma^1_{3,1} + \sigma^2_{3,2} + \sigma^3_{3,3}$ (A.3-50e)

$$= \frac{\partial \sigma^1_3}{\partial q^1} + \begin{Bmatrix} 1 \\ s\ 1 \end{Bmatrix} \sigma^s_3 - \begin{Bmatrix} t \\ 3\ 1 \end{Bmatrix} \sigma^1_t + \frac{\partial \sigma^2_3}{\partial q^2} + \begin{Bmatrix} 2 \\ s\ 2 \end{Bmatrix} \sigma^s_3 - \begin{Bmatrix} t \\ 3\ 2 \end{Bmatrix} \sigma^2_t$$

$$+ \frac{\partial \sigma^3_3}{\partial q^3} + \begin{Bmatrix} 3 \\ s\ 3 \end{Bmatrix} \sigma^s_3 - \begin{Bmatrix} t \\ 3\ 3 \end{Bmatrix} \sigma^3_t. \tag{A.3-50f}$$

For the spherical polar coordinate system we identify

$$q^1 = r, \quad q^2 = \theta, \quad q^3 = \phi. \tag{A.3-51}$$

Making use of equation A.3-20, equation A.3-50 becomes

$$-\frac{\partial P}{\partial r} = \frac{\partial \sigma^r_r}{\partial r} + \frac{\partial \sigma^\theta_r}{\partial \theta} + \frac{\sigma^r_r}{r} - \frac{\sigma^\theta_\theta}{r} + \frac{\sigma^r_r}{r} + \cot\theta \sigma^\theta_r - \frac{\sigma^\phi_\phi}{r}, \tag{A.3-52a}$$

$$-\frac{\partial P}{\partial \theta} = \frac{\partial \sigma^r_\theta}{\partial r} - \frac{\partial \sigma^r_\theta}{\partial r} + \frac{\partial \sigma^\theta_\theta}{\partial \theta} + \frac{\sigma^\theta_\theta}{r} + r\sigma^\theta_r + \frac{\sigma^r_\theta}{r} + \cot\theta(\sigma^\theta_\theta - \sigma^\phi_\phi), \tag{A.3-52b}$$

and $\quad 0 = \dfrac{\partial \sigma^r_\theta}{\partial r} - \dfrac{\sigma^r_\phi}{r} + \dfrac{\partial \sigma^\theta_\phi}{\partial \theta} + \dfrac{\sigma^r_\phi}{r} - \cot\theta \sigma^\theta_\phi + \dfrac{\sigma^r_\phi}{r} + \cot\theta \sigma^\phi_\phi$

$$+ r \sin^2\theta \sigma^\phi_r + \sin\theta \cos\theta \sigma^\phi_\theta. \tag{A.3-52c}$$

Transforming all the covariant and mixed components to physical components, we obtain

$$-\frac{\partial P}{\partial r} = \frac{\partial}{\partial r}(\sigma_{(rr)}) + \frac{\partial}{\partial \theta}\left(\frac{\sigma_{(r\theta)}}{r}\right) + \left(\frac{2\sigma_{(rr)} - \sigma_{(\theta\theta)} - \sigma_{(\phi\phi)}}{r}\right) + \frac{\sigma_{(r\theta)} \cot\theta}{r}, \tag{A.3-53a}$$

$$-\frac{1}{r}\frac{\partial P}{\partial \theta} = \frac{1}{r}\left[\frac{\partial}{\partial r}(r\sigma_{(\theta r)}) + \frac{\partial \sigma_{(\theta\theta)}}{\partial \theta} + 2\sigma_{(\theta r)} + \cot\theta(\sigma_{(\theta\theta)} - \sigma_{(\phi\phi)})\right], \tag{A.3-53b}$$

and $\quad 0 = \dfrac{1}{r \sin\theta}\left[\dfrac{\partial}{\partial r}(r \sin\theta \sigma_{(\phi r)}) + \dfrac{\partial}{\partial \theta}(\sin\phi \sigma_{(\phi\phi)}) + 2\sin\theta \sigma_{(\phi r)} + \cos\theta \sigma_{(\theta\phi)}\right].$ (A.3-53c)

Note that in equation A.3-53 every term has the same dimensions.

A.4 Integral Transforms

The divergence theorem, which transforms a volume integral to a surface integral, and Stokes' theorem, which transforms a surface integral to a line integral, can be extended to higher order tensors and higher dimensional spaces. In this section, we state the divergence theorem and Stokes' theorem for a first- and second-order tensor in a generalized coordinate system.

We recall that the divergence theorem for a vector $\underline{u}$ is

$$\int_V \text{div } \underline{u} \, dV = \int_S \underline{u} \cdot \underline{n} \, dS \qquad \text{(A.4-1a)}$$

or

$$\int_V u^j{}_{,j} \, dV = \int_S u^j n_j \, dS, \qquad \text{(A.4-1b)}$$

where V is the volume enclosed by the surface S whose outward unit normal is $\underline{n}$.

To extend the theorem to a second order tensor $\underline{\underline{T}}$, we may replace $\underline{u}$ by $\underline{\underline{T}}$ in equation A.4-1a, and in component form we obtain

$$\int_V T^{ij}{}_{,j} \, dV = \int_S T^{ij} n_j \, dS. \qquad \text{(A.4-2)}$$

We note that equation A.4-2 is a vector equation. Both sides contain one free index (i) and the equation represents three equations for $i = 1, 2,$ and 3. Equation A.4-1b is a scalar equation, containing no free index.

Stokes' theorem may be written as

$$\int_S (\varepsilon^{ijk} u_{k,j}) n_i \, dS = \oint_C u_i \, dq^i, \qquad \text{(A.4-3)}$$

where, as usual, C is the closed curve bounding the surface S. For a second-order tensor $\underline{\underline{T}}$, equation A.4-3 becomes

$$\int_S (\varepsilon^{ijk} T_{kl,j}) n_i \, dS = \oint_C T_{kl} \, dq^k. \qquad \text{(A.4-4)}$$

Again, equation A.4-4 represents three equations ($l = 1, 2,$ and 3).

Example A.4-1 Volume-Surface Integral Transformation
Show that

$$\int_V \text{curl } \underline{u} \, dV = \int_S \underline{n} \times \underline{u} \, dS. \qquad \text{(A.4-5)}$$

Solution
Let

$$T^{ij} = \varepsilon^{ijk} u_k. \qquad \text{(A.4-6)}$$

On substituting equation A.4-6 into A.4-2 and noting that ε^{ijk} is a constant with respect to covariant differentiation, we obtain the following

$$\int_V (\varepsilon^{ijk} u_k)_{,j} dV = \int_S \varepsilon^{ijk} u_k n_j dS \qquad (A.4\text{-}7a)$$

$$\text{or} \quad \int_V \varepsilon^{ijk} u_{k,j} dV = \int_S \varepsilon^{ijk} n_j u_k dS. \qquad (A.4\text{-}7b)$$

Equation A.4-7b is the component form of equation A.4-5.

Example A.4-2 Equation of Motion

Applying the law of conservation of momentum to a volume of a continuous medium in motion, Bird et al. (1987) have obtained the equation

$$\frac{d}{dt} \int_V \rho \underline{V} dV = -\int_S [\underline{n} \cdot \rho \underline{V}\underline{V}] dS - \int_S \underline{n} \cdot \underline{\underline{\pi}} dS + \int_V \rho \underline{g} dV, \qquad (A.4\text{-}8)$$

where $\underline{V}$ is the velocity, ρ is the density, $\underline{\underline{\pi}}$ is the stress tensor, and $\underline{g}$ is the gravity force.

Write the equation in component form, and hence deduce the equation of motion at each point in space.

Solution

We choose to write the components as contravariant components, and hence equation A.4-8 becomes

$$\frac{d}{dt} \int_V \rho V^i dV = -\int_S n_j \rho V^j V^i dS - \int_S n_j \pi^{ji} dS + \int_V \rho g^i dV. \qquad (A.4\text{-}9)$$

To obtain the equation of motion at each point, we need to use the divergence theorem to transform the surface integrals to volume integrals. Since the volume V is fixed in space, we may include the time derivative inside the volume V. Equation A.4-9 now becomes

$$\int_V \frac{\partial}{\partial t}(\rho V^i) dV = -\int_V (\rho V^j V^i)_{,j} dV - \int_V \pi^{ji}_{,j} dV + \int_V \rho g^i dV. \qquad (A.4\text{-}10)$$

Since V is an arbitrary volume, equation A.4-10 holds at every point and we obtain

$$\frac{\partial}{\partial t}(\rho V^i) + (\rho V^j V^i)_{,j} = -\pi^{ji}_{,j} + \rho g^i. \qquad (A.4\text{-}11)$$

Expanding the left side of equation A.4-11, we obtain

$$\rho \frac{\partial V^i}{\partial t} + V^i \frac{\partial \rho}{\partial t} + \rho V^j V^i_{,j} + \rho V^j_{,j} V^i + \rho_{,j} V^j V^i$$

$$= \rho \left(\frac{\partial V^i}{\partial t} + V^j V^i_{,j} \right) + V^i \left(\frac{\partial \rho}{\partial t} + V^j \rho_{,j} + \rho V^j_{,j} \right). \qquad (A.4\text{-}12)$$

From the mass balance, we obtain the equation of continuity, which may be written as

$$\frac{\partial \rho}{\partial t} + (\rho V^j)_{,j} = 0. \qquad (A.4\text{-}13)$$

Combining equations A.4-12 and A.4-13, equation A.4-11 becomes

$$\rho\left(\frac{\partial V^i}{\partial t} + V^j V^i_{,j}\right) = -\pi^{ji}_{,j} + \rho g^i. \qquad (A.4\text{-}14)$$

The left side of equation A.4-14 is often written as DV^i/Dt, where D/Dt is known as the substantial (material) derivative. It is the time derivative following a material element.

A.5 Isotropic Tensors, Objective Tensors, and Tensor-Valued Functions

A.5-1 Isotropic Tensors

Many materials are isotropic, that is, their properties are independent of direction. Thus, if these properties are described by tensors, the components of tensors are identical in all rectangular Cartesian coordinate systems. To find out if a tensor is isotropic or not, we express its components in a rectangular Cartesian coordinate system (x^1, x^2, x^3) and rotate the axes to obtain a new coordinate system $(\bar{x}^1, \bar{x}^2, \bar{x}^3)$. If in the new coordinate system the components are identical, the tensor is isotropic. Below we list the isotropic tensors of order zero to four. Here we consider only Cartesian coordinate systems.

(i) All tensors of order zero are isotropic. Since tensors of order zero are scalars and are independent of direction, they are isotropic.

(ii) $\underline{0}$ is the only isotropic tensor of order one. If (u_1, u_2, u_3) are the components of a tensor of order one (a vector) in the (x^i) coordinate system, and $(\bar{u}^i)$ are the components of the same vector in the $(\bar{x}^i)$ coordinate system, we can write

$$\bar{u}_m = l_{mn} u_n. \qquad (A.5\text{-}1)$$

Let $(\bar{x}^i)$ be the coordinate axes obtained by rotating the (x^i) system through π rad about the x^3 axis, then

$$l_{11} = -1, \quad l_{22} = -1, \quad l_{33} = 1, \quad \text{the other } l_{ij} = 0. \qquad (A.5\text{-}2)$$

Combining equations A.5-1 and A.5-2 yields

$$\bar{u}_1 = -u_1, \quad \bar{u}_2 = -u_2, \quad \bar{u}_3 = u_3. \qquad (A.5\text{-}3)$$

From equation A.5-3 we deduce that for $\underline{u}$ to be isotropic,

$$\bar{u}_1 = u_1 = 0, \quad \bar{u}_2 = u_2 = 0. \qquad (A.5\text{-}4)$$

By rotating the axes about the x^1-axis through π radians, we obtain

$$u_3 = 0. \qquad (A.5\text{-}5)$$

Thus $\underline{0}$ is the only isotropic tensor, that is, there is no nonzero isotropic vector.

A.5 Isotropic Tensors, Objective Tensors, and Tensor-Valued Functions

(iii) The Kronecker delta δ_{ij} is isotropic. If T_{ij} is a component of a second-order tensor in the (x^i) coordinate system, and $\bar{T}_{rs}$ is a component of the same tensor in the $(\bar{x}^i)$ coordinate system, then

$$\bar{T}_{rs} = l_{ri}l_{sj}T_{ij}. \tag{A.5-6}$$

Letting T_{ij} be the Kronecker delta, equation A.5-6 becomes

$$\bar{T}_{rs} = l_{ri}l_{sj}\delta_{ij} \tag{A.5-7a}$$
$$= l_{ri}l_{si} \tag{A.5-7b}$$
$$= \delta_{rs}. \tag{A.5-7c}$$

Thus the Kronecker delta transforms into itself and is thus an isotropic tensor.

In fluid mechanics, the isotropic part of the stress tensor $\pi_{ij}^{(0)}$ can be written as

$$\pi_{ij}^{(0)} = P\delta_{ij}, \tag{A.5-8}$$

where P is a scalar.

Any second-order isotropic tensor can be expressed in terms of the Kronecker delta.

(iv) The permutation tensor e_{ijk} is an isotropic tensor of third-order. A useful relation between e_{ijk} and δ_{rs} is

$$e_{ijk}e_{rsk} = \delta_{ir}\delta_{js} - \delta_{is}\delta_{jr}. \tag{A.5-9}$$

(v) Any fourth-order isotropic tensor c_{ijkl} may be expressed as

$$c_{ijkl} = \lambda \delta_{ij}\delta_{kl} + \mu \delta_{ik}\delta_{jl} + \nu \delta_{il}\delta_{jk}, \tag{A.5-10}$$

where λ, μ, and ν are scalars.

Example A.5-1 Constitutive Equation for a Newtonian Fluid

For a Newtonian fluid, the deviatoric stress tensor $\underline{\underline{\sigma}}$ depends linearly on the rate-of-deformation tensor $\underline{\underline{\dot{\gamma}}}$. Obtain the constitutive equation of an isotropic, incompressible Newtonian fluid.

Solution

The constitutive equation of a Newtonian fluid may be written in the most general form as

$$\sigma_{ij} = \pm c_{ijkl}\dot{\gamma}_{kl}. \tag{A.5-11}$$

Since the fluid is isotropic, we obtain, using equation A.5-10

$$\sigma_{ij} = \pm[\lambda \delta_{ij}\delta_{kl} + \mu \delta_{ik}\delta_{jl} + \nu \delta_{il}\delta_{jk}]\dot{\gamma}_{kl} \tag{A.5-12a}$$
$$= \pm[\lambda \dot{\gamma}_{kk}\delta_{ij} + \mu \dot{\gamma}_{il}\delta_{jl} + \nu \dot{\gamma}_{ki}\delta_{jk}] \tag{A.5-12b}$$
$$= \pm[\lambda \dot{\gamma}_{kk}\delta_{ij} + \mu \dot{\gamma}_{ij} + \nu \dot{\gamma}_{ji}] \tag{A.5-12c}$$
$$= \pm(\nu + \mu)\dot{\gamma}_{ij}. \tag{A.5-12d}$$

Equation A.5-12d follows from equation A.5-12c, since the fluid is incompressible. That is, $\dot{\gamma}_{kk}$ is zero. Also $\dot{\gamma}_{ij}$ is symmetric. The coefficient $(\nu + \mu)$ is known as the viscosity of the fluid. Some authors adopt the positive sign in equation A.5-12a, and others [Bird et al. (1987)] adopt the negative sign. The latter is the convention used in this book.

A.5-2 Objective Tensors

The constitutive equation of a material should be independent of the motion of the material. Alternatively, we may state that the constitutive equation should be the same for all observers, irrespective of whether they are at rest or in motion. Quantities which are indifferent to the motion of the observers are known as objective quantities.

Consider two observers, one at rest (coordinate system $\underline{x} = x^i$), and the other in relative motion (coordinate system $x^{*i} = \underline{x}^*$). Since the second observer is both translating and rotating relative to the first one, we can relate these two systems by

$$\underline{x}^* = \underline{c}(t) + \underline{\underline{Q}}(t) \cdot \underline{x}. \qquad (A.5\text{-}13)$$

The vector $\underline{c}(t)$ in equation A.5-13 denotes the translation of the second observer relative to the first observer. The matrix $\underline{\underline{Q}}(t)$ denotes the rotation of the second observer relative to the first one, and the elements of $\underline{\underline{Q}}$ are the l_{ij}, the direction cosines of the axes x^{*i} relative to x^i. Note that in the present transformation, both $\underline{c}$ and $\underline{\underline{Q}}$ are functions of time t. Such a transformation is known as a transformation of frames of reference. $\underline{\underline{Q}}$ is orthogonal at all times.

Objective tensors are thus tensors which are invariant under a change of frame of reference. If we denote the components of a vector $\underline{u}$ relative to the x^i coordinate system as u^i, and the components of the same vector $\underline{u}$ relative to the x^{*i} coordinate system as u^{*i}, then if

$$\underline{u}^* = \underline{\underline{Q}} \cdot \underline{u}, \qquad (A.5\text{-}14)$$

$\underline{u}$ is an objective tensor (vector).

Note that due to the rotation of the axes, the components u^{*i} transform to components u^i under the usual tensor transformation laws.

Equally, a second-order tensor $\underline{\underline{T}}$ is an objective tensor if

$$\underline{\underline{T}}^* = \underline{\underline{Q}} \cdot \underline{\underline{T}} \cdot \underline{\underline{Q}}^+. \qquad (A.5\text{-}15)$$

We now examine the objectivity of some tensors.

(i) The velocity vector $\underline{V}$

Differentiating equation A.5-13, we obtain

$$\underline{V}^* = \dot{\underline{c}}(t) + \underline{\underline{Q}}(t) \cdot \underline{V} + \dot{\underline{\underline{Q}}}(t) \cdot \underline{x}. \qquad (A.5\text{-}16)$$

Since $\dot{\underline{c}}(t)$ and $\dot{\underline{\underline{Q}}}(t)$ do not vanish at all times, $\underline{V}$ is not an objective vector. This observation is a common experience. Sitting in a moving bus and watching the passenger sitting opposite to us, we seem to be at rest, but to an observer standing on the road we are traveling at a finite velocity.

A.5 Isotropic Tensors, Objective Tensors, and Tensor-Valued Functions

(ii) The line element $(ds)^2$

From equation A.5-13 we have

$$d\underline{x}^* = \underline{\underline{Q}} \cdot d\underline{x}, \tag{A.5-17}$$

$$\text{and } (ds^*)^2 = d\underline{x}^{*+} d\underline{x}^* = d\underline{x}^* \cdot d\underline{x}^* \tag{A.5-18a}$$

$$= d\underline{x}^+ \underline{\underline{Q}}^+ \cdot \underline{\underline{Q}} d\underline{x} \tag{A.5-18b}$$

$$= d\underline{x}^+ d\underline{x} \tag{A.5-18c}$$

$$= (ds)^2. \tag{A.5-18d}$$

Equation A.5-18c follows since $\underline{\underline{Q}}$ is orthogonal (i.e., $\underline{\underline{Q}}^+ \cdot \underline{\underline{Q}} = \underline{\underline{I}}$).

Thus, $(ds)^2$ is an objective quantity. The length of an object in non-relativistic mechanics does not depend on the motion of the observer.

(iii) The rate-of-deformation tensor $\underline{\underline{\dot{\gamma}}}$

Let

$$\underline{\underline{L}}^* = \frac{\partial \underline{V}^*}{\partial \underline{x}^*}, \quad \underline{\underline{L}} = \frac{\partial \underline{V}}{\partial \underline{x}}. \tag{A.5-19}$$

Using equation A.5-16, $\underline{\underline{L}}^*$ becomes

$$\underline{\underline{L}}^* = \frac{\partial}{\partial \underline{x}}[\dot{c}(t) + \underline{\underline{Q}} \cdot \underline{V} + \underline{\underline{\dot{Q}}} \cdot \underline{x}] \frac{\partial \underline{x}}{\partial \underline{x}^*}. \tag{A.5-20}$$

Inverting equation A.5-13 we obtain

$$\underline{\underline{Q}}^+ \cdot \underline{x}^* - \underline{\underline{Q}}^+ \cdot \underline{c} = \underline{x}. \tag{A.5-21}$$

Note that $\partial \underline{x}/\partial \underline{x}^* = \underline{\underline{Q}}^+$.

Combining equations A.5-20 and A.5-21 yields

$$\underline{\underline{L}}^* = \underline{\underline{Q}} \cdot \underline{\underline{L}} \cdot \underline{\underline{Q}}^+ + \underline{\underline{\dot{Q}}} \cdot \underline{\underline{Q}}^+. \tag{A.5-22}$$

$\underline{\underline{L}}^*$ is not an objective tensor since $\underline{\underline{\dot{Q}}} \cdot \underline{\underline{Q}}^+$ is not zero at all times.
The rate-of-deformation tensor $\underline{\underline{\dot{\gamma}}}$ is defined as

$$\underline{\underline{\dot{\gamma}}}^* = \underline{\underline{L}}^* + \underline{\underline{L}}^{*+} \tag{A.5-23a}$$

$$= \underline{\underline{Q}} \cdot \underline{\underline{L}} \cdot \underline{\underline{Q}}^+ + \underline{\underline{\dot{Q}}} \cdot \underline{\underline{Q}}^+ + \underline{\underline{Q}} \cdot \underline{\underline{L}}^+ \cdot \underline{\underline{Q}}^+ + \underline{\underline{Q}} \cdot \underline{\underline{\dot{Q}}}^+ \tag{A.5-23b}$$

$$= \underline{\underline{Q}} \cdot (\underline{\underline{L}} + \underline{\underline{L}}^+) \cdot \underline{\underline{Q}}^+ + (\underline{\underline{\dot{Q}}} \cdot \underline{\underline{Q}}^+ + \underline{\underline{Q}} \cdot \underline{\underline{\dot{Q}}}^+) \tag{A.5-23c}$$

$$= \underline{\underline{Q}} \cdot \underline{\underline{\dot{\gamma}}} \cdot \underline{\underline{Q}}^+ + \frac{d}{dt}(\underline{\underline{Q}} \cdot \underline{\underline{Q}}^+) \tag{A.5-23d}$$

$$= \underline{\underline{Q}} \cdot \underline{\underline{\dot{\gamma}}} \cdot \underline{\underline{Q}}^+. \tag{A.5-23e}$$

Equation A.5-23e follows since $\underline{\underline{Q}}$ is orthogonal. Thus $\underline{\underline{\dot{\gamma}}}$ is an objective tensor and is an admissible quantity in a constitutive equation.

Not all equations in physics are objective. The equation of motion is not objective, because velocity and acceleration are not objective quantities. The equation of motion holds only relative to an inertial frame of reference. If the motion of the earth can be neglected, then a frame of reference fixed on the surface of the earth can be considered to be an inertial frame.

A.5-3 Tensor-Valued Functions

As indicated earlier, the invariants of a tensor are scalar quantities, which remain unchanged when the coordinate system is transformed. They have an important role in tensor-valued functions. The scalar product of $\underline{u}$ and $\underline{u}$ ($u^i u_i$) is an invariant. For second-order tensors, we have three principal invariants. These invariants arise naturally when we consider the eigenvalues and eigenvectors of a second-order tensor. A nonzero vector $\underline{u}$ is said to be an eigenvector of a second-order tensor $\underline{\underline{T}}$ if the product $\underline{\underline{T}} \cdot \underline{u}$ is parallel to $\underline{u}$. This can be expressed as

$$\underline{\underline{T}} \cdot \underline{u} = \lambda \underline{u}, \tag{A.5-24}$$

where λ is a scalar and is an eigenvalue of $\underline{\underline{T}}$.

The condition for the existence of a nonzero solution to equation A.5-24 is

$$\det[\underline{\underline{T}} - \lambda \underline{\underline{I}}] = 0. \tag{A.5-25}$$

On expanding the determinant we obtain

$$-\lambda^3 + I_1 \lambda^2 - I_2 \lambda + I_3 = 0, \tag{A.5-26}$$

where $I_1 = \operatorname{tr} \underline{\underline{T}}$, $I_2 = 1/2[(\operatorname{tr} \underline{\underline{T}})^2 - \operatorname{tr}(\underline{\underline{T}}^2)]$, and $I_3 = \det \underline{\underline{T}}$. The trace of tensor $\underline{\underline{T}}$ ($\operatorname{tr} \underline{\underline{T}}$) is the sum of the diagonal elements. The functions I_1, I_2, and I_3 are known as the principal invariants of $\underline{\underline{T}}$, and equation A.5-26 is the characteristic equation of $\underline{\underline{T}}$.

Another set of invariants is defined as

$$I = \operatorname{tr} \underline{\underline{T}} = I_1, \tag{A.5-27a}$$

$$II = \operatorname{tr} \underline{\underline{T}}^2 = I_1^2 - 2I_2, \tag{A.5-27b}$$

$$\text{and} \quad III = \operatorname{tr} \underline{\underline{T}}^3 = \frac{1}{2}(6I_3 + 2I_1^3 - 6I_1 I_2). \tag{A.5-27c}$$

If $\underline{\underline{T}}$ is a symmetric tensor, then its eigenvalues are real and it is diagonalizable. That is, if λ_1, λ_2, and λ_3 are its eigenvalues, $\underline{\underline{T}}$ can be transformed to a diagonal matrix with λ_1, λ_2, and λ_3 as its diagonal elements. Then,

$$I_1 = \lambda_1 + \lambda_2 + \lambda_3, \tag{A.5-28a}$$

$$I_2 = \lambda_1 \lambda_2 + \lambda_1 \lambda_3 + \lambda_2 \lambda_3, \tag{A.5-28b}$$

$$\text{and} \quad I_3 = \lambda_1 \lambda_2 \lambda_3. \tag{A.5-28c}$$

Furthermore, if φ is a scalar function of $\underline{\underline{T}}$, then φ is a function of the invariants of $\underline{\underline{T}}$, which in turn is a function of λ_1, λ_2, and λ_3 in the case of a symmetric tensor $\underline{\underline{T}}$.

A.5 Isotropic Tensors, Objective Tensors, and Tensor-Valued Functions

We also need to consider tensor-valued functions of $\underline{\underline{T}}$, and we write

$$\underline{\underline{S}} = \underline{\underline{F}}(\underline{\underline{T}}) \quad \text{or} \quad S_{ij} = F_{ij}(T_{kl}). \tag{A.5-29}$$

If $\underline{\underline{S}}$ can be expressed as a polynomial in $\underline{\underline{T}}$, then the *Cayley-Hamilton* theorem can be used to simplify the representation of $\underline{\underline{S}}$. The Cayley-Hamilton theorem can be stated as follows: Every matrix satisfies its own characteristic equation.

Thus, $\underline{\underline{T}}$ satisfies equation A.5-26. That is,

$$-\underline{\underline{T}}^3 + I_1\underline{\underline{T}}^2 - I_2\underline{\underline{T}} + I_3\underline{\underline{\delta}} = 0. \tag{A.5-30}$$

Expanding $\underline{\underline{F}}$ as a polynomial in $\underline{\underline{T}}$, we have

$$\underline{\underline{S}} = \alpha_0\underline{\underline{\delta}} + \alpha_1\underline{\underline{T}} + \alpha_2\underline{\underline{T}}^2 + \alpha_3\underline{\underline{T}}^3 + \cdots + \alpha_n\underline{\underline{T}}^n, \tag{A.5-31}$$

where $\alpha_0, \alpha_1, \ldots, \alpha_n$ are constants.

Using the Cayley-Hamilton theorem (equation A.5-30), we find that $\underline{\underline{T}}^3$ can be replaced by $\underline{\underline{T}}^2, \underline{\underline{T}}, \underline{\underline{\delta}}$ and I_1, I_2, and I_3. Similarly, all powers of $\underline{\underline{T}}$ higher than two in equation A.5-31 can be replaced by $\underline{\underline{T}}^2, \underline{\underline{T}}, \underline{\underline{\delta}}$, and the three invariants of $\underline{\underline{T}}$. Thus, equation A.5-31 simplifies to

$$\underline{\underline{S}} = \beta_0\underline{\underline{\delta}} + \beta_1\underline{\underline{T}} + \beta_2\underline{\underline{T}}^2, \tag{A.5-32}$$

where β_0, β_1, and β_2 are functions of the invariants of $\underline{\underline{T}}$.

In continuum mechanics, it is not uncommon to restrict attention to isotropic materials, and $\underline{\underline{F}}$ is then an isotropic function; that is, $\underline{\underline{F}}$ is the same in all Cartesian coordinate systems. Thus, on rotating the axes, we have

$$\underline{\underline{\bar{S}}} = \underline{\underline{Q}} \cdot \underline{\underline{S}} \cdot \underline{\underline{Q}}^+ = \underline{\underline{Q}} \cdot \underline{\underline{F}}(\underline{\underline{T}}) \cdot \underline{\underline{Q}}^+ \tag{A.5-33a}$$

$$\text{and} \quad \underline{\underline{\bar{T}}} = \underline{\underline{Q}} \cdot \underline{\underline{T}} \cdot \underline{\underline{Q}}^+, \tag{A.5-33b}$$

where $\underline{\underline{Q}}$ is the orthogonal matrix defined in equation A.5-13.

Since $\underline{\underline{F}}$ is the same in both coordinate systems,

$$\underline{\underline{\bar{S}}} = \underline{\underline{F}}(\underline{\underline{\bar{T}}}). \tag{A.5-34}$$

Combining the two sets of equations we deduce

$$\underline{\underline{Q}} \cdot \underline{\underline{F}}(\underline{\underline{T}}) \cdot \underline{\underline{Q}}^+ = \underline{\underline{F}}(\underline{\underline{Q}} \cdot \underline{\underline{T}} \cdot \underline{\underline{Q}}^+). \tag{A.5-35}$$

Equation A.5-35 defines an isotropic second-order tensor-valued function $\underline{\underline{F}}$.

If $\underline{\underline{F}}$ is isotropic, and $\underline{\underline{S}}$ and $\underline{\underline{T}}$ are symmetric, equation A.5-32 is a representation of $\underline{\underline{S}}$ without requiring that it can be expressed as a polynomial in $\underline{\underline{T}}$.

We note that $\underline{\underline{T}}^2$ may be written in terms of $\underline{\underline{T}}, \underline{\underline{I}}$, and the inverse of $\underline{\underline{T}}(\underline{\underline{T}}^{-1})$, if it exists. On multiplying equation A.5-30 by $\underline{\underline{T}}^{-1}$, we obtain

$$-\underline{\underline{T}}^2 + I_1\underline{\underline{T}} - I_2\underline{\underline{I}} + I_3\underline{\underline{T}}^{-1} = 0. \tag{A.5-36}$$

Thus, an alternative representation of $\underline{\underline{S}}$ is

$$\underline{\underline{S}} = \gamma_0\underline{\underline{\delta}} + \gamma_1\underline{\underline{T}} + \gamma_{-1}\underline{\underline{T}}^{-1}, \tag{A.5-37}$$

where γ_0, γ_1, and γ_{-1} are functions of I_1, I_2, and I_3.

The function $\underline{\underline{F}}$ can be a function of more than one tensor. In the case where $\underline{\underline{F}}$ is an isotropic function of two symmetric tensors, $\underline{\underline{T}}_1$ and $\underline{\underline{T}}_2$,

$$\begin{aligned}\underline{\underline{S}} &= \underline{\underline{F}}(\underline{\underline{T}}_1, \underline{\underline{T}}_2) \\ &= \psi_0 \underline{\underline{I}} + \psi_1 \underline{\underline{T}}_1 + \psi_2 \underline{\underline{T}}_2 + \psi_3 \underline{\underline{T}}_1^2 + \psi_4 \underline{\underline{T}}_2^2 \\ &\quad + \psi_5(\underline{\underline{T}}_1 \cdot \underline{\underline{T}}_2 + \underline{\underline{T}}_2 \cdot \underline{\underline{T}}_1) + \psi_6(\underline{\underline{T}}_1^2 \cdot \underline{\underline{T}}_2 + \underline{\underline{T}}_2 \cdot \underline{\underline{T}}_1^2) + \psi_7(\underline{\underline{T}}_1 \cdot \underline{\underline{T}}_2^2 + \underline{\underline{T}}_2^2 \cdot \underline{\underline{T}}_1) \\ &\quad + \psi_8(\underline{\underline{T}}_1^2 \cdot \underline{\underline{T}}_2^2 + \underline{\underline{T}}_2^2 \cdot \underline{\underline{T}}_1^2). \end{aligned} \qquad (A.5\text{-}38)$$

The quantities $\psi_0, \ldots, \psi_8$ are functions of the invariants of $\underline{\underline{T}}_1$, $\underline{\underline{T}}_2$, and their products.

The ten principal invariants are

$$\text{tr}(\underline{\underline{T}}_i), \quad \text{tr}(\underline{\underline{T}}_i^2), \quad \text{tr}(\underline{\underline{T}}_i^3), \quad (i = 1, 2)$$
$$\text{tr}(\underline{\underline{T}}_1 \cdot \underline{\underline{T}}_2), \quad \text{tr}(\underline{\underline{T}}_1^2 \cdot \underline{\underline{T}}_2), \quad \text{tr}(\underline{\underline{T}}_1 \cdot \underline{\underline{T}}_2^2), \quad \text{tr}(\underline{\underline{T}}_1^2 \cdot \underline{\underline{T}}_2^2).$$

In extending from one tensor to two tensors, we have increased the number of functions from three to eight. The number of arguments of each function has increased from three to ten.

The results for a function of an arbitrary number of tensors can be found in Truesdell and Noll (1965).

Example A.5-2   Constitutive Equation for a Stokesian Fluid

A Stokesian fluid is a fluid whose deviatoric stress tensor $\underline{\underline{\sigma}}$ depends on the rate-of-deformation tensor $\underline{\underline{\dot{\gamma}}}$. Obtain a representation for $\underline{\underline{\sigma}}$.

Solution

$\underline{\underline{\sigma}}$ is a tensor-valued function of $\underline{\underline{\dot{\gamma}}}$, and if we assume that $\underline{\underline{\sigma}}$ can be expanded as a power series of $\underline{\underline{\dot{\gamma}}}$, we obtain equation A.5-32, which in this case is written as

$$\underline{\underline{\sigma}} = \beta_1 \underline{\underline{\dot{\gamma}}} + \beta_2 \underline{\underline{\dot{\gamma}}}^2. \qquad (A.5\text{-}39)$$

The term $\beta_0 \underline{\underline{\delta}}$ has been dropped since we are considering the deviatoric stress.

A Stokesian fluid is isotropic. Since both $\underline{\underline{\sigma}}$ and $\underline{\underline{\dot{\gamma}}}$ are symmetric, equation A.5-39 is an exact representation of $\underline{\underline{\sigma}}$. If the fluid is incompressible, the first invariant of $\underline{\underline{\dot{\gamma}}}$ is zero, because the equation of continuity has to be satisfied. In some flows, such as shear flows, the third invariant is also zero. Thus β_1 and β_2 are functions of at most one invariant of $\underline{\underline{\dot{\gamma}}}$.

As a matter of historical interest, we observe that Stokes proposed his constitutive equation in 1845, Hamilton stated the Cayley-Hamilton theorem in 1853 for a special class of matrices, generalized by Cayley in 1858, but it was only in 1945 that Reiner combined both results to obtain equation A.5-39. Rivlin obtained equation A.5-39 two years later without requiring the polynomial approximation. It is thus not surprising that equation A.5-39 is known as the constitutive equation of a Reiner-Rivlin fluid.

It is perhaps appropriate to close this Appendix by observing that Lord Kelvin (M.J. Crowe, 1967) did not believe vectors would be of the slightest use to any creature. Can we imagine a present-day physicist not using vectors at all? Many engineers believe tensors are of no use. Are they better prophets than Lord Kelvin?

A.6 Problems

A.6-1ₐ Rotation of Axes

The set of axes $O\bar{x}\bar{y}\bar{z}$ is obtained by rotating the set of axes $Oxyz$ through an angle of 60° about the z-axis, the direction of rotation being from the x-axis to the y-axis. Write down the set of direction cosines which corresponds to this rotation. Hence, show that

$$\begin{bmatrix} \bar{x} \\ \bar{y} \\ \bar{z} \end{bmatrix} = \begin{bmatrix} \dfrac{1}{2} & \dfrac{\sqrt{3}}{2} & 0 \\ -\dfrac{\sqrt{3}}{2} & \dfrac{1}{2} & 0 \\ 0 & 0 & 1 \end{bmatrix} \begin{bmatrix} x \\ y \\ z \end{bmatrix}. \tag{A.6-1}$$

If the coordinates of a point P are $x=1$, $y=2$, $z=3$, find $\bar{x}$, $\bar{y}$, $\bar{z}$. Calculate $\sqrt{x^2+y^2+z^2}$ and $\sqrt{\bar{x}^2+\bar{y}^2+\bar{z}^2}$. Explain your result.

A.6-2ₐ Contraction

Let $m=n$ in equation A.2-41. Using the summation convention, show that T_n^n is a scalar. This process of setting a superscript equal to a subscript is known as the process of contraction.

A.6-3ₐ Quotient Law

Let T_{ij} be a covariant tensor of order two. $v_{(j)}$ is a quantity whose tensorial properties are not known. But the inner product $T_{ij}v_{(j)}$ is known to be a covariant vector (tensor of order one). Show that $v_{(j)}$ is a contravariant vector. That is, $v_{(j)}$ transforms as given by equation A.2-19.

This method of ascertaining whether a quantity is a tensor is known as the quotient law.

A.6-4ₐ Transformation Rule for the Contravariant Components of a Second-Order Tensor

Use the chain rule to prove that

$$\bar{T}^{mn} = \frac{\partial \bar{q}^m}{\partial q^r}\frac{\partial \bar{q}^n}{\partial q^s} T^{rs}. \tag{A.6-2}$$

A.6-5a Christoffel Symbols

Show that the only nonzero Christoffel symbols of the second kind corresponding to the cylindrical polar coordinate system (r, θ, z) are $\begin{Bmatrix} r \\ \theta\,\theta \end{Bmatrix}$ and $\begin{Bmatrix} \theta \\ r\,\theta \end{Bmatrix}$. Determine them.

The transformation from the Cartesian coordinate system (x^1, x^2, x^3) to the cylindrical coordinate system (r, θ, z) is

$$x^1 = r\cos\theta, \quad x^2 = r\sin\theta, \quad x^3 = z.$$

The (r, θ, z) system is orthogonal and the metric tensors are $g_{rr} = g_{zz} = 1$, $g_{\theta\theta} = r^2$.

A.6-6a Cylindrical Coordinates

(i) Obtain the covariant and the contravariant components of a vector $\underline{v}$ and of the metric tensor $\underline{\underline{g}}$. Express your answers in terms of the physical components $v_{(r)}$, $v_{(\theta)}$, and $v_{(z)}$.

(ii) Compute all the Christoffel symbols.

A.6-7a Covariant Derivative

Write the covariant derivatives with respect to q^k of each of the following:

(i) $g_{ij}v^j$ (ii) $g^{ij}v_j$ (iii) $T_{ij}v^j$

A.6-8a Physical Components

Calculate the physical components of grad φ in the cylindrical polar coordinate system (see Problem A.6-5).

A.6-9b Divergence Theorem

The symmetric stress tensor π_{ij} satisfies the equation

$$\pi_{ij,j} = 0. \tag{A.6-3}$$

Show that

$$\frac{1}{2}\int_V \pi_{ij}\dot{\gamma}^{ij}\,dV = \int_S V^i \pi_{ij} n^j\,dS, \tag{A.6-4}$$

where V is the volume enclosed by the surface S, V^i are the velocity components and,

$$\dot{\gamma}_{ij} = V_{i,j} + V_{j,i}. \tag{A.6-5}$$

A.6-10$_b$ Isotropic Tensor

By considering the transformation of coordinates given in Problem A.6-1, show that if T_{ij} is an isotropic tensor, it is necessarily of the form $\lambda \delta_{ij}$.

A.6-11$_b$ Objectivity

The rate-of-deformation tensor $\underline{\underline{\dot{\gamma}}}$ has been shown to be objective. Show that

(i)

$$\frac{D\dot{\gamma}^{ij}}{Dt} = \frac{\partial \dot{\gamma}^{ij}}{\partial t} + V^s \dot{\gamma}^{ij}_{,s} \quad \text{is not objective, and} \tag{A.6-6}$$

(ii)

$$\frac{\delta \dot{\gamma}^{ij}}{\delta t} = \frac{\partial \dot{\gamma}^{ij}}{\partial t} + V^s \frac{\partial \dot{\gamma}^{ij}}{\partial x^s} - \frac{\partial V^i}{\partial x^s} \dot{\gamma}^{sj} - \frac{\partial V^j}{\partial x^s} \dot{\gamma}^{ij} \quad \text{is objective.} \tag{A.6-7}$$

Note that in (ii) we have ordinary partial derivatives, and not covariant derivatives as in (i). Verify that the Christoffel symbols cancel out. The derivative defined in (ii) is known as the Oldroyd contravariant upper-convected derivative.

A.6-12$_a$ Invariants

The rate-of-deformation tensors $\underline{\underline{\dot{\gamma}}}$ for (i) a simple shear flow and (ii) a uniaxial elongational flow, are given respectively by

(i)

$$\underline{\underline{\dot{\gamma}}} = \dot{\gamma} \begin{bmatrix} 0 & 1 & 0 \\ 1 & 0 & 0 \\ 0 & 0 & 0 \end{bmatrix}, \quad \text{and} \tag{A.6-8}$$

(ii)

$$\underline{\underline{\dot{\gamma}}} = \dot{\varepsilon} \begin{bmatrix} -1 & 0 & 0 \\ 0 & -1 & 0 \\ 0 & 0 & 2 \end{bmatrix}. \tag{A.6-9}$$

Compute the principal invariants of $\underline{\underline{\dot{\gamma}}}$ for the two flows. Answer: (i) $2\dot{\gamma}^2$; (ii) $6\dot{\varepsilon}^2$.

A.6-13$_b$ Tensor-Valued Function

The scalar function $\phi(I_1, I_2)$ is a function of the first two principal invariants of a second-order tensor $\underline{\underline{T}}$. Working with Cartesian components, show that

$$\frac{\partial \phi}{\partial T_{rs}} = \frac{\partial \phi}{\partial I_1}\delta_{rs} + \frac{\partial \phi}{\partial I_2}(I_1\delta_{rs} - T_{rs}). \tag{A.6-10}$$

A.6-14$_b$ Elongational Flow

The flow into a die can be considered to be elongational (see Fig. A.6-1). In the case of an axisymmetrical geometry, the velocity gradient in the z-direction is given by

$$\frac{dV_z}{dz} = \dot{\varepsilon}. \tag{A.6-11a}$$

(i) Obtain the components of the velocity gradient tensor for the flow of an incompressible fluid, in terms of $\dot{\varepsilon}$.
(ii) Compute the three invariants of the velocity gradient tensor.

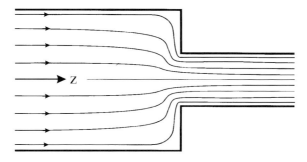

Figure A.6-1 Die entry flow

Appendix B Equations of Change

B.1 The Equation of Continuity in Three Coordinate Systems 487
B.2 The Equation of Motion in Rectangular Coordinates (x, y, z) 487
    B.2-1 In Terms of $\underline{\underline{\sigma}}$ 487
    B.2-2 In Terms of Velocity Gradients for a Newtonian Fluid with Constant ρ and μ . . . 488

B.3 The Equation of Motion in Cylindrical Coordinates (r, θ, z) 488
    B.3-1 In Terms of $\underline{\underline{\sigma}}$ 488
    B.3-2 In Terms of Velocity Gradients for a Newtonian Fluid with Constant ρ and μ . . . 489

B.4 The Equation of Motion in Spherical Coordinates (r, θ, ϕ) 490
    B.4-1 In Terms of $\underline{\underline{\sigma}}$ 490
    B.4-2 In Terms of Velocity Gradients for a Newtonian Fluid with Constant ρ and μ . . . 490

B.1 The Equation of Continuity in Three Coordinate Systems

Rectangular coordinates (x, y, z):

$$\frac{\partial \rho}{\partial t} + \frac{\partial}{\partial x}(\rho V_x) + \frac{\partial}{\partial y}(\rho V_y) + \frac{\partial}{\partial z}(\rho V_z) = 0 \tag{B.1-1}$$

Cylindrical coordinates (r, θ, z):

$$\frac{\partial \rho}{\partial t} + \frac{1}{r}\frac{\partial}{\partial r}(\rho r V_r) + \frac{1}{r}\frac{\partial}{\partial \theta}(\rho V_\theta) + \frac{\partial}{\partial z}(\rho V_z) = 0 \tag{B.1-2}$$

Spherical coordinates (r, θ, ϕ):

$$\frac{\partial \rho}{\partial t} + \frac{1}{r^2}\frac{\partial}{\partial r}(\rho r^2 V_r) + \frac{1}{r\sin\theta}\frac{\partial}{\partial \theta}(\rho V_\theta \sin\theta) + \frac{1}{r\sin\theta}\frac{\partial}{\partial \phi}(\rho V_\phi) = 0 \tag{B.1-3}$$

B.2 The Equation of Motion in Rectangular Coordinates (x, y, z)

B.2-1 In Terms of $\underline{\underline{\sigma}}$

x-component:

$$\rho\left(\frac{\partial V_x}{\partial t} + V_x\frac{\partial V_x}{\partial x} + V_y\frac{\partial V_x}{\partial y} + V_z\frac{\partial V_x}{\partial z}\right) = -\frac{\partial P}{\partial x} - \left(\frac{\partial \sigma_{xx}}{\partial x} + \frac{\partial \sigma_{yx}}{\partial y} + \frac{\partial \sigma_{zx}}{\partial z}\right) + \rho g_x \tag{B.2-1}$$

y-component:

$$\rho\left(\frac{\partial V_y}{\partial t} + V_x\frac{\partial V_y}{\partial x} + V_y\frac{\partial V_y}{\partial y} + V_z\frac{\partial V_y}{\partial z}\right) = -\frac{\partial P}{\partial y} - \left(\frac{\partial \sigma_{xy}}{\partial x} + \frac{\partial \sigma_{yy}}{\partial y} + \frac{\partial \sigma_{zy}}{\partial z}\right) + \rho g_y \quad \text{(B.2-2)}$$

z-component:

$$\rho\left(\frac{\partial V_z}{\partial t} + V_x\frac{\partial V_z}{\partial x} + V_y\frac{\partial V_z}{\partial y} + V_z\frac{\partial V_z}{\partial z}\right) = -\frac{\partial P}{\partial z} - \left(\frac{\partial \sigma_{xz}}{\partial x} + \frac{\partial \sigma_{yz}}{\partial y} + \frac{\partial \sigma_{zz}}{\partial z}\right) + \rho g_z \quad \text{(B.2-3)}$$

B.2-2 In Terms of Velocity Gradients for a Newtonian Fluid with Constant ρ and μ

x-component:

$$\rho\left(\frac{\partial V_x}{\partial t} + V_x\frac{\partial V_x}{\partial x} + V_y\frac{\partial V_x}{\partial y} + V_z\frac{\partial V_x}{\partial z}\right) = -\frac{\partial P}{\partial x} + \mu\left(\frac{\partial^2 V_x}{\partial x^2} + \frac{\partial^2 V_x}{\partial y^2} + \frac{\partial^2 V_x}{\partial z^2}\right) + \rho g_x \quad \text{(B.2-4)}$$

y-component:

$$\rho\left(\frac{\partial V_y}{\partial t} + V_x\frac{\partial V_y}{\partial x} + V_y\frac{\partial V_y}{\partial y} + V_z\frac{\partial V_y}{\partial z}\right) = -\frac{\partial P}{\partial y} + \mu\left(\frac{\partial^2 V_y}{\partial x^2} + \frac{\partial^2 V_y}{\partial y^2} + \frac{\partial^2 V_y}{\partial z^2}\right) + \rho g_y \quad \text{(B.2-5)}$$

z-component:

$$\rho\left(\frac{\partial V_z}{\partial t} + V_x\frac{\partial V_z}{\partial x} + V_y\frac{\partial V_z}{\partial y} + V_z\frac{\partial V_z}{\partial z}\right) = -\frac{\partial P}{\partial z} + \mu\left(\frac{\partial^2 V_z}{\partial x^2} + \frac{\partial^2 V_z}{\partial y^2} + \frac{\partial^2 V_z}{\partial z^2}\right) + \rho g_z \quad \text{(B.2-6)}$$

B.3 The Equation of Motion in Cylindrical Coordinates (r, θ, z)

B.3-1 In Terms of $\underline{\underline{\sigma}}$

r-component:

$$\rho\left(\frac{\partial V_r}{\partial t} + V_r\frac{\partial V_r}{\partial r} + \frac{V_\theta}{r}\frac{\partial V_r}{\partial \theta} - \frac{V_\theta^2}{r} + V_z\frac{\partial V_r}{\partial z}\right)$$
$$= -\frac{\partial P}{\partial r} - \left(\frac{1}{r}\frac{\partial}{\partial r}(r\sigma_{rr}) + \frac{1}{r}\frac{\partial \sigma_{r\theta}}{\partial \theta} - \frac{\sigma_{\theta\theta}}{r} + \frac{\partial \sigma_{rz}}{\partial z}\right) + \rho g_r \quad \text{(B.3-1)}$$

θ-component:

$$\rho\left(\frac{\partial V_\theta}{\partial t} + V_r\frac{\partial V_\theta}{\partial r} + \frac{V_\theta}{r}\frac{\partial V_\theta}{\partial \theta} + \frac{V_r V_\theta}{r} + V_z\frac{\partial V_\theta}{\partial z}\right)$$

$$= -\frac{1}{r}\frac{\partial P}{\partial \theta} - \left(\frac{1}{r^2}\frac{\partial}{\partial r}(r^2\sigma_{r\theta}) + \frac{1}{r}\frac{\partial \sigma_{\theta\theta}}{\partial \theta} + \frac{\partial \sigma_{\theta z}}{\partial z}\right) + \rho g_\theta \quad \text{(B.3-2)}$$

z-component:

$$\rho\left(\frac{\partial V_z}{\partial t} + V_r\frac{\partial V_z}{\partial r} + \frac{V_\theta}{r}\frac{\partial V_z}{\partial \theta} + V_z\frac{\partial V_z}{\partial z}\right)$$

$$= -\frac{\partial P}{\partial z} - \left(\frac{1}{r}\frac{\partial}{\partial r}(r\sigma_{rz}) + \frac{1}{r}\frac{\partial \sigma_{\theta z}}{\partial \theta} + \frac{\partial \sigma_{zz}}{\partial z}\right) + \rho g_z \quad \text{(B.3-3)}$$

B.3-2 In Terms of Velocity Gradients for a Newtonian Fluid with Constant ρ and μ

r-component:

$$\rho\left(\frac{\partial V_r}{\partial t} + V_r\frac{\partial V_r}{\partial r} + \frac{V_\theta}{r}\frac{\partial V_r}{\partial \theta} - \frac{V_\theta^2}{r} + V_z\frac{\partial V_r}{\partial z}\right)$$

$$= -\frac{\partial P}{\partial r} + \mu\left[\frac{\partial}{\partial r}\left(\frac{1}{r}\frac{\partial}{\partial r}(rV_r)\right) + \frac{1}{r^2}\frac{\partial^2 V_r}{\partial \theta^2} - \frac{2}{r^2}\frac{\partial V_\theta}{\partial \theta} + \frac{\partial^2 V_r}{\partial z^2}\right] + \rho g_r \quad \text{(B.3-4)}$$

θ-component:

$$\rho\left(\frac{\partial V_\theta}{\partial t} + V_r\frac{\partial V_\theta}{\partial r} + \frac{V_\theta}{r}\frac{\partial V_\theta}{\partial \theta} + \frac{V_r V_\theta}{r} + V_z\frac{\partial V_\theta}{\partial z}\right)$$

$$= -\frac{1}{r}\frac{\partial P}{\partial \theta} + \mu\left[\frac{\partial}{\partial r}\left(\frac{1}{r}\frac{\partial}{\partial r}(rV_\theta)\right) + \frac{1}{r^2}\frac{\partial^2 V_\theta}{\partial \theta^2} + \frac{2}{r^2}\frac{\partial V_r}{\partial \theta} + \frac{\partial^2 V_\theta}{\partial z^2}\right] + \rho g_\theta \quad \text{(B.3-5)}$$

z-component:

$$\rho\left(\frac{\partial V_z}{\partial t} + V_r\frac{\partial V_z}{\partial r} + \frac{V_\theta}{r}\frac{\partial V_z}{\partial \theta} + V_z\frac{\partial V_z}{\partial z}\right)$$

$$= -\frac{\partial P}{\partial z} + \mu\left[\frac{1}{r}\frac{\partial}{\partial r}\left(r\frac{\partial V_z}{\partial r}\right) + \frac{1}{r^2}\frac{\partial^2 V_z}{\partial \theta^2} + \frac{\partial^2 V_z}{\partial z^2}\right] + \rho g_z \quad \text{(B.3-6)}$$

490   Equations of Change

B.4 The Equation of Motion in Spherical Coordinates (r, θ, ϕ)

B.4-1 In Terms of $\underline{\underline{\sigma}}$

r-component:

$$\rho\left(\frac{\partial V_r}{\partial t} + V_r\frac{\partial V_r}{\partial r} + \frac{V_\theta}{r}\frac{\partial V_r}{\partial \theta} + \frac{V_\phi}{r\sin\theta}\frac{\partial V_r}{\partial \phi} - \frac{V_\theta^2 + V_\phi^2}{r}\right)$$

$$= -\frac{\partial P}{\partial r} - \left(\frac{1}{r^2}\frac{\partial}{\partial r}(r^2\sigma_{rr}) + \frac{1}{r\sin\theta}\frac{\partial}{\partial \theta}(\sigma_{r\theta}\sin\theta) + \frac{1}{r\sin\theta}\frac{\partial \sigma_{r\phi}}{\partial \phi} - \frac{\sigma_{\theta\theta} + \sigma_{\phi\phi}}{r}\right) + \rho g_r \quad \text{(B.4-1)}$$

θ-component:

$$\rho\left(\frac{\partial V_\theta}{\partial t} + V_r\frac{\partial V_\theta}{\partial r} + \frac{V_\theta}{r}\frac{\partial V_\theta}{\partial \theta} + \frac{V_\phi}{r\sin\theta}\frac{\partial V_\theta}{\partial \phi} + \frac{V_r V_\theta}{r} - \frac{V_\phi^2 \cot\theta}{r}\right)$$

$$= -\frac{1}{r}\frac{\partial P}{\partial \theta} - \left(\frac{1}{r^2}\frac{\partial}{\partial r}(r^2\sigma_{r\theta}) + \frac{1}{r\sin\theta}\frac{\partial}{\partial \theta}(\sigma_{\theta\theta}\sin\theta) + \frac{1}{r\sin\theta}\frac{\partial \sigma_{\theta\phi}}{\partial \phi} + \frac{\sigma_{r\theta}}{r} - \frac{\cot\theta}{r}\sigma_{\phi\phi}\right) + \rho g_\theta$$

(B.4-2)

ϕ-component:

$$\rho\left(\frac{\partial V_\phi}{\partial t} + V_r\frac{\partial V_\phi}{\partial r} + \frac{V_\theta}{r}\frac{\partial V_\phi}{\partial \theta} + \frac{V_\phi}{r\sin\theta}\frac{\partial V_\phi}{\partial \phi} + \frac{V_\phi V_r}{r} + \frac{V_\theta V_\phi}{r}\cot\theta\right)$$

$$= -\frac{1}{r\sin\theta}\frac{\partial P}{\partial \phi} - \left(\frac{1}{r^2}\frac{\partial}{\partial r}(r^2\sigma_{r\phi}) + \frac{1}{r}\frac{\partial \sigma_{\theta\phi}}{\partial \theta} + \frac{1}{r\sin\theta}\frac{\partial \sigma_{\phi\phi}}{\partial \phi} + \frac{\sigma_{r\phi}}{r} + \frac{2\cot\theta}{r}\sigma_{\theta\phi}\right) + \rho g_\phi$$

(B.4-3)

B.4-2 In Terms of Velocity Gradients for a Newtonian Fluid with Constant ρ and μ[a]

r-component:

$$\rho\left(\frac{\partial V_r}{\partial t} + V_r\frac{\partial V_r}{\partial r} + \frac{V_\theta}{r}\frac{\partial V_r}{\partial \theta} + \frac{V_\phi}{r\sin\theta}\frac{\partial V_r}{\partial \phi} - \frac{V_\theta^2 + V_\phi^2}{r}\right)$$

$$= -\frac{\partial P}{\partial r} + \mu\left(\nabla^2 V_r - \frac{2}{r^2}V_r - \frac{2}{r^2}\frac{\partial V_\theta}{\partial \theta} - \frac{2}{r^2}V_\theta\cot\theta - \frac{2}{r^2\sin\theta}\frac{\partial V_\phi}{\partial \phi}\right) + \rho g_r \quad \text{(B.4-4)}$$

[a] In these equations

$$\nabla^2 = \frac{1}{r^2}\frac{\partial}{\partial r}\left(r^2\frac{\partial}{\partial r}\right) + \frac{1}{r^2\sin\theta}\frac{\partial}{\partial \theta}\left(\sin\theta\frac{\partial}{\partial \theta}\right) + \frac{1}{r^2\sin^2\theta}\left(\frac{\partial^2}{\partial \phi^2}\right)$$

θ-component:

$$\rho\left(\frac{\partial V_\theta}{\partial t} + V_r\frac{\partial V_\theta}{\partial r} + \frac{V_\theta}{r}\frac{\partial V_\theta}{\partial \theta} + \frac{V_\phi}{r\sin\theta}\frac{\partial V_\theta}{\partial \phi} + \frac{V_r V_\theta}{r} - \frac{V_\phi^2 \cot\theta}{r}\right)$$

$$= -\frac{1}{r}\frac{\partial P}{\partial \theta} + \mu\left(\nabla^2 V_\theta + \frac{2}{r^2}\frac{\partial V_r}{\partial \theta} - \frac{V_\theta}{r^2\sin^2\theta} - \frac{2\cos\theta}{r^2\sin^2\theta}\frac{\partial V_\phi}{\partial \phi}\right) + \rho g_\theta \qquad (B.4\text{-}5)$$

ϕ-component:

$$\rho\left(\frac{\partial V_\phi}{\partial t} + V_r\frac{\partial V_\phi}{\partial r} + \frac{V_\theta}{r}\frac{\partial V_\phi}{\partial \theta} + \frac{V_\phi}{r\sin\theta}\frac{\partial V_\phi}{\partial \phi} + \frac{V_\phi V_r}{r} + \frac{V_\theta V_\phi}{r}\cot\theta\right)$$

$$= -\frac{1}{r\sin\theta}\frac{\partial P}{\partial \phi} + \mu\left(\nabla^2 V_\phi - \frac{V_\phi}{r^2\sin^2\theta} + \frac{2}{r^2\sin\theta}\frac{\partial V_r}{\partial \phi} + \frac{2\cos\theta}{r^2\sin^2\theta}\frac{\partial V_\theta}{\partial \phi}\right) + \rho g_\phi \qquad (B.4\text{-}6)$$

References

Abid, M., Xuereb, C., and Bertrand, J., *Chem. Eng. Res. Des.*, (1992) 70, p. 377
Acharya, A., Mashelkar, R.A., and Ulbrecht, J.J., *Can. J. Chem. Eng.*, (1978) 56, p. 19
Acierno, D., La Mantia, F.P., Marrucci, G., and Titomanlio, G., *J. Non-Newt. Fluid Mech.*, (1976) 1, p. 125 and 147
Acrivos, A., *AIChE J.*, (1960) 6, p. 584
Acrivos, A., Shah, M.J., and Petersen, E.E., *AIChE J.*, (1960) 6, p. 312
Adler, P.M., Nadim, A., and Brenner, H., *Adv. Chem. Eng.*, (1990) 15, p. 1
Agassant, J.F., Avenas, P., Sergent, J.P., and Carreau, P.J., *Polymer Processing: Principles and Modeling* (1991) Hanser Publishers, Munich
Ait-Kadi, A., Carreau, P.J., and Chauveteau, G., *J. Rheol.*, (1987) 31, p. 537
Ait-Kadi, A., Grmela, M., and Carreau, P.J., *Rheol. Acta*, (1988) 27, p. 241
Ait-Kadi, A., Choplin, L., and Carreau, P.J., *Polym. Eng. Sci.*, (1989) 29, p. 1265
Ajji, A., Carreau, P.J., Grmela, M., and Schreiber, H.P., *J. Rheol.*, (1989) 33, p. 401
Al-Farris, T.F., and Pinder, K.L., *Can. J. Chem. Eng.*, (1987) 65, p. 391
Alfrey, T., and Dotey, P., *J. Applied Phys.*, (1945) 16, p. 700
Allen, E., and Uhlherr, P.H.T., *J. Rheol.*, (1989) 33, p. 627
Amato, W.S., and Tien, C., *Int. J. Heat Mass Transf.*, (1976) 19, p. 1257
Ashare, E., *Trans. Soc. Rheol.*, (1968) 12, p. 535
Astarita, G., Marrucci, G., and Palumbo, G., *Ind. Eng. Chem. Fundam.*, (1964) 3, p. 333
Astarita, G., and Appuzzo, G., *AIChE J.*, (1965) 11, p. 815
Astarita, G., *Ind. Eng. Chem. Fundam.*, (1996) 5, p. 14
Astarita, G., *J. Non-Newt. Fluid Mech.*, (1979), 4, p. 285
Atapattu, D.D., Chhabra, R.P., and Uhlherr, P.H.T., *J. Non-Newt. Fluid Mech.*, (1990) 38, p. 31
Attané, P., Le Roy, P., and Turrel, G., *J. Non-Newt. Fluid Mech.*, (1980) 6, p. 269
Attané, P., Le Roy, P., Pierrard, J.M., and Turrel, G., *J. Non- Newt. Fluid Mech.*, (1981) 9, p. 13
Attané, P., *Comportement en cisaillement de solutions polymériques enchevêtrées: adimensionnalisation et tests de lois rhéologiques* (1984) Ph.D. thesis, Institut National Polytechnique de Grenoble, France
Attané, P., Pierrard, J.M., and Turrel, G., *J. Non-Newt. Fluid Mech.*, (1985a) 18, p. 295
Attané, P., Pierrard, J.M., and Turrel, G., *J. Non-Newt. Fluid Mech.*, (1985b) 18, p. 319
Attané, P., Turrel, G., Pierrard, J.M., and Carreau, P.J., *Rheol.*, (1988) 32, p. 23
Ausias, G., Agassant, J.F., Vincent, M., Carreau, P.J., and Lafleur, P.G., *J. Rheol.*, (1992) 36, p. 525
Ayazi Shamlou, P., and Edwards, M.F., *Chem. Eng. Sci.*, (1986) 41, p. 1957
Bagley, E.B., *J. Appl. Phys.*, (1957) 28, p. 264
Bailey, F., and Koleske, J., *Poly(ethylene oxide)* (1976) Academic Press, New York, NY
Baird, D.G., and Ballman, R.L., *J. Rheol.*, (1979) 23, p. 505
Ballenger, I.F., Chen, I.J., Crowder, J.W., Hagler, G.E., Bogue, D.C., and White, J.L., *Trans. Soc. Rheol.*, (1971) 15, p. 195
Banerjee, S., *Chem. Eng. Sci.*, (1992) 47, p. 1793
Barnes, H.A., Hutton, J.F., and Walters, K., *An Introduction to Rheology* (1989) Elsevier, New York, NY
Bartram, E., Goldsmith, H.L., and Mason, S.G., *Rheol. Acta*, (1975) 14, p. 1063
Bates, R.L., Fondy, P.L., and Corpstein, R.R., *Ind. Eng. Chem., Proc. Des. Dev.*, (1963) 2, p. 310
Beckner, J.L., and Smith, J.M., *Trans. Inst. Chem. Engrs.*, (1966) 44, p. 224
Beris, A.N., Tsamopoulos, J., Armstrong, R.C., and Brown, R.A., *J. Fluid Mech.*, (1985) 158, p. 219
Bernstein, B., Kearsley, E.A., and Zapas, L.J., *Trans. Soc. Rheol.*, (1963) 7, p. 391
Bhavaraju, S.M., Mashelkar, R.A., and Blanch, H.W., *AIChE J.*, (1978) 24, p. 1063
Bingham, E.C., *Fluidity and Plasticity* (1922) McGraw-Hill, New York, NY

Bird, R.B., Stewart, W.E., and Lightfoot, E.N., *Transport Phenomena* (1960) John Wiley & Sons, New York, NY
Bird, R.B., and Carreau, P.J., *Chem. Eng. Sci.*, (1968) 23, p. 427
Bird, R.B., Hassager, O., and Abdel-Khalik, S.I., *AIChE J.*, (1974) 20, p. 1041
Bird, R.B., Armstrong, R.C., and Hassager, O., *Dynamics of Polymeric Liquids, Vol. 1*, Fluid Mechanics (1977) Wiley, New York, NY
Bird, R.B., Dotson, P.J., and Johnson, N.L., *J. Non-Newt. Fluid Mech.*, (1980) 7, p. 213
Bird, R.B., Saab, H.H., and Curtiss, C.F., *J. Phys. Chem.*, (1982) 86, p. 1102 and *J. Chem. Phys.*, (1982) 77, p. 4747
Bird, R.B., Armstrong, R.C., and Hassager, O., *Dynamics of Polymeric Liquids, Vol. 1*, Fluid Mechanics, Second edition (1987) Wiley, New York, NY
Bird, R.B., Curtiss, C.F., Armstrong, R.C., and Hassager, O., *Dynamics of Polymeric Liquids, Vol. 2*, Kinetic Theory, Second edition (1987) Wiley, New York, NY
Bird, R.B., and Wiest, J.M., *Annual Review of Fluid Mechanics*, (1995) 27, p. 169
Bizzell, G.D., and Slattery, J.C., *Chem. Eng. Sci.*, (1962) 17, p. 777
Boger, D.V., and Ramamurthy, A.V., *Rheol. Acta*, (1972) 11, p. 61
Boger, D.V., *J. Non-Newt. Fluid Mech.*, (1977) 3, p. 87
Boger, D.V., and Denn, M.M., *J. Non-Newt. Fluid Mech.*, (1980) 6, p. 163
Boger, D.V., and Walters, K., *Rheological Phenomena in Focus* (1993) Elsevier, New York, NY
Bourne, J.R., and Butler, H., *Trans. Inst. Chem. Engrs.*, (1969) 47, p. 11
Bourne, J.R., Buerli, M., and Regenass, W., *Chem. Eng. Sci.*, (1981) 36, p. 782
Bousmina, M., *Etude des corrélations entre la morphologie et les propriétés rhéologiques linéaires et non-linéaires de mélanges de polymères incompatibles à l'état fondu* (1992) Ph.D. thesis, Université Louis-Pasteur de Strasbourg, France
Bousmina, M., and Muller, R., *J. Rheol.*, (1993) 37, p. 663
Bousmina, M., Bataille, P.B., Sapieha, S., and Schreiber, H.P., *J. Rheol.*, (1995) 39, p. 499
Brahimi, B., Ait-Kadi, A., Ajji, A., Jerome, R., and Fayt, R., *J. Rheol.*, (1991) 35, p. 1069
Brinkman, H.C., *Appl. Sci. Res.*, (1947) A1, p. 27. Also see *ibid.*, (1948) A1, p. 81
Brito, E., Leuliet, J.C., Choplin, L., and Tanguy, P., *Chem. Eng. Res. Des.*, (1991) 69, p. 324
Brunn, P.O., *J. Fluid Mech.*, (1977) 82, p. 529
Brunn, P.O., *J. Non-Newt. Fluid Mech.*, (1980) 7, p. 271
Bueche, F., *J. Chem. Phys.*, (1952) 29, p. 1959
Calderbank, P.H., and Moo-Young, M.B., *Trans. Inst. Chem. Engrs.*, (1959) 37, p. 26
Calderbank, P.H., and Moo-Young, M.B., *Trans. Inst. Chem. Engrs.*, (1961) 39, p. 337
Cameron, A., *Principles of lubrication* (1966) John Wiley & Sons, New York, NY
Carman, P.C., *Flow of Gases through porous media* (1956) Butterworths, London, UK
Carreau, P.J., *Trans. Soc. Rheol.*, (1972) 16, p. 99
Carreau, P.J., Devic, M., and Kapellas, M., *Rheol. Acta*, (1974) 13, p. 477
Carreau, P.J., Patterson, I., and Yap, C.Y., *Can. J. Chem. Eng.*, (1976) 54, p. 135
Carreau, P.J., Bui, Q.H., and Leroux, P., *Rheol. Acta*, (1979a) 18, p. 600
Carreau, P.J., De Kee, D., and Daroux, M., *Can. J. Chem. Eng.*, (1979b) 57, p. 135
Carreau, P.J., Choplin, L., and Clermont, J.R., *Polym. Eng. Sci.*, (1985) 25, p. 669
Carreau, P.J., Grmela, M., and Rollin, A., in *Recent Developements in Structured Continua*. D. De Kee and P.N. Kaloni (Eds.) (1990) Longman Scientific and Technical, New York, NY, Chapter 6
Carreau, P.J., in *Transport Processes in Bubbles, Drops and Particles*. Chhabra, R.P. and D. De Kee (Eds.) (1992) Hemisphere, New York, NY, Chapter 8
Carreau, P.J., Paris, J., and Guérin, P., *Can. J. Chem. Eng.*, (1992) 70, p. 1071
Carreau, P.J., Chhabra, R.P., and Cheng, J., *AIChE J.*, (1993) 39, p. 1421
Carreau, P.J., and Lavoie, P.A., *Rheology of coating colors: A Rheologist Point of View* TAPPI (1993) Symposium on Paper Coating, Minneapolis, MN, April 30
Carreau, P.J., Bousmina, M., and Ajji, A., in *Progress in Pacific Polymer Science 3*. K.P. Ghiggin (Ed.) (1994a) Springer-Verlag, New York, NY, p. 25
Carreau, P.J., Paris, J., and Guérin, P., *Can. J. Chem. Eng.*, (1994b) 72, p. 966
Casson, N., in *Rheology of Disperse Systems*. C.C. Mill (Ed.) (1959) Pergamon Press, New York, NY

Chan, P.C.H., and Leal, L.G., *J. Fluid Mech.*, (1979) 92, p. 131
Chan, P.C.H., and Leal, L.G., *Int. J. Multiphase Flow*, (1981) 7, p. 83
Chan Man Fong, C.F., and De Kee, D., *Rheol. Acta*, (1992) 31, p. 490
Chan Man Fong, C.F., De Kee, D., and Gryte, C., *J. Non-Newt. Fluid Mech.*, (1993) 46, p. 111
Chan Man Fong, C.F., and De Kee, D., *J. Non-Newt. Fluid Mech.*, (1995a) 57, p. 39
Chan Man Fong, C.F., and De Kee, D., *Physica A*, (1995b) 218, p. 56
Chang, H., and Lodge, A.S., *Rheol. Acta*, (1972) 11, p. 127
Chang, K.I., Yoo, S.S., and Hartnett, J.P., *Trans. Soc. Rheol.*, (1975) 19, p. 155
Chapman, F.S., and Holland, F.A., *Trans. Inst. Chem. Engrs.*, (1965) 43, p. T131
Chavan, V.V., and Ulbrecht, J.J., *Trans. Inst. Chem. Engrs.*, (1972a) 50, p. 147
Chavan, V.V., and Ulbrecht, J.J., *Chem. Eng. J.*, (1972b) 3, p. 308
Chavan, V.V., and Ulbrecht, J.J., *Ind. Eng. Chem. Proc. Des. Dev.*, (1973a) 12, p. 472. Correction *ibid.*, (1974) 13, p. 309
Chavan, V.V., and Ulbrecht, J.J., *Chem. Eng. J.*, (1973b) 6, p. 213
Chavan, V.V., Arumugam, M., and Ulbrecht, J.J., *AIChE J.*, (1975a) 21, p. 613
Chavan, V.V., Ford, D.E., and Arumugam, M., *Can. J. Chem. Eng.*, (1975b) 53, p. 62
Chavan, V.V., *AIChE J.*, (1983) 29, p. 177
Cheng, D.C.H., *Chem. Eng. Sci.*, (1968) 23, p. 895
Cheng, D.C.H., Schofield, C., and Jane, R.J., *Proc. 1st Engr. Conf. on Mixing and Centrifugal Sep.*, Cambridge, England, BHRA Fluid Eng. (1974) C2-15
Cheng, D.C.H., *Rheol. Acta*, (1986) 25, p. 542
Cheng, J., and Carreau, P.J., *Chem. Eng. Sci.*, (1994a) 49, p. 1965
Cheng, J., and Carreau, P.J., *Can. J. Chem. Eng.*, (1994b) 72, p. 418
Cheng, J., Carreau, P.J., Chhabra, R.P., *AIChE Symposium Series*, (1996), 91, No 305, p. 115
Chhabra, R.P., and Uhlherr, P.H.T., *Rheol. Acta*, (1980) 19, p. 187
Chhabra, R.P., and Uhlherr, P.H.T., in *Encyclopedia of Fluid Mechanics*. N.P. Cheremisinoff (Ed.) (1988) Vol. 7, p. 611
Chhabra, R.P., and Srinivas, B.K., *Powder Technol.*, (1991) 67, p. 15
Chhabra, R.P., and De Kee, D., *Transport Processes in Bubbles, Drops and Particles* (1992) Hemisphere, New York, NY
Chhabra, R.P., Unnikrishnan, A., and Nair, V.R.U., *Can. J. Chem. Eng.*, (1992) 70, p. 716
Chhabra, R.P., *Bubbles, Drops and Particles in non-Newtonian Fluids* (1993a) CRC Press, Boca Raton, FL
Chhabra, R.P., *Adv. Heat Transf.*, (1993b) 23, p. 187
Chhabra, R.P., *Adv. Transport Processes*, (1993c) 9, p. 571
Chhabra, R.P., *Powder Technology*, (1993d) 76, p. 225
Chilton, T.H., and Colburn, A.P., *Ind. Eng. Chem.*, (1934) 26, p. 1183
Cho, Y.I., and Hartnett, J.P., *J. Non-Newt. Fluid Mech.*, (1983) 13, p. 229
Choplin, L., and Carreau, P.J., *J.Non-Newt. Fluid Mech.*, (1981) 9, p. 119
Choplin, L., and Carreau, P.J., *Rheol. Acta*, (1986) 25, p. 95
Christiansen, E.B., and Craig, S.E., *AIChE J.*, (1962) 8, p. 154
Christiansen, E.B., Jensen, G.E., and Tao, F.S., *AIChE J.*, (1966) 12, p. 196
Christiansen, E.B., and Leppard, W.R., *Trans. Soc. Rheol.*, (1974) 18, p. 65
Christopher, R.H., and Middleman, S., *Ind. Eng. Chem. Fundam.*, (1965) 4, p. 422
Clift, R., Grace, J., and Weber, M.E., *Bubbles, Drops and Particles* (1978) Academic, New York, NY
Cohen, M., and Turnbull, D., *J. Chem. Phys.*, (1959) 31, p. 1164
Cohen, Y., and Metzner, A.B., *AIChE J.*, (1981) 27, p. 705
Coleman, B.D., Markovitz, H., and Noll, W., *Viscometric flows of Non-Newtonian Fluids* (1966) Springer, New York, NY
Coleman, B.D., and Noll, W., *Rev. Mod. Phys.*, (1961) 33, p. 239
Collyer, A.A., and Clegg, D.W., *Rheological Measurement* (1988) Elsevier Applied Science, London, UK
Couette, M., *Étude sur le frottement des liquides* (1890) Ph.D. thesis, Faculté des Sciences de Paris
Cox, W., and Merz, E., *J. Polym. Sci.*, (1958) 28, p. 619
Coyle, C.K., Hirschland, H.E., Michel, B.J., and Oldshue, J.Y., *Can. J. Chem. Eng.*, (1970) 48, p. 275

Crawley, R.L., and Graessley, W.W., *Trans. Soc. Rheol.*, (1977) 21, p. 19
Criminale, W.O., Ericksen, J.L., and Filbey, G.L., *Arch. Rat. Mech. Anal.*, (1958) 1, p. 410
Crochet, M.J., Davies, A.R., and Walters, K., *Numerical Simulation of Non-Newtonian Flow* (1984) Elsevier, Amsterdam
Cross, M.M., *J. Colloid Sci.*, (1965) 20, p. 417
Curtiss, C.F., and Bird, R.B., *J. Chem. Phys.*, (1981) 74, p. 2016 and 2026
Dajan, A., (1985) M.A.Sc. thesis, University of Windsor, Windsor, ON, Canada
Davies, M.J., and Walters, K., in *Rheology of lubricants*. T.C. Davenport (Ed.) (1973) Halsted Press, New York, NY
Dealy, J.M., *Rheometers for Molten Plastics* (1982) Van Nostrand Reinhold Company, New York, NY
Dealy, J.M., and Wissbrun, K.F., *Melt Rheology and its Role in Plastics Processing, Theory and Applications* (1990) Van Nostrand Reinhold, New York, NY
De Cleyn, G., and Mewis, J., *J. Non-Newt. Fluid Mech.*, (1979) 9, p. 91
De Gennes, P.G., *J. Chem. Phys.*, (1971) 55, p. 572
De Gennes, P.G., *J. Chem. Phys.*, (1974) 60, p. 5030
De Gennes, P.G., *Scaling Concepts in Polymer Physics* (1979) Cornell Univ. Press, Ithaca, NY
De Kee, D., *Equations rhéologiques pour décrire le comportement des fluides polymériques* (1977) Ph.D. Thesis, Ecole Polytechnique, Montreal, QC, Canada
De Kee, D., and Carreau, P.J., *J. Non-Newt. Fluid Mech.*, (1979) 6, p. 91
De Kee, D., and Turcotte, G., *Chem. Eng. Commun.*, (1980) 6, p. 273
De Kee, D., Code, R., and Turcotte, G., *J. Rheol.*, (1983) 27, p. 581
De Kee, D., and Carreau, P.J., *J. Rheol.*, (1984) 28, p. 109
De Kee, D., and Stastna, J., *J. Rheol.*, (1986) 30, p. 207
De Kee, D., Mohan, P., and Soong, D., *J. Macromol. Sci. Phys.*, (1986a) B25 (1 & 2), p. 153
De Kee, D., Carreau, P.J., and Mordarski, J., *Chem. Eng. Sci.*, (1986b) 41, p. 2273
De Kee, D., and Chhabra, R.P., *Rheol. Acta*, (1988) 27, p. 656
De Kee, D., Schlesinger, M., and Godo, M., *Chem. Eng. Sci.*, (1988) 43, p. 1603
De Kee, D., and Durning, C.J., in *Polymer Rheology and Processing*. A. Collyer and L.A. Utracki (Eds.) (1990) Elsevier, New York, NY
De Kee, D., Chhabra, R.P., and Dajan, A., *J. Non-Newt. Fluid Mech.*, (1990a) 37, p. 1
De Kee, D., Chhabra, R.P., Powley, M.B., and Roy, S., *Chem. Eng. Comm.*, (1990b) 96, p. 229
De Kee, D., and Carreau, P.J., *Can. J. Chem. Eng.*, (1993) 71, p. 183
De Kee, D., and Chan Man Fong, C.F., *Polym. Eng. Sci.*, (1994) 34, p. 438
De Kee, D., and Chhabra, R.P., *Rheol. Acta*, (1994) 33, p. 238
De Kee, D., Chhabra, R.P., and Rodrigue, D., in *Handbook of Applied Polymer Processing*. N.P. Cheremisinoff (Ed.) (1996), Marcel Dekker, New York, NY, Chapter 3
De Witt, T.W., *J. Appl. Phys.*, (1955) 26, p. 889
Del Villar, R., Carreau, P.J., and Patterson, W.I., *Chem. Eng. Commun.*, (1984) 25, p. 321
Den Otter, J.L., *Plast. Polym.*, (1970) 38, p. 155
Den Otter, J.L., *Rhéol. Acta*, (1971) 10, p. 200
Denn, M.M., and Porteous, K.C., *Chem. Eng. J.*, (1971) 2, p. 280
Derdouri, A., *Rhéologie des lubrifiants non-newtoniens: leur comportement dans des systèmes de lubrification* (1985) Ph.D. Thesis, Ecole Polytechnique, Montreal, QC, Canada
Derdouri, A., and Carreau, P.J., *Tribology Trans.*, (1989) 32, p. 161
Desplanches, H., Llinas, J.R., and Chevalier, J.L., *Can. J. Chem. Eng.*, (1980) 58, p. 160
Desplanches, H., Bruxelmane, M., Chevalier, J.L., and Ducla, J., *Chem. Eng. Res. Des.*, (1983) 61, p. 3
Dexter, F.D., *J. Appl. Phys.*, (1954) 25, p. 1124
Dierckes Jr., A.C., and Schowalter, W.R., *Ind. Eng. Chem. Fundam.*, (1966) 5, p. 263
Dlugogorski, B.Z., Grmela, M., and Carreau, P.J., *J. Non-Newt. Fluid Mech.*, (1993) 49, p. 23
Doi, M., and Edwards, S.F., *J. Chem. Soc. Faraday Trans. II*, (1978) 74, p. 1789, 1802, and 1818
Doi, M., and Edwards, S.F., *J. Chem. Soc. Faraday Trans. II*, (1979) 75, p. 38
Doi, M., *J. Polym. Sci.: Polym. Phys.*, (1981) 19, p. 229
Doi, M., and Edwards, S.F., *The Theory of Polymer Dynamics* (1986) Oxford University Press, New York, NY

Doi, M., Graessley, W.W., Helfand, E., and Pearson, D.S., *Macromolecules*, (1987) 20, p. 1900
Doolittle, A.K., *J. Appl. Phys.*, (1951) 22, p. 1031
Ducla, J.M., Desplanches, H., and Chevalier, J.J., *Chem. Eng. Commun.*, (1983) 21, p. 29
Dufresne, A., *Propriétés mécaniques électriques et rhéologiques de matériaux thermoplastiques renforcés de noir de carbone* (1989) M.A.Sc. thesis, Ecole Polytechnique, Montreal, QC, Canada
Dunlap, R.N., and Leal, L.G., *Rheol. Acta*, (1984) 23, p. 238
Durst, F., and Haas, R., *Rheol. Acta*, (1981) 20, p. 179
Durst, R., Haas, R., and Interthal, W., *J. Non-Newt. Fluid Mech.*, (1987) 22, p. 169
Earle, R.L., *Trans. Inst. Chem. Engrs.*, (1959) 37, p. 297
Edney, H.G.S., and Edwards, M.F., *Trans. Inst. Chem. Engrs.*, (1976) 54, p. 160
Edwards, M.F., and Wilkinson, W.L., *The Chem. Engr.*, (1972) 257, p. 310
Edwards, M.F., Godfrey, J.C., and Kashim, M.M., *J. Non-Newt. Fluid Mech.*, (1976) 1, p. 309
Einstein, A., *Ann. Physik*, (1906) 19, p. 289 and (1911) 34, p. 591
Elson, T.P., *Chem. Eng. Comm.*, (1990) 94, p. 143
Eveson, G.F., in *Rheology of Dispersed Systems*. C.C. Mills (Ed.) (1959) Academic Press, New York, NY
Fand, R.M., and Thinakaran, R., *J. Fluids Eng. (ASME)*, (1990) 112, p. 84
Faulkner, D.L., and Schmidt, L.R., *Polym. Eng. Sci.*, (1977) 17, p. 657
Ferry, J.D., and Williams, M.L., *J. Colloid. Sci.*, (1952) 7, p. 347
Ferry, J.D., *Viscoelastic Properties of Polymers*, 2nd Edition (1970) John Wiley and Sons, New York, NY; 3rd Edition (1980)
Fisher, R.J., and Denn, M.M., *AIChE J.*, (1976) 22, p. 236
Flory, P.J., and Fox, T. G., *J. Am. Chem. Soc.*, (1951) 73, p. 1904
Flory, P.J., *Principles of Polymer Chemistry* (1953) Cornell University Press, Ithaca, NY
Flumerfelt, R.W., Pierick, M.W., Cooper, S.L., and Bird, R.B., *Ind. Eng. Chem. Fundam.*, (1969) 8, p. 354
Forschner, P., Krebs, R., and Schneider, T., *Proceeding of the 7th European Conference on Mixing* (1991) Brugge, Belgium, p. 161
Fredrickson, A.G., and Bird, R.B., *Ind. Eng. Chem.*, (1958) 50, p. 347
Fredrickson, A.G., *Principles and Applications of Rheology* (1964) Prentice Hall, Englewood Cliffs, NJ
Froelich, D., Muller, R., and Zang, Y.H., *Rub. Chem. Techn.*, (1986) 59, p. 564
Fuller, C.G., and Leal, L.G., *Rheol. Acta*, (1980) 19, p. 580
Gauthier, F., Goldsmith, H.L., and Mason, S.G., *Rheol. Acta*, (1971) 10, p. 344. Also see *Trans. Soc. Rheol.*, (1971) 15, p. 297
Gerhardt, L.G., and Manke, C.W., *J. Rheol.*, (1994) 38, p. 1227
Ghosh, T., *Rheological Modeling of Complex Polymeric Liquids*, (1993) Ph.D. Thesis, Ecole Polytechnique of Montreal, Montreal, QC, Canada
Ghosh, T., Grmela, M., and Carreau, P.J., *Polym. Compos.*, (1995) 16, p. 144
Ghosh, U.K., Upadhyay, S.N., and Chhabra, R.P., *Adv. Heat Transf.*, (1994) 25, p. 251
Giesekus, H., *Rheol. Acta*, (1965) 4, p. 85
Giesekus, H., *Rheol. Acta*, (1966) 5, p. 29
Giesekus, H., *J. Non-Newt. Fluid Mech.*, (1982) 11, p. 69
Giesekus, H., *J. Non-Newt. Fluid Mech.*, (1985) 17, p. 349
Gleissle, W., in *Rheology*, Vol. 2, G. Astarita, G. Marrucci and L. Nicolais (Eds.) (1980) Plenum Press, New York, NY, p. 457
Gluz, M., and Pavlushenko, I.S., *J. App. Chem.* (U.S.S.R.), (1966) 39, p. 2223
Goddard, J.D., and Miller, C., *Rheol. Acta*, (1966) 5, p. 177
Godfrey, J.C., in *Mixing in the Processing Industries*. N. Harnby, M.F. Edwards and A.W. Nienow (Eds) (1985) Butterworths, London, pp. 185–201
Godleski, E.S., and Smith, J.C., *AIChE J.*, (1962) 8, p. 617
Gordon, R. J., and Schowalter, W. R., *Trans. Soc. Rheol.*, (1972) 16, p. 79
Gorla, R.S.R., *Chem. Eng. Commun.*, (1986) 49, p. 13
Graebling, D., and Muller, R., *J. Rheol.*, (1990) 34, p. 2 and 193
Graebling, D., Muller, R., and Palierne, J.F., *Macromolecules*, (1993) 26, p. 320
Graessley, W.W., *J. Polym. Sci., Polym. Phys. Ed.*, (1980) 18, p. 27
Graetz, L., *Ann. Phys.*, (1885) 25, p. 337

Graham, D.I., and Jones, T.E.R., *J. Non-Newt. Fluid Mech.*, (1994) 54, p. 465
Green, M.S., and Tobolsky, A.V., *J. Chem. Phys.*, (1946) 14, p. 80
Griskey, R.G., and Wiehe, I.A., *AIChE J.*, (1966) 12, p. 308
Grmela, M., *Physics Lett.*, (1985) IIIA, p. 36 and 41
Grmela, M., *Physica D.*, (1986) 21, p. 179
Grmela, M., and Carreau, P.J., *J. Non-Newt. Fluid Mech.*, (1987) 23, p. 271
Grmela, M., and Chhon, L.Y., *Physics Lett. A*, (1987) 120, p. 281
Gu, D., and Tanner, R.I., *J. Non-Newt. Fluid Mech.*, (1985) 17, p. 1
Guérin, P., Carreau, P.J., Patterson, I., and Paris, J., *Can. J. Chem. Eng.*, (1984) 62, p. 301
Gummalam, S., and Chhabra, R.P., *Can. J. Chem. Eng.*, (1987) 65, p. 1004
Gummalam, S., Narayan, K.A., and Chhabra, R.P., *Int. J. Multiphase Flow*, (1988) 14, p. 361
Gupta, S.K., Hamilton, G.M., and Hirst, W., *Proc. Int. Engrs.*, (1969) 184, p. 148
Gupta, R.K., and Seshadri, S.G., *J. Rheol.*, (1986) 30, p. 503
Hagen, G., *Ann. Phys. Chem.*, (1839) 46, p. 423
Hall, K.R., and Godfrey, J.C., *Trans. Inst. Chem. Engrs.*, (1968) 46, p. 205
Han, C.D., *Rheology in Polymer Processing* (1976) Academic Press, New York, NY
Hand, C.L., *J. Fluid Mech.*, (1964) 13, p. 33
Hanks, R.W., and Larsen, K.M., *Ind. Eng. Chem. Fundam.*, (1979) 18, p. 33
Happel, J., and Brenner, H., *Low Reynolds Number Hydrodynamics* (1965) Prentice Hall, New Jersey. (Reprinted 1973)
Harnby, N., Edwards, M.F., and Nienow, A.W., *Mixing in the Processing Industries* (1992) Butterworths, London, 2nd ed.
Harriott, P., *Chem. Eng. Prog.*, (1959) 55, p. 137
Henzler, H. J., *Chem. Ing. Tech.*, (1980) 52, p. 643
Hiem, A., *Int. Chem. Eng.*, (1980) 20, p. 271 and 279
Herschel, W.H., and Bulkley, R., *Kolloid-Z.*, (1926) 34, p. 291
Hewitt, G.F., Shires, G.L., and Bott, T.R., *Process Heat Transfer* (1994) CRC Press, Boca Raton, FL
Hickman, A.D., *Proceedings of 6th. Eur. Conf. Mixing, Pavia, Italy (BHRA Fluid Eng., Cranfield, England) (1988) p. 369*
Hilal, M., Brunjail, D., and Comiti, J., *J. Appl. Electrochem.*, (1991) 17, p. 583
Hinch, E.J., *Colloques Internationaux du CNRS* (1974) No. 233 and 241
Hiraoka, S., Yamada, I., and Mizoguchi, K., *J. Chem. Eng. Japan*, (1978) 11, p. 487
Hiraoka, S., Yamada, I., and Mizoguchi, K., *J. Chem. Eng. Japan*, (1979) 12, p. 56
Hirose, T. and Moo-Young, M., *Can. J. Chem. Eng.*, (1969) 47, p. 265
Hocker, H., Langer, G., and Werner, U., *Ger. Chem. Eng.*, (1981) 4, p. 113
Holland, F.A., and Chapman, F.S., *Liquid Mixing and Processing in Stirred Tanks* (1966) Reinhold, New York, NY
Hudson, N.E., and Jones, T.E.R., *J. Non-Newt. Fluid Mech.*, (1993) 46, p. 69
Huggins, M., *J. Am. Chem. Soc.*, (1942) 64, p. 2716
Huneault, M.A., *L'extrusion des profilés en PVC: rhéologie et conception de filières* (1992) Ph.D. thesis, Ecole Polytechnique of Montreal, Montreal, QC, Canada
Huneault, M.A., Carreau, P.J., Lafleur, P.G., and Gupta, V.P., *ANTEC 92*, (1992) p. 1063
Huppler, J.D., *Trans. Soc. Rheol.*, (1965) 9, p. 273
Hwang, S.J., Lui, C.B., and Lu, W.J., *Chem. Eng. J.*, (1993) 52, p. 131
Ide, Y., and White, J.L., *J. App. Polym. Sci.*, (1974) 18, p. 2997
Ide, Y., and White, J.L., *J. Non-Newt. Fluid Mech.*, (1977) 2, p. 281
Jaiswal, A.K., Sundararajan, T., and Chhabra, R.P., *Can. J. Chem. Eng.*, (1991) 69, p. 1235
Jaiswal, A.K., Sundararajan, T., and Chhabra, R.P., *Int. J. Eng. Sci.*, (1993a) 31, p. 293
Jaiswal, A.K., Sundararajan, T., and Chhabra, R.P., *Can. J. Chem. Eng.*, (1993b) 71, p. 646
Jaiswal, A.K., Sundararajan, T., and Chhabra, R.P., *Can. J. Chem. Eng.*, (1994) 72, p. 352
James, D.F., and Acosta, A.J., *J. Fluid Mech.*, (1970) 42, p. 269
James, D.F., and McLaren, R., *J. Fluid. Mech.*, (1975) 70, p. 733
James, H.M., *J. Chem. Phys.*, (1947) 15, p. 651
Jeffery, G.B., *Proc. Roy. Soc. A*, (1922) 102, p. 161

Johnson, M.W., and Segalman, D., *J. Non-Newt. Fluid Mech.*, (1981) 9, p. 481
Jomha, A.I., and Edwards, M.F., *Chem. Eng. Sci.*, (1990) 45, p. 1389
Jones, W.M., Price, A.H., and Walters, K., *J. Non-Newt. Fluid Mech.*, (1994) 53, p. 175
Joshi, J.B., *Chem. Eng. Res. Des.*, (1983) 61, p. 143
Kai, W., and Shengyao, Y., *Chem. Eng. Sci.*, (1989) 44, p. 33
Kale, D.D., Mashelkar, R.A., and Ulbrecht, J.J., *Chem. Ing. Tech.*, (1974) 46, p. 69
Kaloni, P.N., and Stastna, J., in *Transport Processes in Bubbles, Drops and Particles*. Chhabra R.P. and D. De Kee (Eds) (1992) Hemisphere, New York, NY, Chapter 7
Kamal, M.R., and Mutel, A.T., *J. Polym. Eng.*, (1986) 5, p. 293
Kappel, M., *Int. Chem. Eng.*, (1979) 19, p. 571
Karnis, A., and Mason, S.G., *Trans. Soc. Rheol.*, (1967) 10, p. 571
Kawase, Y., and Ulbrecht, J., *Chem. Eng. Sci.*, (1981) 36, p. 1193
Kawase, Y., and Moo-Young, M., *Chem. Eng. Res. Des.*, (1988) 66, p. 284
Kaye, A., *College of Aeronautics* (1962) Cranfield, UK Note 134
Kaye, A., *British J. Appl. Phys.*, (1966) 17, p. 803
Keirstead, K.F., De Kee, D., and Carreau, P.J., *Can. J. Chem. Eng.*, (1980) 58, p. 549
Kelkar, J.V., Mashelkar, R.A., and Ulbrecht, J.J., *J. Appl. Polym. Sci.*, (1973) 17, p. 3069
Kemblowski, Z., and Dziubinski, M., *Rheol. Acta*, (1978) 17, p. 176
Kemblowski, Z., Dziubinski, M., and Sek, J., *Adv. Transport Processes*, (1987) 5, p. 117
Keunings, R., and Crochet, M.J., *J. Non-Newt. Fluid Mech.*, (1984) 14, p. 279
Khan, A.R., and Richardson, J.F., *Chem. Eng. Commun.*, (1989) 78, p. 111
Khan, S.A., and Larson, R.G., *J. Rheol.*, (1987) 31, p. 207
Khokhlov, A.R., and Semenov, A.N., *Macromolecules*, (1984) 17, p. 2678
Khokhlov, A.R., and Semenov, A.N., *J. Stat. Phys.*, (1985) 38, p. 161
Kozicki, W., and Slegr, H., *J. Non-Newt. Fluid Mech.*, (1994) 53, p. 129
Krieger, I.M., and Maron, S.H., *J. Appl. Phys.*, (1954) 25, p. 72
Kumar, S., and Upadhyay, S.N., *Ind. Eng. Chem. Fundam.*, (1981) 20, p. 186
Kuriyama, M., Arai, K., and Saito, S., *J. Chem. Eng. Jpn.*, (1983) 16, p. 489
Lacroix, C., Bousmina, M., Carreau, P.J., Favis, B.D., and Michel, A., *Polymer*, (1996) 37, p. 2939
Lakdawala, K., and Salovey, R., *Polym. Eng. Sci.*, (1987) 27, p. 1035
Landau, L.D., and Lifshitz, E.M., *Fluid Mechanics* (1959) Pergamon Press, London
Larson, R.G., and Valesano, V.A., *J. Rheol.*, (1986) 30, p. 1093
Larson, R.G., *Constitutive Equations for Polymer Melts and Solutions* (1988) Butterworths, Boston, MA
Laun, H.M., *Rheol. Acta*, (1978) 17, p. 1
Laun, H.M., *J. Rheol.*, (1986) 30, p. 459
Leal, L.G., *J. Fluid Mech.*, (1975) 69, p. 305
Leal, L.G., *Ann. Rev. Fluid Mech.*, (1980) 12, p. 435
Lee, S.Y., and Ames, W.F., *AIChE J.*, (1966) 12, p. 700
Lee, T.L., and Donatelli, A.A., *Ind. Eng. Chem. Res.*, (1989) 28, p. 105
Leonov, A.I., *Rheol. Acta*, (1976) 15, p. 85
Leroy, P., and Pierrard, J.M., *Rheol. Acta*, (1973) 12, p. 449
Lévêque, J., *J. Ann. Mines*, Ser. 12, (1928) 13, p. 201, 305, and 381
Levich, V.G., *Physico-chemical Hydrodynamics* (1965) Prentice Hall, Englewood Cliffs, NJ
Liang, C.H., and Mackay, M.E., *J. Rheol.*, (1993) 37, p. 149
Liew, K.S., and Adelman, M., *Can. J. Chem. Eng.*, (1975) 53, p. 494
Lin, H.T., and Shih, Y.P., *Chem. Eng. Commun.*, (1980) 4, p. 557. Also see *ibid.*, (1980) 7, p. 327
Liu, T.Y., Soong, D.S., and De Kee, D., *Chem. Eng. Commun.*, (1983) 22, p. 273
Liu, Y.J., and Joseph, D.D., *J. Fluid Mech.*, (1993) 255, p. 565
Lockett, T.J., Richardson, S.M., Worraker, W.J., *J. Non-Newt. Fluid Mech.*, (1992) 43, p. 165
Lodge, A.S., *Trans. Faraday Soc.*, (1956) 52, p. 120
Lodge, A.S., *Kolloid-Z.*, (1960) 171, p. 46
Lodge, A.S., *Elastic Liquids* (1964) Academic, New York, NY
Lodge, A.S., *Rheol. Acta*, (1968) 7, p. 379
Lodge, A.S., and Meissner, J., *Rheol. Acta*, (1972) 11, p. 351

Lodge, A.S., *Body Tensor Fields in Continuum Mechanics* (1974) Academic Press, New York, NY
Macdonald, I.F., (1968) Ph.D. Thesis, University of Wisconsin, Madison, WI
Macdonald, I.F., Marsh, B.D., and Ashare, E., *Chem. Eng. Sci.*, (1969) 24, p. 1615
Macdonald, I.F., *Rheol. Acta*, (1975a) 14, p. 801
Macdonald, I.F., *Rheol. Acta*, (1975b) 14, p. 899
Macdonald, I.F., El-Sayed, M.S., Mow, K., and Dullien, F.A.L., *Ind. Eng. Chem. Fund.*, (1979) 18, p. 199
Machac, I., Balcar, M., and Lecjaks, Z., *Chem. Eng. Sci.*, (1986) 41, p. 591
Machon, V., Vlcek, J., Nienow, A.W., and Solomon, J., *Chem. Eng. J.*, (1980) 14, p. 67
Mahalingam, R.L., Chan, S.F., and Coulson, J.M., *Chem. Eng. J.*, (1975b) 9, p. 161
Mahalingam, R.L., Tilton, O., and Coulson, J.M., *Chem. Eng. Sci.*, (1975a) 30, p. 921
Malik, T.M., Carreau, P.J., Grmela, M., and Dufresne, A., *Polym. Compos.*, (1988) 9, p. 412
Malik, T.M., Carreau, P.J., and Chapleau, N., *Polym. Eng. Sci.*, (1989) 29, p. 600
Maron, S.H., and Pierce, P.E., *J. Colloid Sci.*, (1956) 11, p. 80
Marrucci, G., *J. Non-Newt. Fluid Mech.*, (1986) 21, p. 329
Marrucci, G., and Grizzuti, N., *J. Non-Newt. Fluid Mech.*, (1986) 21, p. 319
Mehta, D., and Hawley, M.C., *Ind. Eng. Chem. Proc. Des. Dev.*, (1969) 8, p. 280
Meissner, J., *Rheol. Acta*, (1971) 10, p. 230
Meissner, J., *J. Appl. Polym. Sci.*, (1972) 16, p. 2877
Metzner, A.B., and Reed, J.C., *AIChE J.*, (1955) 1, p. 434
Metzner, A.B., and Otto, R.E., *AIChE J.*, (1957) 3, p. 3
Metzner, A.B., Vaughn, R.D., and Houghton, G.L., *AIChE J.*, (1957) 3, p. 92
Metzner, A.B., *J. Rheol.*, (1985) 29, p. 739
Mewis, J., and Denn, M.M., *J. Non-Newt. Fluid Mech.*, (1983) 12, p. 69
Michele, J., Parzold, R., and Donis, R., *Rheol. Acta*, (1977) 16, p. 317
Miller, C., *Ind. Eng. Chem. Fundam.*, (1972) 11, p. 524
Minagawa, N., and White, J.L., *Polym. Eng. Sci.*, (1975) 15, p. 825
Mitsuishi, N., and Hirai, N.J., *J. Chem. Eng. Japan*, (1969) 2, p. 217
Mohan, V., *AIChE J.*, (1974) 20, p. 180
Mohr, C.M., and Newman, J., *Electrochim. Acta*, (1973) 18, p. 761
Moldenaers, P., and Mewis, J., *J. Rheol.*, (1993) 37, p. 367
Mourniac, Ph., Agasant, J.F., and Vergnes, B., *Rheol. Acta*, (1992) 31, p. 565
Murakami, Y., Hirose, T., and Oshima, M., *Chem. Eng. Prog.*, (1980) 76(5), p. 78
Nagata, S., Yanagimoto, M., and Yokoyama, T., *Memoirs. Fac. Eng., Kyoto Uni.*, (1956) 18, p. 444
Nagata, S., Nishikawa, M., Tada, H., Hirabayashi, H., and Gotoh, S., *J. Chem. Eng. Japan*, (1970) 3, p. 237
Nagata, S., Nishikawa, M., and Kayama, T., *J. Chem. Eng. Jpn.*, (1972a) 5, p. 83
Nagata, S., Nishikawa, M., Kayama, T., and Nakajima, M., *J. Chem. Eng. Jpn.*, (1972b) 5, p. 187
Nagata, S., *Mixing: Principles and Applications* (1975) Wiley, New York, NY
Nakano, Y., and Tien, C., *AIChE J.*, (1970) 16, p. 569
Nazem, F., and Hansen, G., *J. Appl. Polym. Sci.*, (1976) 20, p. 1355
Nielsen, L.E., *Polymer Rheology* (1977) Van Nostrand Reinhold Co., New York, NY
Nienow, A.W., Wisdom, D.J., Solomon, J., Machon, V., and Vlacek, J., *Chem. Eng. Commun.*, (1983) 19, p. 273
Nienow, A.W., and Ulbrecht, J.J., in *Mixing of Liquids by Mechanical Agitation*. J.J. Ulbrecht and G.K. Patterson (Eds) (1985) Gordon and Breach, New York, NY, Chapter 6
Nguyen, Q.D., and Boger, D.V., *Ann. Rev. Fluid Mech.*, (1992) 24, p. 47
Nishikawa, M., Kato, H., and Hashimoto, K., *Ind. Eng. Chem. Proc. Des. Dev.*, (1977) 16, p. 133
Nishikawa, M., Nakamura, M., Yagi, H., and Hashimoto, K., *J. Chem. Eng. Japan*, (1981) 14, p. 219. Also see *ibid.*, (1981) 14, p. 227
Norwood, K.W., and Metzner, A.B., *AIChE J.*, (1960) 6, p. 432
Olbricht, W.L., and Leal, L.G., *J. Fluid Mech.*, (1982) 115, p. 187
Olbricht, W.L., and Leal, L.G., *J. Fluid Mech.*, (1983) 134, p. 329
Oldroyd, J.G., *Proc. Roy. Soc. London*, (1950) A200, p. 523; A202, p. 345
Oldroyd, J.G., *Proc. Roy. Soc. London*, (1953) A218, p. 122
Oldroyd, J.G., *Proc. Roy. Soc. London*, (1955) A232, p. 567

Oldroyd, J.G., *Proc. Roy. Soc. London*, (1958) A245, p. 278
Oldroyd, J.G., *Proc. Roy. Soc. London*, (1965) A283, p. 115
Oldshue, J.Y., *Fluid Mixing Technology* (1983) McGraw Hill, New York, NY
Oliver, D.R., Nienow, A.W., Mitson, R.J., and Terry, K., *Chem. Eng. Res. Des.*, (1984) 62, p. 123
Onogi, S., Masuda, T., and Kitagawa, K., *Macromolecules*, (1970) 3, p. 109
Onsager, L., *Ann. NY Acad. Sci.*, (1949) 51, p. 627
Ortiz, M.E., *Comportement rhéologique du poly(oxyéthylène) en solution* (1992) Ph.D. Thesis, Ecole Polytechnique of Montreal, Montreal, QC, Canada
Ortiz, M.E., De Kee, D., and Carreau, P.J., *J. Rheol.*, (1994) 38, p. 519
Osaki, K., *Proc. Seventh International Congress on Rheology* (1976) Gothenburg, Sweden, p. 104
Osaki, K., Kimura, S., and Kurata, M., *J. Rheol.*, (1981) 25, p. 549
Ostwald, W., *Kolloid Z.*, (1925) 36, p. 99
Otsubo, Y., *J. Colloid and Interf. Sci.*, (1986) 112, p. 380
Ottino, J.M., *Sci. Amer.*, (1989) 260, p. 56
Ottino, J.M., *The Kinematics of Mixing* (1990) Cambridge University Press, Cambridge, UK
Palierne, J.F., *Rheol. Acta*, (1990) 29, p. 204
Papanastasiou, A.C., Scriven, L.E., and Macosko, C., *J. Non Newt. Fluid Mech.*, (1984) 16, p. 53
Park, H.C., Hawley, M.C., and Blanks, R.F., *Polym. Eng. Sci.*, (1975) 15, p. 761
Patterson, W.I., Carreau, P.J., and Yap, C.Y., *AIChE J.*, (1979) 25, p. 508
Peev, G., *Chem. Eng. Sci.*, (1985) 40, p. 1985
Perez, J.F., and Sandall, O.C., *AIChE J.*, (1974) 20, p. 1073
Peterlin, A., *J. Polym. Sci., Polym. Lett.*, (1966) 4B, p. 287
Peters, D.C., and Smith, J.M., *Trans. Inst. Chem. Engrs.*, (1967) 45, p. 360
Pfeffer, R., *Ind. Eng. Chem. Fundam.*, (1964) 3, p. 380
Phan-Thien, N., and Tanner, R.I., *J. Non-Newt. Fluid Mech.*, (1977) 2, p. 353
Philippoff, W., and Gaskins, F.H., *Trans. Soc. Rheol.*, (1958) 2, p. 263
Poggermann, R., Steiff, A., and Weinspach, P.M., *Ger. Chem. Eng.*, (1980) 3, p. 163
Poiseuille, J.L., *Comptes rendus*, (1841) 12, p. 112
Pollard, J., and Kantyka, T.A., *Trans. Inst. Chem. Engrs.*, (1969) 47, p. 21
Poslinski, A.J., Ryan, M.E., Gupta, R.K., Seshadri, S.G., and Frechette, F.J., *J. Rheol.*, (1988) 32, p. 703 and 751
Prud'homme, R.K., and Shaqfeh, E., *AIChE J.*, (1984) 30, p. 485
Quraishi, A.Q., Mashelkar, R.A., and Ulbrecht, J.J., *J. Non-Newt. Fluid Mech.*, (1976) 1, p. 223
Quemada, D., *J. Theor. and Appl. Mech.*, (1985) p. 267
Quemada, D., in *Rheological Modelling: Thermodynamical and Statistical Approaches*. J. Casas-Vasquez and D. Jou (Eds.) (1991) Springer-Verlag, New York, NY, 381, p. 158
Quintana, G., in *Transport Processes in Bubbles, Drops and Particles*. R.P. Chhabra and D. De Kee (Eds.) (1992) Hemisphere, New York, NY, Chapter 4
Rabinowitsch, B., *Z. Physik Chemie*, (1929) A145, p. 1
Ranade, V.R., and Ulbrecht, J.J., *AIChE J.*, (1978) 24, p. 796
Rao, P.T., and Chhabra, R.P., *Powder Technol.*, (1993) 77, p. 171
Rautenbach, R., and Bollenrath, F.M., *Ger. Chem. Eng.*, (1979) 2, p. 18
Richardson, J.F., and Zaki, W.N., *Trans. Inst. Chem. Engrs.*, (1954) 32, p. 35
Riddle, M.J., Narvaez, C., and Bird, R.B., *J. Non-Newt. Fluid Mech.*, (1977) 2, p. 23
Rivlin, R.S., and Ericksen, J.L., *J. Rat. Mech. Anal.*, (1955) 4, p. 323
Rodrigue, D., Chhabra, R.P., and De Kee, D., *Can. J. Chem. Eng.*, (1994) 72, p. 583
Rouse, P.E., *J. Chem. Phys.*, (1953) 21, p. 7 and 1272
Saab, H.H., and Bird, R.B., *J. Chem. Phys.*, (1982) 77, p. 4758
Sadowski, T.J., and Bird, R.B., *Trans. Soc. Rheol.*, (1965) 9, p. 243
Sakai, T., *J. Polym. Sci.*, Part A-2, (1968) 6, p. 1535
Sandall, O.C., and Patel, K.G., *Ind. Eng. Chem. Proc. Des. Dev.*, (1970) 9, p. 139
Sangani, A., and Acrivos, A., *Int. J. Multiphase Flow*, (1982) 8, p. 343
Saunders, P.R., Stern, D.M., Kurath, S.F., Sakoonkim, C., and Ferry, J.D., *Colloid Sci.*, (1959) 14, p. 222
Schieber, J.B., *J. Chem. Phys.*, (1987) 87, p. 4917 and 4928

Schlichting, H., *Boundary Layer Theory, Sixth Edition* (1968) McGraw-Hill, New York, NY
Schowalter, W.R., Chaffey, C.E., and Brenner, H., *J. Coll. Interface Sci.*, (1968) 26, p. 152
Schowalter, W.R., *Mechanics of Non-Newtonian Fluids* (1978) Pergamon Press, New York, NY
Schwarzl, F., and Stavermann, A.J., *Appl. Sci. Research*, (1953) A4, p. 127
Sellin, R.H.J., Hoyt, J.W., and Scrivener, O., *J. Hyd. Res.*, (1982) 20, p. 29. Also see *ibid.*, (1982) 20, p. 235
Shah, M.J., Petersen, E.E., and Acrivos, A., *AIChE J.*, (1962) 8, p. 542
Shah, P.P., and Parsania, P.H., *J. Macromol. Sci. Phys.*, (1984) B23, p. 363
Sharma, M.K., and Chhabra, R.P., *Can. J. Chem. Eng.*, (1992) 70, p. 586
Sheffield, R.E., and Metzner, A.B., *AIChE J.*, (1976) 22, p. 736
Shenoy, A.V., in *Encyclopedia of Fluid Mechanics, Vol. 7* (1988) Gulf Publishing Co, Houston, TX, Chapter 23
Sieder, E.N., and Tate, G.E., *Ind. Eng. Chem.*, (1936) 28, p. 1429
Skelland, A.H.P., *Non-Newtonian Flow and Heat Transfer* (1967) Wiley, New York, NY
Skelland, A.H.P., in *Handbook of Fluids in Motion*. N.P. Cheremisinoff and R. Gupta (Eds) (1983) Ann Arbor, MI
Slattery, J.C., *Momentum, Heat and Energy Transfer in Continua* (1972) McGraw Hill, New York, NY
Smith, J.M., in *Mixing of Liquids by Mechanical Agitation*. J.J. Ulbrecht and G.K. Patterson (Eds) (1985) Gordon and Breach, New York, NY, Chapter 5
Solomon, J., Nienow, A.W., and Pace, G.W., *Fluid Mixing, I. Chem. Engr. Sym.*, Ser. No. 64 (1981) A1
Sorbie, K.S., *Polymer Improved Oil Recovery* (1991) Blackie and Sons, Glasgow, UK
Sridhar, T., Chhabra, R.P., Uhlherr, P.H.T., and Potter, O.E., *Rheol. Acta*, (1978) 17, p. 519
Sridhar, T., *J. Non-Newt. Fluid Mech.*, (1990) 35, p. 85
Srinivas, B.K., and Chhabra, R.P., *Chem. Eng. Processes*, (1991) 29, p. 212
Srinivas, B.K., and Chhabra, R.P., *Int. J. Eng. Fluid Mech.*, (1992) 5, p. 309
Stewart, W.E., *Int. J. Heat Mass Transf.*, (1971) 14, p. 1013
Tallmadge, J.A., *J. Phys. Chem.*, (1971) 75, p. 583
Tanguy, P., Lacroix, A., Bertrand, F., Choplin, L., and De La Fuente, E.B., *AIChE J.*, (1992) 38, p. 939
Tanner, R.I., *Trans. Soc. Rheol.*, (1975) 19, p. 37
Tanner, R.I., *Engineering Rheology* (1985) Clarendon Press, Oxford, UK
Tatterson, G.B., *Fluid Mixing and Gas Dispersion in Agitated Tanks* (1991) McGraw Hill, New York, NY
Tatterson, G.B., *Scaleup and Design of Industrial Mixing Processes* (1994) McGraw Hill, New York, NY
Taylor, G.I., *Proc. Roy. Soc. A*, (1932) 138, p. 41
Thomas, D.G., *J. Colloid Sci.*, (1965) 20, p. 267
Tiefenbruck, G., and Leal, L.G., *J. Non-Newt. Fluid Mech.*, (1980) 6, p. 201
Tiefenbruck, G., and Leal, L.G., *J. Non Newt. Fluid Mech.*, (1982) 10, p. 115
Tirtaatmadja, V., *Measurement of Extensional Viscosity of Polymer Solutions* (1993) Ph.D. Thesis, Monash University, Clayton, Victoria, Australia
Tirtaatmadja, V., and Sridhar, T., *J. Rheol.*, (1993) 37, p. 1081
Tiu, C., Moussa, T., and Carreau, P.J., *Rheol. Acta*, (1995) 34, p. 586
Tobolsky, A.V., *Properties and Structure of Polymers* (1960) John Wiley & Sons, New York, NY
Tomita, Y., *Bull. J.S.M.E.*, (1959) 2, p. 469
Tonini, R.D., *Encyclopedia of Fluid Mech.*, (1987) 6, p. 495
Tran, Q.K., Horsley, R.R., Reizes, J.A., and Ang, H.M., *Theo. & Appl. Rheo.*, (1992) Elsevier, Amsterdam, p. 631
Tripathi, A., Chhabra, R.P., and Sundararajan, T., *Ind. Eng. Chem. Res.*, (1994) 33, p. 403
Tripathi, A., and Chhabra, R.P., *AIChE J.*, (1995) 41, p. 728
Turcotte, G., (1980) M.A.Sc. thesis, Queen's University, Kingston, ON, Canada
Turian, R.M., *Ind. Eng. Chem. Fund.*, (1972) 11, p. 361
Uhl, V.W., and Gray, J.B. (Eds) *Mixing, Vol. 1* (1966) Academic, New York, NY, Chapter 5
Ulbrecht, J.J., *The Chem. Engr.*, (1974) 286, p. 347
Ulbrecht, J.J., and Carreau, P.J., in *Mixing of Liquids by Mechanical Agitation*. J.J. Ulbrecht and G.K. Patterson (Eds) (1985) Gordon and Breach Science Publishers, New York, NY, Chapter 4
Ulbrecht, J.J., and Patterson, G.K. (Eds) *Mixing of Liquids by Mechanical Agitation* (1985) Gordon and Breach, New York, NY

Umeya, K., and Kanno, T., Symp. on Suspensions (1979) 2nd joint meeting U.S. and Japan Soc. Rheol., Hawaii, April 6–8
Utracki, L.A., in *Rheological Measurement*. Collyer, A.A. and Clegg, D.W. (Eds) (1988) Elsevier Applied Science, London and New York
Van't Reit, K., (1975) Ph.D. Thesis, University of Delft, The Netherlands
Virk, P.S., *AIChE J.*, (1975) 21, p. 625
Vradis, G.C., and Protopapas, A.L., *J. Hyd. Engrg. (ASCE)*, (1993) 119, p. 95
Vrentas, C.M., and Graessley, W.W., *J. Non-Newt. Fluid Mech.*, (1981) 9, p. 339
Vrentas, J.S., Venerus, D.C., and Vrentas, C.M., *Rheol. Acta*, (1990) 29, p. 298
Wagner, M.G., and Slattery, J.C., *AIChE J.*, (1971) 17, p. 1198
Wagner, M.H., *Rheol. Acta*, (1976) 15, p. 136
Wagner, M.H., *Rheol. Acta*, (1977) 16, p. 43
Wagner, M.H., *Rheol. Acta*, (1978a) 17, p. 138
Wagner, M.H., *J. Non-Newt. Fluid Mech.*, (1978b) 4, p. 39
Wagner, M.H., and Meissner, J., *Makromol. Chem.*, (1980) 181, p. 1533
Wall, F.T., *J. Chem. Phys.*, (1942) 10, p. 485
Walters, K., and Waters, N.D., in *Polymer Systems: Deformation and Flow*. R.E. Wetton and R.W. Whorlow (Eds) (1968) Macmillan, London, UK
Walters, K., *Rheometry* (1975) Chapman and Hall, London, UK
Walters, K. (Ed.) *Rheometry: Industrial Applications* (1980) Wiley, New York, NY
Walters, K., and Tanner, R.I., in *Transport Processes in Bubbles, Drops and Particles*. Chhabra R.P. and De Kee D. (Eds) (1992) Hemisphere, New York, NY, Chapter 3
Warner, H.R., Jr, *Ind. Eng. Chem. Fund.*, (1972) 11, p. 379
Wasserman, M.L., and Slattery, J.C., *AIChE J.*, (1964) 10, p. 383
Weissenberg, K., *Nature*, (1947) 159, p. 310
Westerberg, K.W., and Finlayson, B.A., *Num. Heat Transf.*, (1990) 17A, p. 329
White, A., *Nature*, (1967) 216, p. 994
White, J.L., and Metzner, A.B., *J. Appl. Polym. Sci.*, (1963) 7, p. 1867
Wichterle, K., and Wein, O., *Int. Chem. Eng.*, (1981) 21, p. 116
Williams, D.J., *Polymer Science and Engineering* (1971) Prentice Hall, Englewood Cliffs, NJ
Williams, M.L., Landel, R.F., and Ferry, J.D., *J. Am. Chem. Soc.*, (1955) 77, p. 3701
Wissler, E.H., *Ind. Eng. Chem. Fundam.*, (1971) 10, p. 411
Wronski, S., and Szembek-Stoeger, M., *Inzinieria Chem. Proc.*, (1988) 4, p. 627
Yagi, H., and Yoshida, F., *Ind. Eng. Chem., Proc. Des. Dev. I*, (1975) 14, p. 488
Yamamoto, M., *J. Phys. Soc. Japan*, (1956) 11, p. 413
Yamamoto, M., *J. Phys. Soc. Japan*, (1957) 12, p. 1148
Yamamoto, M., *J. Phys. Soc. Japan*, (1958) 13, p. 1200
Yamamoto, M., *Trans. Soc. Rheol.*, (1971) 15, p. 331
Yamanaka, A., and Mitsuishi, N., *J. Chem. Eng. Japan*, (1977) 10, p. 370
Yap, Y.C., Patterson, W.I., and Carreau, P.J., *AIChE J.*, (1979) 25, p. 516
Yasuda, K. (1979) Ph.D. thesis, Massachussetts Institute of Technology, Cambridge, MA
Zhu, J., *Rheol. Acta*, (1990) 29, p. 409
Zhu, W., Lai, W.M., and Mow, Van C., *J. Biomechanics*, (1991) 24, p. 1007
Zdilar, A.M., and Tanner, R.I., *J. Rheol.*, (1994) 38, p. 909
Zimm, B., *J. Chem. Phys.*, (1956) 24, p. 2 and 269

Notation*

A	area (m^2)
A_{ij}	Rouse matrix (–)
$A = U - TS$	Helmholtz free energy (J)
a	parameter (–), interfacial area per unit volume of bed (m^{-1}), eccentricity (m)
a_T	shift factor for time-temperature superposition, defined by eq. 2.5-3 (–)
B_{kv}, B_{kv}^*	transformation matrices (–)
b	parameter in De Kee-Chan Man Fong model (eqs. 2.3-1 and 2.3-2) (–)
b	extensibility parameter (eq. 7.4-12) (–)
b_n	constant for Gaussian segments, defined by eq. 7.2-2 (–)
C	dimensionless concentration (–)
C_d	drag coefficient for the flow about a particle (–)
C_f	drag coefficient for the flow over a plate (–)
C_n	concentration of n-segments in network theories (m^{-3})
$\hat{C}_p$	Heat capacity at constant pressure, per unit mass (J/kg·K)
C^*	critical polymer concentration for interactions (kg/m^3)
$\underline{\underline{C}}^{-1}$	Finger strain tensor defined by eq. 6.1-8 (–)
$\underline{\underline{C}}$	Cauchy-Green strain tensor defined by eq. 6.1-9 (–)
C_1, C_2	WLF constants (eq. 5.4-3)
c	total molar or mass concentration (mol/m^3 kg/m^3)
$c(n)$	shear-thinning function defined by eq. 4.5-21 (–)
c	parameter in De Kee-Chan Man Fong model (eqs. 2.3-1 and 2) (–)
c_0	initial concentration (mol/m^3)
c^*	solubility (mol/m^3)
$\underline{\underline{c}}$	conformation tensor (m^2)
D	diameter of vessel (m)
D_A, D_L	diffusivity coefficient of A or gas diffusivity in a liquid (m^2/s)
D_c	diameter of column, diameter of cavity in mixing (m)
D_t	diameter of draft tube in mixing (m)

---

* Dimensions are given in SI units. Vectors and tensors have one underbar and two underbars respectively. Superscripts in vector and tensor components refer to contravariant components; subscripts refer to covariant components. Some symbols that appear infrequently are not listed.

504   Notation

$\dfrac{D}{Dt} = \dfrac{\partial}{\partial t} + \underline{V}\cdot\nabla$	substantial derivative (s^{-1})
$\dfrac{\mathcal{D}}{\mathcal{D}t}(\ )$	Jaumann derivative defined by eq. 6.1-21 or 6.1-22 (s^{-1})
d	diameter of impeller (m)
d_{co}	outside diameter of heat transfer coil (m)
d_e	equivalent impeller diameter in mixing (m)
d_p	particle or drop diameter (m)
d_v	volume average diameter (m)
E	activation energy in eq. 2.5-2 (J)
E_p	tensile moduli in generalized Maxwell model (Pa)
E_r	relaxation modulus in tension (Pa)
E_v	total rate of viscous dissipation of mechanical energy (W)
e	electrostatic parameter (eq. 7.4-13) (–)
e_c	entrance Couette correction (–)
e_0	Bagley correction defined by eq. 3.1-18 (–)
F	force acting on a plate (N)
F_b	buoyant force (N)
F_D	drag force (N)
$\underline{F}$	force vector of a fluid on an adjacent solid (N)
$\underline{F}_k$	force acting on the kth bead (N)
$\underline{F}_1^d$	drag force vector acting on bead (1) (N)
$\underline{F}_1^b$	force vector due to Brownian motion acting on bead (1) (N)
F_i^c	component i of connector force (N)
f	friction factor or drag coefficient, defined by eq. 1.2-1 (–)
f, f_0, f_∞	wall factors for the motion of a sphere (eq. 8.3-13) (–)
f_p	function of $II_{\dot{\gamma}}$ in the network theory (–)
f_1, f_2	shear rate functions in De Kee-Chan Man Fong model (eqs. 2.3-1 and 2)
G	elastic modulus (Pa)
$G(t)$	relaxation modulus, defined by eq. 2.1-22 (Pa)
G_e	relaxation modulus at equilibrium (Pa)
G_p	shear moduli in rheological models (Pa)
$G'(\omega)$	storage or elastic modulus, defined by eq. 2.1-12 (Pa)
$G'_m(\omega), G'_s(\omega)$	elastic modulus of matrix, of suspension (Pa)
$G''(\omega)$	loss or viscous modulus (Pa)
$G''_m(\omega), G''_s(\omega)$	loss modulus of matrix, of suspension (Pa)
$G_N(t)$	normal stress relaxation modulus defined by eq. 6.4-11 (Pa)
G_N^0	plateau modulus (Pa)
$G^*(\omega)$	complex modulus defined by eq. 2.1-12 (Pa)
G_0	shear modulus (Pa)
G_1, G_2	characteristic moduli (Pa)
$\underline{g}^i, \underline{\hat{g}}^i$	reciprocal base vectors in nonorthogonal coordinates, in embedded coordinates

$\underline{g}_i, \hat{\underline{g}}_i$	base vectors in nonorthogonal coordinates, in embedded coordinates
$g_{ij}, \hat{g}_{ij}$	convariant components of metric tensor, in embedded coordinates
$g^{ij}, \hat{g}^{ij}$	contravariant components of metric tensor, in embedded coordinates
g_p	function of $II_{\dot\gamma}$ in the network theory (–)
H	proportionality constant in dumbbell models (N/m)
H	parameter in Quemada model, eq. 2.3-4 (–)
H	height of fluid in mixing vessel, half-thickness of channel (m)
H_0	Hookean constant in dumbbell models (N/m)
h, h_x	heat transfer coefficient (W/m² · K)
h	height of impeller or gap in concentric disk geometry (m)
$h(t)$	Heaviside function defined by eq. 2.1-14 (–)
$h(\ )$	damping function in Wagner model (–)
h_a	average heat transfer coefficient (W/m² · K)
h_{co}	heat transfer coefficient from coil (W/m² · K)
h_{loc}	local heat transfer coefficient (W/m² · K)
$I_{\dot\gamma} = \text{tr}\,\underline{\underline{\dot\gamma}}$	first invariant of rate-of-deformation tensor (s⁻¹)
$II_{\dot\gamma} = \underline{\underline{\dot\gamma}} : \underline{\underline{\dot\gamma}}$	second invariant of rate-of-deformation tensor (s⁻²)
$III_{\dot\gamma} = \det\underline{\underline{\dot\gamma}}$	third invariant of rate of deformation tensor (s⁻³)
$i = \sqrt{-1}$	imaginary number
J_e, J_e^0	equilibrium compliance (eq. 5.2-57) or steady-state recoverable compliance defined by eq. 3.3-30 (Pa⁻¹)
j	mass transfer factor defined by eq. 8.7-8 (–)
j_H	Chilton-Colburn j-factor for heat transfer, defined by eq. 1.2-3 (–)
K	Mark-Houwink constant defined by eq. 3.1-27 (m³/kg)
$K = 8^{n-1}m$	in equation 4.3-40 (Pa · sⁿ)
$K_P = PoRe$	power consumption constant (–)
K_{ij}	Kramer's matrix (–)
k	thermal conductivity (W/m · K)
k_a	average mass transfer coefficient (mol/s · m²)
k_B	Boltzmann constant (1.381 × 10⁻²³ J/K)
k_c	structural parameter in De Kee-Chan Man Fong model (–)
k_L	mass transfer coefficient (mol/s · m²)
k_{loc}	local mass transfer coefficient (mol/s · m²)
k_s	Metzner-Otto constant (–)
k'	Huggins' constant in eq. 3.1-25 (–)
k''	Kramer's constant in eq. 3.1-26 (–)
k, k_0, k_∞	parameters in Quemada model (eqs. 2.3-4 and 2.3-5) (–)
L	length of tube, contour length of polymeric chain, or other characteristic length (m)
L_c	characteristic length in heat transfer (m)
L_e	effective length of packed bed (m)
$\underline{\underline{L}}$	effective velocity gradient tensor (eq. 6.3-50) (s⁻¹)

l	chain segment (m)
l_c	mean circulation length in mixing (m)
l_G	Giesekus parameter in eq. 7.4-44 (–)
M	molar molecular weight (kg/kmol), number of elements in rheological equation (–)
$M(t - t', II_{\dot{\gamma}})$	memory functional in Carreau equation 6.4-14 (Pa·s^{-1})
M_c	critical molecular weight (kg/kmol)
M_e	molecular weight of a polymeric chain segment between entanglements (kg/kmol)
M_v	viscosity average molecular weight (kg/kmol)
M_w	weight average molecular weight (kg/kmol)
m	parameter in power-law model (eq. 2.2-2) (Pa·s^n)
m	mass of a bead (kg)
$\dot{m}$	mass flow rate (kg/s)
m'	power-law parameter for normal stress data (Pa·s^n)
$m(\ ), m_1(\ ), m_2(\ )$	memory functions (Pa·s^{-1})
N	rate of rotation of an impeller (rev/s)
N	number of elements, or beads in Rouse, Zimm, and reptation models
N_n, N_{n0}	number density of segments of n links and number density at equilibrium (eq. 7.2-31) (m^{-3})
N_{1w}	primary normal stress difference evaluated at the wall (Pa)
$N_1 = -(\sigma_{11} - \sigma_{22})$	primary normal stress difference (Pa)
$N_2 = -(\sigma_{22} - \sigma_{33})$	secondary normal stress difference (Pa)
n	molecular concentration or number density (m^{-3})
n	parameter in power-law model, defined by eq. 2.2-2 (–)
n'	slope defined by eq. 3.1-8 (–), or power-law index for normal stress data (–)
$\underline{n}$	unit vector normal to surface (–)
n_i	component i of unit vector (–)
P	power consumption in mixing (W)
P	fluid pressure (Pa)
P_a	atmospheric pressure (Pa)
P_g	power consumption under aeration in mixing (W)
$P(t)$	probability function in reptation theories
P_0, P_L	fluid pressure at 0, at L (Pa)
P^p	polymer contribution to the pressure (Pa)
P^s	solvent contribution to the pressure (Pa)
p	pitch of impeller (m)
Q	volumetric flow rate (m^3/s)
Q	rate of energy flow across a surface (W)
q	heat flux (W/m^2)
q_r	radial component of heat flux vector (W/m^2)
R	gas constant (8.314 J/mol·K)
R	radius of sphere, polymer coil, or cylinder (m)

Notation

$\underline{R}$	end-to-end vector of dumbbell or macromolecule conformation (m)
R_f	radial dimension of disk-shaped mold (m)
R_g	radius of gyration (m)
R_h	hydraulic radius (m)
$\underline{R}_k$	connector vector in bead-spring model (m)
R_0	end-to-end vector in rigid-in-average chains, radius of circular die (m)
R_1	characteristic radius (m)
R_∞	maximum value of end-to-end vector in FENE models, or radius of cell model sphere (m)
r	radial distance in both cylindrical and spherical coordinates (m)
$\underline{r}_c$	position vector of center of mass (m)
$\underline{r}_v$	position vector of bead v (m)
$\underline{\dot{r}}_1$	velocity vector of bead (1) in dumbbell models (m/s)
$\underline{\ddot{r}}_1$	acceleration vector of bead (1) in dumbbell models (m²/s)
S	cross-section area (m²)
S	entropy (J/K)
S_B	Brownian motion parameter in conformation tensor models (–)
S_L	flexibility parameter in conformation tensor models, defined by eq. 7.4-18 (–)
S_R	recoverable shear defined by eq. 3.1-20 (- -)
S_0	Onsager parameter in conformation tensor model, in eq. 7.4-43 (–)
s	Laplace transform variable, distance from wall (m)
$s = t - t'$	time interval (s)
$T = L_e/L$	tortuosity factor in packed beds (–)
T	temperature (K) or torque (N·m)
T_b	bulk temperature (K)
$T_f = \dfrac{T_W + T_b}{2}$	film temperature (K)
T_g	glass transition temperature (K)
T_W	wall temperature (K)
T_0	reference temperature (K)
t	time (s)
t_a	average temperature (K)
t_c	circulation time (s)
t_m	mixing time or matrix characteristic time (s)
t_r	reduced time constant (eq. 8.2-36) (–), delay time of signal in rheometry, residence time (s)
t_s	response time of instrument in rheometry (s)
t_1	parameter or characteristic time (s)
t'	past time (s)
$\tan\delta = G''(\omega)/G'(\omega)$	loss tangent (–)
$\mathrm{tr}(\ )$	trace of tensor

U	intramolecular energy (J)
U_x	separation in the x-direction (m)
u	strain energy function in eq. 6.4-38 (W/m^3)
$\underline{u}$	unit vector tangent to a polymeric chain (–)
V	volume (m^3) or characteristic velocity (m/s)
$\underline{V}$	velocity vector (m/s)
V_c	mean circulation velocity of liquid in mixing vessel (m/s)
$\langle V \rangle$	average velocity across flow section (m/s)
V_g	superficial gas velocity (m/s)
V_f	free volume (m^3)
V_{mf}	minimum fluidization velocity (m/s)
V_t	terminal velocity (m/s)
V_s	slip velocity in Mooney correction (m/s)
V_x, V_y, V_z	velocity components (m/s)
V_r, V_θ, V_ϕ	velocity components (m/s)
V_0	free stream or superficial velocity (m/s), volume at 0 K (m^3)
V_∞	terminal velocity or velocity at $r \to \infty$ (m/s)
V_z'	derivative of V_z with respect to z (s^{-1})
W	width of channel or plate (m), load (N)
$\dot{W}$	energy dissipated per unit volume (W/m^3)
w	width of impeller blades (m)
X	correction factor for the motion of sphere (eq. 8.3-11) (–)
x	rectangular coordinate (m)
x'	coordinate at time t'
x^i	component i of general coordinate system
x_i	component i of Cartesian coordinates
$\hat{x}^i$	component i in embedded coordinates
y	rectangular coordinate (m)
Z	nonlinear operator defined by eq. 7.1-48b (–)
$Z(\ )$	Riemann zeta function
z	rectangular coordinate (m)

Greek Letters

α	interfacial tension (N/m)
α	shear strain angle (rad)
α	parameter (–)
α_0	parameter in De Kee-Chan Man Fong model (eqs. 2.3-1 and 2.3-2) (–)
$\alpha = k/\rho\hat{C}_p$	thermal diffusivity (m^2/s)
B	Lagrange multiplier
β	conformation parameter defined by eq. 7.4-11 (–)
$\beta = V_s/\sigma_R$	slip coefficient in Mooney correction (m/s·Pa)

$\beta_1, \beta_2, \beta_3$	rheological parameters
Γ	pressure dependence coefficient defined by eq. 2.5-8 (Pa^{-1})
$\Gamma(\ )$	gamma function
$\Gamma_{ij} = C_{ij} - \delta_{ij}$	relative Cauchy-Green tensor (–)
$\Gamma_{ij}^{-1} = C_{ij}^{-1} - \delta_{ij}$	relative Finger tensor (–)
$\underline{\underline{\Gamma}}$	corotational rate-of-strain tensor (s^{-1})
γ_0	strain in relaxation experiments (–)
γ	shear deformation or strain (–)
γ_R	shear recovery (eq. 5.2-27a) (–)
γ_{yx}	shear component of the linear strain tensor (–)
$\dot{\gamma}, \dot{\gamma}_{21}, \dot{\gamma}_{yx}$	shear rate (s^{-1})
$\dot{\gamma}_c$	critical shear rate in Quemada model (eq. 2.3-5) (s^{-1})
$\dot{\gamma}_e$	effective rate of deformation (s^{-1})
$\dot{\gamma}_R$	wall-shear rate in Poiseuille flow (s^{-1})
$\dot{\gamma}_w$	wall-shear rate (s^{-1})
$\dot{\gamma}_0$	initial shear rate in relaxation experiments (s^{-1})
$\dot{\gamma}^0$	shear rate amplitude in dynamic experiments (s^{-1})
$\dot{\gamma}_\infty$	applied shear rate in stress growth experiments (s^{-1})
$\dot{\underline{\underline{\gamma}}} = \nabla \underline{V} + \nabla \underline{V}^+$	rate-of-deformation tensor (s^{-1})
$\bar{\dot{\gamma}} = \sqrt{\tfrac{1}{2} II_\gamma}$	effective deformation rate (s^{-1})
$\Delta a = a_2 - a_1$	differences in a, referred to two control surfaces
δ	falling film or boundary thickness (m)
δ	loss tangent angle, $\tan \delta = G''(\omega)/G'(\omega)$ (–)
δ_t	thermal boundary layer thickness (m)
$\delta(t - t')$	Dirac delta function (–)
$\underline{\underline{\delta}}$	unit tensor (–)
$\underline{\delta}_i$	base vector in Cartesian coordinates (–)
δ_{ij}	Kronecker delta
$\dfrac{\delta}{\delta t}(\ )$	convected derivatives (eqs. 6.1-13 and 6.1-14)
$\dfrac{\bar{\delta}}{\delta t}(\ )$	Gordon-Schowalter derivative (eq. 6.3-49)
ε	relative eccentricity (–)
$\dot{\varepsilon}$	elongational rate (s^{-1})
$\dot{\varepsilon}_B$	biaxial extensional rate (s^{-1})
$\dot{\varepsilon}_\infty$	applied elongational rate in stress growth experiments (s^{-1})
ε	parameter in the Carreau rheological equation (eq. 6.4-13) (–)
ε	link tension coefficient in the Curtiss-Bird model (Section 7.3-3) (–)
ε	fractional void space or bed porosity (–)
ε_{mf}	bed porosity at minimum fluidization (–)
η	non-Newtonian viscosity (Pa·s)
$[\eta]$	intrinsic velocity defined by eq. 3.1-25 or 3.1-26 (m^3/kg)
η_a	apparent viscosity defined by eq. 3.1-10 (Pa·s)

η_B	biaxial extensional viscosity defined by eq. 2.1-32 (Pa·s)
η_E	uniaxial elongational viscosity (Pa·s)
η_e	shear viscosity at equilibrium (Pa·s)
η_{eff}	effective viscosity in mixing (Pa·s)
$\eta_r = \eta/\eta_s$	reduced viscosity (–)
η_p	parameter in Quemada model, eq. 2.3-4 or viscosity parameter in generalized Maxwell model (Pa·s)
$\eta_{sp} = \eta_r - 1$	specific viscosity (–)
η_s	solvent viscosity (Pa·s)
η_0	zero-shear viscosity or parameter (Pa·s)
η_∞	infinite shear viscosity or parameter (Pa·s)
$\eta^+(t, \dot{\gamma}_\infty)$	shear stress growth function defined by eq. 2.1-15 (Pa·s)
$\eta^-(t, \dot{\gamma}_0)$	shear stress relaxation function defined by eq. 2.1-19 (Pa·s)
$\eta^*(\omega)$	complex viscosity defined by eq. 2.1-9 (Pa·s)
Θ	velocity distribution function defined by eq. 7.1-15 (–)
θ	dimensionless temperature (–), angle in cylindrical or spherical coordinates (rad)
θ_0	angle of cone-and-plate geometry (rad)
$\underline{\underline{\kappa}} = \nabla \underline{V}^+$	transpose of velocity gradient tensor (s^{-1})
$\underline{\underline{\Lambda}}$	mobility tensor (–)
Λ_0	constant mobility (–)
λ	characteristic relaxation time, longest relaxation time (s)
λ_d	reptation time constant, defined by eq. 7.3-6 (s)
λ_e	relaxation time in reptation theory and characteristic relaxation time in FENE conformation model, defined by eq. 7.4-16 (s)
λ_p	characteristic relaxation time in multi-mode rheological equations (s)
λ_R	Rouse relaxation time, defined by eq. 7.3-3 (s)
λ_w	weight average characteristic time, defined by eq. 3.3-30 or 7.4-42 (s)
λ_β	characteristic relaxation time, defined by eq. 7.4-14 (s)
λ_1	relaxation time (s)
λ_2	retardation time (s)
μ	Newtonian viscosity (Pa·s)
μ_0	parameter in Bingham model (Pa·s)
ν	number of entanglements per unit volume (m^{-3})
$\nu = \mu/\rho$	kinematic viscosity (m^2/s)
ξ	dimensionless position variable or slip parameter (–)
$\underline{\underline{\pi}} = \underline{\underline{\sigma}} + P\underline{\underline{\delta}}$	total stress tensor (Pa)
$\underline{\underline{\pi}}^s$	contribution of the solvent to the total stress tensor (Pa)
ρ	fluid density (kg/m^3)
ρ_s	solvent density (kg/m^3)
σ_H	wall-shear stress in planar Poiseuille flow (Pa)
σ_R	wall-shear stress in Poiseuille flow (Pa)
σ_w	wall-shear stress (Pa)

σ_δ	wall-shear stress in the flow on an inclined plane (Pa)
σ_0	applied shear stress in creep experiments (Pa)
σ_0	parameter in Bingham model, magnitude of yield stress (Pa)
σ^0	shear stress amplitude in dynamic experiments (Pa)
$\sigma_{1/2}$	parameter in the Ellis model (Pa)
$\sigma_{21}, \sigma_{yx}, \sigma_{rz}$	shear stress (Pa)
$\underline{\underline{\sigma}}$	extra stress tensor (Pa)
$\underline{\underline{\sigma}}^p$	contribution of the polymer to the extra stress tensor (Pa)
$\underline{\underline{\sigma}}^s$	contribution of solvent to the extra stress tensor (Pa)
τ	injection period (s)
τ_n	characteristic time of normal thrust response (s)
τ_t	characteristic time of torque response (s)
ϕ	holdup, volume fraction of dispersed phase (–)
ϕ_i	volume fraction of droplets of diameter d_i (–)
ϕ_m	volume solid fraction at maximum packing (–)
Φ, φ, φ_n	distribution functions (–)
χ	dimensionless variable, reduced extension in conformation tensor model of Section 7.4-2 (–)
$\psi(R, t)$	distribution function in dumbbell models (–)
ψ_n	distribution function (–)
ψ_1	primary normal stress coefficient defined by eq. 2.1-7 (Pa·s^2)
ψ_2	secondary normal stress coefficient defined by eq. 2.1-8 (Pa·s^2)
$\psi_1^+(t, \dot{\gamma}_\infty)$	shear primary normal stress growth function defined by eq. 2.1-16 (Pa·s^2)
$\psi_2^+(t, \dot{\gamma}_\infty)$	shear secondary normal stress growth function defined by eq. 2.1-17 (Pa·s^2)
$\psi_1^-(t, \dot{\gamma}_0)$	shear primary normal stress relaxation function defined by eq. 2.1-20 (Pa·s^2)
$\psi_2^-(t, \dot{\gamma}_0)$	shear secondary normal stress relaxation function defined by eq. 2.1-21 (Pa·s^2)
Ω	characteristic angular velocity (s^{-1})
ω	angular velocity or frequency (s^{-1})
ω'	derivative of angular velocity with respect to r (m^{-1}·s^{-1})
$\underline{\underline{\omega}}$	vorticity tensor (s^{-1})
ζ	dimensionless position variable (–)
ζ	drag coefficient (–)
$\underline{\underline{\zeta}}$	drag tensor (–)
ζ_0	drag coefficient on a chain segment (–)

Dimensionless Numbers

Bm	Bingham number (eq. 9.3-17)
$Cu = t_1 V_\infty / R$	Carreau number

De	Deborah number (eq. 4.1-1)
DDR	Draw down ratio (eq. 6.3-33c)
Fr	Froude number (eq. 9.3-15)
Gr	Grashof number for heat transfer (eq. 9.8-2c)
Gz	Graetz number (eq. 4.3-3 or 4.3-34)
N	Force ratio (eq. 6.3-32)
Nt_m	dimensionless mixing time
Nu	Nusselt number for heat transfer (eq. 1.2-3)
Nu_c	Nusselt number based on the circulation concept (eq. 9.8-17)
Nu_x	Nusselt number for the flow over a plate (eq. 4.5-27)
Pr	Prandtl number (eq. 1.2-5 or 9.8-2b)
Pr_c	Prandtl number based on the circulation concept (eq. 9.8-16)
$Po = \dfrac{P}{\rho N^3 d^5}$	power number in mixing
Re	Reynolds number (eq. 1.2-2)
Re_c	Reynolds number based on the circulation concept (eq. 9.8-15)
Re_x, Re_L	Reynolds numbers for the flow over a plate (eqs. 4.5-12, 4.5-14)
Sc	Schmidt number (eq. 4.4-33)
Sh_{loc}, Sh_a	local or average Sherwood number (eq. 4.4-28)
So	Sommerfeld number (eq. 6.2-35)
$Vi = \dfrac{\mu_W}{\mu_b}$	viscosity number
We	Weber number (eq. 9.3-16)
Wi	Weissenberg number (eq. 4.1-3 or 4.1-4, eq. 9.3-18)

Index

Admissible equations *209, 259*
Agitators (*see impellers*)
Affine deformation *270, 305, 308, 325*
Axial velocities in mixing tank *419*

Baffles *436*
Bagley correction *68, 71*
    elasticity determination *71*
Bead spring models *264*
Bird-Carreau model *245*
Biaxial elongation *35, 259*
Bingham model *41, 59, 357*
Bingham number *394*
Blasius equation *11*
Boger fluid *24, 25, 406, 427*
Boltzmann parameter *324, 325*
Boundary layer flows *144–151*
Brownian force *269*
Bubbles (*see gas bubbles*)
Bulk temperature *136*

Capillary rheometers *62–63*
Carreau model *38–39, 241, 243, 356, 360*
Carreau constitutive equation *239, 261, 289*
Carreau-Yasuda model *41*
Casson model *42*
Cauchy-Green tensor *198, 200, 201, 207*
Cayley-Hamilton theorem *481*
Chaotic distortion *7*
Characteristic time *113, 164, 228, 249, 273, 282, 294, 302, 308*
Charged macromolecules *307*
Chilton-Colburn analogy *12*
Creep deformation *175*
Christoffel symbol *463*
    of the first kind *463*
    of the second kind *463*
    properties of *464*
Circulation time (in mixing) *420*
Circulation velocity *434*
Clausius-Duhem inequality *305*
Coil conformation *54*

Complex modulus *26*
    emulsions *338*
    suspensions *346*
Complex compliance *192*
Complex viscosity *25*
Compliance in creep *181–183*
Compression *205*
Configuration space *290*
Conformation model *304, 320, 328*
Conformation tensor *304, 317*
Connector force *266, 270*
Constitutive equations (*see rheological equations*)
Contraction *484*
Contravariant components *196, 445–454*
Contravariant transformation *449*
Coulombic potential *307*
Covariant components *196, 445–454*
Covariant differentiation *462*
Covariant transformation *451*
Cox-Merz relation *50*
Creep experiment *175*
Creeping flow
    around a bubble *362, 384*
    around a droplet *366*
    around a sphere *352, 384*
    in fluidized beds *377*
    in packed beds *368*
    of power-law fluids *368*
    of viscoelastic fluids *373*
    of viscoplastic fluids *356*
Criminale-Ericksen-Filbey equation *217*
Criteria for
    linear viscoelasticity *163*
    measurement in concentric disk geometry *99*
    neglecting elastic effects *113–114*
    scale-up in mixing *391–395, 411–412*
    transient experiments *94–98*
Critical concentration *75*
Critical molecular weight *52, 296*
Cross-Williamson model *39, 47, 59*

514   Index

Curtiss-Bird kinetic theory *300*
Curtiss-Bird model *302*

Damping function *250, 253*
De Kee model *40, 59, 247*
De Kee-Chan Man Fong model *46*
De Kee-Turcotte model *42*
De Witt model *221*
Deborah number *113, 227*
Definitions
    base vectors *446*
    biaxial extensional viscosity *35*
    bulk temperature *136*
    Cauchy-Green tensor *198*
    complex compliance *192*
    complex modulus *26*
    complex viscosity *25*
    Deborah number *113*
    drag coefficient on a sphere *353*
    elongational viscosity *33*
    Finger tensor *197*
    generalized Newtonian fluid *36*
    Graetz number *127, 143*
    intrinsic viscosity *73*
    invariant of rate-of-deformation tensor *44*
    Jaumann derivatives *199*
    loss modulus *26*
    lower convected derivative *198*
    metric tensor *196, 454, 457*
    non-Newtonian viscosity *21*
    Nusselt number *11*
    power number *396*
    Prandtl number *11*
    primary normal stress coefficient *21*
    primary normal stress difference *22*
    rate-of-deformation tensor *20*
    Reynolds number *8*
    rheology, rheological behavior *1–3*
    secondary normal stress coefficient *22*
    secondary normal stress difference *22*
    specific viscosity *73*
    storage modulus *26*
    unit step function *26*
    upper-convected derivative *198*
    vorticity tensor *199*
    Weissenberg number *113*
Deformation *197, 201*
Deformation rate *197, 201*
Determinism *209*

Differential constitutive equations *220*
Dilatant material *37, 354*
Distribution function *267–268*
Divergence theorem *474*
Doi-Edwards model *296, 299*
Drag on a sphere
    coefficient *353*
    dilatant fluids *354*
    dilute solutions *360*
    power-law fluids *354*
    viscoelastic fluids *357*
    viscoplastic fluids *357*
    wall effects *357–358*
Drag reduction *7, 10, 360*
Drag tensor *300*
Draw-dawn ratio *228*
Droplets
    drag *366*
    power-law fluids *366*
Dumbbell *265*

Equations of continuity *488*
Equations of motions *475, 488–492*
Effective deformation rate in mixing *398–412, 440*
Effective viscosity in mixing *398–412, 426, 440*
Einstein equation *334*
Elastic liquids *288*
Elastic time constant (see also *characteristic time*) *71*
Elastic Modulus (see *storage modulus*)
Elasticity of suspensions *346–347*
Ellis model *37, 58*
Elongational stress growth *235*
Elongational viscosity *33, 213, 258, 260*
    biaxial elongation *34–35, 259*
    planar elongation *204*
    uniaxial elongation *32–33, 200*
Embedded coordinates *196, 201*
Emulsions
    complex modulus *334*
    elasticity *338, 383*
    Oldroyd model *336*
    Palierne model *338*
    rheology *334*
    viscosity *335*
End effect corrections
    Bagley correction *68*

Couette geometry *82*
Expansion *205*
Extensional viscosity (*see elongational viscosity*)
Extra stress tensor *305*
Extruders *437, 439*
Extrudate swell *5–6*

Fading memory *7*
FENE charged macromolecules *307*
FENE dumbbell *272*
FENE-P dumbbell *273*
Finger tensor *198, 201, 207*
First normal difference *21*
Flexibility parameter *312*
Flow
    about a sphere *352*
    about a gas bubble *362*
    about a droplet *366*
    between coaxial cylinders *155*
    drainage *157*
    generalized Couette *156*
    helical *119*
    in a disk shaped mold *122*
    in a journal bearing *217*
    in a thin slit *118*
    in a tube *114*
    on inclined plane *116*
    past particles *352*
Flow patterns in mixing vessels *414*
    class I agitators *415*
    class II agitators *416*
    class III agitators *418*
    viscoelastic systems *416, 418*
    viscoplastic systems *415*
Fluidized beds
    bed expansion *380*
    fluidizing velocity *378*
    liquid-solid mass transfer *382*
    power-law fluids *377*
    Sherwood number *382*
Friction factor
    definition *8*
    in Poiseuille flow *152*
    packed beds *369, 384*
Froude number *394*

Gaussian distribution *286, 288*
Gaussian segments *285*

Gas bubbles *15, 362, 384*
    coalescence *16, 365*
    drag *362, 384*
    power-law liquids *362, 384*
    rise velocity *363*
    shapes *15, 363, 365*
    size *428*
    viscoelastic systems *364*
Gas dispersion mechanisms *423*
Generalized Newtonian fluid *36, 113, 215*
Generalized Maxwell model *170, 173*
Giesekus equation *270–271, 282*
Giesekus parameter *324–325*
Goddard-Miller equation *254*
Gordon-Schowalter derivative *233*
Grashof number *430*
Graetz number
    heat transfer *127–132*
    mass transfer *143*

Heat generation in Poiseuille flow *134–138*
    equilibrium regime *135*
    transition regime *136*
    adiabatic regime *138*
Heat transfer in
    boundary layer *160*
    falling films *159*
    mixing vessels *430–435*
    packed beds *373*
    slit flow *158*
    tube flow *126*
Helmholtz free energy *284, 286, 288, 305*
Herschel-Bulkley model *42*
Hookean elastic dumbbells *265, 270, 327*
Hooke's law *1*
Hold-up *428*
Hole pressure error *13*
Huggins equation *73*
Hydrodynamic forces *270*
Hydrodynamic interactions *283*

Impellers *402, 437, 438*
Independent alignment approximation *299*
Indifference (principle) *209*
Inelastic material *3*
Inertial frame of reference *480*
Inextensibility in average *324*
Integral constitutive equations *234*
Integral transforms *474*

Interactive particles
    effects on viscosity *347–351*
Intrinsic viscosity *73, 284*
Invariants *480–481*
    rate-of-deformation tensor *44*
Inviscid fluid *2*
Isotropic mobility *306*
Isotropic tensor *476*
Isothermal flow (*see flow*)

Jaumann derivative *199*
Jeffreys model *178, 190*
Journal bearing *217*

Kaye-BKZ constitutive equation *247*
Kramers expression *268, 273*
Kramers matrix *280–281*
Kronecker delta *196, 443*
Kuhn segment *312*

Lagrange multilier *313–314*
LeRoy-Pierrard equation *254*
Lévêque analysis *127*
Lifshitz contribution *312*
Linear deformation tensor *199*
Linear viscoelasticity *162*
Lodge-Meissner relation *237*
Lodge model *235*
Loss modulus *26*
Lower-convected derivative *198*
Lower-convected Maxwell model *211–213*

Mark-Houwink equation *74, 284*
Maron-Pierce equation *344*
Marrucci model *230, 261*
Mass transfer
    gas in boundary layer *160*
    gas-liquid systems *429*
    in falling films *160*
    in fluidized beds *382*
    power-law fluids *138*
Material functions *19*
    biaxial extensional viscosity *35*
    complex modulus *26*
    complex viscosity *25*
    Cox-Merz relation *50*
    dynamic rigidity *25*
    dynamic viscosity *25*
    elongational viscosity *33*
    loss modulus *26*
    non-Newtonian viscosity *21*
    primary normal stress coefficient *21*
    relations between material functions *49–50, 249, 252, 255*
    relaxation modulus *31*
    secondary normal stress coefficient *22*
    storage modulus *26*
    stress growth functions *28*
    stress relaxation functions *30*
    stress tensor *20*
    uniaxial elongational viscosity *33*
Material objectivity *209*
Maxwell (linear) models *164–178, 189*
Maxwell convected models *210–215, 261, 272*
Maxwell extruder *110*
Maxwellian distribution function *267*
Melt (defect) fracture *6, 7*
Melt index *64*
Metric tensor *196, 454, 457*
Metzner-Otto concept *398*
Mild curvature assumption *301*
Minimum fluidization velocity *378–379*
Mixing *14, 387*
    bubble size *428–429*
    effective shear rate *398, 400, 409*
    equipment *436, 438*
    extruders *437*
    flow patterns *414, 418*
    gas-liquid systems *423*
    heat transfer with coil and jackets *430–435*
    laminar *389*
    liquid systems *387*
    mass transfer *429*
    mechanical agitation *436*
    mechanisms *388*
    power consumption *396–412*
    power-law liquids *411*
    rate of *420*
    scale-up *391, 411*
    similarity criteria *391, 411*
    time *420*
    turbulent flow *391*
    viscoelastic systems *412*
Mixing phenomena *14*
Mixing time *420*
    inelastic liquids *420–421*
    viscoelastic systems *422*

Mobility *308, 321*
Mobility tensor *305, 324*
Molecular models *264*
    bead-spring model *264*
    Curtiss-Bird kinetic theory *300*
    conformation models *304*
    Doi-Edwards model *296*
    FENE model *272*
    Hookean elastic dumbbell *265*
    network theories *284*
    reptation models *294*
    Rouse-Zimm models *276*
    rubber-like solid *286*
Mooney correction *72*
Multiphase systems *330*
    concentrated suspensions *343–351*
    dilute suspensions *331*
    emulsions *334–343, 383*
    fluidization *377–382*
    mixing in gas-liquid systems *423–429*
    packed beds *368*
    suspensions *334–351*

Network theories *284, 327*
Newtonian law *2*
Newtonian viscosity *2, 37*
Non-linear deformations *195*
Non-Newtonian phenomena *3*
    bamboo effect *7*
    bubbles *15*
    chaotic distortion *7*
    drag reducer *7, 10, 12*
    entry flow *5*
    extrudate swell *5–6*
    hole pressure error *13*
    melts defects, fracture *6–7*
    mixing *14*
    recoil *7, 9, 10*
    sharkskin *6–7*
    slip-stick *7*
    spheres *17*
    vibrating nozzle *8*
    Weissenberg effect *4*
Non-Newtonian viscosity *21, 36*
Non-spherical particles *359, 375, 381*
    drag *359*
    fluidized beds *381*
    packed beds *375*
Normal force corrections *92—94*

Normal stress determination from
    co-axial cylinders *82–84*
    concentric disk geometry *101–102*
    cone-and-plate geometry *88–90*
    exit pressure *109–110*
Normal stress relaxation modulus *237*
Nusselt number, definition *11, 430*
    local *130*
    average *132*
    over a plate *150–151*

Objective tensors *478*
Objectivity *209*
    Oldroyd
    models *222, 260*
    emulsion model *336–337*
Onsager parameter *312*
Ostwald-De Waele model *37*
Ostwald viscometer *62*
Outer product *465*

Packed beds
    capillary model *368*
    effect of particle shape *375*
    flow in *368*
    porosity *369*
    power-law fluids *368*
    pressure drop *368, 384*
    submerged objects model *376*
    viscoelastic effects *374*
    wall effects *374*
Palierne emulsion model *338–341*
Particle shape
    effect on drag *359*
    effect on pressure drop *375, 381*
Peterlin pre-averaging approximation *273, 310*
Permutation tensor *456*
Phan-Thien Tanner model *232*
Physical components *458*
Planar elongation *204, 258*
Plateau modulus *96, 297, 300, 303*
Power consumption in mixing *396–414*
    gas-liquid systems *423–428*
    helical impellers *403–410*
    high viscosity systems *397*
    low viscosity liquids *396*
    viscoelastic systems *412*
Power-law fluids

droplets *366*
fluidization *377*
gas bubbles *362*
mixing *397*
model *37*
packed beds *368*
sphere *352*
Power number *396, 412*
Prandtl number *11, 430*
Primary normal stress coefficient *21*
Primary normal stress difference *22*
Pseudoplastic material *3, 37*
Pure deformation *200*
Purely elastic material *3*
Purely viscous fluid *3, 36*

Quemada model *47*
Quotient law *484*

Rabinowitsch analysis *64–67, 106–107*
Radius of gyration *74–75*
Rate-of-deformation tensor *20, 198, 201*
Recoil *7, 9, 10, 168*
Recoverable compliance *298*
Recoverable shear *71*
Relations between material functions *49, 50, 249, 252, 255*
Relative deformation *197*
Relaxation experiment *29, 31, 173*
Relaxation function *296*
Relaxation modulus *31, 193, 237, 297, 302*
Relaxation spectrum *184*
Relaxation time (see *characteristic time*)
Reptation theory *294*
Reynolds number definition *8*
Rheological equations (models)
    Carreau A et B models *241, 243*
    Carreau constitutive equation *239*
    Conformation model *304, 320, 328*
    Criminale Ericksen-Filbey equation *217*
    Curtiss-Bird model *302*
    De Kee model *247*
    De Witt model *221*
    Differential constitutive equations *220*
    Doi-Edwards reptation model *296, 299*
    Dumbbell models *272–276*
    Generalized Newtonian model *36, 215*
    Goddard-Miller equation *254*
    Integral constitutive equations *234*
    Jeffreys model *178*
    Kaye-BKZ constitutive equation *247*
    LeRoy-Pierrard equation *254*
    Lodge model *235*
    Marrucci model *230*
    Maxwell convected models *210–215*
    Maxwell (linear) models *164–165, 171, 173*
    Network constitutive equation *288–289*
    Oldroyd models *222, 260, 336*
    Palierne model *338–341*
    Phan-Thien Tanner model *232*
    Rivlin-Ericksen equation *216*
    Rouse constitutive equation *282*
    Viscosity models *37–48*
    Voigt-Kelvin model *180*
    Wagner model *250*
    White-Metzner model *223, 260*
Rheological functions (see *material functions*)
Rheology
    definitions *1*
    summary of behavior *2*
Rheology of
    concentrated suspensions *343*
    dilute suspensions *331, 334*
    emulsions *334*
    interacting systems *347*
    non-interacting systems *343*
    suspensions *331*
Rheometers, rheometry
    capillary viscometers *62–64*
    co-axial cylinder rheometers *76*
    concentric disk geometry *98*
    cone-and-plate geometry *84, 107*
    Couette rheometers *76, 77*
    Dexter's viscometer *77*
    falling cylinder rheometer *108*
    parallel plate rheometer *98, 108*
Rheopexy *44, 46*
Riemann Christoffel tensor *466*
Rigid in average *313*
Rivlin-Ericksen fluid *216*
Rod-like macromolecule *312, 328*
Rouse constants *282*
Rouse matrix *280*
Rouse model *276, 282*
Rubber-like solids *286*

S-shaped behavior *310–312*

Scale-up of mixing equipment *411*
Schmidt number *382*
Schowalter emulsion model *335*
Second-order fluid *216, 260*
Secondary normal stress coefficient *22*
Secondary normal stress difference *13, 22*
Sedimentation
    concentrated suspensions *361*
    single sphere *352, 356*
Settling velocity *354, 361*
Simple fluid *209*
Sharkskin *6–7*
Shear thickening *37, 310, 311*
Shear-thinning material *3, 37*
Sherwood number *143, 382*
Shift factor *187*
Short contact time analysis
    in heat transfer *127–133*
    in mass transfer *139–144*
Simple shear flow *19, 205*
Sinusoidal shear flow *25*
Slip parameter *233, 305*
Slip-stick *7*
Slip velocity, coefficient *72*
Sommerfeld number *218*
Specific viscosity *73*
Spheres *15, 352*
Spinning of a fiber *224*
Stokes' law *55, 269*
Stokes' theorem *474*
Storage modulus *26*
Stress growth experiment and functions *28*
Stress relaxation experiment *29, 31, 173*
Stress relaxation following a sudden deformation *31*
Stress relaxation function *30*
Stress tensor *20, 216*
Suspensions *331*
    concentrated *343*
    dilute *331, 334*
    Einstein viscosity equation *334*
    elasticity *346*
    interacting systems *347*
    Maron-Pierce equation *344*
    maximum packing fraction *344*
    non-interacting systems *343*
    relaxation time *344*
    rheology *331*

Tanner equation *273, 327*
Taylor equation *334*
Tensor-valued function *480*
Thermal boundary layer *149*
Time-temperature superposition *186*
Thermal diffusivity *128*
Thixotropy *44*
Torque correction *91*
Total stress tensor *216, 333*
Transient shear flows *26*
    criteria *94–98*
Trouton relation *33, 215*
Tube model *294*

Ubbelohde viscometer *62*
Uniaxial elongation *200*
Unit step function *26*
Unit tensor *198*
Upper-convected derivative *198*
Upper-convected Maxwell model *211–215, 272*

Velocity controller *157*
Vibrating nozzle *8*
Viscoelastic effects
    drag on particles *352*
    emulsions *336–343*
    flow patterns in mixing tanks *416, 418*
    in journal bearing *217*
    in oscillatory flow *260, 262*
    in packed beds *373*
    in spinning *224*
    in suspensions *346–347*
    power consumption *412*
Viscoelastic material *3*
Viscometers (*see rheometers*)
Viscometric functions *19*
Viscosity
    corrections for temperature in heat transfer *133*
    effect of molecular weight *52–53*
    effect of pressure *52–53*
    effect of temperature *50–51*
    intrinsic viscosity *73*
Viscosity average molecular weight *74*
Viscosity determination from
    capillary *64–70*
    co-axial cylinders *77–81*

concentric disk geometry *99–100*
cone-and-plate geometry *86*
Viscosity of
   concentrated suspensions of interacting particles *347–351*
   concentrated suspensions of non-interacting particles *343–347*
   dilute systems *332, 334*
   emulsions *334–343, 383*
   suspensions *332*
Viscosity models
   Bingham model *41*
   Carreau 3-, 4-parameter models *38–39*
   Carreau-Yasuda model *41*
   Casson equation *42*
   Cross-Williamson model *39, 59*
   De Kee model *40*
   De Kee-Chan Man Fong model *46*
   De Kee-Turcotte model *42*
   Ellis model *37*
   Herschel-Bulkley model *42*
   Power-law model *37*
   Quemada model *47*
Viscous dissipation *134–138*
Voigt-Kelvin model *180, 191*

Vorticity tensor *199*

Wagner model *250*
Wall effects
   flow of a sphere *357–359*
   in packed columns *374–375*
   sedimentation of glass beads *361–362*
Weber number *394*
Weissenberg effect *4, 109*
Weissenberg number *113, 394*
Weissenberg postulate *13*
White-Metzner model *223, 260*
Wire coating *154*
Worm-like macromolecules *312*

Yamamato relation *255*
Yield stress *41–42*
   determination *103, 111*
   sphere drag *385*
   suspensions *347*

Zimm hydrodynamic interaction matrix *283*
Zimm model *276, 283*